KV-703-968

PROFESSOR W. G. CHALONER, FRS
Biology Department
Royal Holloway University of London
Egham Hill, Egham
Surrey TW20 0EX

Sept 26 '94
(Complimentary)

INTERNATIONAL CODE OF BOTANICAL NOMENCLATURE
(TOKYO CODE)
1994

Regnum Vegetabile

a series of publications for the use of plant taxonomists published under the auspices of the International Association for Plant Taxonomy

edited by Werner Greuter

Volume 131

ISSN 0080-0694

International Code of Botanical Nomenclature

(Tokyo Code)

Adopted by the Fifteenth International Botanical Congress, Yokohama, August-September 1993

Prepared and edited by

W. GREUTER, Chairman
F. R. BARRIE, H. M. BURDET, W. G. CHALONER,
V. DEMOULIN, D. L. HAWKSWORTH, P. M. JØRGENSEN,
D. H. NICOLSON, P. C. SILVA, P. TREHANE, Members
J. MCNEILL, Secretary
of the Editorial Committee

1994

Koeltz Scientific Books
D-61453 Königstein, Germany

ISBN 3-87429-367-X or 1-878762-66-4 or 80-901699-1-0

CONTENTS

Preface		vii
Key to the numbering of the Articles, Notes and Recommendations		xvi
Important dates in the Code		xviii
Preamble	(1-11)	1
Division I	Principles (I-VI)	3
Division II	Rules and Recommendations (Art. 1-76)	4
Chapter I	Ranks of taxa (Art. 1-5)	4
Chapter II	Names of taxa (general provisions) (Art. 6-15)	6
Section 1	Definitions (Art. 6)	6
Section 2	Typification (Art. 7-10)	7
Section 3	Priority (Art. 11 and 12)	15
Section 4	Limitation of the principle of priority (Art. 13-15)	19
Chapter III	Nomenclature of taxa according to their rank (Art. 16-28)	25
Section 1	Names of taxa above the rank of family (Art. 16 and 17)	25
Section 2	Names of families and subfamilies, tribes and subtribes (Art. 18 and 19)	27
Section 3	Names of genera and subdivisions of genera (Art. 20-22)	29
Section 4	Names of species (Art. 23)	33
Section 5	Names of taxa below the rank of species (infraspecific taxa) (Art. 24-27)	36
Section 6	Names of plants in cultivation (Art. 28)	39
Chapter IV	Effective and valid publication (Art. 29-50)	40
Section 1	Conditions and dates of effective publication (Art. 29-31)	40
Section 2	Conditions and dates of valid publication of names (Art. 32-45)	42
Section 3	Citation of authors' names for purposes of precision (Art. 46-50)	56
Section 4	General recommendations on citation (Rec. 50A-F)	62
Chapter V	Rejection of names (Art. 51-58)	64
Chapter VI	Names of fungi with a pleomorphic life cycle (Art. 59)	71

Chapter VII	Orthography of names and epithets and gender of generic names (Art. 60-62)	73
Section 1	Orthography of names and epithets (Art. 60 and 61)	
Section 2	Gender of generic names (Art. 62)	80
Division III	Provisions for modification of the Code	82
Appendix I	Names of hybrids (Art. H.1-H.12)	84
Appendix IIA	Nomina familiarum algarum, fungorum et pteridophytorum conservanda et rejicienda	91
Appendix IIB	Nomina familiarum bryophytorum et spermatophytorum conservanda	98
Appendix IIIA	Nomina generica conservanda et rejicienda	118
A. Algae	1. Bacillariophyceae (incl. foss.)	119
	2. Bodonophyceae	123
	3. Chlorophyceae	123
	4. Chrysophyceae	127
	5. Cyanophyceae	128
	6. Dinophyceae	129
	7. Euglenophyceae	130
	8. Phaeophyceae	130
	9. Rhodophyceae	134
	10. Trichomonadophyceae	141
	11. Xanthophyceae	142
B. Fungi		142
C. Bryophyta	1. Hepaticae	167
	2. Musci	172
D. Pteridophyta		179
E. Spermatophyta		183
F. Plantae fossiles (exl. Bacillarioph.)		324
Appendix IIIB	Nomina specifica conservanda et rejicienda	326
B. Fungi		326
C. Bryophyta		327
E. Spermatophyta		327
Appendix IV	Nomina utique rejicienda	330
B. Fungi		330
C. Bryophyta		331
E. Spermatophyta		331
Appendix V	Opera utique oppressa	333
Index to Appendix IIIA		336
Index of scientific names		354
Subject index		369

PREFACE

This new edition of the *International code of botanical nomenclature,* like its immediate predecessor, is in English only (British rather than American English, in cases of discrepancy). French, German, and Japanese editions of the text of the *Berlin Code* were produced separately and it is anticipated that at least French and German editions of the *Tokyo Code* will be prepared.

The *Tokyo Code,* is, however, significantly different from the *Berlin Code.* There are two reasons for this difference: one relates to the arrangement of the book and other technical matters, the other to its content.

Considering arrangement first, the Tokyo Congress agreed to delete the major part of Chapter V dealing with "Retention, choice, and rejection of names and epithets", transferring such material as was not already covered elsewhere in the *Code* to other Articles, notably Art. 11. With five other Articles deleted from the latter part of the *Code* at previous Congresses (two in Leningrad in 1975, one in Sydney in 1981, and two in Berlin in 1987), only 11 Articles remained in the *Code* following Art. 50, yet the numbering extended to 76. For this reason and because a completely new subject index was being prepared for this edition, the Editorial Committee decided that a renumbering of the Articles in the latter part of the *Tokyo Code* was essential for clarity. This renumbering was executed in the spirit of the Nomenclature Section's usual instruction "to preserve the numbering of Articles and Recommendations in so far as possible", in that only the Articles following Art. 50 have been renumbered, and, by a rearrangement of the ordering of the Articles, the commonly cited Art. 59, on names of fungi with a pleomorphic life cycle, has retained its traditional number. The final Article is now Art. 62, an extra Article having been created by the division of the old Art. 69 (see below).

The Editorial Committee also took the opportunity to clarify the rules on typification and effective publication by creating a more logical arrangement of Art. 7-10 and 29-31, respectively. Art. 7 now deals with general matters of typification, Art. 8 with typification of names of species and infraspecific taxa, Art. 9 with the various categories of types applicable to such names, and Art. 10 with the typification of supraspecific names. Art. 29 now addresses the general issue of effective publication, Art. 30 special cases, and Art. 31 the date of effective publication.

Among the more familiar Articles whose numbering has changed are the former Art. 63 on superfluous names, which is now Art. 52, and the former Art. 69 on *nomina rejicienda*, which now forms Art. 56 and 57 (see below). A tabular key is provided, setting out the changes in numbering of Articles and Recommendations, as well as their individual paragraphs and Notes, between the *Tokyo Code* and the *Berlin Code*. The numbering of the Articles remained virtually unchanged between the *Seattle Code* adopted in 1969 and the *Berlin Code*. The *Seattle Code* includes a "Key to the numbering of the Articles and Recommendations" for the five editions of the *Code* from Stockholm (1952) to Seattle (1972), and the *Stockholm Code* includes a similar key comparing its arrangement with that of the *Cambridge Rules* (1935).

In this edition, the body of the *Code* is printed in three different type sizes, the Recommendations and Notes being set in somewhat smaller type than are the rules, whereas the type of the Examples is even smaller. This reflects the distinction between the rules (primarily the Articles), the complementary and advisory material (the Notes and Recommendations), and the mainly explanatory material (the Examples). The nature of Articles, Recommendations and Examples is generally well understood, particularly now that voted Examples are designated as such (see below), but the role of Notes is less clear. Like an Article, a Note in the *Code* states something that is mandatory; it differs from an Article, however, in that a Note does not introduce any new rule or concept, but merely spells out something that may not be evident to the user but is provided for elsewhere in the *Code*, either explicitly or by implication.

Content, however, is much more important than arrangement or format, although the latter are sometimes the more noticeable. The Tokyo Congress was noteworthy in that conservation of species names and rejection of any name that would cause a disadvantageous nomenclatural change were both accepted by an overwhelming majority on a show of hands. Those who remember the narrow majorities by which conservation of names of species of major economic importance and names which represent the type of a conserved generic name were approved at Sydney and Berlin, respectively, will recognize the fundamental change that took place in botanical nomenclature at Yokohama. The *Code* is no longer a handicap but an encouragement to the maintenance of nomenclatural stability (see also Greuter & Nicolson in Taxon 42: 925-927. 1993).

The restrictions on conservation of specific names have, therefore, been removed from Art. 14.2, and conservation of the names of species, as for families and genera, now simply "aims at retention of those names which best serve stability of nomenclature". With the adoption by the Tokyo Congress of an amendment to the former Art. 69 that provides for the rejection of any name "that would cause a disadvantageous nomenclatural change", the Article came to cover two distinctly different situations and has been divided into two. The first, the new Art. 56, deals with the general case (i.e. any disadvantageous nomenclatural change) and includes the mechanisms by which names can be rejected as in the previous Art. 69.2. The second, the new Art. 57, relates to the more restricted case, to which the former Art. 69 was hitherto confined, i.e. names that have been widely and persistently used for a taxon or taxa not including their type. Such names continue not to be available for use in a sense that conflicts with current usage, unless and until a proposal to deal with them under the conservation provisions of Art. 14 or the rejection provisions of the new Art. 56 has been submitted and rejected. The separation of Art. 56 and 57 makes even clearer the requirement of the *Code* (formerly in Art. 69.4) not to use such a name in a sense that conflicts with current usage unless the appropriate Committee has authorized its use.

It is of particular note that the Nomenclature Section at Yokohama, recognizing the significance of this increased scope for conservation and rejection of names in ensuring nomenclatural stability, adopted a resolution in the following terms: "The Section urges the General Committee and through it all Permanent Committees to make full use of the options that the *Code* now provides in order to ensure nomenclatural clarity and stability." Individual users of the *Code* also have a responsibility to help ensure nomenclatural clarity and stability by making appropriate proposals for conservation or rejection rather than change names for purely nomenclatural reasons (see also the Congress Resolution, below).

One entirely new concept is incorporated in this edition of the *Code*, that of interpretative types to serve when an established type cannot be reliably identified for purposes of precise application of a name. The original proposal had used the term "protype" for such a specimen or illustration, but the Nomenclature Section asked the Editorial Committee to determine the most appropriate term and the Committee adopted "epitype" as better expressing the meaning ("on top of the type") and because protype had been used in other senses in the past. The provision appears in Art. 9.7.

Other additions of note decided by the Tokyo Congress and now incorporated in the *Code* include provision for the use of the term *"phylum"* as an alternative to *divisio* (Art. 4.2); the requirement that, on or after 1 January 2000 (subject to the approval of the XVI International Botanical Congress) names be registered (Art. 32.1); the designation of "suppressed works" in which names of certain categories are ruled to be not validly published (Art. 32.8 and the new App. V); and the requirement that in order to be validly published a name of a new taxon of fossil plants must, on or after 1 January 1996, be accompanied by a description or diagnosis in Latin or English, or reference to such, not in any language as before (Art. 36.3). An extensive revision of Art. 46 has clarified the situation in which "ex" may be used in the citation of authors' names and confirmed that the preposition "in", and what may follow, is bibliographic and not part of the author citation. One of the Permanent Committees listed in Div. II was abolished (the Committee for Hybrids) and one was renamed (the Committee for Fungi and Lichens, now Committee for Fungi).

From time to time Nomenclature Sections have accepted specific Examples ("voted Examples") in order to legislate nomenclatural practice where the corresponding Article of the *Code* is open to divergent interpretation or may not even cover the matter at all. One such Example, adopted by the Tokyo Congress, appears as Art. 8 Ex. 1, making clear what had hitherto been controversial, namely that "cultures permanently preserved in a metabolically inactive state" are to be considered "preserved permanently" (Art. 8.2), and, although in a sense "living plants or cultures", are eligible as types, regularizing a procedure adopted by yeast taxonomists in particular. Whereas the Editorial Committee normally has the power to delete, modify or add Examples in order to clarify the *Code*, this power does not extend to voted Examples, which the Editorial Committee is obligated to retain, whether or not they actually exemplify the rules. In response to a suggestion, made at the Yokohama meetings, that voted Examples should be clearly indicated, an asterisk (*) has been inserted against each one.

Although proposals made to the Tokyo Congress to provide for the granting of protection to names (or some aspect of names, e.g. types) on approved lists (the "NCU proposals") did not receive the 60 % majority necessary for their adoption, the Section was particularly impressed by the utility of the list of species names in *Trichocomaceae* (incl. *Aspergillus* and *Penicillium*) in ensuring nomenclatural stability in that group. Accordingly the Section adopted the following resolution that authorizes users of names in that family to suspend application of the *Code* where necessary: "The Nomenclature Section, noting that the 'List of Names in Current Use in the *Trichocomaceae*' (Regnum Veg. 128: 13-57. 1994) has already been approved by the International Commission on *Penicillium* and *Aspergillus* of the International Union of Microbiological Societies (IUMS), urges taxonomists not to adopt names that would compete with or change the application of any names on that list."

Apart from the addition of the new App. V referred to above, the Appendices remain those established in the *Berlin Code*. App. I deals with the naming of hybrids; App. IIA lists conserved names of families of algae, fungi, and pteridophytes, which are only conserved against listed rejected names, and App. IIB those of bryophytes and spermatophytes, which are conserved against all competing names not themselves included in the list; App. IIIA lists conserved generic names and the corresponding rejected names, and App. IIIB likewise for species names; App. IV lists names rejected under what is now Art. 56 (formerly Art. 69). Within these Appendices, some restructuring became necessary, partly because of the growing number of algal classes involved in conservation (two being added in this *Code*) and partly because of the present and foreseeable expansion of some of the smaller Appendices (IIIB and IV in ·rticular). Six major groups are now recognized for the purposes of all these

Appendices, and are designated by identical capital letters in each Appendix: A for algae, B for fungi, C for bryophytes, D for pteridophytes, E for spermatophytes, and F for fossil plants. In App. IIIA, subheadings with Arabic numerals are used for the algal and bryophyte classes. Under these headings and subheadings, conserved names are listed alphabetically, except in the case of genera of spermatophytes, for which the Dalla Torre & Harms numbering system and family definitions have been retained once more. It has been made explicit that, in App. III, all names of diatoms, whether with fossil or recent types and whether or not the genera include recent species, are listed under *Bacillariophyceae* rather than with the other fossil plants.

The procedures for producing this edition of the *Code* have followed the pattern outlined in Div. III of the *Code* and traditions well established since the Paris Congress of 1954. Published proposals for amendment, with comments by the Rapporteurs, were assembled in a "Synopsis of proposals" (Taxon 42: 191-271. 1993). Results of the Preliminary Mail Vote on these proposals, a strictly advisory but very helpful expression of opinion, were made available at registration for the Nomenclature Section of the Tokyo Congress, in the Congress Center of Pacifico, Yokohama, Japan. The Section met from 23 to 27 August, just before the regular sessions of the Congress, and acted on the 321 proposals before it, accepting some 82 and referring another 42 to the Editorial Committee for modification of the *Code*. The Section's decisions were sanctioned by resolution of the closing plenary session of the Congress on 3 September 1993 (see below) and became official at that time. A list of them appeared along with the results of the preliminary mail vote (in Taxon 42: 907-922. 1993). A preliminary transcript of the complete tape records of the nomenclature sessions, prepared by Fred Barrie, Werner Greuter, and John McNeill, was available to all members of the Editorial Committee at their meeting in January 1994. The full report of the Section's proceedings, including the essence of the debates and comments made during the deliberations, has since been published as a separate volume (Englera 14. 1994).

It is the duty of the Editorial Committee, elected by the Section (and, by tradition, from among those present for the discussions), to incorporate the decisions of the Congress into the *Code* and to make whatever strictly editorial changes are desirable for smooth, consistent, accurate, and clear reading. The composition of the Editorial Committee usually changes slightly at each Congress, and this was also the case on this occasion. Although the positions of Chairman and Secretary were unchanged in that Werner Greuter had continued as Rapporteur-général for the Tokyo Congress and John McNeill as Vice-rapporteur, the Committee lost three former members, Riclef Grolle, Frans Stafleu and Ed Voss, the two latter having served for several terms on the Editorial Committee. Frans Stafleu had been Vice-rapporteur and Secretary of the Editorial Committee from 1954 to 1964, Rapporteur-général and Chairman of the Committee from 1964 to 1979, and as President of the Nomenclature Section at Berlin in 1989 he had returned to the Committee for the preparation of the *Berlin Code;* while Ed Voss had served continuously on the Committee since 1964, being Vice-rapporteur and Secretary of the Committee from 1964 to 1979, Rapporteur-général and Chairman of the Committee for the 1981 Sydney Congress, and a member of the Committee for the *Berlin Code*. Although their experience was missed, the Committee was very well served by their replacements: Fred Barrie, Missouri Botanical Garden (currently on assignment at the Field Museum, Chicago); Per Magnus Jørgensen, University of Bergen; and Piers Trehane, Wimborne, Dorset, U.K. (co-opted to replace Alan Leslie, Royal Horticultural Society, Wisley, U.K., elected to the Committee in Yokohama but unable to serve); all of whom joined most effectively in the work of the Committee.

After circulation of a first draft of the text of the new *Code*, the Editorial Committee met at the "Botanischer Garten und Botanisches Museum Berlin-Dahlem" from 2-7 January 1994. All 11 members of the Committee were present for this traditional and essential meeting that explores as exhaustively as possible the clearest and most concise way to express in the *Code* the decisions of the Nomenclature Section. A myriad of editorial details must be addressed to ensure that the resulting work is unambiguous to all users, regardless of their primary language. The Editorial Committee recognizes that complete clarity and consistency are hardly achievable. Some instances of imprecise, conflicting, or otherwise unsatisfactory wording may still remain in the *Code*. Any effort to resolve certain points would result in extending the operation of the *Code* or restricting it, depending upon one's reading of the present text, and would not, therefore, be covered by the Editorial Committee's mandate.

The method by which some or all scientific names are set off in printed text varies substantially between different countries and language traditions. Perhaps as a result, there has been an unevenness in this regard in different editions of the *Code*. In an attempt to achieve uniformity, the *Sydney Code* and the *Berlin Code* italicized all scientific names at the rank of family and below, i.e. those for which priority is mandatory. The present Editorial Committee recognized that this policy was rather illogical, and, in the *Tokyo Code,* all scientific names falling under the provisions of the *Code* are italicized, whereas informal designations appear in Roman type. For example, in Art. 13.1 *(d)* the ordinal names *Uredinales, Ustilaginales,* etc. are italicized, whereas the informal group name "fungi" is not. The Editorial Committee considers this to be the most appropriate form of presentation in a code of nomenclature but does not aim to impose this as a standard to be followed in other publications, which may have different editorial traditions, often of long standing.

Consistency within the *Code* regarding bibliographic style and details, in a manner which is non-confusable and agreeable to all users, has been among our prime editorial concerns. Standardization aids are now available which did not exist years ago, providing an almost complete coverage of high standard for various categories of data. We have consistently used "TL-2" (Stafleu & Cowan, *Taxonomic literature*, ed. 2; with Suppl. 1 & 2, by Stafleu & Mennega) for book title abbreviations, "B-P-H" *(Botanico-Periodicum-Huntianum)* and its Supplement for the citation of journal titles, and Brummitt & Powell's *Authors of plant names* (Royal Botanic Gardens, Kew, 1992), for author citations. The style of citation used in the Names in Current Use (NCU) lists, and explained there in detail (see Regnum Veg. 126: 912. 1993), has been followed throughout. For the countless nomenclatural citations in the Appendices this standardization was a major undertaking. Whereas new-style entries were readily available for most of the conserved names of families (from Regnum Veg. 126) and genera (from Regnum Veg. 129), they had to be established anew for *nomina rejicienda*. Standardized author citations were introduced by Paul Kirk, International Mycological Institute, Egham, for all such names that had been downloaded from the *Index nominum genericorum* database by Ellen Farr, Smithsonian Institution, Washington – which were only those that are explicitly marked *"nom. rej."* in that database (perhaps slightly more than one half of the total). Norbert Kilian, Berlin, working under an IAPT contract, carried out all the remaining standardizations manually. Brigitte Zimmer, Berlin, did extensive consistency checks.

The updating of App. II-IV, meaning not only the additions and changes reflecting adopted proposals but also extensive verification of extant entries which in several cases resulted in substantial editorial corrections, was done in conjunction with experts acting either in their capacity of Editorial Committee members or Permanent Committee secretaries. Paul Silva took care of the algal entries; Vincent Demoulin, assisted by Walter Gams, of the fungal ones; Gea Zijlstra not only provided the bryophyte additions but also many updates and amendments resulting from her work with the *Index nominum genericorum;* Dan Nicolson prepared the additions for pteridophytes and spermatophytes, for the latter of which Dick Brummitt provided countless careful contributions; and Bill Chaloner took responsibility for fossil plants. The great and generous help of them all is gratefully acknowledged.

Whereas the index to App. III, completely computer-generated from its content, has not undergone any major change, the main index has been completely restructured and divided into two halves (scientific names and subject index) in an effort to make the content of the *Code* more readily accessible to teachers and students of botany as well as to those who apply the *Code* on a regular basis. This complete restructuring and rewriting, which we hope will be seen as a significant improvement on previous versions, was carried out by Piers Trehane. It proved to be a demanding and exhausting chore, for which he deserves all our gratitude.

The final editing of the whole text, including the Appendices and Indices, was undertaken by Werner Greuter in close contact with other members of the

Editorial Committee, a circumstance made easier by the development of fax transmission and electronic communication. Brigitte Zimmer, assisted by Norbert Kilian for the Appendices portion, produced camera-ready copy of the text.

In addition to those who have helped produce this new edition of the *Code*, botanical nomenclature depends on the scores of botanists who serve on the Permanent Nomenclature Committees that work continuously between Congresses, dealing principally with proposals for conservation or rejection, and those others who are members of Special Committees, reviewing and seeking solutions to the problems assigned to them by the Nomenclature Section of the previous Congress. Botanical Nomenclature is remarkable for the large number of taxonomists who voluntarily work so effectively and for such long hours, to the immeasurable benefit of all their colleagues who must use plant names and on whose behalf this word of sincere thanks is expressed.

Ultimately, however, plant nomenclature is not governed by a bureaucracy of committees but, in an open and democratic manner, by the community of its users represented by the enrolled members of International Botanical Congresses. The user-driven process by which plant nomenclature is regulated is of utmost importance for a *Code* which, having no "teeth" in the way of penalties for infringements, entirely depends on user consensus for its universal application and implementation.

The *International code of botanical nomenclature* is therefore published under the ultimate authority of the International Botanical Congresses. The Tokyo Congress at its final plenary session adopted the following resolution relating to nomenclature:

"Considering the great importance of a stable system of scientific names of plants for use in the pure and applied sciences and in many other domains of public life and economy;

"noting with satisfaction recent important improvements in the *International code of botanical nomenclature* and ongoing efforts to explore new avenues for increased stability and security in the application of plant names;

"the XV International Botanical Congress urges plant taxonomists, while such work continues, to avoid displacing well established names for purely nomenclatural reasons, whether by change in their application or by resurrection of long-forgotten names;

"resolves that the decisions of the Nomenclature Section with respect to the *International code of botanical nomenclature,* as well as the appointment of officers and members of the nomenclature committees, made by that section during its meetings, 22–27 August, be accepted."

This resolution goes far beyond the traditional act of ratification of nomenclature actions and Permanent Committee nominations by Congress. By it, and through the International Union of Biological Sciences under whose auspices those Congresses are held, plant taxonomists are urged to become the champions of nomenclatural stability. Name changes made for purely nomenclatural reasons (as opposed to those that result from changing taxonomic concepts, hopefully reflecting a progress of our science) are to be avoided.

Does this mean that the present *Code* is a document of little consequence, to be set aside each time its application leads to results felt (by some) to be disagreeable? Certainly not. The *Code* now offers generous new ways to avoid nomenclatural changes by proposing the conservation or rejection of names, and these opportunities are to be used. Should these not suffice, new provisions may have to be devised and incorporated in the future.

The *Code* is a living and adapting body of law, and as long as it keeps evolving in tune with changing needs and new challenges it will keep its authority and strength. The *Tokyo Code* is, we believe, a significant landmark in this ongoing adaptive process.

May 1994

Werner Greuter
John McNeill

Key to the numbering of the Articles, Notes and Recommendations

The key includes only those items that either are new, or newly deleted, or whose numbering has changed. Examples are omitted since they can easily be traced via the indexes, through the plant names mentioned.

1. Berlin Code to Tokyo Code

Berlin	Tokyo
Pre. 7	Pre. 7 + 8
Pre. 8-10	Pre. 9-11
4.1	4.1 + 4.2
4.3-3	4.3-4
7.3	9.1
7.4	9.9
7.4 footnote	9.7 footnote
7.5	9.2 + 9 Note 3
7.6-9	9.3-6
7.10	9.11
7.11	7.3
7.12	7.4 p.p.
7.13	7.5
7.14	7.7 p.p.
7.15	–
7.16	7.7 p.p.
7.17	14.8
7.18	8.4
7.19	7.9
7.20	7.8
7.21	7.6
7 Note 1	9 Note 1
7 Note 2	7 Note 1
7B.1	–
7B.2-4	9A.1-3
7B.5-6	9A.5-6
7B.7	10A
7C	9B
8.1	9.13 + 10.5
8.1 footnote	9.4 footnote
8.2-3	7.10-11
8.4	9.14
8.5	9.12
9.1	8.1
9.2	9.10
9.3	8.3
9.4	8A.3 + 8.5
9.5	8.2
9A.1	8B.1
10.3	10.4
10.4-5	10.6-7
10 Note 1	10 Note 3
10A	10 Note 2
11.2-3	11.3-4
11.4	11.9
13.1(d) p.p.	15.1 p.p.
14.8	14.9
14.9	14.10 + 15.2
14.10-11	14.11-12
14.12	14.13 + 15.1 p.p.
14 Note 1	–
14 Note 2	14Note 3 + 15 Note 1
15	14.14
15A	14A
16 Note 1	16 Note 2
19.2-3	19.3-4
19.4-5	–
19.6-7	19.6-7
19A.1	19A.2
22.2-5	22.3-6
23.6(a)+(b)	23.6(b)
23.6(c)	23.6(a)+(c)
23B.1	23A.3
24.5	24 Note 1
24 Note 1	24 Note 2
26.2	26.3
29.2-4	30.1-3
29A	30A
30.1-2	31.1-2
30A	31A
31.1	30.4
31 Note 1	30 Note 1
32.2-6	32.3-7
32 Note 1	–
33.2	33.2 + 33.3
33.2 footnote	33 Note 1
33.3-5	33.4-6
33 Note 1	33 Note 2
34.1 + 34.2	34.1
34.3	34.2
40.2	11.8
42.2	42.3
42 Note 1	42.2
42 Note 2	42.4
45.2-4	45.3-5
46.2	46 Note 1
46.3	46.4 + 46.5
50A.2	–
51-54	–
55.1	–
55.2	7.4 p.p.
56	–
57.1	–
57.2-3	11.5-6
57 Note 1	11 Note 2
57A	–
58.1	11.7
60.1	11.2
61	–
61A.1	19A.1
61A.2	21B.3
61A.3	24B.2
61A.4	–
62-63	51-52
64.1-3	53.1-3
64.3 footnote	53.4
64.4	53.5 + 53 Note 2
64.5	53.6
64 Note 1	53 Note 1
65	54
68	55
69.1-2	56.1-2
69.3	–
69.4	57.1
72.1-2	58.1-2
72 Note1	58.3
72A	58A
73.1-9	60.1-9
73.10	60.11
73 Notes 1-3	60 Notes 1-3
73A-I	60A-I
75.1	61.1
75.2-4	61.3-5
75 Note 1	61.2
75 Note 2	61 Note 1
76	62
76A	62A
App. IIIB p.p.	14 Note 2

2. Tokyo Code to Berlin Code

Tokyo	Berlin
Pre. 7 + 8	Pre. 7
Pre. 9-11	Pre. 8-10
4.1 + 4.2	4.1
4.3-4	4.2-3
7.3	7.11
7.4	7.12 + 55.2
7.5	7.13
7.6	7.21
7.7	7.16 + 7.14
7-8	7.20
7.9	7.19
7.10-11	8.2-3
7 Note 1	7 Note 2
8.1	9.1
8.2	9.5
8.3	9.3
8.4	7.18
8.5	9.4 p.p.
8A.1 + 8A.2	–
8A.3	9.4 p.p.
8B.1	9A.1
8B.2	–
9.1	7.3
9.2	7.5 p.p.
9.3-9.4	7.6-7.7
9.4 footnote	8.1 footnote
9.5-9.6	7.8-7.9
9.7 + 9.8	–
9.9	7.4
9.10	9.2
9.11	7.10
9.12	8.5
9.13	8.1 p.p.
9.14	8.4
9 Note 1	7 Note 1
9 Note 2	–
9 Note 3	7.5 p.p.
9A.1-3	7B.2-4
9A.4	–
9A.5-6	7B.5-6
9B	7C
10.3	–
10.4	10.3
10.5	8.1 p.p.
10.6-7	10.4-5
10 Note 1	–
10 Note 2	10A
10 Note 3	10 Note 1
10A	7B.7
11.2	60.1
11.3-4	11.2-3
11.5-6	57.2-3
11.7	58.1
11.8	40.2
11.9	11.4
11 Note 1	–
11 Note 2	57 Note 1
14.8	7.17
14.9-13	14.8-12
14.14	15
14 Note 1	–
14 Note 2	App. IIIB p.p.
14 Note 3	14 Note 2 p.p.
14A	15A
15.1	13.1(d) p.p. + 14.12 p.p.
15.2	14.9 p.p.
15.3+4+5+6	–
15 Note 1	14 Note 2 p.p.
16 Note 1	–
16 Note 2	16 Note 1
19.2	–
19.3-4	19.2-3
19.5	–
19.6-7	19.6-7
19A.1	61A.1
19A.2	19A.1
21B.3	61A.2
22.2	–
22.3-6	22.2-5
23.6(a)	23.6(c) p.p.
23.6(b)	23.6(a)+(b)
23.6(c)	23.6(c) p.p.
23.7 + 23.8	–
23A.3	23B.1
24 Note 1	24.5
24 Note 2	24 Note 1
24B.2	61A.3
26.2	–
26.3	26.2
30.1-3	29.2-4
30.4	31.1
30 Note 1	31 Note 1
30A	29A
31.1-2	30.1-2
31A	30A
32.2	–
32.3-7	32.2-6
32.8 + 32.9	–
32 Note 1	–
32F	–
33.2 + 33.3	33.2
33.4-6	33.3-5
33 Note 1	33.2 footnote
33 Note 2	33 Note 1
34.1	34.1 + 34.2
34.2	34.3
36.3	–
42.2	42 Note 1
42.3	42.2
42.4	42 Note 2
45.2	–
45.3-5	45.2-4
46.2 + 46.3	–
46.4 + 46.5	46.3
46.6	–
46 Note 1	46.2
46 Note 2	–
46A Note 1	–
49 Note 1	–
51-52	62-63
53.1-3	64.1-3
53.4	64.3 footnote
53.5	64.4 p.p.
53.6	64.5
53 Note 1	64 Note 1
53 Note 2	64.4 p.p.
54	65
55	68
56.1-2	69.1-2
57.1	69.4
58.1-2	72.1-2
58.3	72 Note 1
58A	72A
60.1-9	73.1-9
60.10	–
60.11	73.10
60 Notes 1-3	73 Notes 1-3
60A-I	73A-I
61.1	75.1
61.2	75 Note 1
61.3-5	75.2-4
61 Note	175 Note 2
62	76
62A	76A
App. V	–

IMPORTANT DATES IN THE CODE

DATES UPON WHICH PARTICULAR PROVISIONS OF THE CODE BECOME EFFECTIVE

1 May 1753	Art. 13.1(a), (c), (d), (e)
1753	Art. 7.7
1 Jan 1801	Art. 13.1(b)
31 Dec 1801	Art. 13.1(d)
31 Dec 1820	Art. 13.1(f)
1 Jan 1821	Art. 13.1(d)
1 Jan 1848	Art. 13.1(e)
1 Jan 1886	Art. 13.1(e)
1 Jan 1890	Art. 35.3
1 Jan 1892	Art. 13.1(e)
1 Jan 1900	Art. 13.1(e)
1 Jan 1908	Art. 42.3; 44.1
1 Jan 1912	Art. 20.2; 38.1
1 Jan 1935	Art. 36.1
1 Jan 1953	Art. 30.1; 30.3; 30.4; 32.4; 33.2; 34.2; 35.1; 35.2
1 Jan 1958	Art. 36.2; 37.1; 39.1
1 Jan 1959	Art. 28 Note 2
1 Jan 1973	Art. 30.3; 45.1
1 Jan 1990	Art. 9.14; 37.4; 37.5
1 Jan 1996	Art. 36.3
1 Jan 2000	Art. 32.1; 45.2

ARTICLES INVOLVING DATES APPLICABLE TO THE MAIN TAXONOMIC GROUPS

All groups	Art. 9.14; 20.2; 30.1; 30.3; 30.4; 32.1; 32.4; 33.2; 34.2; 35.1; 35.2; 35.3; 37.1; 37.4; 37.5; 42.3; 44.1; 45.1; 45.2
All except algae and fossils	Art. 36.1
Spermatophyta	Art. 13.1(a)
Pteridophyta	Art. 13.1(a)
Bryophyta	Art. 7.7; 13.1(b), (c)
Fungi (incl. Myxomycetes)	Art. 13.1(d)
Algae	Art. 7.7; 13.1(f); 36.2; 39.1
Fossil plants	Art. 7.7; 13.1(f); 36.3; 38.1
Cultivated plants	Art. 28 Note 2

ARTICLES DEFINING THE DATES OF CERTAIN WORKS

Art. 13.1(a)-(f); 13.5

INTERNATIONAL CODE OF BOTANICAL NOMENCLATURE

PREAMBLE

1. Botany requires a precise and simple system of nomenclature used by botanists in all countries, dealing on the one hand with the terms which denote the ranks of taxonomic groups or units, and on the other hand with the scientific names which are applied to the individual taxonomic groups of plants[1]. The purpose of giving a name to a taxonomic group is not to indicate its characters or history, but to supply a means of referring to it and to indicate its taxonomic rank. This *Code* aims at the provision of a stable method of naming taxonomic groups, avoiding and rejecting the use of names which may cause error or ambiguity or throw science into confusion. Next in importance is the avoidance of the useless creation of names. Other considerations, such as absolute grammatical correctness, regularity or euphony of names, more or less prevailing custom, regard for persons, etc., notwithstanding their undeniable importance, are relatively accessory.

2. The Principles form the basis of the system of botanical nomenclature.

3. The detailed Provisions are divided into Rules, set out in the Articles, and Recommendations. Examples (Ex.) are added to the rules and recommendations to illustrate them.

4. The object of the Rules is to put the nomenclature of the past into order and to provide for that of the future; names contrary to a rule cannot be maintained.

5. The Recommendations deal with subsidiary points, their object being to bring about greater uniformity and clearness, especially in future nomenclature; names contrary to a recommendation cannot, on that account, be rejected, but they are not examples to be followed.

6. The provisions regulating the modification of this *Code* form its last division.

---

[1] In this *Code,* unless otherwise indicated, the word "plant" means any organism traditionally studied by botanists (see Pre. 7).

7. The rules and recommendations apply to all organisms traditionally treated as plants, whether fossil or non-fossil[1], e.g., blue-green algae *(Cyanobacteria)*[2]; fungi, including chytrids, oomycetes, and slime moulds; photosynthetic protists and taxonomically related non-photosynthetic groups.

8. Special provisions are needed for certain groups of plants: The *International code of nomenclature for cultivated plants-1980* was adopted by the International Commission for the Nomenclature of Cultivated Plants; provisions for the names of hybrids appear in App. I.

9. The only proper reasons for changing a name are either a more profound knowledge of the facts resulting from adequate taxonomic study or the necessity of giving up a nomenclature that is contrary to the rules.

10. In the absence of a relevant rule or where the consequences of rules are doubtful, established custom is followed.

11. This edition of the *Code* supersedes all previous editions.

---

[1] In this *Code,* the term "fossil" is applied to a taxon when its name is based on a fossil type and the term "non-fossil" is applied to a taxon when its name is based on a non-fossil type (see Art. 13.3).

[2] For the nomenclature of other prokaryotic groups, see the *International code of nomenclature of bacteria.*

DIVISION I. PRINCIPLES

Principle I

Botanical nomenclature is independent of zoological and bacteriological nomenclature. The *Code* applies equally to names of taxonomic groups treated as plants whether or not these groups were originally so treated (see Pre. 7).

Principle II

The application of names of taxonomic groups is determined by means of nomenclatural types.

Principle III

The nomenclature of a taxonomic group is based upon priority of publication.

Principle IV

Each taxonomic group with a particular circumscription, position, and rank can bear only one correct name, the earliest that is in accordance with the Rules, except in specified cases.

Principle V

Scientific names of taxonomic groups are treated as Latin regardless of their derivation.

Principle VI

The Rules of nomenclature are retroactive unless expressly limited.

DIVISION II. RULES AND RECOMMENDATIONS

CHAPTER I. RANKS OF TAXA

Article 1

1.1. Taxonomic groups of any rank will, in this *Code,* be referred to as *taxa* (singular: *taxon*).

Article 2

2.1. Every individual plant is treated as belonging to an indefinite number of taxa of consecutively subordinate rank, among which the rank of species *(species)* is basic.

Article 3

3.1. The principal ranks of taxa in descending sequence are: kingdom *(regnum),* division or phylum *(divisio, phylum),* class *(classis),* order *(ordo),* family *(familia),* genus *(genus),* and species *(species)*. Thus, except for some fossil plants (see Art. 3.3), each species is assignable to a genus, each genus to a family, etc.

3.2. The principal ranks of nothotaxa (hybrid taxa) are nothogenus and nothospecies. These are the same ranks as genus and species, only the terms denoting the ranks differing in order to indicate the hybrid character (see App. I).

3.3. Because of the fragmentary nature of the specimens on which the species of some fossil plants are based, the genera to which they are assigned are not assignable to a family, although they may be referable to a taxon of higher rank. Such genera are known as form-genera *(forma-genera).*

Ex. 1. Form-genera: *Dadoxylon* Endl. *(Coniferopsida), Pecopteris* (Brongn.) Sternb. *(Pteridopsida), Stigmaria* Brongn. *(Lepidodendrales), Spermatites* Miner (seed-bearing plants).

Ex. 2. The following are, however, not form-genera: *Lepidocarpon* D. H. Scott *(Lepidocarpaceae), Mazocarpon* M. J. Benson *(Sigillariaceae), Siltaria* Traverse *(Fagaceae).*

Note 1. Art. 59 provides for form-taxa for asexual forms (anamorphs) of certain pleomorphic fungi, at any rank.

3.4. As in the case of certain pleomorphic fungi, the provisions of this *Code* do not prevent the publication and use of names of form-genera of fossils.

Article 4

4.1. The secondary ranks of taxa in descending sequence are tribe *(tribus)* between family and genus, section *(sectio)* and series *(series)* between genus and species, and variety *(varietas)* and form *(forma)* below species.

4.2. If a greater number of ranks of taxa is desired, the terms for these are made by adding the prefix *sub-* to the terms denoting the principal or secondary ranks. A plant may thus be assigned to taxa of the following ranks (in descending sequence): *regnum, subregnum, divisio* or *phylum, subdivisio* or *subphylum, classis, subclassis, ordo, subordo, familia, subfamilia, tribus, subtribus, genus, subgenus, sectio, subsectio, series, subseries, species, subspecies, varietas, subvarietas, forma, subforma.*

4.3. Further ranks may also be intercalated or added, provided that confusion or error is not thereby introduced.

4.4. The subordinate ranks of nothotaxa are the same as the subordinate ranks of non-hybrid taxa, except that nothogenus is the highest rank permitted (see App. I).

Note 1. Throughout this *Code* the phrase "subdivision of a family" refers only to taxa of a rank between family and genus and "subdivision of a genus" refers only to taxa of a rank between genus and species.

Note 2. For the designation of certain variants of species in cultivation, see Art. 28 Notes 1 and 2.

Note 3. In classifying parasites, especially fungi, authors who do not give specific, subspecific, or varietal value to taxa characterized from a physiological standpoint but scarcely or not at all from a morphological standpoint may distinguish within the species special forms *(formae speciales)* characterized by their adaptation to different hosts, but the nomenclature of special forms is not governed by the provisions of this *Code*.

Article 5

5.1. The relative order of the ranks specified in Art. 3 and 4 must not be altered (see Art. 33.5 and 33.6).

CHAPTER II. NAMES OF TAXA (GENERAL PROVISIONS)

SECTION 1. DEFINITIONS

Article 6

6.1. Effective publication is publication in accordance with Art. 29-31.

6.2. Valid publication of names is publication in accordance with Art. 32-45 or H.9 (see also Art. 61).

6.3. A legitimate name is one that is in accordance with the rules.

6.4. An illegitimate name is one that is designated as such in Art. 18.3, 19.5, or 52-54 (see also Art. 21 Note 1 and Art. 24 Note 2). A name which according to this *Code* was illegitimate when published cannot become legitimate later unless it is conserved or sanctioned.

6.5. The correct name of a taxon with a particular circumscription, position, and rank is the legitimate name which must be adopted for it under the rules (see Art. 11).

Ex. 1. The generic name *Vexillifera* Ducke (1922), based on the single species *V. micranthera,* is legitimate because it is in accordance with the rules. The same is true of the generic name *Dussia* Krug & Urb. ex Taub. (1892), based on the single species *D. martinicensis.* Both generic names are correct when the genera are thought to be separate. Harms (in Repert. Spec. Nov. Regni Veg. 19: 291. 1924), however, united *Vexillifera* and *Dussia* in a single genus; the latter name is the correct one for the genus with this particular circumscription. The legitimate name *Vexillifera* may therefore be correct or incorrect according to different concepts of the taxa.

6.6. In this *Code,* unless otherwise indicated, the word "name" means a name that has been validly published, whether it is legitimate or illegitimate (see Art. 12).

6.7. The name of a taxon below the rank of genus, consisting of the name of a genus combined with one or two epithets, is termed a combination (see Art. 21, 23, and 24).

Ex. 2. Combinations: *Mouriri* subg. *Pericrene, Arytera* sect. *Mischarytera, Gentiana lutea, Gentiana tenella* var. *occidentalis, Equisetum palustre* var. *americanum, Equisetum palustre* f. *fluitans.*

6.8. Autonyms are such names as can be established automatically under Art. 22.3 and 26.3, whether or not they appear in print in the publication in which they are created (see Art. 32.7).

SECTION 2. TYPIFICATION

Article 7

7.1. The application of names of taxa of the rank of family or below is determined by means of nomenclatural types (types of names of taxa). The application of names of taxa in the higher ranks is also determined by means of types when the names are ultimately based on generic names (see Art. 10.7).

7.2. A nomenclatural type *(typus)* is that element to which the name of a taxon is permanently attached, whether as a correct name or as a synonym. The nomenclatural type is not necessarily the most typical or representative element of a taxon.

7.3. A new name published as an avowed substitute *(nomen novum)* for an older name is typified by the type of the older name (see Art. 33.2; but see Art. 33 Note 2).

Ex. 1. Myrcia lucida McVaugh (1969) was published as a *nomen novum* for *M. laevis* O. Berg (1862), an illegitimate homonym of *M. laevis* G. Don (1832). The type of *M. lucida* is therefore the type of *M. laevis* O. Berg (non G. Don), namely, Spruce 3502.

7.4. A new name formed from a previously published legitimate name *(stat. nov., comb. nov.)* is, in all circumstances, typified by the type of the basionym, even though it may have been applied erroneously to a taxon now considered not to include that type (but see Art. 48.1 and 59.6).

Ex. 2. Pinus mertensiana Bong. was transferred to the genus *Tsuga* by Carrière, who, however, as is evident from his description, erroneously applied the new combination *T. mertensiana* to another species of *Tsuga,* namely *T. heterophylla* (Raf.) Sarg. The combination *Tsuga mertensiana* (Bong.) Carrière must not be applied to *T. heterophylla* but must be retained for *P. mertensiana* when that species is placed in *Tsuga;* the citation in parentheses (under Art. 49) of the name of the original author, Bongard, indicates the type of the name.

Ex. 3. Iridaea splendens (Setch. & N. L. Gardner) Papenf., *I. cordata* var. *splendens* (Setch. & N. L. Gardner) I. A. Abbott (in Syesis 4: 55. 1972), and *Gigartina cordata* var. *splendens* (Setch. & N. L. Gardner) D. H. Kim (in Nova Hedwigia 27: 40. 1976) all have the same type as their basionym, *Iridophycus splendens* Setch. & N. L. Gardner, namely, Gardner 7781 (UC No. 539565).

7.5. A name which, under Art. 52, was illegitimate when published is either automatically typified by the type of the name which ought to have been adopted under the rules, or by a different type designated or definitely indicated by the author of the illegitimate name. Automatic typification does not apply to names sanctioned under Art. 15.

7.6. The type of an autonym is the same as that of the name from which it is derived.

7.7. A name validly published by reference to a previously and effectively published description or diagnosis (Art. 32.1(c)) is to be typified by an element selected from the context of the validating description or diagnosis, unless the validating author has definitely designated a different type (but see Art. 10.2). However, the type of a name of a taxon assigned to a group with a nomenclatural starting-point later than 1753 (see Art. 13.1) is to be determined in accordance with the indication or descriptive and other matter accompanying its valid publication (see Art. 32-45).

Ex. 4. Since the name *Adenanthera bicolor* Moon (1824) is validated solely by reference to Rumphius (Herb. Amboin. 3: t. 112. 1743), the type of the name, in the absence of the specimen from which it was figured, is the illustration referred to. It is not the specimen, at Kew, collected by Moon and labelled *"Adenanthera bicolor"*, since Moon did not definitely designate the latter as the type.

Ex. 5. *Echium lycopsis* L. (Fl. Angl.: 12. 1754) was published without a description or diagnosis but with reference to Ray (Syn. Meth. Stirp. Brit., ed. 3: 227. 1724), in which a *"Lycopsis"* species was discussed with no description or diagnosis but with citation of earlier references, including Bauhin (Pinax: 255. 1623). The accepted validating description of *E. lycopsis* is that of Bauhin, and the type must be chosen from the context of his work. Consequently the Sherard specimen in the Morison herbarium (OXF), selected by Klotz (in Wiss. Z. Martin-Luther-Univ. Halle-Wittenberg Math.-Naturwiss. Reihe 9: 375-376. 1960), although probably consulted by Ray, is not eligible as type. The first acceptable choice is that of the illustration, cited by both Ray and Bauhin, of *"Echii altera species"* in Dodonaeus (Stirp. Hist. Pempt.: 620. 1583), suggested by Gibbs (in Lagascalia 1: 60-61. 1971) and formally made by Stearn (in Ray Soc. Publ. 148, Introd.: 65. 1973).

7.8. Typification of names adopted in one of the works specified in Art. 13.1(d), and thereby sanctioned (Art. 15), may be effected in the light of anything associated with the name in that work.

7.9. The typification of names of form-genera of plant fossils (Art. 3.3), of fungal anamorphs (Art. 59), and of any other analogous genera or lower taxa does not differ from that indicated above.

Note 1. See also Art. 59 for details regarding typification of names in certain pleomorphic fungi.

7.10. For purposes of priority (Art. 9.13 and 10.5), designation of a type is achieved only by effective publication (Art. 29-31).

7.11. For purposes of priority (Art. 9.13 and 10.5), designation of a type is achieved only if the type is definitely accepted as such by the typifying author, and if the type element is clearly indicated by direct citation including the term "type" or an equivalent.

Ex. 6. *Chlorosarcina* Gerneck (1907) originally comprised two species, *C. minor* and *C. elegans*. Vischer (1933) transferred the former to *Chlorosphaera* G. A. Klebs and retained the latter in *Chlorosarcina*. He did not, however, use the term "type" or an equivalent, so that his action does not constitute typification of *Chlorosarcina*. The first to designate a type, as "LT.", was Starr (in ING Card No. 16528, Nov 1962), who selected *Chlorosarcina elegans*.

**Ex. 7.* The phrase "standard species" as used by Hitchcock & Green (in Anon., Nomencl. Prop. Brit. Botanists: 110-199. 1929) is now treated as equivalent to "type", and hence type designations in this work are acceptable.

Recommendation 7A

7A.1. It is strongly recommended that the material on which the name of a taxon is based, especially the holotype, be deposited in a public herbarium or other public collection with a policy of giving *bona fide* botanists open access to deposited material, and that it be scrupulously conserved.

Article 8

8.1. The type of a name of a species or infraspecific taxon is a single specimen or illustration except in the following case: for small herbaceous plants and for most non-vascular plants, the type may consist of more than one individual, which ought to be conserved permanently on one herbarium sheet or in one equivalent preparation (e.g., box, packet, jar, microscope slide).

8.2. Type specimens of names of taxa must be preserved permanently and cannot be living plants or cultures.

**Ex. 1.* The strain CBS 7351, given as the type of the name *Candida populi* Hagler & al. (in Int. J. Syst. Bacteriol. 39: 98. 1989), is acceptable as a nomenclatural type as it is permanently preserved in a metabolically inactive state by lyophilization (see also Rec. 8B.2).

8.3. If it is impossible to preserve a specimen as the type of a name of a species or infraspecific taxon of non-fossil plants, or if such a name is without a type specimen, the type may be an illustration.

8.4. The type of the name of a taxon of fossil plants of the rank of species or below is the specimen whose figure either accompanies or is cited in the valid publication of the name (see Art. 38). If figures of more than one specimen were given or cited when the name was validly published, one of those specimens must be chosen as the type.

8.5. One whole specimen used in establishing a taxon of fossil plants is to be considered the nomenclatural type.

Recommendation 8A

8A.1. When a holotype, a lectotype or a neotype is an illustration (see Art. 8.3), the specimen or specimens upon which that illustration is based should be used to help determine the application of the name.

8A.2. When it is impossible to preserve a type specimen and an illustration is designated as the type of the name of a new taxon (see Art. 8.3), the collection data of the illustrated material should be given (see also Rec. 32D.2).

---

* Here and elsewhere in the *Code*, a prefixed asterisk denotes a "voted Example" (see Preface, p. x).

8A.3. If the type specimen of a name of a fossil plant is cut into pieces (sections of fossil wood, pieces of coal-ball plants, etc.), all parts originally used in establishing the diagnosis ought to be clearly marked.

Recommendation 8B

8B.1. Whenever practicable a living culture should be prepared from the holotype material of the name of a newly described taxon of fungi or algae and deposited in at least two institutional culture or genetic resource collections. (Such action does not obviate the requirement for a holotype specimen under Art. 8.2.)

8B.2. In cases where the nomenclatural type is a culture permanently preserved in a metabolically inactive state (see Art. 8 Ex. 1), any living isolates obtained from that should be referred to as "ex-type" *(ex typo)*, "ex-holotype" *(ex holotypo)*, "ex-isotype" *(ex isotypo)*, etc., in order to make it clear they are derived from the type but are not themselves the nomenclatural type.

Article 9

9.1. A holotype of a name of a species or infraspecific taxon is the one specimen or illustration used by the author, or designated by the author as the nomenclatural type. As long as a holotype is extant, it fixes the application of the name concerned (see also Art. 10).

Note 1. Any designation made by the original author, if definitely expressed at the time of the original publication of the name of the taxon, is final (but see Art. 9.9). If the author included only one element, that one must be accepted as the holotype. If a new name is based on a previously published description or diagnosis of the taxon, the same considerations apply to material included by the earlier author (see Art. 7.7 and 7.8).

9.2. A lectotype is a specimen or illustration (see Art. 8.3) designated as the nomenclatural type, in conformity with Art. 9.9, when no holotype was indicated at the time of publication, when the holotype is found to belong to more than one taxon, or as long as it is missing.

9.3. An isotype is any duplicate[1] of the holotype; it is always a specimen.

9.4. A syntype is any one of two or more specimens cited in the protologue[2] when no holotype was designated, or any one of two or more specimens simultaneously designated as types.

---

[1] Here and elsewhere, the word duplicate is given its usual meaning in herbarium curatorial practice. It is part of a single gathering of a single species or infraspecific taxon made by a collector at one time. The possibility of a mixed gathering must always be considered by an author choosing a lectotype, and corresponding caution used.

[2] Protologue (from the Greek *protos,* first; *logos,* discourse): everything associated with a name at its valid publication, i.e., description or diagnosis, illustrations, references, synonymy, geographical data, citation of specimens, discussion, and comments.

9.5. A paratype is a specimen cited in the protologue that is neither the holotype nor an isotype, nor one of the syntypes if two or more specimens were simultaneously designated as types.

Ex. 1. The holotype of the name *Rheedia kappleri* Eyma, which applies to a polygamous species, is a male specimen collected by Kappler (593a in U). The author designated a hermaphroditic specimen collected by the Forestry Service of Surinam as a paratype (B. W. 1618 in U).

Note 2. In most cases in which no holotype was designated there will also be no paratypes, since all the cited specimens will be syntypes. However, when an author designated two or more specimens as types (Art. 9.4), any remaining cited specimens are paratypes and not syntypes.

9.6. A neotype is a specimen or illustration (see Art. 8.3) selected to serve as nomenclatural type as long as all of the material on which the name of the taxon was based is missing (see also Art. 9.11).

9.7. An epitype is a specimen or illustration selected to serve as an interpretative type when the holotype, lectotype or previously designated neotype, or all original material[1] associated with a validly published name, is demonstrably ambiguous and cannot be critically identified for purposes of the precise application of the name of a taxon. When an epitype is designated, the holotype, lectotype or neotype that the epitype supports must be explicitly cited.

9.8. The use of a term defined in the *Code* (Art. 9.1-9.7) as denoting a type, in a sense other than that in which it is so defined, is treated as an error to be corrected (for example, the use of the term lectotype to denote what is in fact a neotype).

Ex. 2. Borssum Waalkes (in Blumea 14: 198. 1966) cited Herb. Linnaeus No. 866.7 (LINN) as the holotype of *Sida retusa* L. (1763). The term is incorrectly used because illustrations in Plukenet (Phytographia: t. 9, f. 2. 1691) and Rumphius (Herb. Amboin. 6: t. 19. 1750) were cited by Linnaeus in the protologue of *S. retusa*. Since all three elements are original material (Art. 9.9, footnote), Borssum Waalkes's use of holotype is an error to be corrected to lectotype.

9.9. If no holotype was indicated by the author of a name of a species or infraspecific taxon, or when the holotype has been lost or destroyed, or when the material designated as type is found to belong to more than one taxon, a lectotype or, if permissible (Art. 9.6), a neotype as a substitute for it may be designated (Art. 7.10 and 7.11). A lectotype always takes precedence over a neotype, except as provided by Art. 9.11. An isotype, if such exists, must be chosen as the lectotype. If no isotype exists, the lectotype must be chosen from among the syntypes, if such exist. If neither an isotype nor a syntype nor an

[1] For the purposes of this *Code*, "original material" comprises *(a)* those specimens and illustrations (both unpublished and published either prior to or together with the protologue) upon which it can be shown that the description or diagnosis validating the name was based; *(b)* the holotype and those specimens which, even if not seen by the author of the description or diagnosis validating the name, were indicated as types (syntypes or paratypes) of the name at its valid publication; and *(c)* the isotypes or isosyntypes of the name irrespective of whether such specimens were seen by either the author of the validating description or diagnosis, or the author of the name.

isosyntype (duplicate of syntype) nor any other part of the original material is extant, a neotype may be selected.

Note 3. When two or more specimens have been designated as types by the author of a name (e.g. male and female, flowering and fruiting, etc.), the lectotype must be chosen from among them (see Art. 9.4).

9.10. When a type specimen (herbarium sheet or equivalent preparation) contains parts belonging to more than one taxon (see Art. 9.9), the name must remain attached to that part which corresponds most nearly with the original description or diagnosis.

Ex. 3. The type of the name *Tillandsia bryoides* Griseb. ex Baker (1878) is Lorentz 128 in BM; this, however, proved to be a mixture. Smith (in Proc. Amer. Acad. Arts 70: 192. 1935) acted in accordance with this rule in designating one part of Lorentz's gathering as the lectotype.

9.11. When a holotype or a previously designated lectotype has been lost or destroyed and it can be shown that all the other original material differs taxonomically from the destroyed type, a neotype may be selected to preserve the usage established by the previous typification (see also Art. 9.12).

9.12. A neotype selected under Art. 9.11 may be superseded if it can be shown to differ taxonomically from the holotype or lectotype that it replaced.

9.13. The author who first designates a lectotype or a neotype must be followed, but his choice is superseded if *(a)* the holotype or, in the case of a neotype, any of the original material is rediscovered; it may also be superseded if *(b)* it can be shown that it is in serious conflict with the protologue and another element is available which is not in conflict with the protologue, or *(c)* that it is contrary to Art. 9.10.

9.14. On or after 1 January 1990, lectotypification or neotypification of a name of a species or infraspecific taxon by a specimen or unpublished illustration (see Art. 8.3) is not effected unless the herbarium or institution in which the type is conserved is specified.

Recommendation 9A

9A.1. Typification of names for which no holotype was designated should only be carried out with an understanding of the author's method of working; in particular it should be realized that some of the material used by the author in describing the taxon may not be in the author's own herbarium or may not even have survived, and conversely, that not all the material surviving in the author's herbarium was necessarily used in describing the taxon.

9A.2. Designation of a lectotype should be undertaken only in the light of an understanding of the group concerned. In choosing a lectotype, all aspects of the protologue should be considered as a basic guide. Mechanical methods, such as the automatic selection of the first species or specimen cited or of a specimen collected by the person after whom a species is named, should be avoided as unscientific and productive of possible future confusion and further changes.

9A.3. In choosing a lectotype, any indication of intent by the author of a name should be given preference unless such indication is contrary to the protologue. Such indications are manuscript notes, annotations on herbarium sheets, recognizable figures, and epithets such as *typicus, genuinus,* etc.

9A.4. When a single collection is cited in the protologue, but a particular institution housing this is not designated, it should be assumed that the specimen housed in the institution where the author is known to have worked is the holotype, unless there is evidence that he used further material of the same collection.

9A.5. When two or more heterogeneous elements were included in or cited with the original description or diagnosis, the lectotype should be so selected as to preserve current usage. In particular, if another author has already segregated one or more elements as other taxa, the residue or part of it should be designated as the lectotype provided that this element is not in conflict with the original description or diagnosis (see Art. 9.13).

9A.6. For the name of a fossil species, the lectotype, when one is needed, should, if possible, be a specimen illustrated at the time of the valid publication of the name (see Art. 8.4).

Recommendation 9B

9B.1. In selecting a neotype, particular care and critical knowledge should be exercised because the reviewer usually has no guide except personal judgement as to what best fits the protologue, and if this selection proves to be faulty, it will inevitably result in further change.

Article 10

10.1. The type of a name of a genus or of any subdivision of a genus is the type of a name of a species (except as provided by Art. 10.4). For purposes of designation or citation of a type, the species name alone suffices, i.e., it is considered as the full equivalent of its type.

Note 1. Terms such as "holotype", "syntype", and "lectotype", as presently defined in Art. 9, although not applicable, strictly speaking, to the types of names in ranks higher than species, are so used by analogy.

10.2. If in the protologue of the name of a genus or of any subdivision of a genus the holotype or lectotype of one or more previously or simultaneously published species name(s) is definitely included (see Art. 10.3), the type must be chosen (Art. 7.10 and 7.11) from among these types unless the type was indicated (Art. 22.5, 22.6, and 37.2) or designated by the author of the name. If no type of a previously or simultaneously published species name was definitely included, a type must be otherwise chosen, but the choice is to be superseded if it can be demonstrated that the selected type is not conspecific with any of the material associated with the protologue.

Ex. 1. The genus *Anacyclus,* as originally circumscribed by Linnaeus (1753), comprised three validly named species. Cassini (in Cuvier, Dict. Sci. Nat. 34: 104. 1825) designated *Anthemis valentina* L. (1753) as type of *Anacyclus,* but this was not an original element of the genus. Green (in Anon.,

Nomencl. Prop. Brit. Botanists: 182. 1929) designated *Anacyclus valentinus* L. (1753), "the only one of the three original species still retained in the genus", as the "standard species" (see Art. 7 Ex. 7), and her choice must be followed (Art. 10.5). Humphries (in Bull. Brit. Mus. (Nat. Hist.) Bot. 7: 109. 1979) designated a specimen in the Clifford Herbarium (BM) as lectotype of *Anacyclus valentinus,* and that specimen thereby became the ultimate type of the generic name.

Ex. 2. *Castanella* Spruce ex Benth. & Hook. f. (1862) was described on the basis of a single specimen and without mention of a species name. Swart (in ING Card No. 2143. 1957) was the first to designate a type (as "T."): *C. granatensis* Triana & Planch. (1862), based on a Linden collection. As long as the Spruce specimen is considered to be conspecific with Linden's collection Swart's type designation cannot be superseded, even though the Spruce specimen became the type of *Paullinia paullinioides* Radlk. (1896), because the latter is not a "previously or simultaneously published species name".

10.3. For the purposes of Art. 10.2, definite inclusion of the type of a name of a species is effected by citation of, or reference (direct or indirect) to, a validly published name, whether accepted or synonymized by the author, or by citation of the holotype or lectotype of a previously or simultaneously published name of a species.

Ex. 3. The protologue of *Elodes* Adans. (1763) included references to *"Elodes"* of Clusius (1601), *"Hypericum"* of Tournefort (1700), and *Hypericum aegypticum* L. (1753). The latter is the only reference to a validly published name of a species, and neither of the other elements is the type of a name of a species. The type of *H. aegypticum* is therefore the type of *Elodes,* even though subsequent authors designated *H. elodes* L. (1759) as the type (see Robson in Bull. Brit. Mus. (Nat. Hist.), Bot. 5: 305, 336. 1977).

10.4. By and only by conservation (Art. 14.9), the type of a name of a genus may be a specimen or illustration, preferably used by the author in the preparation of the protologue, other than the type of a name of an included species.

Ex. 4. *Physconia* Poelt has been conserved with the specimen "*'Lichen pulverulentus',* Germania, Lipsia in *Tilia,* 1767, Schreber (M)" as the type.

Note 2. If the element designated under Art. 10.4 is the type of a species name, that name may be cited as the type of the generic name. If the element is not the type of a species name, a parenthetical reference to the correct name of the type element may be added.

10.5. The author who first designates a type of a name of a genus or subdivision of a genus must be followed, but the choice may be superseded if *(a)* it can be shown that it is in serious conflict with the protologue and another element is available which is not in conflict with the protologue, or *(b)* that it was based on a largely mechanical method of selection.

Ex. 5. Fink (in Contr. U.S. Natl. Herb. 14(1): 2. 1910) specified that he was "stating the types of the genera according to the 'first species' rule". His type designations may therefore be superseded.

**Ex. 6.* Authors following the *American code of botanical nomenclature,* Canon 15 (in Bull. Torrey Bot. Club 34: 172. 1907), designated as the type "the first binomial species in order" eligible under certain provisions. This method of selection is to be considered as largely mechanical. Thus the first type designation for *Delphinium* L., by Britton (in Britton & Brown, Ill. Fl. N. U.S., ed. 2, 2: 93. 1913), who followed the *American code* and chose *D. consolida* L., has been superseded under Art. 10.5(b) by the designation of *D. peregrinum* L. by Green (in Anon., Nomencl. Prop. Brit. Botanists: 162. 1929). The unicarpellate *D. consolida* could not have been superseded as type by the tricarpellate *D. peregrinum* under Art. 10.5(a), however, because it is not in serious conflict with the generic protologue, which specifies "germina tria vel unum", the assignment of the genus to "Polyandria Trigynia" by Linnaeus notwithstanding.

10.6. The type of a name of a family or of any subdivision of a family is the same as that of the generic name on which it is based (see Art. 18.1). For purposes of designation or citation of a type, the generic name alone suffices. The type of a name of a family or subfamily not based on a generic name is the same as that of the corresponding alternative name (Art. 18.5 and 19.8).

10.7. The principle of typification does not apply to names of taxa above the rank of family, except for names that are automatically typified by being based on generic names (see Art. 16). The type of such a name is the same as that of the generic name on which it is based.

Note 3. For the typification of some names of subdivisions of genera see Art. 22.5 and 22.6.

Recommendation 10A

10A.1. When a combination in a rank of subdivision of a genus has been published under a generic name that has not yet been typified, the type of the generic name should be selected from the subdivision of the genus that was designated as nomenclaturally typical, if that is apparent.

SECTION 3. PRIORITY

Article 11

11.1. Each family or taxon of lower rank with a particular circumscription, position, and rank can bear only one correct name, special exceptions being made for 9 families and 1 subfamily for which alternative names are permitted (see Art. 18.5 and 19.7). However, the use of separate names for the form-taxa of fungi and for form-genera of fossil plants is allowed under Art. 3.3 and 59.5.

11.2. In no case does a name have priority outside the rank in which it is published (but see Art. 53.5).

Ex. 1. Campanula sect. *Campanopsis* R. Br. (Prodr.: 561. 1810) when treated as a genus is called *Wahlenbergia* Roth (1821), a name conserved against the taxonomic synonym *Cervicina* Delile (1813), and not *Campanopsis* (R. Br.) Kuntze (1891).

Ex. 2. Magnolia virginiana var. *foetida* L. (1753) when raised to specific rank is called *M. grandiflora* L. (1759), not *M. foetida* (L.) Sarg. (1889).

Ex. 3. Lythrum intermedium Ledeb. (1822) when treated as a variety of *L. salicaria* L. (1753) is called *L. salicaria* var. *glabrum* Ledeb. (Fl. Ross. 2: 127. 1843), not *L. salicaria* var. *intermedium* (Ledeb.) Koehne (in Bot. Jahrb. Syst. 1: 327. 1881).

Ex. 4. When the two varieties constituting *Hemerocallis lilioasphodelus* L. (1753), var. *flava* L. and var. *fulva* L., are considered to be distinct species, the one not including the lectotype of the species name is called *H. fulva* (L.) L. (1762), but the other one bears the name *H. lilioasphodelus* L., which in the rank of species has priority over *H. flava* (L.) L. (1762).

11.3. For any taxon from family to genus inclusive, the correct name is the earliest legitimate one with the same rank, except in cases of limitation of priority by conservation (see Art. 14) or where Art. 11.7, 15, 19.4, 56, 57, or 59 apply.

Ex. 5. When *Aesculus* L. (1753), *Pavia* Mill. (1754), *Macrothyrsus* Spach (1834) and *Calothyrsus* Spach (1834) are referred to a single genus, its name is *Aesculus* L.

11.4. For any taxon below the rank of genus, the correct name is the combination of the final epithet[1] of the earliest legitimate name of the taxon in the same rank, with the correct name of the genus or species to which it is assigned, except *(a)* in cases of limitation of priority under Art. 14, 15, 56 or 57, or *(b)* if the resulting combination would be invalid under Art. 32.1(b) or illegitimate under Art. 53, or *(c)* if Art. 11.7, 22.1, 26.1, or 59 rule that a different combination is to be used.

Ex. 6. Primula sect. *Dionysiopsis* Pax (in Jahresber. Schles. Ges. Vaterländ. Kultur 87: 20. 1909) when transferred to *Dionysia* Fenzl becomes *D.* sect. *Dionysiopsis* (Pax) Melch. (in Mitt. Thüring. Bot. Vereins 50: 164-168. 1943); the substitute name *D.* sect. *Ariadna* Wendelbo (in Bot. Not. 112: 496. 1959) is illegitimate.

Ex. 7. Antirrhinum spurium L. (1753) when transferred to *Linaria* Mill. is called *L. spuria* (L.) Mill. (1768).

Ex. 8. When transferring *Serratula chamaepeuce* L. (1753) to *Ptilostemon* Cass., Cassini illegitimately named the species *P. muticus* Cass. (1826). In that genus, the correct name is *P. chamaepeuce* (L.) Less. (1832).

Ex. 9. Spartium biflorum Desf. (1798) when transferred to *Cytisus* Desf. could not be called *C. biflorus* because of the previously and validly published *C. biflorus* L'Hér. (1791); the substitute name *C. fontanesii* Spach (1849) was therefore correctly proposed.

Ex. 10. Spergula stricta Sw. (1799) when transferred to *Arenaria* L. is called *A. uliginosa* Schleich. ex Schltdl. (1808) because of the existence of the name *A. stricta* Michx. (1803), based on a different type; but on further transfer to the genus *Minuartia* the epithet *stricta* is again available and the species is called *M. stricta* (Sw.) Hiern (1899).

Ex. 11. Arum dracunculus L. (1753) when transferred to *Dracunculus* Mill. is named *D. vulgaris* Schott (1832), as use of the Linnaean epithet would result in a tautonym.

Ex. 12. Cucubalus behen L. (1753) when transferred to *Behen* Moench was legitimately renamed *B. vulgaris* Moench (1794) to avoid the tautonym *"B. behen"*. In *Silene* L., the epithet *behen* is unavailable because of the existence of *S. behen* L. (1753). Therefore, the substitute name *S. cucubalus* Wibel (1799) was proposed. This, however, is illegitimate since the specific epithet *vulgaris* was available. In *Silene,* the correct name of the species is *S. vulgaris* (Moench) Garcke (1869).

Ex. 13. Helianthemum italicum var. *micranthum* Gren. & Godr. (Fl. France 1: 171. 1847) when transferred as a variety to *H. penicillatum* Thibaud ex Dunal retains its varietal epithet and is named *H. penicillatum* var. *micranthum* (Gren. & Godr.) Grosser (in Engler, Pflanzenr. 14: 115. 1903).

11.5. When, for any taxon of the rank of family or below, a choice is possible between legitimate names of equal priority in the corresponding rank, or between available final epithets of names of equal priority in the corresponding rank, the first such choice to be effectively published (Art. 29-31) establishes the priority of the chosen name, and of any legitimate combination with the

[1] Here and elsewhere in this *Code,* the phrase "final epithet" refers to the last epithet in sequence in any particular combination, whether in the rank of a subdivision of a genus, or of a species, or of an infraspecific taxon.

same type and final epithet at that rank, over the other competing name(s) (but see Art. 11.6).

Note 1. A choice as provided for in Art. 11.5 is effected by adopting one of the competing names, or its final epithet in the required combination, and simultaneously rejecting or relegating to synonymy the other(s), or nomenclatural synonyms thereof.

Ex. 14. When *Dentaria* L. (1753) and *Cardamine* L. (1753) are united, the resulting genus is called *Cardamine* because that name was chosen by Crantz (Cl. Crucif. Emend.: 126. 1769), who first united them.

Ex. 15. When *Entoloma* (Fr. ex Rabenh.) P. Kumm. (1871), *Leptonia* (Fr. : Fr.) P. Kumm. (1871), *Eccilia* (Fr. : Fr.) P. Kumm. (1871), *Nolanea* (Fr. : Fr.) P. Kumm. (1871), and *Claudopus* Gillet (1876) are united, one of the generic names simultaneously published by Kummer must be used for the combined genus. Donk, who did so (in Bull. Jard. Bot. Buitenzorg, ser. 3, 18(1): 157. 1949), selected *Entoloma*, which is therefore treated as having priority over the other names.

Ex. 16. Brown (in Tuckey, Narr. Exp. Congo: 484. 1818) was the first to unite *Waltheria americana* L. (1753) and *W. indica* L. (1753). He adopted the name *W. indica* for the combined species, and this name is accordingly treated as having priority over *W. americana.*

Ex. 17. Baillon (in Adansonia 3: 162. 1863), when uniting for the first time *Sclerocroton integerrimus* Hochst. (1845) and *S. reticulatus* Hochst. (1845), adopted the name *Stillingia integerrima* (Hochst.) Baill. for the combined taxon. Consequently *Sclerocroton integerrimus* is treated as having priority over *S. reticulatus* irrespective of the genus *(Sclerocroton, Stillingia, Excoecaria, Sapium)* to which the species is assigned.

Ex. 18. Linnaeus (1753) simultaneously published the names *Verbesina alba* and *V. prostrata.* Later (1771), he published *Eclipta erecta,* an illegitimate name because *V. alba* was cited in synonymy, and *E. prostrata,* based on *V. prostrata.* The first author to unite these taxa was Roxburgh (Fl. Ind. 3: 438. 1832), who adopted the name *E. prostrata* (L.) L. Therefore *V. prostrata* is treated as having priority over *V. alba.*

Ex. 19. *Donia speciosa* and *D. formosa,* which were simultaneously published by Don (1832), were illegitimately renamed *Clianthus oxleyi* and *C. dampieri* by Lindley (1835). Brown (1849) united both in a single species, adopting the illegitimate name *C. dampieri* and citing *D. speciosa* and *C. oxleyi* as synonyms; his choice is not of the kind provided for by Art. 11.5. *C. speciosus* (D. Don) Asch. & Graebn. (1909), published with *D. speciosa* and *C. dampieri* listed as synonyms, is an illegitimate later homonym of *C. speciosus* (Endl.) Steud. (1840); again, conditions for a choice under Art. 11.5 were not satisfied. Ford & Vickery (1950) published the legitimate combination *C. formosus* (D. Don) Ford & Vickery and cited *D. formosa* and *D. speciosa* as synonyms, but since the epithet of the latter was unavailable in *Clianthus* a choice was not possible and again Art. 11.5 does not apply. Thompson (1990) was the first to effect an acceptable choice when publishing the combination *Swainsona formosa* (D. Don) Joy Thomps. and indicating that *D. speciosa* was a synonym of it.

11.6. An autonym is treated as having priority over the name or names of the same date and rank that established it.

Note 2. When the final epithet of an autonym is used in a new combination under the requirements of Art. 11.6, the basionym of that combination is the name from which the autonym is derived, or its basionym if it has one.

Ex. 20. *Heracleum sibiricum* L. (1753) includes *H. sibiricum* subsp. *lecokii* (Godr. & Gren.) Nyman (Consp. Fl. Europ.: 290. 1879) and *H. sibiricum* subsp. *sibiricum* automatically established at the same time. When *H. sibiricum* is included in *H. sphondylium* L. (1753) as a subspecies, the correct name for the taxon is *H. sphondylium* subsp. *sibiricum* (L.) Simonk. (Enum. Fl. Transsilv.: 266. 1887), not subsp. *lecokii,* whether or not subsp. *lecokii* is treated as distinct.

Ex. 21. The publication of *Salix tristis* var. *microphylla* Andersson (Salices Bor.-Amer.: 21. 1858) created the autonym *S. tristis* Aiton (1789) var. *tristis.* If *S. tristis,* including var. *microphylla,* is recognized as a variety of *S. humilis* Marshall (1785), the correct name is *S. humilis* var. *tristis* (Aiton)

Griggs (in Proc. Ohio Acad. Sci. 4: 301. 1905). However, if both varieties of *S. tristis* are recognized as varieties of *S. humilis*, then the names *S. humilis* var. *tristis* and *S. humilis* var. *microphylla* (Andersson) Fernald (in Rhodora 48: 46. 1946) are both used.

Ex. 22. In the classification adopted by Rollins and Shaw, *Lesquerella lasiocarpa* (Hook. ex A. Gray) S. Watson (1888) is composed of two subspecies, subsp. *lasiocarpa* (which includes the type of the name of the species and is cited without an author) and subsp. *berlandieri* (A. Gray) Rollins & E. A. Shaw. The latter subspecies is composed of two varieties. In that classification the correct name of the variety which includes the type of subsp. *berlandieri* is *L. lasiocarpa* var. *berlandieri* (A. Gray) Payson (1922), not *L. lasiocarpa* var. *berlandieri* (cited without an author) or *L. lasiocarpa* var. *hispida* (S. Watson) Rollins & E. A. Shaw (1972), based on *Synthlipsis berlandieri* var. *hispida* S. Watson (1882), since publication of the latter name established the autonym *S. berlandieri* A. Gray var. *berlandieri* which, at varietal rank, is treated as having priority over var. *hispida*.

11.7. Names of plants (algae excepted) based on a non-fossil type are treated as having priority over names of the same rank based on a fossil (or subfossil) type.

Ex. 23. If *Platycarya* Siebold & Zucc. (1843), a non-fossil genus, and *Petrophiloides* Bowerb. (1840), a fossil genus, are united, the name *Platycarya* is accepted for the combined genus, although it is antedated by *Petrophiloides*.

Ex. 24. The generic name *Metasequoia* Miki (1941) was based on the fossil type of *M. disticha* (Heer) Miki. After discovery of the non-fossil species *M. glyptostroboides* Hu & W. C. Cheng, conservation of *Metasequoia* Hu & W. C. Cheng (1948) as based on the non-fossil type was approved. Otherwise, any new generic name based on *M. glyptostroboides* would have had to be treated as having priority over *Metasequoia* Miki.

11.8. For purposes of priority, names in Latin form given to hybrids are subject to the same rules as are those of non-hybrid taxa of equivalent rank.

Ex. 25. The name ×*Solidaster* H. R. Wehrh. (1932) antedates the name ×*Asterago* Everett (1937) for the hybrid *Aster* L. × *Solidago* L.

Ex. 26. The name ×*Gaulnettya* Marchant (1937) antedates the name ×*Gaulthettya* Camp (1939) for the hybrid *Gaultheria* L. × *Pernettya* Gaudich.

Ex. 27. *Anemone* ×*hybrida* Paxton (1848) antedates *A.* ×*elegans* Decne. (1852), pro sp., as the binomial for the hybrids derived from *A. hupehensis* (Lemoine & E. Lemoine) Lemoine & E. Lemoine × *A. vitifolia* Buch.-Ham. ex DC.

Ex. 28. Camus (in Bull. Mus. Natl. Hist. Nat. 33: 538. 1927) published the name ×*Agroelymus* A. Camus as the name of a nothogenus, without a Latin description or diagnosis, mentioning only the names of the parents involved (*Agropyron* Gaertn. and *Elymus* L.). Since this name was not validly published under the *Code* then in force (Stockholm 1953), Rousseau (in Mém. Jard. Bot. Montréal 29: 10-11. 1952), published a Latin diagnosis. However, the date of valid publication of the name ×*Agroelymus* under this *Code* (Art. H.9) is 1927, not 1952, and the name also antedates ×*Elymopyrum* Cugnac (in Bull. Soc. Hist. Nat. Ardennes 33: 14. 1938) which is accompanied by a statement of parentage and a description in French but not Latin.

11.9. The principle of priority is not mandatory for names of taxa above the rank of family (but see Rec. 16B).

Article 12

12.1. A name of a taxon has no status under this *Code* unless it is validly published (see Art. 32-45).

SECTION 4. LIMITATION OF THE PRINCIPLE OF PRIORITY

Article 13

13.1. Valid publication of names for plants of the different groups is treated as beginning at the following dates (for each group a work is mentioned which is treated as having been published on the date given for that group):

Non-fossil plants:

(a) SPERMATOPHYTA and PTERIDOPHYTA, 1 May 1753 (Linnaeus, *Species plantarum,* ed. 1).

(b) MUSCI (the *Sphagnaceae* excepted), 1 January 1801 (Hedwig, *Species muscorum*).

(c) SPHAGNACEAE and HEPATICAE, 1 May 1753 (Linnaeus, *Species plantarum,* ed. 1).

(d) FUNGI (including slime moulds and lichen-forming fungi), 1 May 1753 (Linnaeus, *Species plantarum,* ed. 1). Names in the *Uredinales, Ustilaginales,* and *Gasteromycetes* (s. l.) adopted by Persoon (*Synopsis methodica fungorum,* 31 December 1801) and names of other fungi (excluding slime moulds) adopted by Fries (*Systema mycologicum,* vol. 1 (1 January 1821) to 3, with additional *Index* (1832), and *Elenchus fungorum,* vol. 1-2), are sanctioned (see Art. 15). For nomenclatural purposes names given to lichens shall be considered as applying to their fungal component.

(e) ALGAE, 1 May 1753 (Linnaeus, *Species plantarum,* ed. 1). Exceptions:

NOSTOCACEAE HOMOCYSTEAE, 1 January 1892 (Gomont, "Monographie des Oscillariées", in Ann. Sci. Nat., Bot., ser. 7, 15: 263-368; 16: 91-264). The two parts of Gomont's "Monographie", which appeared in 1892 and 1893 respectively, are treated as having been published simultaneously on 1 January 1892.

NOSTOCACEAE HETEROCYSTEAE, 1 January 1886 (Bornet & Flahault, "Révision des Nostocacées hétérocystées", in Ann. Sci. Nat., Bot., ser. 7, 3: 323-381; 4: 343-373; 5: 51-129; 7: 177-262). The four parts of the "Révision", which appeared in 1886, 1886, 1887, and 1888 respectively, are treated as having been published simultaneously on 1 January 1886.

DESMIDIACEAE (s. l.), 1 January 1848 (Ralfs, *British* Desmidieae).

OEDOGONIACEAE, 1 January 1900 (Hirn, "Monographie und Iconographie der Oedogoniaceen", in Acta Soc. Sci. Fenn. 27(1)).

Fossil plants:

(f) ALL GROUPS, 31 December 1820 (Sternberg, *Flora der Vorwelt, Versuch* 1: 1-24, t. 1-13). Schlotheim's *Petrefactenkunde* (1820) is regarded as published before 31 December 1820.

13.2. The group to which a name is assigned for the purposes of this Article is determined by the accepted taxonomic position of the type of the name.

Ex. 1. The genus *Porella* and its single species, *P. pinnata,* were referred by Linnaeus (1753) to the *Musci;* since the type specimen of *P. pinnata* is now accepted as belonging to the *Hepaticae,* the names were validly published in 1753.

Ex. 2. The lectotype of *Lycopodium* L. (1753) is *L. clavatum* L. (1753) and the type specimen of this is currently accepted as a pteridophyte. Accordingly, although the genus is listed by Linnaeus among the *Musci,* the generic name and the names of the pteridophyte species included by Linnaeus under it were validly published in 1753.

13.3. For nomenclatural purposes, a name is treated as pertaining to a non-fossil taxon unless its type is fossil in origin. Fossil material is distinguished from non-fossil material by stratigraphic relations at the site of original occurrence. In cases of doubtful stratigraphic relations, provisions for non-fossil taxa apply.

13.4. Generic names which first appear in Linnaeus's *Species plantarum,* ed. 1 (1753) and ed. 2 (1762-1763), are associated with the first subsequent description given under those names in Linnaeus's *Genera plantarum,* ed. 5 (1754) and ed. 6 (1764). The spelling of the generic names included in *Species plantarum,* ed. 1, is not to be altered because a different spelling has been used in *Genera plantarum,* ed. 5.

13.5. The two volumes of Linnaeus's *Species plantarum,* ed. 1 (1753), which appeared in May and August, 1753, respectively, are treated as having been published simultaneously on 1 May 1753.

Ex. 3. The generic names *Thea* L. (Sp. Pl.: 515. 24 Mai 1753), and *Camellia* L. (Sp. Pl.: 698. 16 Aug 1753; Gen. Pl., ed. 5: 311. 1754), are treated as having been published simultaneously on 1 May 1753. Under Art. 11.5 the combined genus bears the name *Camellia,* since Sweet (Hort. Suburb. Lond.: 157. 1818), who was the first to unite the two genera, chose that name, and cited *Thea* as a synonym.

13.6. Names of anamorphs of fungi with a pleomorphic life cycle do not, irrespective of priority, affect the nomenclatural status of the names of the correlated holomorphs (see Art. 59.4).

Article 14

14.1. In order to avoid disadvantageous changes in the nomenclature of families, genera, and species entailed by the strict application of the rules, and especially of the principle of priority in starting from the dates given in Art. 13, this *Code* provides, in App. II and III, lists of names that are conserved *(nomina conservanda)* and must be retained as useful exceptions. Conserved names are legitimate even though initially they may have been illegitimate.

14.2. Conservation aims at retention of those names which best serve stability of nomenclature (see Rec. 50E).

14.3. The application of both conserved and rejected names is determined by nomenclatural types. The type of the specific name cited as the type of a conserved generic name may, if desirable, be conserved and listed in App. IIIA.

14.4. A conserved name of a family or genus is conserved against all other names in the same rank based on the same type (nomenclatural synonyms, which are to be rejected) whether these are cited in the corresponding list of rejected names or not, and against those names based on different types (taxonomic synonyms) that are cited in that list[1]. A conserved name of a species is conserved against all names listed as rejected, and against all combinations based on the rejected names.

Note 1. The *Code* does not provide for conservation of a name against itself, i.e. against the same name with the same type but with a different place and date of valid publication than is given in the relevant entry in App. II or III, and perhaps with a different authorship[2] (but see Art. 14.9).

Note 2. A species name listed as conserved or rejected in App. IIIB may have been published as the name of a new taxon, or as a combination based on an earlier name. Rejection of a name based on an earlier name does not in itself preclude the use of the earlier name since that name is not "a combination based on a rejected name" (Art. 14.4).

Ex. 1. Rejection of *Lycopersicon lycopersicum* (L.) H. Karst. in favour of *L. esculentum* Mill. does not preclude the use of the homotypic *Solanum lycopersicum* L.

14.5. When a conserved name competes with one or more names based on different types and against which it is not explicitly conserved, the earliest of the competing names is adopted in accordance with Art. 11, except for some conserved family names (App. IIB), which are conserved against unlisted names.

Ex. 2. If *Weihea* Spreng. (1825) is united with *Cassipourea* Aubl. (1775), the combined genus will bear the prior name *Cassipourea,* although *Weihea* is conserved and *Cassipourea* is not.

Ex. 3. If *Mahonia* Nutt. (1818) is united with *Berberis* L. (1753), the combined genus will bear the prior name *Berberis,* although *Mahonia* is conserved and *Berberis* is not.

Ex. 4. *Nasturtium* R. Br. (1812) was conserved only against the homonym *Nasturtium* Mill. (1754) and the nomenclatural synonym *Cardaminum* Moench (1794); consequently if reunited with *Rorippa* Scop. (1760) it must bear the name *Rorippa.*

14.6. When a name of a taxon has been conserved against an earlier name based on a different type, the latter is to be restored, subject to Art. 11, if it is

---

[1] The *International code of zoological nomenclature* and the *International code of nomenclature of bacteria* use the terms "objective synonym" and "subjective synonym" for nomenclatural and taxonomic synonym, respectively.

[2] As a temporary exception, the Tokyo Congress has authorized maintenance of the current entries in App. IIB even though, for many of the listed family names, places of earlier valid publication, by different authors, have come to light.

considered the name of a taxon at the same rank distinct from that of the *nomen conservandum,* except when the earlier rejected name is a homonym of the conserved name.

Ex. 5. The generic name *Luzuriaga* Ruiz & Pav. (1802) is conserved against the earlier names *Enargea* Banks ex Gaertn. (1788) and *Callixene* Comm. ex Juss. (1789). If, however, *Enargea* is considered to be a separate genus, the name *Enargea* is retained for it.

14.7. A rejected name, or a combination based on a rejected name, may not be restored for a taxon which includes the type of the corresponding conserved name.

Ex. 6. Enallagma Baill. (1888) is conserved against *Dendrosicus* Raf. (1838), but not against *Amphitecna* Miers (1868); if *Enallagma* and *Amphitecna* are united, the combined genus must bear the name *Amphitecna,* although the latter is not explicitly conserved against *Dendrosicus.*

14.8. The listed type of a conserved name may not be changed except by the procedure outlined in Art. 14.12.

Ex. 7. Bullock & Killick (in Taxon 6: 239. 1957) published a proposal that the listed type of *Plectranthus* L'Hér. be changed from *P. punctatus* (L. f.) L'Hér. to *P. fruticosus* L'Hér. This proposal was approved by the appropriate Committees and by an International Botanical Congress.

14.9. A name may be conserved with a different type from that designated by the author or determined by application of the *Code* (see also Art. 10.4). Such a name may be conserved either from its place of valid publication (even though the type may not then have been included in the named taxon) or from a later publication by an author who did include the type as conserved. In the latter case the original name and the name as conserved are treated as if they were homonyms (Art. 53), whether or not the name as conserved was accompanied by a description or diagnosis of the taxon named.

Ex. 8. Bromus sterilis L. (1753) has been conserved from its place of valid publication even though its conserved type, a specimen (Hubbard 9045, E) collected in 1932, was not originally included in Linnaeus's species.

Ex. 9. Protea L. (1753) did not include the conserved type of the generic name, *P. cynaroides* (L.) L. (1771), which in 1753 was placed in the genus *Leucadendron. Protea* was therefore conserved from the 1771 publication, and *Protea* L. (1771), although not designed to be a new generic name and still including the original type elements, is treated as if it were a validly published homonym of *Protea* L. (1753).

14.10. A conserved name, with its corresponding autonyms, is conserved against all earlier homonyms. An earlier homonym of a conserved name is not made illegitimate by that conservation but is unavailable for use; if legitimate, it may serve as basionym of another name or combination based on the same type (see also Art. 55.3).

Ex. 10. The generic name *Smithia* Aiton (1789), conserved against *Damapana* Adans. (1763), is thereby conserved automatically against the earlier homonym *Smithia* Scop. (1777).

14.11. A name may be conserved in order to preserve a particular orthography or gender. A name so conserved is to be attributed without change of priority to the author who validly published it, not to an author who later introduced the conserved spelling or gender.

Ex. 11. The spelling *Rhodymenia,* used by Montagne (1839), has been conserved against the original spelling *Rhodomenia,* used by Greville (1830). The name is to be cited as *Rhodymenia* Grev. (1830).

Note 3. The date of conservation does not affect the priority (Art. 11) of a conserved name, which is determined only on the basis of the date of valid publication (Art. 32-45).

14.12. The lists of conserved names will remain permanently open for additions and changes. Any proposal of an additional name must be accompanied by a detailed statement of the cases both for and against its conservation. Such proposals must be submitted to the General Committee (see Div. III), which will refer them for examination to the committees for the various taxonomic groups.

14.13. Entries of conserved names may not be deleted.

14.14. When a proposal for the conservation (or rejection under Art. 56) of a name has been approved by the General Committee after study by the Committee for the taxonomic group concerned, retention (or rejection) of that name is authorized subject to the decision of a later International Botanical Congress.

Recommendation 14A

14A.1. When a proposal for the conservation or rejection of a name has been referred to the appropriate Committee for study, authors should follow existing usage as far as possible pending the General Committee's recommendation on the proposal.

Article 15

15.1. Names sanctioned under Art. 13.1(d) are treated as if conserved against earlier homonyms and competing synonyms. Such names, once sanctioned, remain sanctioned even if elsewhere in the sanctioning works the sanctioning author does not recognize them.

Ex. 1. Agaricus ericetorum Fr. was accepted by Fries in *Systema mycologicum* (1821), but later (1828) regarded by him as a synonym of *A. umbelliferus* L. and not included in his *Index* (1832) as an accepted name. Nevertheless *A. ericetorum* is a sanctioned name.

15.2. An earlier homonym of a sanctioned name is not made illegitimate by that sanctioning but is unavailable for use; if legitimate, it may serve as a basionym of another name or combination based on the same type (see also Art. 55.3).

Ex. 2. Patellaria Hedw. (1794) is an earlier homonym of the sanctioned generic name *Patellaria* Fr. (1822). Hedwig's name is legitimate but unavailable for use. *Lecanidion* Endl. (1830), based on the same type as *Patellaria* Fr. : Fr. non Hedw., is illegitimate.

Ex. 3. Agaricus cervinus Schaeff. (1774) is an earlier homonym of the sanctioned *A. cervinus* Hoffm. (1789) : Fr.; Schaeffer's name is unavailable for use, but it may serve as basionym for combinations in other genera. In *Pluteus* Fr. the combination should be cited as *P. cervinus* (Schaeff.) P. Kumm. and has priority over the heterotypic synonym *P. atricapillus* (Batsch) Fayod, based on *A. atricapillus* Batsch (1786).

15.3. When, for a taxon from family to genus inclusive, two or more sanctioned names compete, Art. 11.3 governs the choice of the correct name (see also Art. 15.5).

15.4. When, for a taxon below the rank of genus, two or more sanctioned names and/or two or more names with the same final epithet and type as a sanctioned name compete, Art. 11.4 governs the choice of the correct name.

Note 1. The date of sanctioning does not affect the priority (Art. 11) of a sanctioned name, which is determined only on the basis of valid publication. In particular, when two or more homonyms are sanctioned only the earliest of them can be used, the later being illegitimate under Art. 53.2.

Ex. 4. Fries (Syst. Mycol. 1: 41. 1821) accepted *Agaricus flavovirens* Pers. (1801), treating *A. equestris* L. (1753) as a synonym. Later (Elench. Fung. 1: 6. 1828) he stated "Nomen prius et aptius arte restituendum" and accepted *A. equestris*. Both names are sanctioned, but when they are considered synonyms *A. equestris*, having priority, is to be used.

15.5. A name which neither is sanctioned nor has the same type and final epithet as a sanctioned name in the same rank may not be applied to a taxon which includes the type of a sanctioned name in that rank the final epithet of which is available for the required combination (see Art. 11.4(b)).

15.6. Conservation (Art. 14) and explicit rejection (Art. 56.1) override sanctioning.

CHAPTER III. NOMENCLATURE OF TAXA ACCORDING TO THEIR RANK

SECTION 1. NAMES OF TAXA ABOVE THE RANK OF FAMILY

Article 16

16.1. Names of taxa above the rank of family are automatically typified if they are based on generic names (see Art. 10.7); for such automatically typified names, the name of a subdivision or subphylum which includes the type of the adopted name of a division or phylum, the name of a subclass which includes the type of the adopted name of a class, and the name of a suborder which includes the type of the adopted name of an order, are to be based on the generic name equivalent to that type.

Note 1. The terms "divisio" and "phylum", and their equivalents in modern languages, are treated as referring to one and the same rank. When "divisio" and "phylum" are used simultaneously to denote different ranks, this usage is contrary to Art. 5, and the corresponding names are not validly published (Art. 33.5).

16.2. Where one of the word elements *-monado-*, *-cocco-*, *-nemato-*, or *-clado-* as the second part of a generic name has been omitted before the termination *-phyceae* or *-phyta,* the shortened class name or division or phylum name is regarded as based on the generic name in question if such derivation is obvious or is indicated at establishment of the group name.

Ex. 1. Raphidophyceae Chadef. ex P. C. Silva (1980) was indicated by its author to be based on *Raphidomonas* F. Stein (1878).

Note 2. The principle of priority is not mandatory for names of taxa above the rank of family (Art. 11.9).

Recommendation 16A

16A.1. The name of a division or phylum is taken either from distinctive characters of the division or phylum (in descriptive names) or from a name of an included genus; it should end in *-phyta,* unless it is a division or phylum of fungi, in which case it should end in *-mycota.*

16A.2. The name of a subdivision or subphylum is formed in a similar manner; it is distinguished from a divisional name by an appropriate prefix or suffix or by the termination *-phytina,* unless it is a subdivision or subphylum of fungi, in which case it should end in *-mycotina.*

16A.3. The name of a class or of a subclass is formed in a similar manner and should end as follows:

(a) In the algae: *-phyceae* (class) and *-phycidae* (subclass);

(b) In the fungi: *-mycetes* (class) and *-mycetidae* (subclass);

(c) In other groups of plants: *-opsida* (class) and *-idae* (subclass).

16A.4. When a name has been published with a Latin termination not agreeing with this recommendation, the termination may be changed to accord with it, without change of author's name or date of publication.

Recommendation 16B

16B.1. In choosing among typified names for a taxon above the rank of family, authors should generally follow the principle of priority.

Article 17

17.1. The name of an order or suborder is taken either from distinctive characters of the taxon (descriptive name) or from a legitimate name of an included family based on a generic name (automatically typified name). An ordinal name of the second category is formed by replacing the termination *-aceae* by *-ales* . A subordinal name of the second category is similarly formed, with the termination *-ineae.*

Ex. 1. Descriptive names of orders: *Centrospermae, Parietales, Farinosae;* of a suborder: *Enantioblastae.*

Ex. 2. Automatically typified names: *Fucales, Polygonales, Ustilaginales; Bromeliineae, Malvineae.*

17.2. Names intended as names of orders, but published with their rank denoted by a term such as "cohors", "nixus", "alliance", or "Reihe" instead of "order", are treated as having been published as names of orders.

17.3. When the name of an order or suborder based on a name of a genus has been published with an improper Latin termination, this termination must be changed to accord with the rule, without change of the author's name or date of publication.

Recommendation 17A

17A.1. Authors should not publish new names of orders for taxa of that rank which include a family from whose name an existing ordinal name is derived.

SECTION 2. NAMES OF FAMILIES AND SUBFAMILIES, TRIBES AND SUBTRIBES

Article 18

18.1. The name of a family is a plural adjective used as a substantive; it is formed from the genitive singular of a legitimate name of an included genus by replacing the genitive singular inflection (Latin *-ae, -i, -us, -is;* transliterated Greek *-ou, -os, -es, -as,* or *-ous,* including the latter's equivalent *-eos*) with the termination *-aceae.* For generic names of non-classical origin, when analogy with classical names is insufficient to determine the genitive singular, *-aceae* is added to the full word. For generic names with alternative genitives the one implicitly used by the original author must be maintained.

Ex. 1. Family names based on a generic name of classical origin: *Rosaceae* (from *Rosa, Rosae*), *Salicaceae* (from *Salix, Salicis*), *Plumbaginaceae* (from *Plumbago, Plumbaginis*), *Rhodophyllaceae* (from *Rhodophyllus, Rhodophylli*), *Rhodophyllidaceae* (from *Rhodophyllis, Rhodophyllidos*), *Sclerodermataceae* (from *Scleroderma, Sclerodermatos*), *Aextoxicaceae* (from *Aextoxicon, Aextoxicou*), *Potamogetonaceae* (from *Potamogeton, Potamogetonos*).

Ex. 2. Family names based on a generic name of non-classical origin: *Nelumbonaceae* (from *Nelumbo, Nelumbonis,* declined by analogy with *umbo, umbonis*), *Ginkgoaceae* (from *Ginkgo,* indeclinable).

18.2. Names intended as names of families, but published with their rank denoted by one of the terms "order" *(ordo)* or "natural order" *(ordo naturalis)* instead of "family", are treated as having been published as names of families (see also Art. 19.2).

Ex. 3. *Cyperaceae* Juss. (1789) and *Xylomataceae* Fr. (1820) were published as "ordo *Cyperoideae*" and "ordo *Xylomaceae*".

18.3. A name of a family based on an illegitimate generic name is illegitimate unless conserved. Contrary to Art. 32.1(b) such a name is validly published if it complies with the other requirements for valid publication.

Ex. 4. *Caryophyllaceae* Juss., *nom. cons.* (from *Caryophyllus* Mill. non L.); *Winteraceae* Lindl., *nom. cons.* (from *Wintera* Murray, an illegitimate synonym of *Drimys* J. R. Forst. & G. Forst.).

18.4. When a name of a family has been published with an improper Latin termination, the termination must be changed to conform with the rule, without change of the author's name or date of publication (see Art. 32.6).

Ex. 5. "*Coscinodisceae*" (Kützing 1844) is to be accepted as *Coscinodiscaceae* Kütz. and not attributed to De Toni, who first used the correct spelling (in Notarisia 5: 915. 1890).

Ex. 6. "*Atherospermeae*" (Brown 1814) is to be accepted as *Atherospermataceae* R. Br. and not attributed to Airy Shaw (in Willis, Dict. Fl. Pl., ed. 7: 104. 1966), who first used the correct spelling, or to Lindley (Veg. Kingd.: 300. 1846), who used the spelling "*Atherospermaceae*".

Ex. 7. However, Tricholomées (Roze in Bull. Soc. Bot. France 23: 49. 1876) is not to be accepted as "*Tricholomataceae* Roze", because it has a French rather than a Latin termination. The name *Tricholomataceae* was later validated by Pouzar (1983; see App. IIA).

18.5. The following names, of long usage, are treated as validly published: *Palmae* (*Arecaceae*; type, *Areca* L.); *Gramineae* (*Poaceae*; type, *Poa* L.); *Cru-*

ciferae (*Brassicaceae*; type, *Brassica* L.); *Leguminosae* (*Fabaceae*; type, *Faba* Mill. [= *Vicia* L.]); *Guttiferae* (*Clusiaceae*; type, *Clusia* L.); *Umbelliferae* (*Apiaceae*; type, *Apium* L.); *Labiatae* (*Lamiaceae*; type, *Lamium* L.); *Compositae* (*Asteraceae*; type, *Aster* L.). When the *Papilionaceae* (*Fabaceae*; type, *Faba* Mill.) are regarded as a family distinct from the remainder of the *Leguminosae*, the name *Papilionaceae* is conserved against *Leguminosae*.

18.6. The use, as alternatives, of the names indicated in parentheses in Art. 18.5 is authorized.

Article 19

19.1. The name of a subfamily is a plural adjective used as a substantive; it is formed in the same manner as the name of a family (Art. 18.1) but by using the termination *-oideae* instead of *-aceae*.

19.2. Names intended as names of subfamilies, but published with their rank denoted by the term "suborder" *(subordo)* instead of subfamily, are treated as having been published as names of subfamilies (see also Art. 18.2).

19.3. A tribe is designated in a similar manner, with the termination *-eae,* and a subtribe similarly with the termination *-inae*.

19.4. The name of any subdivision of a family that includes the type of the adopted, legitimate name of the family to which it is assigned is to be based on the generic name equivalent to that type.

Ex. 1. The type of the family name *Rosaceae* Juss. is *Rosa* L. and hence the subfamily and tribe which include *Rosa* are to be called *Rosoideae* Endl. and *Roseae* DC.

Ex. 2. The type of the family name *Poaceae* Barnhart (nom. alt., *Gramineae* Juss. – see Art. 18.5) is *Poa* L. and hence the subfamily and tribe which include *Poa* are to be called *Pooideae* Asch. and *Poëae* R. Br.

Note 1. This provision applies only to the names of those subordinate taxa that include the type of the adopted name of the family (but see Rec. 19A.2).

Ex. 3. The subfamily including the type of the family name *Ericaceae* Juss. (*Erica* L.), irrespective of priority, is to be called *Ericoideae* Endl., and the tribe including this type is called *Ericeae* D. Don. However, the correct name of the tribe including both *Rhododendron* L., the type of the subfamily name *Rhododendroideae* Endl., and *Rhodora* L. is *Rhodoreae* D. Don (1834) not *Rhododendreae* Brongn. (1843).

Ex. 4. The subfamily of the family *Asteraceae* Dumort. (nom. alt., *Compositae* Giseke) including *Aster* L., the type of the family name, is irrespective of priority to be called *Asteroideae* Asch., and the tribe and subtribe including *Aster* are to be called *Astereae* Cass. and *Asterinae* Less., respectively. However, the correct name of the tribe including both *Cichorium* L., the type of the subfamily name *Cichorioideae* W. D. J. Koch (1837), and *Lactuca* L. is *Lactuceae* Cass. (1815), not *Cichorieae* D. Don (1829), while that of the subtribe including both *Cichorium* and *Hyoseris* L. is *Hyoseridinae* Less. (1832), not *Cichoriinae* Sch. Bip. (1841) (unless the *Cichoriaceae* Juss. are accepted as a family distinct from *Compositae*).

family automatically establishes the corresponding autonym (see also Art. 11.6 and 32.7).

19.6. A name of a subdivision of a family based on an illegitimate generic name that is not the base of a conserved family name is illegitimate. Contrary to Art. 32.1(b) such a name is validly published if it complies with the other requirements for valid publication.

Ex. 5. *Caryophylloideae* (Juss.) Rabeler & Bittrich, based on *Caryophyllaceae* Juss., *nom. cons.*, is legitimate although it is derived from the illegitimate *Caryophyllus* Mill. non L.

19.7. When a name of a taxon assigned to one of the above categories has been published with an improper Latin termination, such as *-eae* for a subfamily or *-oideae* for a tribe, the termination must be changed to accord with the rule, without change of the author's name or date of publication (see Art. 32.6).

Ex. 6. The subfamily name *"Climacieae"* Grout (Moss Fl. N. Amer. 3: 4. 1928) is to be changed to *Climacioideae* with rank and author's name unchanged.

19.8. When the *Papilionaceae* are included in the family *Leguminosae* (nom. alt., *Fabaceae*; see Art. 18.5) as a subfamily, the name *Papilionoideae* may be used as an alternative to *Faboideae.*

Recommendation 19A

19A.1. When a family is changed to the rank of a subdivision of a family, or the inverse change occurs, and no legitimate name is available in the new rank, the name should be retained, and only its termination *(-aceae, -oideae, -eae, -inae)* altered.

Ex. 1. The subtribe *Drypetinae* Pax (1890) *(Euphorbiaceae)* when raised to the rank of tribe was named *Drypeteae* (Pax) Hurus. (1954); the subtribe *Antidesmatinae* Pax (1890) *(Euphorbiaceae)* when raised to the rank of subfamily was named *Antidesmatoideae* (Pax) Hurus. (1954).

19A.2. When a subdivision of a family is changed to another such rank, and no legitimate name is available in the new rank, its name should be based on the same generic name as the name in the former rank.

Ex. 2. Three tribes of the family *Ericaceae,* none of which includes the type of that family name (*Erica* L.), are *Pyroleae* D. Don, *Monotropeae* D. Don, and *Vaccinieae* D. Don. The later names *Pyroloideae* (D. Don) A. Gray, *Monotropoideae* (D. Don) A. Gray, and *Vaccinioideae* (D. Don) Endl. are based on the same generic names.

SECTION 3. NAMES OF GENERA AND SUBDIVISIONS OF GENERA

Article 20

20.1. The name of a genus is a substantive in the singular, or a word treated as such, and is written with a capital initial letter (see Art. 60.2). It may be taken from any source whatever, and may even be composed in an absolutely arbitrary manner.

Ex. 1. Rosa, Convolvulus, Hedysarum, Bartramia, Liquidambar, Gloriosa, Impatiens, Rhododendron, Manihot, Ifloga (an anagram of *Filago*).

20.2. The name of a genus may not coincide with a technical term currently used in morphology unless it was published before 1 January 1912 and accompanied by a specific name published in accordance with the binary system of Linnaeus.

Ex. 2. "Radicula" (Hill, 1756) coincides with the technical term "radicula" (radicle) and was not accompanied by a specific name in accordance with the binary system of Linnaeus. The name *Radicula* is correctly attributed to Moench (1794), who first combined it with specific epithets.

Ex. 3. Tuber F. H. Wigg. : Fr., when published in 1780, was accompanied by a binary specific name (*Tuber gulosorum* F. H. Wigg.) and is therefore validly published.

Ex. 4. The intended generic names *"Lanceolatus"* (Plumstead, 1952) and *"Lobata"* (Chapman, 1952) coincide with technical terms and are therefore not validly published.

Ex. 5. Words such as "radix", "caulis", "folium", "spina", etc., cannot now be validly published as generic names.

20.3. The name of a genus may not consist of two words, unless these words are joined by a hyphen.

Ex. 6. "Uva ursi", as originally published by Miller (1754), consisted of two separate words unconnected by a hyphen, and is therefore not validly published (Art. 32.1(b)); the name is correctly attributed to Duhamel (1755) as *Uva-ursi* (hyphenated when published).

Ex. 7. However, names such as *Quisqualis* L. (formed by combining two words into one when originally published), *Sebastiano-schaueria* Nees, and *Neves-armondia* K. Schum. (both hyphenated when originally published) are validly published.

Note 1. The names of intergeneric hybrids are formed according to the provisions of Art. H.6.

20.4. The following are not to be regarded as generic names:

(a) Words not intended as names.

Ex. 8. The designation *"Anonymos"* was applied by Walter (Fl. Carol.: 2, 4, 9, etc. 1788) to 28 different genera to indicate that they were without names.

Ex. 9. "Schaenoides" and *"Scirpoides"*, as used by Rottbøll (Descr. Pl. Rar. Progr.: 14, 27. 1772) to indicate unnamed genera resembling *Schoenus* and *Scirpus* which he stated (on p. 7) he intended to name later, are token words and not generic names. They were later legitimately named *Kyllinga* Rottb. and *Fuirena* Rottb.

(b) Unitary designations of species.

Note 2. Examples such as *"Leptostachys"* and *"Anthopogon"*, listed in earlier editions of the *Code*, were from publications now listed in App. V.

Recommendation 20A

20A.1. Authors forming generic names should comply with the following suggestions:

(a) To use Latin terminations insofar as possible.

(b) To avoid names not readily adaptable to the Latin language.

(c) Not to make names which are very long or difficult to pronounce in Latin.

(d) Not to make names by combining words from different languages.

(e) To indicate, if possible, by the formation or ending of the name the affinities or analogies of the genus.

(f) To avoid adjectives used as nouns.

(g) Not to use a name similar to or derived from the epithet in the name of one of the species of the genus.

(h) Not to dedicate genera to persons quite unconnected with botany or at least with natural science.

(i) To give a feminine form to all personal generic names, whether they commemorate a man or a woman (see Rec. 60B).

(j) Not to form generic names by combining parts of two existing generic names, because such names are likely to be confused with nothogeneric names (see Art. H.6).

Ex. 1. Hordelymus (K. Jess.) K. Jess. derives from a subgeneric epithet that was formed by combining parts of the generic names *Hordeum* L. and *Elymus* L. (see also Art. H.3 Ex. 2).

Article 21

21.1. The name of a subdivision of a genus is a combination of a generic name and a subdivisional epithet connected by a term (subgenus, sectio, series, etc.) denoting its rank.

21.2. The epithet is either of the same form as a generic name, or a plural adjective agreeing in gender with the generic name and written with a capital initial letter (see Art. 32.6 and 60.2).

21.3. The epithet in the name of a subdivision of a genus is not to be formed from the name of the genus to which it belongs by adding the prefix *Eu-*.

Ex. 1. Costus subg. *Metacostus; Ricinocarpos* sect. *Anomodiscus; Valeriana* sect. *Valerianopsis; Euphorbia* sect. *Tithymalus; Euphorbia* subsect. *Tenellae; Sapium* subsect. *Patentinervia; Arenaria* ser. *Anomalae;* but not *Carex* sect. *Eucarex.*

Note 1. The use within the same genus of the same epithet in names of subdivisions of the genus, even in different ranks, based on different types is illegitimate under Art. 53.

Note 2. The names of hybrids with the rank of a subdivision of a genus are formed according to the provisions of Art. H.7.

Recommendation 21A

21A.1. When it is desired to indicate the name of a subdivision of the genus to which a particular species belongs in connection with the generic name and specific epithet, the subdivisional epithet should be placed in parentheses between the two; when desirable, the subdivisional rank may also be indicated.

Ex. 1. Astragalus (Cycloglottis) contortuplicatus; A. (Phaca) umbellatus; Loranthus (sect. *Ischnanthus*) *gabonensis.*

Recommendation 21B

21B.1. The epithet in the name of a subgenus or section is preferably a substantive, that of a subsection or lower subdivision of a genus preferably a plural adjective.

21B.2. Authors, when proposing new epithets for names of subdivisions of genera, should avoid those in the form of a substantive when other co-ordinate subdivisions of the same genus have them in the form of a plural adjective, and vice-versa. They should also avoid, when proposing an epithet for a name of a subdivision of a genus, one already used for a subdivision of a closely related genus, or one which is identical with the name of such a genus.

21B.3. When a section or a subgenus is raised to the rank of genus, or the inverse change occurs, the original name or epithet should be retained unless the resulting name would be contrary to this *Code*.

Article 22

22.1. The name of any subdivision of a genus that includes the type of the adopted, legitimate name of the genus to which it is assigned is to repeat that generic name unaltered as its epithet, but not followed by an author's name (see Art. 46). Such names are termed autonyms (Art. 6.8; see also Art. 7.6).

Note 1. This provision applies only to the names of those subordinate taxa that include the type of the adopted name of the genus (but see Rec. 22A).

22.2. A name of a subdivision of a genus that includes the type (i.e. the original type or all elements eligible as type or the previously designated type) of the adopted, legitimate name of the genus is not validly published unless its epithet repeats the generic name unaltered. For the purposes of this provision, explicit indication that the nomenclaturally typical element of the genus is included is considered as equivalent to inclusion of the type, whether or not it has been previously designated (see also Art. 21.3).

Ex. 1. "*Dodecatheon* sect. *Etubulosa*" (Knuth in Engler, Pflanzenr. 22: 234. 1905) was not validly published since it was proposed for a section that included *D. meadia* L., the original type of the generic name *Dodecatheon* L.

Ex. 2. *Cactus* [unranked] *Melocactus* L. (Gen. Pl., ed. 5: 210. 1754) was proposed for one of four unranked (Art. 35.2), named subdivisions of the genus *Cactus*, comprising *C. melocactus* L. (its type under Art. 22.5) and *C. mammillaris* L. It is validly published, even though *C. melocactus* was subsequently designated as the type of *Cactus* L. (by Britton & Millspaugh, Bahama Fl.: 294. 1920) and, later still, *C. mammillaris* became the conserved type of the generic name (by the way in which the family name *Cactaceae* Juss. was conserved).

22.3. The first instance of valid publication of a name of a subdivision of a genus that does not include the type of the adopted, legitimate name of the genus automatically establishes the corresponding autonym (see also Art. 11.6 and 32.7).

Ex. 3. The subgenus of *Malpighia* L. that includes the lectotype of the generic name (*M. glabra* L.) is called *M.* subg. *Malpighia,* not *M.* subg. *Homoiostylis* Nied.; and the section of *Malpighia* including the lectotype of the generic name is called *M.* sect. *Malpighia,* not *M.* sect. *Apyrae* DC.

Ex. 4. However, the correct name of the section of the genus *Rhododendron* L. that includes *R. luteum* Sweet, the type of *R.* subg. *Anthodendron* (Rchb.) Rehder, is *R.* sect. *Pentanthera* G. Don, the oldest legitimate name for the section, and not *R.* sect. *Anthodendron.*

22.4. The epithet in the name of a subdivision of a genus may not repeat unchanged the correct name of the genus, unless the two names have the same type.

22.5. When the epithet in the name of a subdivision of a genus is identical with or derived from the epithet of one of its constituent species, the type of the name of the subdivision of the genus is the same as that of the species name, unless the original author of the subdivisional name designated another type.

Ex. 5. The type of *Euphorbia* subg. *Esula* Pers. is *E. esula* L.; the designation of *E. peplus* L. as lectotype by Croizat (in Revista Sudamer. Bot. 6: 13. 1939) has no standing.

22.6. When the epithet in the name of a subdivision of a genus is identical with or derived from the epithet in a specific name that is a later homonym, its type is the type of that later homonym, whose correct name necessarily has a different epithet.

Recommendation 22A

22A.1. A section including the type of the correct name of a subgenus, but not including the type of the correct name of the genus, should, where there is no obstacle under the rules, be given a name with the same epithet and type as the subgeneric name.

22A.2. A subgenus not including the type of the correct name of the genus should, where there is no obstacle under the rules, be given a name with the same epithet and type as the correct name of one of its subordinate sections.

Ex. 1. Instead of using a new epithet at the subgeneric level, Brizicky raised *Rhamnus* sect. *Pseudofrangula* Grubov to the rank of subgenus as *R.* subg. *Pseudofrangula* (Grubov) Brizicky. The type of both names is the same, *R. alnifolia* L'Hér.

SECTION 4. NAMES OF SPECIES

Article 23

23.1. The name of a species is a binary combination consisting of the name of the genus followed by a single specific epithet in the form of an adjective, a noun in the genitive, or a word in apposition, or several words, but not a phrase name of one or more descriptive substantives and associated adjectives in the ablative (see Art. 23.6(a)), nor certain other irregularly formed designations (see Art. 23.6(c)). If an epithet consists of two or more words, these are to be united or hyphenated. An epithet not so joined when originally published is not to be rejected but, when used, is to be united or hyphenated, as specified in Art. 60.9.

23.2. The epithet in the name of a species may be taken from any source whatever, and may even be composed arbitrarily (but see Art. 60.1).

Ex. 1. Cornus sanguinea, Dianthus monspessulanus, Papaver rhoeas, Uromyces fabae, Fumaria gussonei, Geranium robertianum, Embelia sarasiniorum, Atropa bella-donna, Impatiens noli-tangere, Adiantum capillus-veneris, Spondias mombin (an indeclinable epithet).

23.3. Symbols forming part of specific epithets proposed by Linnaeus do not invalidate the relevant names but must be transcribed.

Ex. 2. Scandix pecten ♀ L. is to be transcribed as *Scandix pecten-veneris; Veronica anagallis* ▽ L. is to be transcribed as *Veronica anagallis-aquatica.*

23.4. The specific epithet may not exactly repeat the generic name with or without the addition of a transcribed symbol (tautonym).

Ex. 3. "Linaria linaria" and *"Nasturtium nasturtium-aquaticum"* are contrary to this rule and cannot be validly published.

23.5. The specific epithet, when adjectival in form and not used as a substantive, agrees grammatically with the generic name (see Art. 32.6).

Ex. 4. Helleborus niger L., *Brassica nigra* (L.) W. D. J. Koch, *Verbascum nigrum* L.; *Vinca major* L., *Tropaeolum majus* L.; *Rubus amnicola* Blanch. *("amnicolus"),* the specific epithet being a Latin substantive; *Peridermium balsameum* Peck, but also *Gloeosporium balsameae* Davis, both derived from the epithet of *Abies balsamea* (L.) Mill., treated as a substantive in the second example.

23.6. The following designations are not to be regarded as specific names:

(a) Descriptive designations consisting of a generic name followed by a phrase name (Linnaean *nomen specificum legitimum*) of one or more descriptive substantives and associated adjectives in the ablative.

Ex. 5. Smilax "caule inermi" (Aublet, Hist. Pl. Guiane 2, Tabl.: 27. 1775) is an abbreviated descriptive reference to an imperfectly known species which is not given a binomial in the text but referred to merely by a phrase name cited from Burman.

(b) Other designations of species consisting of a generic name followed by one or more words not intended as specific epithets.

Ex. 6. Viola "qualis" (Krocker, Fl. Siles. 2: 512, 517. 1790); *Urtica "dubia?"* (Forsskål, Fl. Aegypt.-Arab.: cxxi. 1775), the word "dubia?" being repeatedly used in Forsskål's work for species which could not be reliably identified.

Ex. 7. Atriplex "nova" (Winterl, Index Hort. Bot. Univ. Hung.: fol. A [8] recto et verso. 1788), the word "nova" (new) being here used in connection with four different species of *Atriplex.* However, in *Artemisia nova* A. Nelson (in Bull. Torrey Bot. Club 27: 274. 1900), *nova* was intended as a specific epithet, the species having been newly distinguished from others.

Ex. 8. Cornus "gharaf" (Forsskål, Fl. Aegypt.-Arab.: xci, xcvi. 1775) is an interim designation not intended as a species name. An interim designation in Forsskål's work is an original designation (for an accepted taxon and thus not a "provisional name" as defined in Art. 34.1(b)) with an epithet-like vernacular which is not used as an epithet in the "Centuriae" part of the work. *Elcaja "roka"* (Forsskål, Fl. Aegypt.-Arab.: xcv. 1775) is another example of such an interim designation; in other parts of the work (p. c, cxvi, 127) this species is not named.

Ex. 9. In *Agaricus "octogesimus nonus"* and *Boletus "vicesimus sextus"* (Schaeffer, Fung. Bavar. Palat. Nasc. 1: t. 100. 1762; 2: t. 137. 1763), the generic names are followed by ordinal adjectives used for enumeration. The corresponding species were given valid names, *A. cinereus* Schaeff. and *B. ungulatus* Schaeff., in the final volume of the same work (1774).

(c) Designations of species consisting of a generic name followed by two or more adjectival words in the nominative case.

Ex. 10. Salvia "africana coerulea" (Linnaeus, Sp. Pl.: 26. 1753) and *Gnaphalium "fruticosum flavum"* (Forsskål, Fl. Aegypt.-Arab.: cxix. 1775) are generic names followed by two adjectival words in the nominative case. They are not to be regarded as species names.

Ex. 11. However, *Rhamnus "vitis idaea"* Burm. f. (Fl. Ind.: 61. 1768) is to be regarded as a species name, since the generic name is followed by a substantive and an adjective, both in the nominative case; these words are to be hyphenated (*R. vitis-idaea*) under the provisions of Art. 23.1 and Art. 60.9. In *Anthyllis "Barba jovis"* L. (Sp. Pl.: 720. 1753) the generic name is followed by substantives in the nominative and in the genitive case respectively, and they are to be hyphenated (*A. barba-jovis*). Likewise, *Hyacinthus "non scriptus"* L. (Sp. Pl.: 316. 1753), where the generic name is followed by a negative particle and a past participle used as an adjective, is corrected to *H. non-scriptus*, and *Impatiens "noli tangere"* L. (Sp. Pl.: 938. 1753), where the generic name is followed by two verbs, is corrected to *I. noli-tangere*.

Ex. 12. Similarly, in *Narcissus "Pseudo Narcissus"* L. (Sp. Pl.: 289. 1753) the generic name is followed by an independent prefix and a substantive in the nominative case, and the name is to be corrected to *N. pseudonarcissus* under the provisions of Art. 23.1 and Art. 60.9.

(d) Formulae designating hybrids (see Art. H.10.3).

23.7. Phrase names used by Linnaeus as specific epithets *(nomina trivialia)* are to be treated as orthographical errors to be corrected in accordance with later usage by Linnaeus himself.

Ex. 13. Apocynum "fol. [foliis] androsaemi" L. is to be cited as *A. androsaemifolium* L. (Sp. Pl.: 213. 1753 [corr. L., Syst. Nat., ed. 10, 2: 946. 1759]); and *Mussaenda "fr. [fructu] frondoso"* L., as *M. frondosa* L. (Sp. Pl.: 177. 1753 [corr. L., Syst. Nat., ed. 10, 2: 931. 1759]).

23.8. Where the status of a designation of a species is uncertain under Art. 23.6, established custom is to be followed (Pre. 10).

**Ex. 14. Polypodium "F. mas", P. "F. femina"*, and *P. "F. fragile"* (Linnaeus, Sp. Pl.: 1090-1091. 1753) are, in accordance with established custom, to be treated as *P. filix-mas* L., *P. filix-femina* L., and *P. fragile* L., respectively. Likewise, *Cambogia "G. gutta"* is to be treated as *C. gummi-gutta* L. (Gen. Pl.: [522]. 1754). The intercalations *"Trich." [Trichomanes]* and *"M." [Melilotus]* in the names of Linnaean species of *Asplenium* and *Trifolium*, respectively, are to be deleted, so that names in the form *Asplenium "Trich. dentatum"* and *Trifolium "M. indica"*, for example, are treated as *A. dentatum* L. and *T. indicum* L. (Sp. Pl.: 765, 1080. 1753).

Recommendation 23A

23A.1. Names of persons and also of countries and localities used in specific epithets should take the form of substantives in the genitive *(clusii, porsildiorum, saharae)* or of adjectives *(clusianus, dahuricus)* (see also Art. 60, Rec. 60C and D).

23A.2. The use of the genitive and the adjectival form of the same word to designate two different species of the same genus should be avoided (e.g. *Lysimachia hemsleyana* Oliv. and *L. hemsleyi* Franch.).

23A.3. In forming specific epithets, authors should comply also with the following suggestions:

(a) To use Latin terminations insofar as possible.

(b) To avoid epithets which are very long and difficult to pronounce in Latin.

(c) Not to make epithets by combining words from different languages.

(d) To avoid those formed of two or more hyphenated words.

(e) To avoid those which have the same meaning as the generic name (pleonasm).

(f) To avoid those which express a character common to all or nearly all the species of a genus.

(g) To avoid in the same genus those which are very much alike, especially those which differ only in their last letters or in the arrangement of two letters.

(h) To avoid those which have been used before in any closely allied genus.

(i) Not to adopt epithets from unpublished names found in correspondence, travellers' notes, herbarium labels, or similar sources, attributing them to their authors, unless these authors have approved publication (see Rec. 34A).

(j) To avoid using the names of little-known or very restricted localities, unless the species is quite local.

SECTION 5. NAMES OF TAXA BELOW THE RANK OF SPECIES (INFRASPECIFIC TAXA)

Article 24

24.1. The name of an infraspecific taxon is a combination of the name of a species and an infraspecific epithet connected by a term denoting its rank.

Ex. 1. Saxifraga aizoon subf. *surculosa* Engl. & Irmsch. This can also be cited as *Saxifraga aizoon* var. *aizoon* subvar. *brevifolia* f. *multicaulis* subf. *surculosa* Engl. & Irmsch.; in this way a full classification of the subforma within the species is given.

24.2. Infraspecific epithets are formed like specific epithets and, when adjectival in form and not used as substantives, they agree grammatically with the generic name (see Art. 32.6).

Ex. 2. Solanum melongena var. *insanum* Prain (Bengal Pl.: 746. 1903, *"insana"*).

24.3. Infraspecific names with final epithets such as *typicus, originalis, originarius, genuinus, verus,* and *veridicus,* purporting to indicate the taxon containing the type of the name of the next higher taxon, are not validly published unless are autonyms (Art. 26).

Ex. 3. Lobelia spicata "var. *originalis*" (McVaugh in Rhodora 38: 308. 1936) was not validly published (see Art. 26 Ex. 1).

24.4. The use of a binary combination instead of an infraspecific epithet is not admissible. Contrary to Art. 32.1(b), names so constructed are validly published but are to be altered to the proper form without change of the author's name or date of publication.

Ex. 4. Salvia grandiflora subsp. *"S. willeana"* (Holmboe in Bergens Mus. Skr., ser. 2, 1(2): 157. 1914) is to be cited as *S. grandiflora* subsp. *willeana* Holmboe.

Ex. 5. Phyllerpa prolifera var. *"Ph. firma"* (Kützing, Sp. Alg.: 495. 1849) is to be altered to *P. prolifera* var. *firma* Kütz.

Note 1. Infraspecific taxa within different species may bear names with the same final epithet; those within one species may bear names with the same final epithet as the names of other species (but see Rec. 24B.1).

Ex. 6. Rosa glutinosa var. *leioclada* H. Christ (in Boissier, Fl. Orient. Suppl.: 222. 1888) and *Rosa jundzillii* f. *leioclada* Borbás (in Math. Term. Közlem. 16: 376, 383. 1880) are both permissible, as is *Viola tricolor* var. *hirta* Ging. (in Candolle, Prodr. 1: 304. 1824), in spite of the previous existence of a species named *Viola hirta* L.

Note 2. Infraspecific taxa within the same species, even if they differ in rank, may not bear names with the same final epithet but different types (Art. 53.5).

Recommendation 24A

24A.1. Recommendations made for forming specific epithets (Rec. 23A) apply equally for infraspecific epithets.

Recommendation 24B

24B.1. Authors proposing new infraspecific names should avoid epithets previously used as specific epithets in the same genus.

24B.2. When an infraspecific taxon is raised to the rank of species, or the inverse change occurs, the final epithet of its name should be retained unless the resulting combination would be contrary to this *Code*.

Article 25

25.1. For nomenclatural purposes, a species or any taxon below the rank of species is regarded as the sum of its subordinate taxa, if any. In fungi, a holomorph also includes its correlated form-taxa (see Art. 59).

Ex. 1. When *Montia parvifolia* (DC.) Greene is treated as comprising two subspecies, one must write *M. parvifolia* subsp. *parvifolia* for that part of the species that includes the nomenclatural type and excludes the type of the name of the other subspecies, *M. parvifolia* subsp. *flagellaris* (Bong.) Ferris. The name *M. parvifolia* applies to the species in its entirety.

Article 26

26.1. The name of any infraspecific taxon that includes the type of the adopted, legitimate name of the species to which it is assigned is to repeat the specific epithet unaltered as its final epithet, but not followed by an author's name (see Art. 46). Such names are termed autonyms (Art. 6.8; see also Art. 7.6).

Ex. 1. The variety which includes the type of the name *Lobelia spicata* Lam. is to be named *Lobelia spicata* Lam. var. *spicata* (see also Art. 24 Ex. 3).

Note 1. This provision applies only to the names of those subordinate taxa that include the type of the adopted name of the species (but see Rec. 26A).

26.2. A name of an infraspecific taxon that includes the type (i.e. the holotype or all syntypes or the previously designated type) of the adopted, legitimate name of the species to which it is assigned is not validly published unless its final epithet repeats the specific epithet unaltered. For the purpose of this provision, explicit indication that the nomenclaturally typical element of the species is included is considered as equivalent to inclusion of the type, whether or not it has been previously designated (see also Art. 24.3).

Ex. 2. Linnaeus (Sp. Pl.: 779-781. 1753) included 13 named varieties under *Medicago polymorpha.* Since *M. polymorpha* L. has no holotype and since no syntypes are cited, all varietal names are validly published irrespective of the fact that the lectotype subsequently chosen (by Heyn in Bull. Res. Council Israel, Sect. D, Bot., 7: 163. 1959) can be attributed to *M. polymorpha* var. *hispida* L.

Ex. 3. The intended combination "*Vulpia myuros* subsp. *pseudomyuros* (Soy.-Will.) Maire & Weiller" was not validly published in Maire (Fl. Afrique N. 3: 177. 1955) because it included "*F. myuros* L., Sp. 1, p. 74 (1753) sensu stricto" in synonymy, *Festuca myuros* L. being the basionym of *Vulpia myuros* (L.) C. C. Gmel.

26.3. The first instance of valid publication of a name of an infraspecific taxon that does not include the type of the adopted, legitimate name of the species automatically establishes the corresponding autonym (see also Art. 32.7 and 11.6).

Ex. 4. The publication of the name *Lycopodium inundatum* var. *bigelovii* Tuck. (in Amer. J. Sci. Arts 45: 47. 1843) automatically established the name of another variety, *L. inundatum* L. var. *inundatum,* the type of which is that of the name *L. inundatum* L.

Ex. 5. Utricularia stellaris L. f. (1781) includes *U. stellaris* var. *coromandeliana* A. DC. (Prodr. 8: 3. 1844) and *U. stellaris* L. f. var. *stellaris* (1844) automatically established at the same time. When *U. stellaris* is included in *U. inflexa* Forssk. (1775) as a variety, the correct name of that variety, under Art. 11.6, is *U. inflexa* var. *stellaris* (L. f.) P. Taylor (1961).

Recommendation 26A

26A.1. A variety including the type of the correct name of a subspecies, but not including the type of the correct name of the species, should, where there is no obstacle under the rules, be given a name with the same final epithet and type as the subspecies name.

26A.2. A subspecies not including the type of the correct name of the species should, where there is no obstacle under the rules, be given a name with the same final epithet and type as a name of one of its subordinate varieties.

26A.3. A taxon of rank lower than variety which includes the type of the correct name of a subspecies or variety, but not the type of the correct name of the species, should, where there is no obstacle under the rules, be given a name with the same final epithet and type as the name of the subspecies or variety. On the other hand, a subspecies or variety which does not include the type of the correct name of the species should not be given a name with the same final epithet as a name of one of its subordinate taxa below the rank of variety.

Ex. 1. Fernald treated *Stachys palustris* subsp. *pilosa* (Nutt.) Epling (in Repert. Spec. Nov. Regni Veg. Beih. 8: 63. 1934) as composed of five varieties, for one of which (that including the type of *S. palustris* subsp. *pilosa*) he made the combination *S. palustris* var. *pilosa* (Nutt.) Fernald (in Rhodora 45: 474. 1943), there being no legitimate varietal name available.

Ex. 2. There being no legitimate name available at the rank of subspecies, Bonaparte made the combination *Pteridium aquilinum* subsp. *caudatum* (L.) Bonap. (Notes Ptérid. 1: 62. 1915), using the same final epithet that Sadebeck had used earlier in the combination *P. aquilinum* var. *caudatum* (L.) Sadeb. (in Jahrb. Hamburg. Wiss. Anst. Beih. 14(3): 5. 1897), both combinations being based on *Pteris caudata* L. Each name is legitimate, and both can be used, as by Tryon (in Rhodora 43: 52-54. 1941), who treated *P. aquilinum* var. *caudatum* as one of four varieties under subsp. *caudatum* (see Art. 34.2).

Article 27

27.1. The final epithet in the name of an infraspecific taxon may not repeat unchanged the epithet of the correct name of the species to which the taxon is assigned unless the two names have the same type.

SECTION 6. NAMES OF PLANTS IN CULTIVATION

Article 28

28.1. Plants brought from the wild into cultivation retain the names that are applied to the same taxa growing in nature.

28.2. Hybrids, including those arising in cultivation, may receive names as provided in App. I (see also Art. 11.8, 40, and 50).

Note 1. Additional, independent designations for plants used in agriculture, forestry, and horticulture (and arising either in nature or cultivation) are dealt with in the *International code of nomenclature for cultivated plants,* where regulations are provided for their formation and use. However, nothing precludes the use for cultivated plants of names published in accordance with the requirements of the present *Code*.

Note 2. Epithets in names published in conformity with this *Code* may be used as cultivar epithets under the rules of the *International code of nomenclature for cultivated plants-1980,* when this is considered to be the appropriate status for the groups concerned. The *International code of nomenclature for cultivated plants-1980,* in its Art. 27, requires new cultivar epithets published on or after 1 January 1959 to be fancy names markedly different from epithets of names in Latin form governed by the present *Code*.

Ex. 1. Cultivar names: *Taxus baccata* 'Variegata' or *Taxus baccata* cv. Variegata (based on *T. baccata* var. *variegata* Weston, Bot. Univ. 1: 292, 347. 1770), *Phlox drummondii* 'Sternenzauber', *Viburnum* ×*bodnantense* 'Dawn'.

CHAPTER IV. EFFECTIVE AND VALID PUBLICATION

SECTION 1. CONDITIONS AND DATES OF EFFECTIVE PUBLICATION

Article 29

29.1. Publication is effected, under this *Code,* only by distribution of printed matter (through sale, exchange, or gift) to the general public or at least to botanical institutions with libraries accessible to botanists generally. It is not effected by communication of new names at a public meeting, by the placing of names in collections or gardens open to the public, or by the issue of microfilm made from manuscripts, type-scripts or other unpublished material.

Ex. 1. Cusson announced his establishment of the genus *Physospermum* in a memoir read at the Société des Sciences de Montpellier in 1770, and later in 1782 or 1783 at the Société de Médecine de Paris, but its effective publication dates from 1787 (in Hist. Soc. Roy. Méd. 5(1): 279).

Article 30

30.1. Publication by indelible autograph before 1 January 1953 is effective.

Ex. 1. Salvia oxyodon Webb & Heldr. was effectively published in a printed autograph catalogue placed on sale (Webb & Heldreich, *Catalogus plantarum hispanicarum ... ab A. Blanco lectarum,* Paris, Jul 1850, folio).

30.2. For the purpose of this Article, indelible autograph is handwritten material reproduced by some mechanical or graphic process (such as lithography, offset, or metallic etching).

Ex. 2. H. Léveillé, *Flore du Kouy Tchéou* (1914-1915), is a work lithographed from a handwritten text.

30.3. Publication on or after 1 January 1953 in trade catalogues or non-scientific newspapers, and on or after 1 January 1973 in seed-exchange lists, does not constitute effective publication.

30.4. The distribution on or after 1 January 1953 of printed matter accompanying exsiccata does not constitute effective publication.

Note 1. If the printed matter is also distributed independently of the exsiccata, it is effectively published.

Ex. 3. Works such as Lundell & Nannfeldt, *Fungi exsiccati suecici* ..., Uppsala 1-..., 1934-..., distributed independently of the exsiccata, whether published before or after 1 January 1953, are effectively published.

Recommendation 30A

30A.1. It is strongly recommended that authors avoid publishing new names and descriptions or diagnoses of new taxa in ephemeral printed matter of any kind, in particular that which is multiplied in restricted and uncertain numbers, where the permanence of the text may be limited, where the effective publication in terms of number of copies is not obvious, or where the printed matter is unlikely to reach the general public. Authors should also avoid publishing new names and descriptions or diagnoses in popular periodicals, in abstracting journals, or on correction slips.

Article 31

31.1. The date of effective publication is the date on which the printed matter became available as defined in Art. 29 and 30. In the absence of proof establishing some other date, the one appearing in the printed matter must be accepted as correct.

Ex. 1. Individual parts of Willdenow's *Species plantarum* were published as follows: 1(1), Jun 1797; 1(2), Jul 1798; 2(1), Mar 1799; 2(2), Dec 1799; 3(1), 1800; 3(2), Nov 1802; 3(3), Apr-Dec 1803 (and later than Michaux's *Flora boreali-americana*); 4(1), 1805; 4(2), 1806; these dates, which are partly in disagreement with the years on the title-pages of the volumes, are presently accepted as the correct dates of effective publication.

Ex. 2. T. M. Fries, "Lichenes arctoi", was first published as an independently paginated preprint in 1860, which predates the identical version published in a journal (Nova Acta Reg. Soc. Sci. Upsal. 3(3): 103-398. 1861).

31.2. When separates from periodicals or other works placed on sale are issued in advance, the date on the separate is accepted as the date of effective publication unless there is evidence that it is erroneous.

Ex. 3. Publication in separates issued in advance: the names of the *Selaginella* species published by Hieronymus (in Hedwigia 51: 241-272) were effectively published on 15 October 1911, since the volume in which the paper appeared, though dated 1912, states (p. ii) that the separate appeared on that date.

Recommendation 31A

31A.1. The date on which the publisher or his agent delivers printed matter to one of the usual carriers for distribution to the public should be accepted as its date of effective publication.

SECTION 2. CONDITIONS AND DATES OF VALID PUBLICATION OF NAMES

Article 32

32.1. In order to be validly published, a name of a taxon (autonyms excepted) must: *(a)* be effectively published (see Art. 29-31) on or after the starting-point date of the respective group (Art. 13.1); *(b)* have a form which complies with the provisions of Art. 16-27 (but see Art. 18.3 and 19.6), and Art. H.6 and H.7; *(c)* be accompanied by a description or diagnosis or by a reference to a previously and effectively published description or diagnosis (except as provided in Art. 42.3, 44.1, and H.9); and *(d)* comply with the special provisions of Art. 33-45 (see also Art. 61). In addition, subject to the approval of the XVI International Botanical Congress, names (autonyms excepted) published on or after 1 January 2000 must be registered.

Ex. 1. "*Egeria*" (Néraud in Gaudichaud, Voy. Uranie, Bot.: 25, 28. 1826), published without a description or a diagnosis or a reference to a former one, was not validly published.

Ex. 2. "*Loranthus macrosolen* Steud." originally appeared without a description or diagnosis on the printed labels issued about the year 1843 with Sect. II, No. 529, 1288, of Schimper's herbarium specimens of Abyssinian plants; the name was not validly published, however, until Richard (Tent. Fl. Abyss. 1: 340. 1847) supplied a description.

**Ex. 3.* In Don, *Sweet's Hortus britannicus*, ed. 3 (1839), for each listed species the flower colour, the duration of the plant, and a translation into English of the specific epithet are given in tabular form. In many genera the flower colour and duration may be identical for all species and clearly their mention is not intended as a validating description or diagnosis. New names appearing in that work are therefore not validly published, except in some cases where reference is made to earlier descriptions or diagnoses or to validly published basionyms.

32.2. Registration is effected by sending the printed matter that includes the protologue(s), with the name(s) to be registered clearly identified, to any registering office designated by the International Association for Plant Taxonomy.

32.3. A diagnosis of a taxon is a statement of that which in the opinion of its author distinguishes the taxon from others.

32.4. For the purpose of valid publication of a name, reference to a previously and effectively published description or diagnosis may be direct or indirect (Art. 32.5). For names published on or after 1 January 1953 it must, however, be full and direct as specified in Art. 33.2.

32.5. An indirect reference is a clear indication, by the citation of the author's name or in some other way, that a previously and effectively published description or diagnosis applies.

Ex. 4. "*Kratzmannia*" (Opiz in Berchtold & Opiz, Oekon.-Techn. Fl. Böhm. 1: 398. 1836) was published with a diagnosis but was not definitely accepted by the author and therefore was not validly published. *Kratzmannia* Opiz (Seznam: 56. 1852), lacking description or diagnosis, is however definitely accepted, and its citation as "*Kratzmannia* O." constitutes indirect reference to the diagnosis published in 1836.

Ex. 5. Opiz published the name of the genus *Hemisphace* (Benth.) Opiz (1852) without a description or diagnosis, but as he wrote *"Hemisphace* Benth." he indirectly referred to the previously effectively published description by Bentham (Labiat. Gen. Spec.: 193. 1833) of *Salvia* sect. *Hemisphace.*

Ex. 6. The new combination *Cymbopogon martini* (Roxb.) W. Watson (1882) is validated by the addition of the number "309", which, as explained at the top of the same page, is the running-number of the species *(Andropogon martini* Roxb.) in Steudel (Syn. Pl. Glumac. 1: 388. 1854). Although the reference to the basionym *Andropogon martini* is indirect, it is unambiguous (see also Rec. 60C.2).

Ex. 7. Miller (1768), in the preface to *The gardeners dictionary,* ed. 8, stated that he had "now applied Linnaeus's method entirely except in such particulars ...", of which he gave examples. In the main text, he often referred to Linnaean genera under his own generic headings, e.g., to *Cactus* L. [pro parte] under *Opuntia* Mill. Therefore an implicit reference to a Linnaean binomial may be assumed when this is appropriate, and Miller's binomials accepted as new combinations (e.g., *Opuntia ficus-indica* (L.) Mill., based on *Cactus ficus-indica* L.) or avowed substitutes (e.g., *Opuntia vulgaris* Mill., based on *Cactus opuntia* L., where both names have the reference to *"Opuntia vulgo herbariorum"* of Bauhin & Cherler in common).

Ex. 8. In Kummer's *Führer in die Pilzkunde* (1871) the statement that the author intended to adopt at generic rank the subdivisions of *Agaricus* then in use, which at the time were those of Fries, and the general arrangement of the work, which faithfully follows that of Fries, provide indirect reference to Fries's earlier names of "tribes". Therefore, names such as *Hypholoma* (Fr. : Fr.) P. Kumm. are accepted as being based on the corresponding Friesian names (here: *A.* "tribus" *Hypholoma* Fr. : Fr.) although Kummer did not explicitly refer to Fries.

32.6. Names published with an incorrect Latin termination but otherwise in accordance with this *Code* are regarded as validly published; they are to be changed to accord with Art. 17-19, 21, 23, and 24, without change of the author's name or date of publication (see also Art. 60.11).

32.7. Autonyms (Art. 6.8) are accepted as validly published names, dating from the publication in which they were established (see Art. 22.3 and 26.3), whether or not they appear in print in that publication.

32.8. Names in specified ranks included in publications listed as suppressed works *(opera utique oppressa;* App. V) are not validly published. Proposals for the addition of publications to App. V must be submitted to the General Committee (see Div. III), which will refer them for examination to the committees for the various taxonomic groups (see Rec. 32F; see also Art. 14.14 and Rec. 14A).

32.9. When a proposal for the suppression of a publication has been approved by the General Committee after study by the committees for the taxonomic groups concerned, treating that publication as suppressed is authorized subject to the decision of a later International Botanical Congress.

Note 1. For valid publication of names of plant taxa that were originally not treated as plants, see Art. 45.5.

Recommendation 32A

32A.1. A name should not be validated solely by a reference to a description or diagnosis published before 1753.

Recommendation 32B

32B.1. The description or diagnosis of any new taxon should mention the points in which the taxon differs from its allies.

Recommendation 32C

32C.1. Authors should avoid adoption of a name which has been previously but not validly published for a different taxon.

Recommendation 32D

32D.1. In describing or diagnosing new taxa, authors should, when possible, supply figures with details of structure as an aid to identification.

32D.2. In the explanation of the figures, authors should indicate the specimen(s) on which they are based (see also Rec. 8A.2).

32D.3. Authors should indicate clearly and precisely the scale of the figures which they publish.

Recommendation 32E

32E.1. Descriptions or diagnoses of parasitic plants should always be followed by indication of the hosts, especially those of parasitic fungi. The hosts should be designated by their scientific names and not solely by names in modern languages, the applications of which are often doubtful.

Recommendation 32F

32F.1. When a proposal for the suppression of a publication under Art. 32.8 has been referred to the appropriate committees for study, authors should follow existing usage as far as possible pending the General Committee's recommendation on the proposal.

Article 33

33.1. A combination (autonyms excepted) is not validly published unless the author definitely associates the final epithet with the name of the genus or species, or with its abbreviation.

Ex. 1. Combinations validly published: In Linnaeus's *Species plantarum* the placing of the epithet in the margin opposite the name of the genus clearly associates the epithet with the name of the genus. The same result is attained in Miller's *Gardeners dictionary,* ed. 8, by the inclusion of the epithet in parentheses immediately after the name of the genus, in Steudel's *Nomenclator botanicus* by the arrangement of the epithets in a list headed by the name of the genus, and in general by any typographical device which associates an epithet with a particular generic or specific name.

Ex. 2. Combinations not validly published: Rafinesque's statement under *Blephilia* that "Le type de ce genre est la *Monarda ciliata* Linn." (in J. Phys. Chim. Hist. Nat. Arts 89: 98. 1819) does not constitute valid publication of the combination *Blephilia ciliata,* since Rafinesque did not definitely associate the epithet *ciliata* with the generic name *Blephilia.* Similarly, the combination *Eulophus peucedanoides* is not to be attributed to Bentham & Hooker (Gen. Pl. 1: 885. 1867) on the basis of their listing of "*Cnidium peucedanoides,* H. B. et K." under *Eulophus.*

33.2. A new combination, or an avowed substitute *(nomen novum),* published on or after 1 January 1953, for a previously and validly published name is not validly published unless its basionym (name-bringing or epithet-bringing synonym) or the replaced synonym (when a new name is proposed) is clearly indicated and a full and direct reference given to its author and place of valid publication with page or plate reference and date.

Ex. 3. In transferring *Ectocarpus mucronatus* D. A. Saunders to *Giffordia,* Kjeldsen & Phinney (in Madroño 22: 90. 27 Apr 1973) cited the basionym and its author but without reference to its place of valid publication. They later (in Madroño 22: 154. 2 Jul 1973) validated the binomial *G. mucronata* (D. A. Saunders) Kjeldsen & Phinney by giving a full and direct reference to the place of valid publication of the basionym.

Note 1. For the purpose of this *Code,* a page reference (for publications with a consecutive pagination) is a reference to the page or pages on which the basionym was validly published or on which the protologue is printed, but not to the pagination of the whole publication unless it is coextensive with that of the protologue.

Ex. 4. When proposing *"Cylindrocladium infestans",* Peerally (in Mycotaxon 40: 337. 1991) cited the basionym as "*Cylindrocladiella infestans* Boesw., Can. J. Bot. 60: 2288-2294. 1982". As this refers to the pagination of Boeswinkel's entire paper, not of the protologue of the intended basionym alone, the combination was not validly published by Peerally.

33.3. Errors of bibliographic citation and incorrect forms of author citation (see Art. 46) do not invalidate publication of a new combination or avowed substitute.

Ex. 5. Aronia arbutifolia var. *nigra* (Willd.) F. Seym. (Fl. New England: 308. 1969) was published as a new combination "Based on *Mespilus arbutifolia* L. var. *nigra* Willd., in Sp. Pl. 2: 1013. 1800." Willdenow treated these plants in the genus *Pyrus,* not *Mespilus,* and publication was in 1799, not 1800; these errors are treated as bibliographic errors of citation and do not invalidate the new combination.

Ex. 6. The combination *Trichipteris kalbreyeri* was proposed by Tryon (in Contr. Gray Herb. 200: 45. 1970) with a full and direct reference to *Alsophila kalbreyeri* C. Chr. (Index Filic.: 44. 1905). This, however, was not the place of valid publication of the basionym, which had previously been published, with the same type, by Baker (Summ. New Ferns: 9. 1892). Tryon's bibliographic error of citation does not invalidate this new combination, which is to be cited as *T. kalbreyeri* (Baker) R. M. Tryon.

Ex. 7. The combination *Lasiobelonium corticale* was proposed by Raitviir (1980) with a full and direct reference to *Peziza corticalis* in Fries (Syst. Mycol. 2: 96. 1822). This, however, was not the place of valid publication of the basionym, which, under the *Code* operating in 1980, was in Mérat (Nouv. Fl. Env. Paris, ed. 2, 1: 22. 1821), and under the present *Code* is in Persoon (Observ. Mycol. 1: 28. 1796). Raitviir's bibliographic error of citation does not invalidate the new combination, which is to be cited as *L. corticale* (Pers.) Raitv.

33.4. Mere reference to the *Index kewensis,* the *Index of fungi,* or any work other than that in which the name was validly published does not constitute a full and direct reference to the original publication of a name.

Ex. 8. Ciferri (in Mycopathol. Mycol. Appl. 7: 86-89. 1954), in proposing 142 new combinations in *Meliola,* omitted references to places of publication of basionyms, stating that they could be found in Petrak's lists or in the *Index of fungi;* none of these combinations was validly published. Similarly, Grummann (Cat. Lich. Germ.: 18. 1963) introduced a new combination in the form *Lecanora campestris* f. "*pseudistera* (Nyl.) Grumm. c.n. – *L. p* Nyl., Z 5: 521", in which "Z 5" referred to Zahlbruckner

(Cat. Lich. Univ. 5: 521. 1928), who gave the full citation of the basionym, *Lecanora pseudistera* Nyl.; Grummann's combination was not validly published.

Note 2. The publication of a name for a taxon previously known under a misapplied name must be valid under Art. 32-45. This procedure is not the same as publishing an avowed substitute (nomen novum) for a validly published but illegitimate name (Art. 58.1(b)), the type of which is necessarily the same as that of the name which it replaced (Art. 7.3).

Ex. 9. Sadleria hillebrandii Rob. (1913) was introduced as a "nom. nov." for "*Sadleria pallida* Hilleb. Fl. Haw. Is. 582. 1888. Not Hook. & Arn. Bot. Beech. 75. 1832." Since the requirements of Art. 32-45 were satisfied (for valid publication, prior to 1935, simple reference to a previous description or diagnosis in any language was sufficient), the name is validly published. It is, however, to be considered the name of a new species, validated by Hillebrand's description of the taxon to which he misapplied the name *S. pallida* Hook. & Arn., and not a *nomen novum* as stated by Robinson; hence, Art. 7.3 does not apply.

Ex. 10. Juncus bufonius "var. *occidentalis*" (Hermann in U.S. Forest Serv., Techn. Rep. RM-18: 14. 1975) was published as a "nom. et stat. nov." for *J. sphaerocarpus* "auct. Am., non Nees". Since there is no Latin diagnosis, designation of type, or reference to any previous publication providing these requirements, the name is not validly published.

33.5. A name given to a taxon whose rank is at the same time, contrary to Art. 5, denoted by a misplaced term is not validly published. Such misplacements include forms divided into varieties, species containing genera, and genera containing families or tribes.

Ex. 11. "Sectio *Orontiaceae*" was not validly published by Brown (Prodr.: 337. 1810) since he misapplied the term "sectio" to taxa of a rank higher than genus.

Ex. 12. "Tribus *Involuta*" and "tribus *Brevipedunculata*" (Huth in Bot. Jahrb. Syst. 20: 365, 368. 1895) are not validly published names, since Huth misapplied the term "tribus" to a taxon of a rank lower than section, within the genus *Delphinium.*

33.6. An exception to Art. 33.5 is made for names of the subdivisions of genera termed tribes *(tribus)* in Fries's *Systema mycologicum,* which are treated as validly published names of subdivisions of genera.

Ex. 13. Agaricus "tribus" *Pholiota* Fr. (Syst. Mycol. 1: 240. 1821), sanctioned in the same work, is the validly published basionym of the generic name *Pholiota* (Fr. : Fr.) P. Kumm. (1871) (see Art. 32 Ex. 8).

Recommendation 33A

33A.1. The full and direct reference to the place of publication of the basionym or replaced synonym should immediately follow a proposed new combination or nomen novum. It should not be provided by mere cross-reference to a bibliography at the end of the publication or to other parts of the same publication, e.g. by use of the abbreviations *"loc. cit."* or *"op. cit."*

Article 34

34.1. A name is not validly published *(a)* when it is not accepted by the author in the original publication; *(b)* when it is merely proposed in anticipation of the future acceptance of the group concerned, or of a particular circumscription, position, or rank of the group (so-called provisional name), except as provided for in Art. 59; *(c)* when it is merely cited as a synonym; *(d)* by the mere mention of the subordinate taxa included in the taxon concerned. Art. 34.1(a) does not apply to names published with a question mark or other indication of taxonomic doubt, yet published and accepted by the author.

Ex. 1. (a) *"Sebertia"*, proposed by Pierre (ms.) for a monotypic genus, was not validly published by Baillon (in Bull. Mens. Soc. Linn. Paris 2: 945. 1891) because he did not accept the genus. Although he gave a description of it, he referred its only species "*Sebertia acuminata* Pierre (ms.)" to the genus *Sersalisia* R. Br. as *Sersalisia* ? *acuminata*, which he thereby validly published under the provision of Art. 34.1, last sentence. The name *Sebertia* was validly published by Engler (1897).

Ex. 2. (a) The designations listed in the left-hand column of the Linnaean thesis *Herbarium amboinense* defended by Stickman (1754) were not names accepted by Linnaeus upon publication and are not validly published.

Ex. 3. (a) (b) The designation *"Conophyton"*, suggested by Haworth (Rev. Pl. Succ.: 82. 1821) for *Mesembryanthemum* sect. *Minima* Haw. (Rev. Pl. Succ.: 81. 1821) in the words "If this section proves to be a genus, the name of *Conophyton* would be apt", was not a validly published generic name since Haworth did not adopt it or accept the genus. The name was validly published as *Conophytum* N. E. Br. (1922).

Ex. 4. (c) "*Acosmus* Desv.", cited by Desfontaines (Cat. Pl. Hort. Paris.: 233. 1829) as a synonym of the generic name *Aspicarpa* Rich., was not validly published thereby.

Ex. 5. (c) *"Ornithogalum undulatum* hort. Bouch." (in Kunth, Enum. Pl. 4: 348. 1843), cited as a synonym under *Myogalum boucheanum* Kunth, was not validly published thereby; when transferred to *Ornithogalum* L., this species is to be called *O. boucheanum* (Kunth) Asch. (1866).

Ex. 6. (c) "*Erythrina micropteryx* Poepp." was not validly published by being cited as a synonym of *Micropteryx poeppigiana* Walp. (1850); the species concerned, when placed under *Erythrina* L., is to be called *E. poeppigiana* (Walp.) O. F. Cook (1901).

Ex. 7. (d) The family designation *"Rhaptopetalaceae"* (Pierre in Bull. Mens. Soc. Linn. Paris 2: 1296. Mai 1897), which was accompanied merely by mention of constituent genera, *Brazzeia* Baill., *"Scytopetalum"*, and *Rhaptopetalum* Oliv., was not validly published, as Pierre gave no description or diagnosis; the family bears the name *Scytopetalaceae* Engl. (Oct 1897), accompanied by a description.

Ex. 8. (d) The generic designation *"Ibidium"* (Salisbury in Trans. Hort. Soc. London 1: 291. 1812) was published merely with the mention of four included species. As Salisbury supplied no generic description or diagnosis, it is not a validly published name.

34.2. When, on or after 1 January 1953, two or more different names are proposed simultaneously for the same taxon by the same author (so-called alternative names), none of them is validly published. This rule does not apply in those cases where the same combination is simultaneously used at different ranks, either for infraspecific taxa within a species or for subdivisions of a genus within a genus (see Rec. 22A.1 and 22A.2, 26A.1-3).

Ex. 9. The species of *Brosimum* Sw. described by Ducke (in Arch. Jard. Bot. Rio de Janeiro 3: 23-29. 1922) were published with alternative names under *Piratinera* Aubl. added in a footnote (pp. 23-24). The publication of both sets of names, being effected before 1 January 1953, is valid.

Ex. 10. *"Euphorbia jaroslavii"* (Poljakov in Bot. Mater. Gerb. Bot. Inst. Komarova Akad. Nauk SSSR 15: 155. 1953) was published with an alternative designation, *"Tithymalus jaroslavii"*. Neither was validly published. However, one name, *Euphorbia yaroslavii* (with a different transliteration of the initial letter), was validly published by Poljakov (1961), who effectively published it with a new reference to the earlier publication and simultaneously rejected the other name.

Ex. 11. Description of *"Malvastrum bicuspidatum* subsp. *tumidum* S. R. Hill var. *tumidum,* subsp. et var. nov." (in Brittonia 32: 474. 1980) simultaneously validated both *M. bicuspidatum* subsp. *tumidum* S. R. Hill and *M. bicuspidatum* var. *tumidum* S. R. Hill.

Ex. 12. Hitchcock (in Univ. Wash. Publ. Biol. 17(1): 507-508. 1969) used the name *Bromus inermis* subsp. *pumpellianus* (Scribn.) Wagnon and provided a full and direct reference to its basionym, *B. pumpellianus* Scribn. Within that subspecies, he recognized varieties one of which he named *B. inermis* var. *pumpellianus* (without author citation but clearly based on the same basionym and type). In so doing, he met the requirements for valid publication of *B. inermis* var. *pumpellianus* (Scribn.) C. L. Hitchc.

Note 1. The name of a fungal holomorph and that of a correlated anamorph (see Art. 59), even if validated simultaneously, are not alternative names in the sense of Art. 34.2. They have different types, and the circumscription of the holomorph is considered to include the anamorph, but not vice versa.

Ex. 13. Lasiosphaeria elinorae Linder (1929), the name of a fungal holomorph, and the simultaneously published name of a correlated anamorph, *Helicosporium elinorae* Linder, are both valid, and both can be used under Art. 59.5.

Recommendation 34A

34A.1. Authors should avoid mentioning in their publications previously unpublished names which they do not accept, especially if the persons responsible for these unpublished names have not formally authorized their publication (see Rec. 23A.3(i)).

Article 35

35.1. A new name or combination published on or after 1 January 1953 without a clear indication of the rank of the taxon concerned is not validly published.

35.2. A new name or combination published before 1 January 1953 without a clear indication of rank is validly published provided that all other requirements for valid publication are fulfilled; it is, however, inoperative in questions of priority except for homonymy (see Art. 53.5). If it is a new name, it may serve as a basionym or replaced synonym for subsequent combinations or avowed substitutes in definite ranks.

Ex. 1. The groups *"Soldanellae", "Sepincoli", "Occidentales",* etc., were published without any indication of rank under *Convolvulus* L. by House (in Muhlenbergia 4: 50. 1908). The names *C.* [unranked] *Soldanellae,* etc., are validly published but they are not in any definite rank and have no status in questions of priority except that they may act as homonyms.

Ex. 2. In *Carex* L., the epithet *Scirpinae* was used in the name of an infrageneric taxon of no stated rank by Tuckerman (Enum. Meth. Caric.: 8. 1843); this was assigned sectional rank by Kükenthal (in Engler, Pflanzenr. 38: 81. 1909) and may be cited as *Carex* sect. *Scirpinae* (Tuck.) Kük. (*C.* [unranked] *Scirpinae* Tuck.).

35.3. If in a given publication prior to 1 January 1890 only one infraspecific rank is admitted, it is considered to be that of variety unless this would be contrary to the statements of the author himself in the same publication.

35.4. In questions of indication of rank, all publications appearing under the same title and by the same author, such as different parts of a flora issued at different times (but not different editions of the same work), must be considered as a whole, and any statement made therein designating the rank of taxa included in the work must be considered as if it had been published together with the first instalment.

Article 36

36.1. In order to be validly published, a name of a new taxon of plants, the algae and all fossils excepted, published on or after 1 January 1935 must be accompanied by a Latin description or diagnosis or by a reference to a previously and effectively published Latin description or diagnosis (but see Art. H.9).

Ex. 1. *Arabis* "Sekt. *Brassicoturritis* O. E. Schulz" and "Sekt. *Brassicarabis* O. E. Schulz" (in Engler & Prantl, Nat. Pflanzenfam., ed. 2, 17b: 543-544. 1936), published with German but no Latin descriptions or diagnoses, are not validly published names.

Ex. 2. *"Schiedea gregoriana"* (Degener, Fl. Hawaiiensis, fam. 119. 9 Apr 1936) was not accompanied by a Latin description or diagnosis, and is accordingly not a validly published name. *S. kealiae* Caum & Hosaka (in Occas. Pap. Bernice Pauahi Bishop Mus. 11(23): 3. 10 Apr 1936), the type of which is part of the material used by Degener, is provided with a Latin description and is validly published.

Ex. 3. *Alyssum flahaultianum* Emb., first published without a Latin description or diagnosis (in Bull. Soc. Hist. Nat. Maroc 15: 199. 1936), was validly published posthumously when a Latin translation of Emberger's original French description was provided (in Willdenowia 15: 62-63. 1985).

36.2. In order to be validly published, a name of a new taxon of non-fossil algae published on or after 1 January 1958 must be accompanied by a Latin description or diagnosis or by a reference to a previously and effectively published Latin description or diagnosis.

Ex. 4. Although *Neoptilota* Kylin (Gatt. Rhodophyc.: 392. 1956) was accompanied by only a German description, it is a validly published name since it applies to an alga and was published before 1958.

36.3. In order to be validly published, a name of a new taxon of fossil plants published on or after 1 Jan 1996 must be accompanied by a Latin or English description or diagnosis or by a reference to a previously and effectively published Latin or English description or diagnosis.

Recommendation 36A

36A.1. Authors publishing names of new taxa of non-fossil plants should give or cite a full description in Latin in addition to the diagnosis.

Article 37

37.1. Publication on or after 1 January 1958 of the name of a new taxon of the rank of genus or below is valid only when the type of the name is indicated (see Art. 7-10; but see Art. H.9 Note 1 for the names of certain hybrids).

37.2. For the name of a new genus or subdivision of a genus, reference (direct or indirect) to one species name only, or the citation of the holotype or lectotype of one previously or simultaneously published species name only, constitutes indication of the type (Art. 10 Note 1; see also Art. 22.5; but see Art. 37.4).

37.3. For the name of a new species or infraspecific taxon, citation of a single element is acceptable as indication of the holotype (but see Art. 37.4). Mere citation of a locality without concrete reference to a specimen does not however constitute indication of a holotype. Citation of the collector's name and/or collecting number and/or date of collection and/or reference to any other detail of the type specimen or illustration is required.

37.4. For the name of a new taxon of the rank of genus or below published on or after 1 January 1990, indication of the type must include one of the words "typus" or "holotypus", or its abbreviation, or its equivalent in a modern language.

37.5. For the name of a new species or infraspecific taxon published on or after 1 January 1990 whose type is a specimen or unpublished illustration, the herbarium or institution in which the type is conserved must be specified.

Note 1. Specification of the herbarium or institution may be made in an abbreviated form, e.g. as given in *Index herbariorum,* part I (Regnum Veg. 120).

Recommendation 37A

37A.1. The indication of the nomenclatural type should immediately follow the description or diagnosis and should use the Latin word "typus" or "holotypus".

Article 38

38.1. In order to be validly published, a name of a new taxon of fossil plants of specific or lower rank published on or after 1 January 1912 must be accompanied by an illustration or figure showing the essential characters, in addition to the description or diagnosis, or by a reference to a previously and effectively published illustration or figure.

Article 39

39.1. In order to be validly published, a name of a new taxon of non-fossil algae of specific or lower rank published on or after 1 January 1958 must be accompanied by an illustration or figure showing the distinctive morphological features, in addition to the Latin description or diagnosis, or by a reference to a previously and effectively published illustration or figure.

Recommendation 39A

39A.1. The illustration or figure required by Art. 39 should be prepared from actual specimens, preferably including the holotype.

Article 40

40.1. In order to be validly published, names of hybrids of specific or lower rank with Latin epithets must comply with the same rules as names of non-hybrid taxa of the same rank.

Ex. 1. "*Nepeta ×faassenii*" (Bergmans, Vaste Pl. Rotsheesters, ed. 2: 544. 1939, with a description in Dutch; Lawrence in Gentes Herb. 8: 64. 1949, with a diagnosis in English) is not validly published, not being accompanied by or associated with a Latin description or diagnosis. The name *Nepeta ×faassenii* Bergmans ex Stearn (1950) is validly published, being accompanied by a Latin description.

Ex. 2. "*Rheum ×cultorum*" (Thorsrud & Reisaeter, Norske Plantenavr.: 95. 1948), being there a nomen nudum, is not validly published.

Ex. 3. "*Fumaria ×salmonii*" (Druce, List Brit. Pl.: 4. 1908) is not validly published, because only the presumed parentage *F. densiflora* × *F. officinalis* is stated.

Note 1. For names of hybrids of the rank of genus or subdivision of a genus, see Art. H.9.

Article 41

41.1. In order to be validly published, a name of a family or subdivision of a family must be accompanied *(a)* by a description or diagnosis of the taxon, or *(b)* by a reference (direct or indirect) to a previously and effectively published description or diagnosis of a family or subdivision of a family.

Ex. 1. "*Pseudoditrichaceae* fam. nov." (Steere & Iwatsuki in Canad. J. Bot. 52: 701. 1974) was not a validly published name of a family as there was no Latin description or diagnosis nor reference to either, but only mention of the single included genus and species (see Art. 34.1(d)), "*Pseudoditrichum mirabile* gen. et sp. nov.", for both of which the name was validated under Art. 42 by a single Latin diagnosis.

41.2. In order to be validly published, a name of a genus or subdivision of a genus must be accompanied *(a)* by a description or diagnosis of the taxon (but see Art. 42), or *(b)* by a reference (direct or indirect) to a previously and effectively published description or diagnosis of a genus or subdivision of a genus.

Ex. 2. Validly published generic names: *Carphalea* Juss., accompanied by a generic description; *Thuspeinanta* T. Durand, replacing the name of the previously described genus *Tapeinanthus* Boiss. ex Benth. (non Herb.); *Aspalathoides* (DC.) K. Koch, based on the name of a previously described section, *Anthyllis* sect. *Aspalathoides* DC.; *Scirpoides* Ség. (Pl. Veron. Suppl.: 73. 1754), accepted there but without a generic description or diagnosis, validated by indirect reference (through the title of the book and a general statement in the preface) to the generic diagnosis and further direct references in Séguier (Pl. Veron. 1: 117. 1745).

Note 1. An exception to Art. 41.2 is made for the generic names first published by Linnaeus in *Species plantarum,* ed. 1 (1753) and ed. 2 (1762-1763), which are treated as having been validly published on those dates (see Art. 13.4).

Note 2. In certain circumstances, an illustration with analysis is accepted as equivalent to a generic description or diagnosis (see Art. 42.3).

41.3. In order to be validly published, a name of a species or infraspecific taxon must be accompanied *(a)* by a description or diagnosis of the taxon (but see Art. 42 and 44), or *(b)* by a reference to a previously and effectively published description or diagnosis of a species or infraspecific taxon. A name of a species may also be validly published *(c),* under certain circumstances, by reference to a genus whose name was previously and validly published simultaneously with its description or diagnosis. A reference as mentioned under (c) is acceptable only if neither the author of the name of the genus nor the author of the name of the species indicate that more than one species belongs to the genus in question.

Ex. 3. Trilepisium Thouars (1806) was validated by a generic description but without mention of a name of a species. *T. madagascariense* DC. (1828) was subsequently proposed without a description or diagnosis of the species. Neither author gave any indication that there was more than one species in the genus. Candolle's specific name is therefore validly published.

Article 42

42.1. The names of a genus and a species may be simultaneously validated by provision of a single description *(descriptio generico-specifica)* or diagnosis, even though this may have been intended as only generic or specific, if all of the following conditions obtain: *(a)* the genus is at that time monotypic; *(b)* no other names (at any rank) have previously been validly published based on the same type; and *(c)* the names of the genus and species otherwise fulfil the requirements for valid publication. Reference to an earlier description or diagnosis is not accepted as provision of such a description or diagnosis.

Ex. 1. Nylander (1879) described the new species *"Anema nummulariellum"* in a new genus *"Anema"* without providing a generic description or diagnosis. Since at the same time he also transferred *Omphalaria nummularia* Durieu & Mont. to *"Anema"*, none of his names were validly published. They were later validated by Forsell (1885).

42.2. For the purpose of Art. 42, a monotypic genus is one for which a single binomial is validly published, even though the author may indicate that other species are attributable to the genus.

Ex. 2. The names *Kedarnatha* P. K. Mukh. & Constance (1986) and *K. sanctuarii* P. K. Mukh. & Constance, the latter designating the only species in the new genus, are both validly published although a Latin description is provided only under the generic name.

Ex. 3. Piptolepis phillyreoides Benth. (1840) was a new species assigned to the monotypic new genus *Piptolepis* published with a combined generic and specific description, and both names are validly published.

Ex. 4. In publishing *"Phaelypea"* without a generic description or diagnosis, P. Browne (Civ. Nat. Hist. Jamaica: 269. 1756) included and described a single species, but he gave the species a phrase-name and did not provide a valid binomial. Art. 42 does not therefore apply and *"Phaelypea"* is not a validly published name.

42.3. Prior to 1 January 1908 an illustration with analysis, or for non-vascular plants a single figure showing details aiding identification, is acceptable, for the purpose of this Article, in place of a written description or diagnosis.

42.4. For the purpose of Art. 42, an analysis is a figure or group of figures, commonly separate from the main illustration of the plant (though usually on the same page or plate), showing details aiding identification, with or without a separate caption.

Ex. 5. The generic name *Philgamia* Baill. (1894) was validly published, as it appeared on a plate with analysis of the only included species, *P. hibbertioides* Baill., and was published before 1 January 1908.

Article 43

43.1. A name of a taxon below the rank of genus is not validly published unless the name of the genus or species to which it is assigned is validly published at the same time or was validly published previously.

Ex. 1. Binary designations for six species of *"Suaeda"*, including *"S. baccata"* and *"S. vera"*, were published with descriptions and diagnoses by Forsskål (Fl. Aegypt.-Arab.: 69-71. 1775), but he provided no description or diagnosis for the genus: these were not therefore validly published names.

Ex. 2. Müller (in Flora 63: 286. 1880) published the new genus *"Phlyctidia"* with the species *"P. hampeana* n. sp.", *"P. boliviensis"* (= *Phlyctis boliviensis* Nyl.), *"P. sorediiformis"* (= *Phlyctis sorediiformis* Kremp.), *"P. brasiliensis"* (= *Phlyctis brasiliensis* Nyl.), and *"P. andensis"* (= *Phlyctis andensis* Nyl.). These were not, however, validly published specific names in this place, because the intended generic name *"Phlyctidia"* was not validly published; Müller gave no generic description or diagnosis but only a description and a diagnosis of the new species *"P. hampeana"*. This description and diagnosis did not validate the generic name as a descriptio generico-specifica under Art. 42 since the new genus was not monotypic. Valid publication of the name *Phlyctidia* was by Müller (1895), who provided a short generic diagnosis and explicitly included only two species, *P. ludoviciensis* Müll. Arg. and *P. boliviensis* (Nyl.) Müll. Arg. The latter names were validly published in 1895.

Note 1. This Article applies also when specific and other epithets are published under words not to be regarded as generic names (see Art. 20.4).

Ex. 3. The binary designation *"Anonymos aquatica"* (Walter, Fl. Carol.: 230. 1788) is not a validly published name. The correct name for the species concerned is *Planera aquatica* J. F. Gmel. (1791), and the date of the name, for purposes of priority, is 1791. The name must not be cited as "*P. aquatica* (Walter) J. F. Gmel."

Ex. 4. Despite the existence of the generic name *Scirpoides* Ség. (1754), the binary designation *"S. paradoxus"* (Rottbøll, Descr. Pl. Rar.: 27. 1772) is not validly published since *"Scirpoides"* in Rottbøll's context was a word not intended as a generic name. The first validly published name for this species is *Fuirena umbellata* Rottb. (1773).

Article 44

44.1. The name of a species or of an infraspecific taxon published before 1 January 1908 is validly published if it is accompanied only by an illustration with analysis (as defined in Art. 42.4).

Ex. 1. *Panax nossibiensis* Drake (1896) was validly published on a plate with analysis.

44.2. Single figures of non-vascular plants showing details aiding identification are considered as illustrations with analysis (see also Art. 42.4).

Ex. 2. *Eunotia gibbosa* Grunow (1881), a name of a diatom, was validly published by provision of a figure of a single valve.

Article 45

45.1. The date of a name is that of its valid publication. When the various conditions for valid publication are not simultaneously fulfilled, the date is that on which the last is fulfilled. However, the name must always be explicitly accepted in the place of its validation. A name published on or after 1 January 1973 for which the various conditions for valid publication are not simultaneously fulfilled is not validly published unless a full and direct reference (Art. 33.2) is given to the places where these requirements were previously fulfilled.

Ex. 1. *"Clypeola minor"* first appeared in the Linnaean thesis *Flora monspeliensis* (1756), in a list of names preceded by numerals but without an explanation of the meaning of these numerals and without any other descriptive matter; when the thesis was reprinted in vol. 4 of the *Amoenitates academicae* (1759), a statement was added explaining that the numbers referred to earlier descriptions published in Magnol's *Botanicon monspeliense*. However, *"Clypeola minor"* was absent from the reprint, being no longer accepted by Linnaeus, and the name was not therefore validly published.

Ex. 2. When proposing *"Graphis meridionalis"* as a new species, Nakanishi (in J. Sci. Hiroshima Univ., ser. B(2), 11: 75. 1966) provided a Latin description but failed to designate a holotype. *G. meridionalis* Nakan. was validly published in 1967 (in J. Sci. Hiroshima Univ., ser. B(2), 11: 265) when he designated the holotype of the name and provided a full and direct reference to the previous publication.

45.2. After 1 January 2000, when one or more of the other conditions for valid publication have not been met prior to registration, the name must be resubmitted for registration after these conditions have been met.

45.3. A correction of the original spelling of a name (see Art. 32.6 and 60) does not affect its date of valid publication.

Ex. 3. The correction of the orthographical error in *Gluta "benghas"* (Linnaeus, Mant.: 293. 1771) to *Gluta renghas* L. does not affect the date of publication of the name even though the correction dates only from 1883 (Engler in Candolle & Candolle, Monogr. Phan. 4: 225).

45.4. For purposes of priority only legitimate names are taken into consideration (see Art. 11, 52-54). However, validly published earlier homonyms, whether legitimate or not, shall cause rejection of their later homonyms, unless the latter are conserved or sanctioned (but see Art. 15 Note 1).

45.5. If a taxon originally assigned to a group not covered by this *Code* is treated as belonging to a group of plants other than algae, the authorship and date of any of its names are determined by the first publication that satisfies the requirements for valid publication under this *Code*. If the taxon is treated as belonging to the algae, any of its names need satisfy only the requirements of the pertinent non-botanical *Code* for status equivalent to valid publication under the present *Code* (but see Art. 54, regarding homonymy).

Ex. 4. Amphiprora Ehrenb. (1843) is an available[1] name for a genus of animals first treated as belonging to the algae by Kützing (1844). *Amphiprora* has priority in botanical nomenclature from 1843, not 1844.

Ex. 5. Petalodinium Cachon & Cachon-Enj. (in Protistologia 5: 16. 1969) is available under the *International code of zoological nomenclature* as the name of a genus of dinoflagellates. When the taxon is treated as belonging to the algae, its name retains its original authorship and date even though the original publication lacked a Latin diagnosis.

Ex. 6. Labyrinthodyction Valkanov (in Progr. Protozool. 3: 373. 1969), although available under the *International code of zoological nomenclature* as the name of a genus of rhizopods, is not valid when the taxon is treated as belonging to the fungi because the original publication lacked a Latin diagnosis.

Ex. 7. Protodiniferaceae Kof. & Swezy (in Mem. Univ. Calif. 5: 111. 1921, *"Protodiniferidae"*), available under the *International code of zoological nomenclature*, is validly published as a name of a family of algae with its original authorship and date but with the original termination changed in accordance with Art. 18.4 and 32.6.

Recommendation 45A

45A.1. Authors using new names in works (floras, catalogues, etc.) written in a modern language should simultaneously comply with the requirements of valid publication.

Recommendation 45B

45B.1. Authors should indicate precisely the dates of publication of their works. In a work appearing in parts the last-published sheet of the volume should indicate the precise dates on which the different fascicles or parts of the volume were published as well as the number of pages and plates in each.

Recommendation 45C

45C.1. On separately printed and issued copies of works published in a periodical, the name of the periodical, the number of its volume or parts, the original pagination, and the date (year, month, and day) should be indicated.

---

[1] The word "available" in the *International code of zoological nomenclature* is equivalent to "validly published" in the present *Code*.

SECTION 3. CITATION OF AUTHORS' NAMES FOR PURPOSES OF PRECISION

Article 46

46.1. For the indication of the name of a taxon to be accurate and complete, and in order that the date may be readily verified, it is necessary to cite the name of the author(s) who validly published the name concerned unless the provisions for autonyms apply (Art. 22.1 and 26.1).

Ex. 1. *Rosaceae* Juss., *Rosa* L., *Rosa gallica* L., *Rosa gallica* var. *eriostyla* R. Keller, *Rosa gallica* L. var. *gallica*.

46.2. A name of a new taxon must be attributed to the author or authors to whom both the name and the validating description or diagnosis were ascribed, even when authorship of the publication is different. A new combination or a *nomen novum* must be attributed to the author or authors to whom it was ascribed when, in the publication in which it appears, it is explicitly stated that they contributed in some way to that publication. Art. 46.4 notwithstanding, authorship of a new name or combination must always be accepted as ascribed, even when it differs from authorship of the publication, when at least one author is common to both.

Ex. 2. The name *Viburnum ternatum* was published in Sargent (Trees & Shrubs 2: 37. 1907). It was ascribed to "Rehd.", and the whole account of the species was signed "Alfred Rehder" at the foot of the article. The name is therefore cited as *V. ternatum* Rehder.

Ex. 3. In a paper by Hilliard & Burtt (1986) names of new species of *Schoenoxiphium*, including *S. altum*, were ascribed to Kukkonen, preceded by a statement "The following diagnostic descriptions of new species have been supplied by Dr. I. Kukkonen in order to make the names available for use". The name is therefore cited as *S. altum* Kukkonen.

Ex. 4. In Torrey & Gray (1838) the names *Calyptridium* and *C. monandrum* were ascribed to "Nutt. mss.", and the descriptions were enclosed in double quotes indicating that Nuttall wrote them, as acknowledged in the preface. The names are therefore cited as *Calyptridium* Nutt. and *C. monandrum* Nutt.

Ex. 5. The name *Brachystelma* was published by Sims (1822) who by implication ascribed it to Brown and added "Brown, Mscr." at the end of the generic diagnosis, indicating that Brown wrote it. The name is therefore cited as *Brachystelma* R. Br.

Ex. 6. Green (1985) ascribed the new combination *Neotysonia phyllostegia* to Paul G. Wilson and elsewhere in the same publication acknowledged his assistance. The name is therefore cited as *N. phyllostegia* (F. Muell.) Paul G. Wilson.

Ex. 7. The authorship of *Steyerbromelia discolor* L. B. Sm. & H. Rob. (1984) is accepted as originally ascribed, although the new species was described in a paper authored by Smith alone. The same applies to the new combination *Sophora tomentosa* subsp. *occidentalis* (L.) Brummitt (in Kirkia 5: 265. 1966), thus ascribed, published in a paper authored jointly by Brummitt & Gillett.

Note 1. When authorship of a name differs from authorship of the publication in which it was validly published, both are sometimes cited, connected by the word "in". In such a case, "in" and what follows are part of a bibliographic citation and are better omitted unless the place of publication is being cited.

46.3. For the purposes of this Article, ascription is the direct association of the name of a person or persons with a new name or description or diagnosis of a taxon. Mention of an author's name in a list of synonyms is not ascription, nor is reference to a basionym or a replaced synonym, including bibliographic errors, nor is reference to a homonym, nor is a formal error.

Ex. 8. Hypnum crassinervium Wilson (1833) was not ascribed to Taylor by Wilson's citing "*Hypnum crassinervium* Dr. Taylor's MS" in the list of synonyms.

Ex. 9. Lichen debilis Sm. (1812) was not ascribed to Turner and Borrer by Smith's citing "*Calicium debile* Turn. and Borr. Mss." as a synonym.

Ex. 10. When Opiz (1852) wrote "*Hemisphace* Bentham" he did not ascribe the generic name to Bentham but provided an indirect reference to the basionym, *Salvia* sect. *Hemisphace* Benth. (see Art. 32 Ex. 5).

Ex. 11. When Brotherus (1907) published "*Dichelodontium nitidulum* Hooker & Wilson" he provided an indirect reference to the basionym, *Leucodon nitidulus* Hook. f. & Wilson, and did not ascribe the new combination to Hooker and Wilson. He did, however, ascribe to them the simultaneously published name of his new genus, *Dichelodontium.*

Ex. 12. When Sirodot (1872) wrote "*Lemanea* Bory" he in fact published a later homonym (see Art. 48 Ex. 1). His reference to Bory is not therefore ascription of the later homonym, *Lemanea* Sirodot, to Bory.

46.4. A name of a new taxon must be attributed to the author or authors of the publication in which it appears when only the name but not the validating description or diagnosis was ascribed to a different author or different authors. A new combination or a *nomen novum* must be attributed to the author or authors of the publication in which it appears, although it was ascribed to a different author or to different authors, when no separate statement was made that they contributed in some way to that publication. However, in both cases authorship as ascribed, followed by "ex", may be inserted before the name(s) of the publishing author(s).

Ex. 13. Seemann (1865) published *Gossypium tomentosum* "Nutt. mss.", followed by a validating description not ascribed to Nuttall; the name may be cited as *Gossypium tomentosum* Nutt. ex Seem. or *G. tomentosum* Seem.

Ex. 14. The name *Lithocarpus polystachyus* published by Rehder (1919) was based on *Quercus polystachya* A. DC. (1864), ascribed by Candolle to "Wall.! list n. 2789" but formerly a *nomen nudum;* Rehder's combination may be cited as *L. polystachyus* (Wall. ex A. DC.) Rehder or *L. polystachyus* (A. DC.) Rehder.

Ex. 15. Lilium tianschanicum was described by Grubov (1977) as a new species and its name was ascribed to Ivanova; since there is no indication that Ivanova provided the validating description, the name may be cited as *L. tianschanicum* N. A. Ivanova ex Grubov or *L. tianschanicum* Grubov.

Ex. 16. In a paper by Boufford, Tsi and Wang (1990) the name *Rubus fanjingshanensis* was ascribed to Lu with no indication that he provided the description; the name should be attributed to Boufford & al. or to L. T. Lu ex Boufford & al.

Ex. 17. Green (1985) ascribed the new combination *Tersonia cyathiflora* to "(Fenzl) A. S. George"; since Green nowhere mentioned that George had contributed in any way, the combining author must be cited as A. S. George ex J. W. Green or just J. W. Green.

46.5. The citation of an author who published the name before the starting point of the group concerned may be indicated by the use of the word "ex". For groups with a starting point later than 1753, when a pre-starting point name was changed in rank or taxonomic position by the first author who validly published it, the name of the pre-starting point author may be added in parentheses, followed by "ex".

Ex. 18. Linnaeus (1754) ascribed the name *Lupinus* to the pre-starting-point author Tournefort; the name may be cited as *Lupinus* Tourn. ex L. (1753) or *Lupinus* L.

Ex. 19. *Lyngbya glutinosa* C. Agardh (Syst. Alg.: 73. 1824) was taken up by Gomont in the publication which marks the starting point of the *"Nostocaceae heterocysteae"* (in Ann. Sci. Nat., Bot., ser. 7, 15: 339. 1892) as *Hydrocoleum glutinosum*. This may be cited as *H. glutinosum* (C. Agardh) ex Gomont.

46.6. In determining the correct author citation, only internal evidence in the publication (as defined in Art. 35.4) where the name was validly published is to be accepted, including ascription of the name, statements in the introduction, title, or acknowledgements, and typographical or stylistic distinctions in the text.

Ex. 20. Names first published in Britton & Brown's *Illustrated flora of the northern United States* (1896-1898; ed. 2, 1913) must, unless ascribed to Britton alone (see Art. 46.2), be attributed to "Britton & A. Br.", since the title page attributes the whole work to both, even though it is generally accepted that A. Brown did not participate in writing it.

Ex. 21. Although the descriptions in Aiton's *Hortus kewensis* (1789) are generally considered to have been written by Solander or Dryander, the names of new taxa published there must be attributed to Aiton, the stated author of the work, except where a name and description were both ascribed in that work to somebody else.

Ex. 22. The name *Andreaea angustata* was published in a work of Limpricht (1885) with the ascription "nov. sp. Lindb. in litt. ad Breidler 1884", but there is no internal evidence that Lindberg had supplied the validating description. Authorship is therefore to be cited as "Limpr." or "Lindb. ex Limpr."

Note 2. Authors publishing new names and wishing to establish that other persons' names followed by "ex" may precede theirs in authorship citation may adopt the "ex" citation in the protologue.

Ex. 23. In validating the name *Nothotsuga,* Page (1989) cited it as "*Nothotsuga* H.-H. Hu ex C. N. Page", noting that in 1951 Hu had published it as a *nomen nudum;* the name may be attributed to Hu ex C. N. Page or just C. N. Page.

Ex. 24. Atwood (1981) ascribed the name of a new species, *Maxillaria mombachoënsis,* to "Heller ex Atwood", with a note stating that it was originally named by Heller, then deceased; the name may be attributed to A. H. Heller ex J. T. Atwood or just J. T. Atwood.

Recommendation 46A

46A.1. Authors' names placed after names of plants may be abbreviated, unless they are very short. For this purpose, particles should be suppressed unless they are an inseparable part of the name, and the first letters should be given without any omission (Lam. for J. B. P. A. Monet Chevalier de Lamarck, but De Wild. for E. De Wildeman).

46A.2. If a name of one syllable is long enough to make it worthwhile to abridge it, the first consonants only should be given (Fr. for Elias Magnus Fries); if the name has two or more syllables, the first syllable and the first letter of the following one should be taken, or the two first when both are consonants (Juss. for Jussieu, Rich. for Richard).

46A.3. When it is necessary to give more of a name to avoid confusion between names beginning with the same syllable, the same system should be followed. For instance, two syllables should be given together with the one or two first consonants of the third; or one of the last characteristic consonants of the name be added (Bertol. for Bertoloni, to distinguish it from Bertero; Michx. for Michaux, to distinguish it from Micheli).

46A.4. Given names or accessory designations serving to distinguish two botanists of the same name should be abridged in the same way (A. Juss. for Adrien de Jussieu, Burm. f. for Burman filius, J. F. Gmel. for Johann Friedrich Gmelin, J. G. Gmel. for Johann Georg Gmelin, C. C. Gmel. for Carl Christian Gmelin, S. G. Gmel. for Samuel Gottlieb Gmelin, Müll. Arg. for Jean Müller of Aargau).

46A.5. When it is a well-established custom to abridge a name in another manner, it is advisable to conform to it (L. for Linnaeus, DC. for Augustin Pyramus de Candolle, St.-Hil. for Saint-Hilaire, R. Br. for Robert Brown).

Note 1. Brummitt & Powell's *Authors of plant names* (1992) provides unambiguous standard abbreviations, in conformity with the present Recommendation, for a large number of authors of plant names, and these abbreviations have been used for author citations throughout the present *Code*.

Recommendation 46B

46B.1. In citing the author of the scientific name of a taxon, the romanization of the author's name given in the original publication should normally be accepted. Where an author failed to give a romanization, or where an author has at different times used different romanizations, then the romanization known to be preferred by the author or that most frequently adopted by the author should be accepted. In the absence of such information the author's name should be romanized in accordance with an internationally available standard.

46B.2. Authors of scientific names whose personal names are not written in Roman letters should romanize their names, preferably (but not necessarily) in accordance with an internationally available standard and, as a matter of typographical convenience, without diacritical signs. Once authors have selected the romanization of their personal names, they should use it consistently thereafter. Whenever possible, authors should not permit editors or publishers to change the romanization of their personal names.

Recommendation 46C

46C.1. After a name published jointly by two authors, the names of both authors should be cited, linked by the word "et" or by an ampersand (&).

Ex. 1. Didymopanax gleasonii Britton et Wilson (or Britton & Wilson).

46C.2. After a name published jointly by more than two authors, the citation should be restricted to the name of the first author followed by "et al." or "& al.", except in the original publication.

Ex. 2. Lapeirousia erythrantha var. *welwitschii* (Baker) Geerinck, Lisowski, Malaisse & Symoens (in Bull. Soc. Roy. Bot. Belgique 105: 336. 1972) should be cited as *L. erythrantha* var. *welwitschii* (Baker) Geerinck & al.

Recommendation 46D

46D.1. Authors should cite their own names after each new name they publish rather than refer to themselves by expressions such as "nobis" *(nob.)* or "mihi" *(m.)*.

Article 47

47.1. An alteration of the diagnostic characters or of the circumscription of a taxon without the exclusion of the type does not warrant a change of author citation for the name of the taxon.

Ex. 1. When the original material of *Arabis beckwithii* S. Watson (1887) is attributed to two different species, as by Munz (1932), that species not including the lectotype must bear a different name (*A. shockleyi* Munz) but the other one is still named *A. beckwithii* S. Watson.

Ex. 2. Myosotis as revised by Brown differs from the genus as originally circumscribed by Linnaeus, but the generic name remains *Myosotis* L. since the type of the name is still included in the genus (it may be cited as *Myosotis* L. emend. R. Br.: see Rec. 47A).

Ex. 3. The variously defined species that includes the types of *Centaurea jacea* L. (1753), *C. amara* L. (1763) and a variable number of other species names is still called *C. jacea* L. (or L. emend. Coss. & Germ., L. emend. Vis., or L. emend. Godr., as the case may be: see Rec. 47A).

Recommendation 47A

47A.1. When an alteration as mentioned in Art. 47 has been considerable, the nature of the change may be indicated by adding such words, abbreviated where suitable, as "emendavit" *(emend.)* (followed by the name of the author responsible for the change), "mutatis characteribus" *(mut. char.),* "pro parte" *(p. p.),* "excluso genere" or "exclusis generibus" *(excl. gen.),* "exclusa specie" or "exclusis speciebus" *(excl. sp.),* "exclusa varietate" or "exclusis varietatibus" *(excl. var.),* "sensu amplo" *(s. ampl.),* "sensu lato" *(s. l.),* "sensu stricto" *(s. str.),* etc.

Ex. 1. Phyllanthus L. emend. Müll. Arg.; *Globularia cordifolia* L. excl. var. (emend. Lam.).

Article 48

48.1. When an author adopts an existing name but definitely excludes its original type, a later homonym that must be attributed solely to that author is considered to have been published. Similarly, when an author who adopts a name refers to an apparent basionym but explicitly excludes its type, a new name is considered to have been published that must be attributed solely to that author. Exclusion can be effected by simultaneous explicit inclusion of the type in a different taxon by the same author (see also Art. 59.6).

Ex. 1. Sirodot (1872) placed the type of *Lemanea* Bory (1808) in *Sacheria* Sirodot (1872); hence *Lemanea,* as treated by Sirodot (1872), is to be cited as *Lemanea* Sirodot non Bory and not as *Lemanea* Bory emend. Sirodot.

Ex. 2. The name *Amorphophallus campanulatus* Decne. (1834) was apparently based on the illegitimate *Arum campanulatum* Roxb. (1819). However, the type of the latter was explicitly excluded by Decaisne, and his name is therefore a legitimate name of a new species, to be attributed solely to him.

Ex. 3. Cenomyce ecmocyna Ach. (1810) is a superfluous name for *Lichen gracilis* L. (1753), and so is *Scyphophora ecmocyna* Gray (1821), the type of *L. gracilis* still being included. However, when proposing the combination *Cladonia ecmocyna,* Leighton (1866) explicitly excluded that type and thereby published a new, legitimate name, *Cladonia ecmocyna* Leight.

Note 1. Misapplication of a new combination to a different taxon, but without explicit exclusion of the type of the basionym, is dealt with under Art. 7.4.

Note 2. Retention of a name in a sense that excludes its original type, or its type designated under Art. 7-10, can be effected only by conservation (see Art. 14.9).

Article 49

49.1. When a genus or a taxon of lower rank is altered in rank but retains its name or the final epithet in its name, the author of the earlier, name- or epithet-bringing legitimate name (the author of the basionym) must be cited in parentheses, followed by the name of the author who effected the alteration (the author of the new name). The same holds when a taxon of lower rank than genus is transferred to another genus or species, with or without alteration of rank.

Ex. 1. Medicago polymorpha var. *orbicularis* L. (1753) when raised to the rank of species becomes *M. orbicularis* (L.) Bartal. (1776).

Ex. 2. Anthyllis sect. *Aspalathoides* DC. (1825) raised to generic rank, retaining the epithet *Aspalathoides* as its name, is cited as *Aspalathoides* (DC.) K. Koch (1853).

Ex. 3. Cineraria sect. *Eriopappus* Dumort. (Fl. Belg.: 65. 1827) when transferred to *Tephroseris* (Rchb.) Rchb. is cited as *T.* sect. *Eriopappus* (Dumort.) Holub (in Folia Geobot. Phytotax. 8: 173. 1973).

Ex. 4. Cistus aegyptiacus L. (1753) when transferred to *Helianthemum* Mill. is cited as *H. aegyptiacum* (L.) Mill. (1768).

Ex. 5. Fumaria bulbosa var. *solida* L. (1753) was elevated to specific rank as *F. solida* (L.) Mill. (1771). The name of this species when transferred to *Corydalis* DC. is to be cited as *C. solida* (L.) Clairv. (1811), not *C. solida* (Mill.) Clairv.

Ex. 6. However, *Pulsatilla montana* var. *serbica* W. Zimm. (in Feddes Repert. Spec. Nov. Regni Veg. 61: 95. 1958), originally placed under *P. montana* subsp. *australis* (Heuff.) Zämelis, retains the same author citation when placed under *P. montana* subsp. *dacica* Rummelsp. (see Art. 24.1) and is not cited as var. *serbica* "(W. Zimm.) Rummelsp." (in Feddes Repert. 71: 29. 1965).

Ex. 7. Salix subsect. *Myrtilloides* C. K. Schneid. (Ill. Handb. Laubholzk. 1: 63. 1904), originally placed under *S.* sect. *Argenteae* W. D. J. Koch, retains the same author citation when placed under *S.* sect. *Glaucae* Pax and is not cited as *S.* subsect. *Myrtilloides* "(C. K. Schneid.) Dorn" (in Canad. J. Bot. 54: 2777. 1976).

Note 1. Art. 46.5 provides for the use of parenthetical author citations preceding the word "ex", after some names in groups with a starting point later than 1753.

Article 50

50.1. When a taxon at the rank of species or below is transferred from the non-hybrid category to the hybrid category of the same rank (Art. H.10.2), or vice versa, the author citation remains unchanged but may be followed by an indication in parentheses of the original category.

Ex. 1. Stachys ambigua Sm. (1809) was published as the name of a species. If regarded as applying to a hybrid, it may be cited as *Stachys ×ambigua* Sm. (pro sp.).

Ex. 2. The binary name *Salix* ×*glaucops* Andersson (1868) was published as the name of a hybrid. Later, Rydberg (in Bull. New York Bot. Gard. 1: 270. 1899) considered the taxon to be a species. If this view is accepted, the name may be cited as *Salix glaucops* Andersson (pro hybr.).

SECTION 4. GENERAL RECOMMENDATIONS ON CITATION

Recommendation 50A

50A.1. In the citation of a name invalidly published as a synonym, the words "as synonym" or "pro syn." should be added.

Recommendation 50B

50B.1. In the citation of a nomen nudum, its status should be indicated by adding the words "nomen nudum" or "nom. nud."

Ex. 1. "Carex bebbii" (Olney, Car. Bor.-Am. 2: 12. 1871), published without a description or diagnosis, should be cited as *Carex bebbii* Olney, nomen nudum (or nom. nud.).

Recommendation 50C

50C.1. The citation of a later homonym should be followed by the name of the author of the earlier homonym preceded by the word "non", preferably with the date of publication added. In some instances it will be advisable to cite also any other homonyms, preceded by the word "nec".

Ex. 1. Ulmus racemosa Thomas in Amer. J. Sci. Arts 19: 170. 1831, non Borkh. 1800; *Lindera* Thunb., Nov. Gen. Pl.: 64. 1783, non Adans. 1763; *Bartlingia* Brongn. in Ann. Sci. Nat. (Paris) 10: 373. 1827, non Rchb. 1824 nec F. Muell. 1882.

Recommendation 50D

50D.1. Misidentifications should not be included in synonymies but added after them. A misapplied name should be indicated by the words "auct. non" followed by the name of the original author and the bibliographic reference of the misidentification.

Ex. 1. Ficus stortophylla Warb. in Ann. Mus. Congo Belge, B, Bot., ser. 4, 1: 32. 1904. *F. irumuënsis* De Wild., Pl. Bequaert. 1: 341. 1922. *F. exasperata* auct. non Vahl: De Wildeman & Durand in Ann. Mus. Congo Belge, B, Bot., ser. 2, 1: 54. 1899; De Wildeman, Miss. Em. Laurent: 26. 1905; Durand & Durand, Syll. Fl. Congol.: 505. 1909.

Recommendation 50E

50E.1. If a generic or specific name is accepted as a *nomen conservandum* (see Art. 14 and App. III) the abbreviation *"nom. cons."* should be added in a full citation.

Ex. 1. Protea L., Mant. Pl.: 187. 1771, *nom. cons.*, non L. 1753; *Combretum* Loefl. (1758), *nom. cons.* [= *Grislea* L. 1753].

50E.2. If it is desirable to indicate the sanctioned status of the names of fungi adopted by Persoon or Fries (see Art. 13.1(d)), ": Pers." or ": Fr." should be added to the citation.

Ex. 2. Boletus piperatus Bull. (Herb. Francc: t. 451, f. 2. 1790) was accepted in Fries (Syst. Mycol. 1: 388. 1821) and was thereby sanctioned. It may thus be cited as *B. piperatus* Bull. : Fr.

Recommendation 50F

50F.1. If a name is cited with alterations from the form as originally published, it is desirable that in full citations the exact original form should be added, preferably between single or double quotation marks.

Ex. 1. Pyrus calleryana Decne. *(P. mairei* H. Lév. in Repert. Spec. Nov. Regni Veg. 12: 189. 1913, *"Pirus").*

Ex. 2. Zanthoxylum cribrosum Spreng., Syst. Veg. 1: 946. 1825, *"Xanthoxylon". (Z. caribaeum* var. *floridanum* (Nutt.) A. Gray in Proc. Amer. Acad. Arts 23: 225. 1888, *"Xanthoxylum").*

Ex. 3. Spathiphyllum solomonense Nicolson in Amer. J. Bot. 54: 496. 1967, *"solomonensis".*

CHAPTER V. REJECTION OF NAMES

Article 51

51.1. A legitimate name must not be rejected merely because it, or its epithet, is inappropriate or disagreeable, or because another is preferable or better known (but see Art. 56.1), or because it has lost its original meaning, or (in pleomorphic fungi with names governed by Art. 59) because the generic name does not accord with the morph represented by its type.

Ex. 1. The following changes are contrary to the rule: *Staphylea* to *Staphylis, Tamus* to *Thamnos, Thamnus,* or *Tamnus, Mentha* to *Minthe, Tillaea* to *Tillia, Vincetoxicum* to *Alexitoxicum;* and *Orobanche rapum* to *O. sarothamnophyta, O. columbariae* to *O. columbarihaerens, O. artemisiae* to *O. artemisiepiphyta.*

Ex. 2. Ardisia quinquegona Blume (1825) is not to be changed to *A. pentagona* A. DC. (1834), although the specific epithet *quinquegona* is a hybrid word (Latin and Greek) (contrary to Rec. 23A.3(c)).

Ex. 3. The name *Scilla peruviana* L. (1753) is not to be rejected merely because the species does not grow in Peru.

Ex. 4. The name *Petrosimonia oppositifolia* (Pall.) Litv. (1911), based on *Polycnemum oppositifolium* Pall. (1771), is not to be rejected merely because the species has leaves only partly opposite, and partly alternate, although there is another closely related species, *Petrosimonia brachiata* (Pall.) Bunge, having all its leaves opposite.

Ex. 5. Richardia L. (1753) is not to be changed to *Richardsonia,* as was done by Kunth (1818), although the name was originally dedicated to the British botanist, Richardson.

Article 52

52.1. A name, unless conserved (Art. 14) or sanctioned (Art. 15), is illegitimate and is to be rejected if it was nomenclaturally superfluous when published, i.e. if the taxon to which it was applied, as circumscribed by its author, definitely included the type (as qualified in Art. 52.2) of a name which ought to have been adopted, or whose epithet ought to have been adopted, under the rules (but see Art. 52.3).

52.2. For the purpose of Art. 52.1, definite inclusion of the type of a name is effected by citation *(a)* of the holotype under Art. 9.1 or the original type under Art. 10 or all syntypes under Art. 9.4 or all elements eligible as types under Art. 10.2; or *(b)* of the previously designated type under Art. 9.9 or 10.2; or *(c)* of the illustrations of these. It is also effected *(d)* by citation of the name itself, unless the type is at the same time excluded either explicitly or by implication.

Ex. 1. The generic name *Cainito* Adans. (1763) is illegitimate because it was a superfluous name for *Chrysophyllum* L. (1753), which Adanson cited as a synonym.

Ex. 2. Chrysophyllum sericeum Salisb. (1796) is illegitimate, being a superfluous name for *C. cainito* L. (1753), which Salisbury cited as a synonym.

Ex. 3. On the other hand, *Salix myrsinifolia* Salisb. (1796) is legitimate, being explicitly based upon *S. myrsinites* of Hoffmann (Hist. Salic. Ill.: 71. 1787), a misapplication of the name *S. myrsinites* L. (1753).

Ex. 4. Picea excelsa Link (1841) is illegitimate because it is based on *Pinus excelsa* Lam. (1778), a superfluous name for *Pinus abies* L. (1753). Under *Picea* the correct name is *Picea abies* (L.) H. Karst. (1881).

Ex. 5. On the other hand, *Cucubalus latifolius* Mill. and *C. angustifolius* Mill. are not illegitimate names, although Miller's species are now united with the species previously named *C. behen* L. (1753): *C. latifolius* and *C. angustifolius* as circumscribed by Miller (1768) did not include the type of *C. behen* L., which name he adopted for another species.

Ex. 6. Explicit exclusion of type: When publishing the name *Galium tricornutum,* Dandy (in Watsonia 4: 47. 1957) cited *G. tricorne* Stokes (1787) pro parte as a synonym, but explicitly excluded the type of the latter name.

Ex. 7. Exclusion of type by implication: *Cedrus* Duhamel (1755) is a legitimate name even though *Juniperus* L. (1753) was cited as a synonym; only some of Linnaeus's species of *Juniperus* were included in *Cedrus* by Duhamel, and the differences between the two genera were discussed, *Juniperus* (including the type of its name) being recognized in the same work as an independent genus.

Ex. 8. Exclusion of type by implication: *Tmesipteris elongata* P. A. Dang. (in Botaniste 2: 213. 1891) was published as a new species but *Psilotum truncatum* R. Br. was cited as a synonym. However, on the following page, *T. truncata* (R. Br.) Desv. is recognized as a different species and two pages later both are distinguished in a key, thus showing that the meaning of the cited synonym was either "*P. truncatum* R. Br. pro parte" or "*P. truncatum* auct. non R. Br."

Ex. 9. Exclusion of type by implication: *Solanum torvum* Sw. (Prodr.: 47. 1788) was published with a new diagnosis but *S. indicum* L. (1753) was cited as a synonym. In accord with the practice in his *Prodromus,* Swartz indicated where the species was to be inserted in the latest edition [ed. 14, by Murray] of Linnaeus's *Systema vegetabilium. S. torvum* was to be inserted between species 26 *(S. insanum)* and 27 *(S. ferox),* the number of *S. indicum* being 32. *S. torvum* is thus a legitimate name.

Note 1. The inclusion, with an expression of doubt, of an element in a new taxon, e.g. the citation of a name with a question mark, does not make the name of the new taxon nomenclaturally superfluous.

Ex. 10. The protologue of *Blandfordia grandiflora* R. Br. (1810) includes, in synonymy, "Aletris punicea. *Labill. nov. holl.* 1. *p.* 85. *t.* 111 ?", indicating that the new species might be the same as *Aletris punicea* Labill. (1805). *B. grandiflora* is nevertheless a legitimate name.

Note 2. The inclusion, in a new taxon, of an element that was subsequently designated as the type of a name which, so typified, ought to have been adopted, or whose epithet ought to have been adopted, does not in itself make the name of the new taxon illegitimate.

52.3. A name that was nomenclaturally superfluous when published is not illegitimate if its basionym is legitimate, or if it is based on the stem of a legitimate generic name. When published it is incorrect, but it may become correct later.

Ex. 11. Chloris radiata (L.) Sw. (1788), based on *Agrostis radiata* L. (1759), was nomenclaturally superfluous when published, since Swartz also cited *Andropogon fasciculatus* L. (1753) as a synonym. It is, however, the correct name in the genus *Chloris* for *Agrostis radiata* when *Andropogon fascicu-*

latus is treated as a different species, as was done by Hackel (in Candolle & Candolle, Monogr. Phan. 6: 177. 1889).

Ex. 12. The generic name *Hordelymus* (K. Jess.) K. Jess. (1885), based on the legitimate *Hordeum* subg. *Hordelymus* K. Jess. (Deutschl. Gräser: 202. 1863), was superfluous when published because its type, *Elymus europaeus* L., is also the type of *Cuviera* Koeler (1802). *Cuviera* Koeler has since been rejected in favour of its later homonym *Cuviera* DC., and *Hordelymus* can now be used as a correct name for the segregate genus containing *Elymus europaeus* L.

Note 3. In no case does a statement of parentage accompanying the publication of a name for a hybrid make the name illegitimate (see Art. H.5).

Ex. 13. The name *Polypodium* ×*shivasiae* Rothm. (1962) was proposed for hybrids between *P. australe* Fée and *P. vulgare* subsp. *prionodes* (Asch.) Rothm., while at the same time the author accepted *P.* ×*font-queri* Rothm. (1936) for hybrids between *P. australe* and *P. vulgare* L. subsp. *vulgare*. Under Art. H.4.1, *P.* ×*shivasiae* is a synonym of *P.* ×*font-queri;* nevertheless, it is not an illegitimate name.

Article 53

53.1. A name of a family, genus or species, unless conserved (Art. 14) or sanctioned (Art. 15), is illegitimate if it is a later homonym, that is, if it is spelled exactly like a name based on a different type that was previously and validly published for a taxon of the same rank.

Note 1. Even if the earlier homonym is illegitimate, or is generally treated as a synonym on taxonomic grounds, the later homonym must be rejected.

Ex. 1. The name *Tapeinanthus* Boiss. ex Benth. (1848), given to a genus of *Labiatae,* is a later homonym of *Tapeinanthus* Herb. (1837), a name previously and validly published for a genus of *Amaryllidaceae. Tapeinanthus* Boiss. ex Benth. is therefore rejected. It was renamed *Thuspeinanta* T. Durand (1888).

Ex. 2. The name *Amblyanthera* Müll. Arg. (1860) is a later homonym of the validly published *Amblyanthera* Blume (1849) and is therefore rejected, although *Amblyanthera* Blume is now considered to be a synonym of *Osbeckia* L. (1753).

Ex. 3. The name *Torreya* Arn. (1838) is a *nomen conservandum* and is therefore not to be rejected because of the existence of the earlier homonym *Torreya* Raf. (1818).

Ex. 4. *Astragalus rhizanthus* Boiss. (1843) is a later homonym of the validly published name *Astragalus rhizanthus* Royle (1835) and it is therefore rejected, as was done by Boissier who renamed it *A. cariensis* Boiss. (1849).

53.2. A sanctioned name is illegitimate if it is a later homonym of another sanctioned name (see also Art. 15 Note 1).

53.3. When two or more generic, specific, or infraspecific names based on different types are so similar that they are likely to be confused (because they are applied to related taxa or for any other reason) they are to be treated as homonyms.

**Ex. 5.* Names treated as homonyms: *Asterostemma* Decne. (1838) and *Astrostemma* Benth. (1880); *Pleuropetalum* Hook. f. (1846) and *Pleuripetalum* T. Durand (1888); *Eschweilera* DC. (1828) and *Eschweileria* Boerl. (1887); *Skytanthus* Meyen (1834) and *Scytanthus* Hook. (1844).

**Ex. 6.* The three generic names *Bradlea* Adans. (1763), *Bradleja* Banks ex Gaertn. (1790), and *Braddleya* Vell. (1827), all commemorating Richard Bradley, are treated as homonyms because only one can be used without serious risk of confusion.

**Ex. 7.* The names *Acanthoica* Lohmann (1902) and *Acanthoeca* W. N. Ellis (1930), both designating flagellates, are sufficiently alike to be considered homonyms (Taxon 22: 313. 1973).

**Ex. 8.* Epithets so similar that they are likely to be confused if combined under the same generic or specific name: *chinensis* and *sinensis; ceylanica* and *zeylanica; napaulensis, nepalensis,* and *nipalensis; polyanthemos* and *polyanthemus; macrostachys* and *macrostachyus; heteropus* and *heteropodus; poikilantha* and *poikilanthes; pteroides* and *pteroideus; trinervis* and *trinervius; macrocarpon* and *macrocarpum; trachycaulum* and *trachycaulon.*

**Ex. 9.* Names not likely to be confused: *Rubia* L. (1753) and *Rubus* L. (1753); *Monochaetum* (DC.) Naudin (1845) and *Monochaete* Döll (1875); *Peponia* Grev. (1863) and *Peponium* Engl. (1897); *Iris* L. (1753) and *Iria* (Pers.) Hedw. (1806); *Desmostachys* Miers (1852) and *Desmostachya* (Stapf) Stapf (1898); *Symphyostemon* Miers (1841) and *Symphostemon* Hiern (1900); *Gerrardina* Oliv. (1870) and *Gerardiina* Engl. (1897); *Urvillea* Kunth (1821) and *Durvillaea* Bory (1826); *Peltophorus* Desv. (1810; *Gramineae*) and *Peltophorum* (Vogel) Benth. (1840; *Leguminosae*); *Senecio napaeifolius* (DC.) Sch. Bip. (1845, *"napeaefolius"*; see Art. 60 Ex. 12) and *S. napifolius* MacOwan (1890; the epithets being derived respectively from *Napaea* and *Brassica napus*); *Lysimachia hemsleyana* Oliv. (1891) and *L. hemsleyi* Franch. (1895) (see, however, Rec. 23A.2); *Euphorbia peplis* L. (1753) and *E. peplus* L. (1753).

Ex. 10. Names conserved against earlier names treated as homonyms (see App. IIIA): *Lyngbya* Gomont (vs. *Lyngbyea* Sommerf.); *Columellia* Ruiz & Pav. (vs. *Columella* Lour.), both commemorating Columella, the Roman writer on agriculture; *Cephalotus* Labill. (vs. *Cephalotos* Adans.); *Simarouba* Aubl. (vs. *Simaruba* Boehm.).

53.4. When it is doubtful whether names are sufficiently alike to be confused, a request for a decision may be submitted to the General Committee (see Div. III) which will refer it for examination to the committee or committees for the appropriate taxonomic group or groups. A recommendation may then be put forward to an International Botanical Congress, and, if ratified, will become a binding decision.

Ex. 11. Names ruled as likely to be confused, and therefore to be treated as homonyms: *Ficus gomelleira* Kunth (1847) and *F. gameleira* Standl. (1937) (Taxon 42: 111. 1993); *Solanum saltiense* S. Moore (1895) and *S. saltense* (Bitter) C. V. Morton (1944) (Taxon 42: 434. 1993); *Balardia* Cambess. (1829; *Caryophyllaceae*) and *Ballardia* Montrouz. (1860; *Myrtaceae*) (Taxon 42: 434. 1993).

Ex. 12. Names ruled as not likely to be confused: *Cathayeia* Ohwi (1931; *Flacourtiaceae*) and *Cathaya* Chun & Kuang (1962; fossil *Pinaceae*) (Taxon 36: 429. 1987); *Cristella* Pat. (1887; *Fungi*) and *Christella* H. Lév. (1915; *Pteridophyta*) (Taxon 35: 551. 1986); *Coluria* R. Br. (1823; *Rosaceae*) and *Colura* (Dumort.) Dumort. (1835; *Hepaticae*) (Taxon 42: 433. 1993); *Acanthococcus* Hook. f. & Harv. (1845; *Rhodophyta*) and *Acanthococos* Barb. Rodr. (1900; *Palmae*) (Taxon 42: 433. 1993); *Rauia* Nees & Mart. (1823; *Rutaceae*) and *Rauhia* Traub (1957; *Amaryllidaceae*) (Taxon 42: 433. 1993).

53.5. The names of two subdivisions of the same genus, or of two infraspecific taxa within the same species, even if they are of different rank, are treated as homonyms if they have the same epithet and are not based on the same type.

Ex. 13. The names *Andropogon sorghum* subsp. *halepensis* (L.) Hack. and *A. sorghum* var. *halepensis* (L.) Hack. (in Candolle & Candolle, Monogr. Phan. 6: 502. 1889) are legitimate, since both have the same type and the epithet may be repeated under Rec. 26A.1.

Ex. 14. *Anagallis arvensis* var. *caerulea* (L.) Gouan (Fl. Monsp.: 30. 1765), based on *A. caerulea* L. (1759), makes illegitimate the name *A. arvensis* subsp. *caerulea* Hartm. (Sv. Norsk Exc.-Fl.: 32. 1846), based on the later homonym *A. caerulea* Schreber (1771).

Ex. 15. *Scenedesmus armatus* var. *brevicaudatus* (Hortob.) Pankow (in Arch. Protistenk. 132: 153. 1986), based on *S. carinatus* var. *brevicaudatus* Hortob. (in Acta Bot. Acad. Sci. Hung. 26: 318. 1981), is a later homonym of *S. armatus* f. *brevicaudatus* L. S. Péterfi (in Stud. Cercet. Biol. (Bucharest), Ser. Biol. Veg. 15: 25. 1963) even though the two names apply to taxa of different infraspecific rank.

Scenedesmus armatus var. *brevicaudatus* (L. S. Péterfi) E. H. Hegew. (in Arch. Hydrobiol. Suppl. 60: 393. 1982), however, is not a later homonym since it is based on the same type as *S. armatus* f. *brevicaudatus* L. S. Péterfi.

Note 2. The same final epithet may be used in the names of subdivisions of different genera, and of infraspecific taxa within different species.

Ex. 16. Verbascum sect. *Aulacosperma* Murb. (Monogr. Verbascum: 34, 593. 1933) is permissible, although there is an earlier *Celsia* sect. *Aulacospermae* Murb. (Monogr. Celsia: 34, 56. 1926). This, however, is not an example to be followed, since it is contrary to Rec. 21B.2.

53.6. When two or more homonyms have equal priority, the first of them that is adopted in an effectively published text (Art. 29-31) by an author who simultaneously rejects the other(s) is treated as having priority. Likewise, if an author in an effectively published text substitutes other names for all but one of these homonyms, the homonym for the taxon that is not renamed is treated as having priority.

Ex. 17. Linnaeus simultaneously published "10." *Mimosa cinerea* (Sp. Pl.: 517. 1753) and "25." *M. cinerea* (Sp. Pl.: 520. 1753). In 1759, he renamed species 10 *M. cineraria* L. and retained the name *M. cinerea* for species 25, so that the latter is treated as having priority over its homonym.

Ex. 18. Rouy & Foucaud (Fl. France 2: 30. 1895) published the name *Erysimum hieraciifolium* var. *longisiliquum*, with two different types, for two different taxa under different subspecies. Only one of these names can be maintained.

Article 54

54.1. Consideration of homonymy does not extend to the names of taxa not treated as plants, except as stated below:

(a) Later homonyms of the names of taxa once treated as plants are illegitimate, even though the taxa have been reassigned to a different group of organisms to which this *Code* does not apply.

(b) A name originally published for a taxon other than a plant, even if validly published under Art. 32-45 of this *Code,* is illegitimate if it becomes a homonym of a plant name when the taxon to which it applies is first treated as a plant (see also Art. 45.5).

Note 1. The *International code of nomenclature of bacteria* provides that a bacterial name is illegitimate if it is a later homonym of a name of a taxon of bacteria, fungi, algae, protozoa, or viruses.

Article 55

55.1. A name of a species or subdivision of a genus, autonyms excepted (Art. 22.1), may be legitimate even if its epithet was originally placed under an illegitimate generic name.

Ex. 1. Agathophyllum Juss. (1789) is an illegitimate name, being a superfluous substitute for *Ravensara* Sonn. (1782). Nevertheless the name *A. neesianum* Blume (1851) is legitimate. Because Meisner (1864) cited *A. neesianum* as a synonym of his new *Mespilodaphne mauritiana* but did not adopt the epithet *neesiana*, *M. mauritiana* Meisn. is a superfluous name and hence illegitimate.

55.2. An infraspecific name, autonyms excepted (Art. 26.1), may be legitimate even if its final epithet was originally placed under an illegitimate specific name.

55.3. The names of species and of subdivisions of genera assigned to genera whose names are conserved or sanctioned later homonyms, and which had earlier been assigned to the genera under the rejected homonyms, are legitimate under the conserved or sanctioned names without change of authorship or date if there is no other obstacle under the rules.

Ex. 2. *Alpinia languas* J. F. Gmel. (1791) and *Alpinia galanga* (L.) Willd. (1797) are to be accepted although *Alpinia* L. (1753), to which they were assigned by their authors, is rejected and the genus in which they are now placed is named *Alpinia* Roxb. (1810), *nom. cons.*

Article 56

56.1. Any name that would cause a disadvantageous nomenclatural change (Art. 14.1) may be proposed for rejection. A name thus rejected, or its basionym if it has one, is placed on a list of *nomina utique rejicienda* (App. IV). Along with the listed names, all combinations based on them are similarly rejected, and none is to be used.

56.2. The list of rejected names will remain permanently open for additions and changes. Any proposal for rejection of a name must be accompanied by a detailed statement of the cases both for and against its rejection, including considerations of typification. Such proposals must be submitted to the General Committee (see Div. III), which will refer them for examination to the committees for the various taxonomic groups (see also Art. 14.14 and Rec. 14A).

Article 57

57.1. A name that has been widely and persistently used for a taxon or taxa not including its type is not to be used in a sense that conflicts with current usage unless and until a proposal to deal with it under Art. 14.1 or 56.1 has been submitted and rejected.

Article 58

58.1. A name rejected or otherwise unavailable for use under Art. 52-54 or 56-57 is replaced by the name that has priority (Art. 11) in the rank concerned. If none exists in any rank a new name must be chosen: *(a)* the taxon may be treated as new and another name published for it, or *(b)* if the illegitimate name is a later homonym, an avowed substitute *(nomen novum)* based on the same type as the rejected name may be published for it. If a name is available in another rank, one of the above alternatives may be chosen, or *(c)* a new combination, based on the name in the other rank, may be published.

58.2. Similar action is to be taken if transfer of an epithet of a legitimate name would result in a combination that cannot be validly published under Art. 21.3, 22.4, 23.4 or 27, or in a later homonym.

Ex. 1. Linum radiola L. (1753) when transferred to *Radiola* Hill may not be named *"Radiola radiola"*, as was done by Karsten (1882), since that combination is invalid (see Art. 23.4 and 32.1(b)). The next oldest name, *L. multiflorum* Lam. (1779), is illegitimate, being a superfluous name for *L. radiola*. Under *Radiola*, the species has been given the legitimate name *R. linoides* Roth (1788).

58.3. When a new epithet is required, an author may adopt the epithet of a previous illegitimate name of the taxon if there is no obstacle to its employment in the new position or sense; the resultant combination is treated as the name of a new taxon or as a *nomen novum,* as the case may be.

Ex. 2. The name *Talinum polyandrum* Hook. (1855) is illegitimate, being a later homonym of *T. polyandrum* Ruiz & Pav. (1798). When Bentham, in 1863, transferred *T. polyandrum* Hook. to *Calandrinia,* he called it *C. polyandra.* This name is treated as having priority from 1863, and is cited as *C. polyandra* Benth., not *C. polyandra* (Hook.) Benth.

Ex. 3. While describing *Collema tremelloides* var. *cyanescens,* Acharius (Syn. Meth. Lich.: 326. 1814) cited *C. tremelloides* var. *caesium* Ach. (Lichenogr. Universalis: 656. 1810) in synonymy, thus rendering his new name illegitimate. The epithet *cyanescens* was taken up in the combination *Parmelia cyanescens* Schaer. (1842), but this is a later homonym of *P. cyanescens* (Pers.) Ach. (1803). In *Collema,* however, the epithet *cyanescens* was available for use, and the name *C. cyanescens* Rabenh. (1845), based on the same type, is legitimate. The correct author citation for *Leptogium cyanescens,* validated by Körber (1855) by reference to *C. cyanescens* "Schaer.", is therefore (Rabenh.) Körb., not (Ach.) Körb. nor (Schaer.) Körb.

Recommendation 58A

58A.1. Authors should avoid adoption of the epithet of an illegitimate name previously published for the same taxon.

CHAPTER VI. NAMES OF FUNGI WITH A PLEOMORPHIC LIFE CYCLE

Article 59

59.1. In ascomycetous and basidiomycetous fungi (including *Ustilaginales*) with mitotic asexual morphs (anamorphs) as well as a meiotic sexual morph (teleomorph), the correct name covering the holomorph (i.e., the species in all its morphs) is – except for lichen-forming fungi – the earliest legitimate name typified by an element representing the teleomorph, i.e. the morph characterized by the production of asci/ascospores, basidia/basidiospores, teliospores, or other basidium-bearing organs.

59.2. For a binary name to qualify as a name of a holomorph, not only must its type specimen be teleomorphic, but also the protologue must include a description or diagnosis of this morph (or be so phrased that the possibility of reference to the teleomorph cannot be excluded).

59.3. If these requirements are not fulfilled, the name is that of a form-taxon and is applicable only to the anamorph represented by its type, as described or referred to in the protologue. The accepted taxonomic disposition of the type of the name determines the application of the name, no matter whether the genus to which a subordinate taxon is assigned by the author(s) is holomorphic or anamorphic.

59.4. The priority of names of holomorphs at any rank is not affected by the earlier publication of names of anamorphs judged to be correlated morphs of the holomorph.

59.5. The provisions of this article shall not be construed as preventing the publication and use of binary names for form-taxa when it is thought necessary or desirable to refer to anamorphs alone.

Ex. 1. Because the teleomorph of *Gibberella stilboides* W. L. Gordon & C. Booth (1971) is only known from strains of the anamorph *Fusarium stilboides* Wollenw. (1924) mating in culture, and has not been found in nature, it may be thought desirable to use the name of the anamorph for the pathogen of *Coffea*.

Ex. 2. Cummins (1971), in *The rust fungi of cereals, grasses and bamboos*, found it to be neither necessary nor desirable to introduce new names of anamorphs under *Aecidium* Pers. : Pers. and *Uredo* Pers. : Pers., for the aecial and uredinial stages of species of *Puccinia* Pers. : Pers. of which the telial stage (teleomorph) was known.

Note 1. When not already available, specific or infraspecific names for anamorphs may be proposed at the time of publication of the name for the holomorphic fungus or later. The epithets may, if desired, be identical, as long as they are not in homonymous combinations.

59.6. As long as there is direct and unambiguous evidence for the deliberate introduction of a new morph judged by the author(s) to be correlated with the morph typifying a purported basionym, and this evidence is strengthened by fulfilment of all requirements in Art. 32-45 for valid publication of a name of a new taxon, any indication such as "comb. nov." or "nom. nov." is regarded as a formal error, and the name introduced is treated as that of a new taxon, and attributed solely to the author(s) thereof. When only the requirements for valid publication of a new combination (Art. 33 and 34) have been fulfilled, the name is accepted as such and based, in accordance with Art. 7.4, on the type of the declared or implicit basionym.

Ex. 3. The name *Penicillium brefeldianum* B. O. Dodge (1933), based on teleomorphic and anamorphic material, is a valid and legitimate name of a holomorph, in spite of the attribution of the species to a form-genus. It is legitimately combined in a holomorphic genus as *Eupenicillium brefeldianum* (B. O. Dodge) Stolk & D. B. Scott (1967). *P. brefeldianum* is not available for use in a restricted sense for the anamorph alone.

Ex. 4. The name *Ravenelia cubensis* Arthur & J. R. Johnst. (1918), based on a specimen bearing only uredinia (an anamorph), is a valid and legitimate name of an anamorph, in spite of the attribution of the species to a holomorphic genus. It is legitimately combined in a form-genus as *Uredo cubensis* (Arthur & J. R. Johnst.) Cummins (1956). *R. cubensis* is not available for use inclusive of the teleomorph.

Ex. 5. *Mycosphaerella aleuritidis* was published as "(Miyake) Ou comb. nov., syn. *Cercospora aleuritidis* Miyake" but with a Latin diagnosis of the teleomorph. The indication "comb. nov." is taken as a formal error, and *M. aleuritidis* S. H. Ou (1940) is accepted as a validly published new specific name for the holomorph, typified by the teleomorphic material described by Ou.

Ex. 6. *Corticium microsclerotium* was originally published as "(Matz) Weber, comb. nov., syn. *Rhizoctonia microsclerotia* Matz" with a description, only in English, of the teleomorph. Because of Art. 36, this may not be considered as the valid publication of the name of a new species, and so *C. microsclerotium* (Matz) G. F. Weber (1939) must be considered a validly published and legitimate new combination based on the specimen of the anamorph that typifies its basionym. *C. microsclerotium* G. F. Weber (1951), published with a Latin description and a teleomorphic type, is an illegitimate later homonym.

Ex. 7. *Hypomyces chrysospermus* Tul. (1860), presented as the name of a holomorph without the indication "comb. nov." but with explicit reference to *Mucor chrysospermus* (Bull.) Bull. and *Sepedonium chrysospermum* (Bull.) Fr., which are names of its anamorph, is not to be considered as a new combination but as the name of a newly described species, with a teleomorphic type.

Recommendation 59A

59A.1. When a new morph of a fungus is described, it should be published either as a new taxon (e.g., gen. nov., sp. nov., var. nov.) whose name has a teleomorphic type, or as a new anamorph (anam. nov.) whose name has an anamorphic type.

59A.2. When in naming a new morph of a fungus the epithet of the name of a different, earlier described morph of the same fungus is used, the new name should be designated as the name of a new taxon or anamorph, as the case may be, but not as a new combination based on the earlier name.

CHAPTER VII. ORTHOGRAPHY OF NAMES AND EPITHETS AND GENDER OF GENERIC NAMES

SECTION 1. ORTHOGRAPHY OF NAMES AND EPITHETS

Article 60

60.1. The original spelling of a name or epithet is to be retained, except for the correction of typographical or orthographical errors and the standardizations imposed by Art. 60.5 (*u/v* or *i/j* used interchangeably), 60.6 (diacritical signs and ligatures), 60.8 (compounding forms), 60.9 (hyphens), 60.10 (apostrophes), and 60.11 (terminations; see also Art. 32.6), as well as Rec. 60H.

Ex. 1. Retention of original spelling: The generic names *Mesembryanthemum* L. (1753) and *Amaranthus* L. (1753) were deliberately so spelled by Linnaeus and the spelling is not to be altered to *"Mesembrianthemum"* and *"Amarantus"* respectively, although these latter forms are philologically preferable (see Bull. Misc. Inform. Kew 1928: 113, 287. 1928). – *Phoradendron* Nutt. (1848) is not to be altered to *"Phoradendrum"*. – *Triaspis mozambica* A. Juss. (1843) is not to be altered to *"T. mossambica"*, as in Engler (Pflanzenw. Ost-Afrikas C: 232. 1895). – *Alyxia ceylanica* Wight (1848) is not to be altered to *"A. zeylanica"*, as in Trimen (Handb. Fl. Ceylon 3: 127. 1895). – *Fagus sylvatica* L. (1753) is not to be altered to *"F. silvatica"*. The classical spelling *silvatica* is recommended for adoption in the case of a new name (Rec. 60E), but the mediaeval spelling *sylvatica* is not an orthographical error. – *Scirpus cespitosus* L. (1753) is not to be altered to *"S. caespitosus"*.

**Ex. 2.* Typographical errors: *Globba "brachycarpa"* Baker (1890) and *Hetaeria "alba"* Ridl. (1896) are typographical errors for *Globba trachycarpa* Baker and *Hetaeria alta* Ridl. respectively (see J. Bot. 59: 349. 1921).

**Ex. 3.* Orthographical error: *Gluta "benghas"* L. (1771), being an orthographical error for *G. renghas*, should be cited as *G. renghas* L., as by Engler (in Candolle & Candolle, Monogr. Phan. 4: 225. 1883); the vernacular name used as a specific epithet by Linnaeus is "renghas", not "benghas".

Ex. 4. The misspelled *Indigofera "longipednnculata"* Y. Y. Fang & C. Z. Zheng (1983) is presumably a typographical error and is to be corrected to *I. longipedunculata*.

Note 1. Art. 14.11 provides for the conservation of an altered spelling of a generic name.

Ex. 5. *Bougainvillea* (see App. IIIA, Spermatophyta, No. 2350).

60.2. The words "original spelling" in this Article mean the spelling employed when the name was validly published. They do not refer to the use of an initial capital or small letter, this being a matter of typography (see Art. 20.1 and 21.2, Rec. 60F).

60.3. The liberty of correcting a name is to be used with reserve, especially if the change affects the first syllable and, above all, the first letter of the name.

**Ex. 6.* The spelling of the generic name *Lespedeza* Michx. (1803) is not to be altered, although it commemorates Vicente Manuel de Céspedes (see Rhodora 36: 130-132, 390-392. 1934). – *Cereus jamacaru* DC. (1828) may not be altered to *C. "mandacaru"*, even if *jamacaru* is believed to be a corruption of the vernacular name "mandacaru".

60.4. The letters *w* and *y,* foreign to classical Latin, and *k,* rare in that language, are permissible in Latin plant names. Other letters and ligatures foreign to classical Latin that may appear in Latin plant names, such as the German *ß* (double *s*), are to be transcribed.

60.5. When a name has been published in a work where the letters *u, v* or *i, j* are used interchangeably or in any other way incompatible with modern practices (one of those letters is not used or only in capitals), those letters are to be transcribed in conformity with modern botanical usage.

Ex. 7. Uffenbachia Fabr. (1763), not *"Vffenbachia"; Taraxacum* Zinn (1757), not *"Taraxacvm"; Curculigo* Gaertn. (1788), not *"Cvrcvligo".*

Ex. 8. "Geastrvm hygrometricvm" and *"Vredo pvstvlata"* of Persoon (1801) are written respectively *Geastrum hygrometricum* Pers. and *Uredo pustulata* Pers.

60.6. Diacritical signs are not used in Latin plant names. In names (either new or old) drawn from words in which such signs appear, the signs are to be suppressed with the necessary transcription of the letters so modified; for example *ä, ö, ü* become respectively *ae, oe, ue; é, è, ê* become *e,* or sometimes *ae; ñ* becomes *n; ø* becomes *oe; å* becomes *ao.* The diaeresis, indicating that a vowel is to be pronounced separately from the preceding vowel (as in *Cephaëlis, Isoëtes*), is permissible; the ligatures *-æ-* and *-œ-*, indicating that the letters are pronounced together, are to be replaced by the separate letters *-ae-* and *-oe-*.

60.7. When changes made in orthography by earlier authors who adopt personal, geographic, or vernacular names in nomenclature are intentional latinizations, they are to be preserved, except for terminations covered by Art. 60.11.

Ex. 9. Valantia L. (1753), *Gleditsia* L. (1753), and *Clutia* L. (1753), commemorating Vaillant, Gleditsch, and Cluyt respectively, are not to be altered to *"Vaillantia", "Gleditschia",* and *"Cluytia";* Linnaeus latinized the names of these botanists deliberately as Valantius, Gleditsius, and Clutius.

Ex. 10. Zygophyllum "billardierii" was named by Candolle (1824) for J. J. H. de Labillardière (de la Billardière). The intended latinization is "Billardierius" (in nominative), but that termination is not acceptable under Art. 60.11 and the name is correctly spelled *Z. billardierei* DC.

60.8. The use of a compounding form contrary to Rec. 60G in an adjectival epithet is treated as an error to be corrected.

Ex. 11. Pereskia "opuntiaeflora" of Candolle (1828) is to be cited as *P. opuntiiflora* DC. However, in *Andromeda polifolia* L. (1753), the epithet is a pre-Linnean plant name (*"Polifolia"* of Buxbaum) used in apposition and not an adjective; it is not to be corrected to *"poliifolia".*

Ex. 12. Cacalia "napeaefolia" and *Senecio "napeaefolius"* are to be cited as *Cacalia napaeifolia* DC. (1838) and *Senecio napaeifolius* (DC.) Sch. Bip. (1845) respectively; the specific epithet refers to the resemblance of the leaves to those of the genus *Napaea* L. (not *"Napea"*), and the substitute (connecting) vowel *-i* should have been used instead of the genitive singular inflection *-ae*.

60.9. The use of a hyphen in a compound epithet is treated as an error to be corrected by deletion of the hyphen, except if an epithet is formed of words that usually stand independently, or if the letters before and after the hyphen are the same, when a hyphen is permitted (see Art. 23.1 and 23.3).

Ex. 13. Hyphen to be omitted: *Acer pseudoplatanus* L. (1753), not *A. "pseudo-platanus"; Ficus neoëbudarum* Summerh. (1932), not *F. "neo-ebudarum"; Lycoperdon atropurpureum* Vittad. (1842), not *L. "atro-purpureum"; Croton ciliatoglandulifer* Ortega (1797), not *C. "ciliato-glandulifer"; Scirpus* sect. *Pseudoëriophorum* Jurtzev (in Bjull. Moskovsk. Obšč. Ips. Prir., Otd. Biol. 70(1): 132. 1965), not *S.* sect. *"Pseudo-eriophorum"*.

Ex. 14. Hyphen to be used: *Aster novae-angliae* L. (1753), *Coix lacryma-jobi* L. (1753), *Peperomia san-felipensis* J. D. Sm. (1894), *Arctostaphylos uva-ursi* (L.) Spreng. (1825), *Veronica anagallis-aquatica* L. (1753; Art. 23.3), *Athyrium austro-occidentale* Ching (1986).

Note 2. Art. 60.9 refers only to epithets (in combinations), not to names of genera or taxa in higher ranks; a generic name published with a hyphen can be changed only by conservation.

Ex. 15. Pseudo-salvinia Piton (1940) cannot be changed to *"Pseudosalvinia"*, but *"Pseudo-elephantopus"* was changed by conservation to *Pseudelephantopus* Rohr (1792).

60.10. The use of an apostrophe in an epithet is treated as an error to be corrected by deletion of the apostrophe.

Ex. 16. Lycium "o'donellii", Cymbidium "i'ansoni" and *Solanum tuberosum* var. *"muru'kewillu"* are to be corrected to *L. odonellii* F. A. Barkley (1953), *C. iansonii* Rolfe (1900) and *S. tuberosum* var. *murukewillu* Ochoa (in Phytologia 65: 112. 1988), respectively.

60.11. The use of a termination (for example *-i, -ii, -ae, -iae, -anus,* or *-ianus*) contrary to Rec. 60C.1 (but not 60C.2) is treated as an error to be corrected (see also Art. 32.6).

Ex. 17. Rosa "pissarti" (Carrière in Rev. Hort. 1880: 314. 1880) is a typographical error for *R. "pissardi"* (see Rev. Hort. 1881: 190. 1881), which in its turn is treated as an error for *R. pissardii* Carrière (see Rec. 60C.1(b)).

Note 3. If the gender and/or number of a substantival epithet derived from a personal name is inappropriate for the sex and/or number of the person(s) whom the name commemorates, the termination is to be corrected in conformity with Rec. 60C.1.

Ex. 18. Rosa ×*"toddii"* was named by Wolley-Dod (in J. Bot. 69, Suppl.: 106. 1931) for "Miss E. S. Todd"; the name is to be corrected to *R.* ×*toddiae* Wolley-Dod.

Ex. 19. Astragalus "matthewsii", dedicated by Podlech and Kirchhoff (in Mitt. Bot. Staatssamml. München 11: 432. 1974) to Victoria A. Matthews, is to be corrected to *A. matthewsiae* Podlech & Kirchhoff; it is not therefore a later homonym of *A. matthewsii* S. Watson (1883) (see Agerer-Kirchhoff & Podlech in Mitt. Bot. Staatssamml. München 12: 375. 1976).

Ex. 20. Codium "geppii" of O. C. Schmidt (in Biblioth. Bot. 91: 50. 1923), which commemorates "A. & E. S. Gepp", is to be corrected to *C. geppiorum* O. C. Schmidt.

Recommendation 60A

60A.1. When a new name or its epithet is to be derived from Greek, the transliteration to Latin should conform to classical usage.

60A.2. The spiritus asper should be transcribed in Latin as the letter *h*.

Recommendation 60B

60B.1. When a new generic name, or subgeneric or sectional epithet, is taken from the name of a person, it should be formed as follows:

(a) When the name of the person ends with a vowel, the letter *-a* is added (thus *Ottoa* after Otto; *Sloanea* after Sloane), except when the name ends with *-a,* when *-ea* is added (e.g. *Collaea* after Colla), or with *-ea* (as *Correa*), when no letter is added.

(b) When the name of the person ends with a consonant, the letters *-ia* are added, but when the name ends with *-er,* either of the terminations *-ia* and *-a* is appropriate (e.g. *Sesleria* after Sesler and *Kernera* after Kerner).

(c) In latinized personal names ending with *-us* this termination is dropped (e.g. *Dillenia* after Dillenius) before applying the procedure described under (a) and (b).

(d) The syllables not modified by these endings retain their original spelling, unless they contain letters foreign to Latin plant names or diacritical signs (see Art. 60.6).

Note 1. Names may be accompanied by a prefix or a suffix, or be modified by anagram or abbreviation. In these cases they count as different words from the original name.

Ex. 1. Durvillaea Bory (1826) and *Urvillea* Kunth (1821); *Lapeirousia* Pourr. (1788) and *Peyrousea* DC. (1838); *Engleria* O. Hoffm. (1888), *Englerastrum* Briq. (1894), and *Englerella* Pierre (1891); *Bouchea* Cham. (1832) and *Ubochea* Baill. (1891); *Gerardia* L. (1753) and *Graderia* Benth. (1846); *Martia* Spreng. (1818) and *Martiusia* Schult. & Schult. f. (1822).

Recommendation 60C

60C.1. Modern personal names may be given Latin terminations and used to form specific and infraspecific epithets as follows (but see Rec. 60C.2):

(a) If the personal name ends with a vowel or *-er,* substantive epithets are formed by adding the genitive inflection appropriate to the sex and number of the person(s) honoured (e.g., *scopoli-i* for Scopoli (m), *fedtschenko-i* for Fedtschenko (m), *glaziou-i* for Glaziou (m), *lace-ae* for Lace (f), *hooker-orum* for the Hookers), except when the name ends with *-a,* in which case adding *-e* (singular) or *-rum* (plural) is appropriate (e.g. *triana-e* for Triana (m)).

(b) If the personal name ends with a consonant (except *-er*), substantive epithets are formed by adding *-i-* (stem augmentation) plus the genitive inflection appropriate to the sex and number of the person(s) honoured (e.g. *lecard-ii* for Lecard (m), *wilson-iae* for Wilson (f), *verlot-iorum* for the Verlot brothers, *braun-iarum* for the Braun sisters).

(c) If the personal name ends with a vowel, adjectival epithets are formed by adding *-an-* plus the nominative singular inflection appropriate to the gender of the generic name (e.g., *Cyperus heyne-anus* for Heyne, *Vanda lindley-ana* for Lindley, *Aspidium bertero-anum* for Bertero), except when the personal name ends with *-a* in which case *-n-* plus the appropriate inflection is added (e.g. *balansa-nus* (m), *balansa-na* (f), and *balansa-num* (n) for Balansa).

(d) If the personal name ends with a consonant, adjectival epithets are formed by adding *-i-* (stem augmentation) plus *-an-* (stem of adjectival suffix) plus the nominative singular inflection appropriate to the gender of the generic name (e.g. *Rosa webb-iana* for Webb, *Desmodium griffith-ianum* for Griffith, *Verbena hassler-iana* for Hassler).

Note 1. The hyphens in the above examples are used only to set off the total appropriate termination.

60C.2. Personal names already in Greek or Latin, or possessing a well-established latinized form, should be given their appropriate Latin genitive to form substantive epithets (e.g. *alexandri* from Alexander or Alexandre, *augusti* from Augustus or August or Auguste, *martini* from Martinus or Martin, *linnaei* from Linnaeus, *martii* from Martius, *beatricis* from Beatrix or Béatrice, *hectoris* from Hector). Treating modern names as if they were in third declension should be avoided (e.g. *munronis* from Munro, *richardsonis* from Richardson).

60C.3. In forming new epithets based on personal names the original spelling of the personal name should not be modified unless it contains letters foreign to Latin plant names or diacritical signs (see Art. 60.4 and 60.6).

60C.4. Prefixes and particles ought to be treated as follows:

(a) The Scottish patronymic prefix "Mac", "Mc" or "M'", meaning "son of", should be spelled "mac" and united with the rest of the name, e.g. *macfadyenii* after Macfadyen, *macgillivrayi* after MacGillivray, *macnabii* after McNab, *mackenii* after M'Ken.

(b) The Irish patronymic prefix "O" should be united with the rest of the name or omitted, e.g. *obrienii, brienianus* after O'Brien, *okellyi* after O'Kelly.

(c) A prefix consisting of an article, e.g. le, la, l', les, el, il, lo, or containing an article e.g. du, de la, des, del, della, should be united to the name, e.g. *leclercii* after Le Clerc, *dubuyssonii* after DuBuysson, *lafarinae* after La Farina, *logatoi* after Lo Gato.

(d) A prefix to a surname indicating ennoblement or canonization should be omitted, e.g. *candollei* after de Candolle, *jussieui* after de Jussieu, *hilairei* after Saint-Hilaire, *remyi* after St. Rémy; in geographical epithets, however, "St." is rendered as *sanctus* (m) or *sancta* (f), e.g. *sancti-johannis,* of St. John, *sanctae-helenae,* of St. Helena.

(e) A German or Dutch prefix when it is normally treated as part of the family name, as often happens outside its country of origin, e.g. in the United States, may be included in the epithet, e.g. *vonhausenii* after Vonhausen, *vanderhoekii* after Vanderhoek, *vanbruntiae* after Mrs Van Brunt, but should otherwise be omitted, e.g. *iheringii* after von Ihering, *martii* after von Martius, *steenisii* after van Steenis, *strassenii* after zu Strassen, *vechtii* after van der Vecht.

Recommendation 60D

60D.1. An epithet derived from a geographical name is preferably an adjective and usually takes the termination *-ensis, -(a)nus, -inus,* or *-icus.*

Ex. 1. Rubus quebecensis L. H. Bailey (from Quebec), *Ostrya virginiana* (Mill.) K. Koch (from Virginia), *Eryngium amorginum* Rech. f. (from Amorgos), *Polygonum pensylvanicum* L. (from Pennsylvania).

Recommendation 60E

60E.1. The epithet in a new name should be written in conformity with the original spelling of the word or words from which it is derived and in accordance with the accepted usage of Latin and latinization (see also Art. 23.5).

Ex. 1. sinensis (not *chinensis*).

Recommendation 60F

60F.1. All specific and infraspecific epithets should be written with a small initial letter, although authors desiring to use capital initial letters may do so when the epithets are directly derived from the names of persons (whether actual or mythical), or are vernacular (or non-Latin) names, or are former generic names.

Recommendation 60G

60G.1. A compound name or an epithet which combines elements derived from two or more Greek or Latin words should be formed, as far as practicable, in accordance with classical usage (see Art. 60.8). This may be stated as follows:

(a) In a true compound, a noun or adjective in non-final position appears as a compounding form generally obtained by

   (1) removing the case ending of the genitive singular (Latin *-ae, -i, -us, -is;* Greek *-os, -es, -as, -ous* and the latter's equivalent *-eos*) and

   (2) before a consonant, adding a connecting vowel (*-i-* for Latin elements, *-o-* for Greek elements).

   (3) Exceptions are common, and one should review earlier usages of a particular compounding form.

(b) A pseudocompound is a noun or adjectival phrase treated as if it were a single compound word. In a pseudocompound, a noun or adjective in a non-final position appears as a word with a case ending, not as a modified stem. Examples are: *nidus-avis* (nest of bird), *Myos-otis* (ear of mouse), *cannae-folius* (leaf of canna), *albo-marginatus* (margined with white), etc. In epithets where tingeing is expressed, the modifying initial colour often is in the ablative because the preposition *e, ex,* is implicit, e.g., *atropurpureus* (blackish purple) from *ex atro purpureus* (purple tinged with black). Others have been deliberately introduced to reveal etymological differences when different word elements have the same compounding forms, such as *tubi-* from tube *(tubus, tubi,* stem *tubo-)* or from trumpet *(tuba, tubae,* stem *tuba-)* where *tubaeflorus* can only mean trumpet-flowered; also *carici-* is the compounding form from both papaya *(carica, caricae,* stem *carica-)* and sedge *(carex, caricis,* stem *caric-)* where *caricaefolius* can only mean papaya-leaved. The latter use of the genitive singular of the first declension for pseudocompounding is treated as an error to be corrected unless it makes an etymological distinction.

(c) Some common irregular forms are used in compounding. Examples are *hydro-* and *hydr-(Hydro-phyllum)* where the regular noun stem is *hydat-; calli- (Calli-stemon)* where the regular adjective stem is *calo-;* and *meli- (Meli-osma, Meli-lotus)* where the regular noun stem is *melit-*.

Note 1. The hyphens in the above examples are given solely for explanatory reasons. For the use of hyphens in botanical names and their epithets see Art. 20.3, 23.1, and 60.9.

Recommendation 60H

60H.1. Epithets of fungus names derived from the generic name of the host plant are spelled in accordance with the accepted spelling of this name; other spellings are regarded as orthographical variants to be corrected (see Art. 61).

Ex. 1. Phyllachora "anonicola" (Chardon in Mycologia 32: 190. 1940) is to be altered to *P. annonicola* Chardon, since the spelling *Annona* is now accepted in preference to *"Anona"*. – *Meliola "albiziae"* (Hansford & Deighton in Mycol. Pap. 23: 26. 1948) is to be altered to *M. albiziae* Hansf. & Deighton, since the spelling *Albizia* is now accepted in preference to *"Albizzia"*.

Recommendation 60I

60I.1 The etymology of new names and their epithets should be given when the meaning of these is not obvious.

Article 61

61.1. Only one orthographical variant of any one name is treated as validly published, the form which appears in the original publication except as provided in Art. 60 (typographical or orthographical errors and standardizations), Art. 14.11 (conserved spellings), and Art. 32.6 (incorrect Latin terminations).

61.2. For the purpose of this *Code,* orthographical variants are the various spelling, compounding, and inflectional forms of a name or its epithet (including typographical errors), only one type being involved.

61.3. If orthographical variants of a name appear in the original publication, the one that conforms to the rules and best suits the recommendations of Art. 60 is to be retained; otherwise the first author who, in an effectively published text (Art. 29-31), explicitly adopts one of the variants, rejecting the other(s), must be followed.

61.4. The orthographical variants of a name are to be corrected to the validly published form of that name. Whenever such a variant appears in print, it is to be treated as if it were printed in its corrected form.

Note 1. In full citations it is desirable that the original form of a corrected orthographical variant of a name be added (Rec. 50F).

61.5. Confusingly similar names based on the same type are treated as orthographical variants. (For confusingly similar names based on different types, see Art. 53.3.)

Ex. 1. Geaster Fr. (1829) and *Geastrum* Pers. (1794) : Pers. (1801) are similar names with the same type (see Taxon 33: 498. 1984); they are treated as orthographical variants despite the fact that they are derived from two different nouns, *aster (asteris)* and *astrum (astri)*.

SECTION 2. GENDER OF GENERIC NAMES

Article 62

62.1. A generic name retains the gender assigned by botanical tradition, irrespective of classical usage or the author's original usage. A generic name without a botanical tradition retains the gender assigned by its author.

Note 1. Botanical tradition usually maintains the classical gender of a Greek or Latin word, when this was well established.

**Ex. 1.* In accordance with botanical tradition, *Adonis* L., *Atriplex* L., *Diospyros* L., *Hemerocallis* L., *Orchis* L., *Stachys* L., and *Strychnos* L. must be treated as feminine while *Lotus* L. and *Melilotus* L. must be treated as masculine. *Eucalyptus* L'Hér., which lacks a botanical tradition, retains the feminine gender assigned by its author. Although their ending suggests masculine gender, *Cedrus* Trew and *Fagus* L., like most other classical tree names, were traditionally treated as feminine and thus retain that gender; similarly, *Rhamnus* L. is feminine, despite the fact that Linnaeus assigned it masculine gender. *Phyteuma* L. (n), *Sicyos* L. (m), and *Erigeron* L. (m) are other names for which botanical tradition has reestablished the classical gender despite another choice by Linnaeus.

62.2. Compound generic names take the gender of the last word in the nominative case in the compound. If the termination is altered, however, the gender is altered accordingly.

Ex. 2. Compound generic names in which the termination of the last word is altered: *Stenocarpus* R. Br., *Dipterocarpus* C. F. Gaertn., and all other compounds ending in the Greek masculine *-carpos* (or *-carpus*), e.g. *Hymenocarpos* Savi, are masculine; those in *-carpa* or *-carpaea,* however, are feminine, e.g. *Callicarpa* L. and *Polycarpaea* Lam.; and those in *-carpon, -carpum,* or *-carpium* are neuter, e.g. *Polycarpon* L., *Ormocarpum* P. Beauv., and *Pisocarpium* Link.

(a) Compounds ending in *-codon, -myces, -odon, -panax, -pogon, -stemon,* and other masculine words, are masculine.

Ex. 3. Irrespective of the fact that the generic names *Andropogon* L. and *Oplopanax* (Torr. & A. Gray) Miq. were originally treated as neuter by their authors, they are masculine.

(b) Compounds ending in *-achne, -chlamys, -daphne, -mecon, -osma* (the modern transcription of the feminine Greek word *osmê*), and other feminine words, are feminine. An exception is made in the case of names ending in *-gaster,* which strictly speaking ought to be feminine, but which are treated as masculine in accordance with botanical tradition.

Ex. 4. Irrespective of the fact that *Dendromecon* Benth. and *Hesperomecon* Greene were originally treated as neuter, they are feminine.

(c) Compounds ending in *-ceras, -dendron, -nema, -stigma, -stoma,* and other neuter words, are neuter. An exception is made for names ending in *-anthos* (or *-anthus*) and *-chilos (-chilus* or *-cheilos),* which ought to be neuter, since that is the gender of the Greek words *anthos* and *cheilos,* but are treated as masculine in accordance with botanical tradition.

Ex. 5. Irrespective of the fact that *Aceras* R. Br. and *Xanthoceras* Bunge were treated as feminine when first published, they are neuter.

62.3. Arbitrarily formed generic names or vernacular names or adjectives used as generic names, whose gender is not apparent, take the gender assigned to them by their authors. If the original author failed to indicate the gender, the next subsequent author may choose a gender, and his choice, if effectively published (Art. 29-31), is to be accepted.

Ex. 6. *Taonabo* Aubl. (1775) is feminine: Aublet's two species were *T. dentata* and *T. punctata.*

Ex. 7. *Agati* Adans. (1763) was published without indication of gender: the feminine gender was assigned to it by Desvaux (in J. Bot. Agric. 1: 120. 1813), who was the first subsequent author to adopt the name in an effectively published text, and his choice is to be accepted.

Ex. 8. The original gender of *Manihot* Mill. (1754), as apparent from some of the species polynomials, was feminine, and *Manihot* is therefore to be treated as feminine.

62.4. Generic names ending in *-anthes, -oides* or *-odes* are treated as feminine and those ending in *-ites* as masculine, irrespective of the gender assigned to them by the original author.

Recommendation 62A

62A.1. When a genus is divided into two or more genera, the gender of the new generic name or names should be that of the generic name that is retained.

Ex. 1. When *Boletus* L. : Fr. is divided, the gender of the new generic names should be masculine: *Xerocomus* Quél. (1887), *Boletellus* Murrill (1909), etc.

DIVISION III. PROVISIONS FOR MODIFICATION OF THE CODE

Div.III.1. Modification of the *Code.* The *Code* may be modified only by action of a plenary session of an International Botanical Congress on a resolution moved by the Nomenclature Section of that Congress[1].

Div.III.2. Nomenclature Committees. Permanent Nomenclature Committees are established under the auspices of the International Association for Plant Taxonomy. Members of these committees are elected by an International Botanical Congress. The Committees have power to co-opt and to establish subcommittees; such officers as may be desired are elected.

(1) General Committee, composed of the secretaries of the other committees, the rapporteur-général, the president and the secretary of the International Association for Plant Taxonomy, and at least 5 members to be appointed by the Nomenclature Section. The rapporteur-général is charged with the presentation of nomenclature proposals to the International Botanical Congress.

(2) Committee for Spermatophyta.

(3) Committee for Pteridophyta.

(4) Committee for Bryophyta.

(5) Committee for Fungi.

(6) Committee for Algae.

(7) Committee for Fossil Plants.

(8) Editorial Committee, charged with the preparation and publication of the *Code* in conformity with the decisions adopted by the International Botanical Congress. Chairman: the rapporteur-général of the previous Congress, who is charged with the general duties in connection with the editing of the *Code*.

Div.III.3. The Bureau of Nomenclature of the International Botanical Congress. Its officers are: *(1)* the president of the Nomenclature Section, elected by the organizing committee of the International Botanical Congress in question; *(2)* the recorder, appointed by the same organizing committee; *(3)* the rappor-

[1] In the event that there should not be another International Botanical Congress, authority for the *International code of botanical nomenclature* shall be transferred to the International Union of Biological Sciences or to an organization at that time corresponding to it. The General Committee is empowered to define the machinery to achieve this.

teur-général, elected by the previous Congress; *(4)* the vice-rapporteur, elected by the organizing committee on the proposal of the rapporteur-général.

Div.III.4. The voting on nomenclature proposals is of two kinds: *(a)* a preliminary guiding mail vote and *(b)* a final and binding vote at the Nomenclature Section of the International Botanical Congress.

Qualifications for voting:

(a) Preliminary mail vote:

(1) The members of the International Association for Plant Taxonomy.

(2) The authors of proposals.

(3) The members of the nomenclature committees.

Note 1. No accumulation or transfer of personal votes is permissible.

(b) Final vote at the sessions of the Nomenclature Section:

(1) All officially enrolled members of the Section. No accumulation or transfer of personal votes is permissible.

(2) Official delegates or vice-delegates of the institutes appearing on a list drawn up by the Bureau of Nomenclature of the International Botanical Congress and submitted to the General Committee for final approval; such institutes are entitled to 1-7 votes, as specified on the list. No single institution, even in the wide sense of the term, is entitled to more than 7 votes. Transfer of institutional votes to specified vice-delegates is permissible, but no single person will be allowed more than 15 votes, his personal vote included. Institutional votes may be deposited at the Bureau of Nomenclature to be counted in a specified way for specified proposals.

APPENDIX I

NAMES OF HYBRIDS

Article H.1

H.1.1. Hybridity is indicated by the use of the multiplication sign ×, or by the addition of the prefix "notho-"[1] to the term denoting the rank of the taxon.

Article H.2

H.2.1. A hybrid between named taxa may be indicated by placing the multiplication sign between the names of the taxa; the whole expression is then called a hybrid formula.

Ex. 1. *Agrostis* L. × *Polypogon* Desf.; *Agrostis stolonifera* L. × *Polypogon monspeliensis* (L.) Desf.; *Salix aurita* L. × *S. caprea* L.; *Mentha aquatica* L. × *M. arvensis* L. × *M. spicata* L.; *Polypodium vulgare* subsp. *prionodes* (Asch.) Rothm. × subsp. *vulgare.*

Recommendation H.2A

H.2A.1. It is usually preferable to place the names or epithets in a formula in alphabetical order. The direction of a cross may be indicated by including the sexual symbols (♀: female; ♂: male) in the formula, or by placing the female parent first. If a non-alphabetical sequence is used, its basis should be clearly indicated.

Article H.3

H.3.1. Hybrids between representatives of two or more taxa may receive a name. For nomenclatural purposes, the hybrid nature of a taxon is indicated by placing the multiplication sign × before the name of an intergeneric hybrid or before the epithet in the name of an interspecific hybrid, or by prefixing the term "notho-" (optionally abbreviated "n-") to the term denoting the rank of the taxon (see Art. 3.2 and 4.4). All such taxa are designated nothotaxa.

---

[1] From the Greek *nothos,* meaning hybrid.

Ex. 1. (The putative or known parentage is found in Art. H.2 Ex. 1.) ×*Agropogon* P. Fourn. (1934); ×*Agropogon littoralis* (Sm.) C. E. Hubb. (1946); *Salix* ×*capreola* Andersson (1867); *Mentha* ×*smithiana* R. A. Graham (1949); *Polypodium vulgare* nothosubsp. *mantoniae* (Rothm.) Schidlay (in Futák, Fl. Slov. 2: 225. 1966).

H.3.2. A nothotaxon cannot be designated unless at least one parental taxon is known or can be postulated.

H.3.3. The epithet in the name of a nothospecies is termed a collective epithet.

H.3.4. For purposes of homonymy and synonymy the multiplication sign and the prefix "notho-" are disregarded.

Ex. 2. ×*Hordelymus* Bachteev & Darevsk. (1950) (= *Elymus* L. × *Hordeum* L.) is a later homonym of *Hordelymus* (K. Jess.) K. Jess. (1885).

Note 1. Taxa which are believed to be of hybrid origin need not be designated as nothotaxa.

Ex. 3. The true-breeding tetraploid raised from the artificial cross *Digitalis grandiflora* L. × *D. purpurea* L. may, if desired, be referred to as *D. mertonensis* B. H. Buxton & C. D. Darl. (1931); *Triticum aestivum* L. (1753) is treated as a species although it is not found in nature and its genome has been shown to be composed of those of *T. dicoccoides* (Körn.) Körn., *T. speltoides* (Tausch) Gren. ex K. Richt., and *T. tauschii* (Coss.) Schmalh.; the taxon known as *Phlox divaricata* subsp. *laphamii* (A. W. Wood) Wherry (in Morris Arbor. Monogr. 3: 41. 1955) is believed by Levin (Evolution 21: 92-108. 1967) to be a stabilized product of hybridization between *P. divaricata* L. subsp. *divaricata* and *P. pilosa* subsp. *ozarkana* Wherry; *Rosa canina* L. (1753), a polyploid believed to be of ancient hybrid origin, is treated as a species.

Note 2. The term "collective epithet" is used in the *International code of nomenclature for cultivated plants-1980* to include also epithets in modern language.

Recommendation H.3A

H.3A.1. The multiplication sign in the name of a nothotaxon should be placed against the initial letter of the name or epithet. However, if the mathematical symbol is not available and the letter "x" is used instead, a single letter space may be left between it and the epithet if this helps to avoid ambiguity. The letter "x" should be in lower case.

Article H.4

H.4.1. When all the parent taxa can be postulated or are known, a nothotaxon is circumscribed so as to include all individuals (as far as they can be recognized) derived from the crossing of representatives of the stated parent taxa (i.e. not only the F_1 but subsequent filial generations and also back-crosses and combinations of these). There can thus be only one correct name corresponding to a particular hybrid formula; this is the earliest legitimate name (see Art. 6.3) in the appropriate rank (Art. H.5), and other names to which the same hybrid formula applies are synonyms of it.

Ex. 1. The names *Oenothera* ×*wienii* Renner ex Rostański (1977) and *O.* ×*hoelscheri* Renner ex Rostański (1968) are both considered to apply to the hybrid *O. rubricaulis* Kleb. × *O. depressa* Greene the types of the two nothospecific names are known to differ by a whole gene complex; nevertheless, the later name is treated as a synonym of the earlier.

Note 1. Variation within nothospecies and nothotaxa of lower rank may be treated according to Art. H.12 or, if appropriate, according to the *International code of nomenclature for cultivated plants-1980.*

Article H.5

H.5.1. The appropriate rank of a nothotaxon is that of the postulated or known parent taxa.

H.5.2. If the postulated or known parent taxa are of unequal rank the appropriate rank of the nothotaxon is the lowest of these ranks.

Note 1. When a taxon is designated by a name in a rank inappropriate to its hybrid formula, the name is incorrect in relation to that hybrid formula but may nevertheless be correct, or may become correct later (see also Art. 52 Note 3).

Ex. 1. The combination *Elymus* ×*laxus* (Fr.) Melderis & D. C. McClint. (1983), based on *Triticum laxum* Fr. (1842), was published for hybrids with the formula *E. farctus* subsp. *boreoatlanticus* (Simonet & Guin.) Melderis × *E. repens* (L.) Gould, so that the combination is in a rank inappropriate to the hybrid formula. It is, however, the correct name applicable to all hybrids between *E. farctus* (Viv.) Melderis and *E. repens*.

Ex. 2. Radcliffe-Smith incorrectly published the nothospecific name *Euphorbia* ×*cornubiensis* Radcl.-Sm. (1985) for *E. amygdaloides* L. × *E. characias* subsp. *wulfenii* (W. D. J. Koch) Radcl.-Sm., although the correct designation for hybrids between *E. amygdaloides* and *E. characias* L. is *E.* ×*martini* Rouy (1900); later, he remedied his mistake by publishing the combination *E.* ×*martini* nothosubsp. *cornubiensis* (Radcl.-Sm.) Radcl.-Sm. (in Taxon 35: 349. 1986). However, the name *E.* ×*cornubiensis* is potentially correct for hybrids with the formula *E. amygdaloides* × *E. wulfenii* W. D. J. Koch.

Recommendation H.5A

H.5A.1. When publishing a name of a new nothotaxon at the rank of species or below, authors should provide any available information on the taxonomic identity, at lower ranks, of the known or postulated parent plants of the type of the name.

Article H.6

H.6.1. A nothogeneric name (i.e. the name at generic rank for a hybrid between representatives of two or more genera) is a condensed formula or is equivalent to a condensed formula.

H.6.2. The nothogeneric name of a bigeneric hybrid is a condensed formula in which the names adopted for the parental genera are combined into a single word, using the first part or the whole of one, the last part or the whole of the other (but not the whole of both) and, optionally, a connecting vowel.

Ex. 1. ×*Agropogon* P. Fourn. (1934) (= *Agrostis* L. × *Polypogon* Desf.); ×*Gymnanacamptis* Asch. & Graebn. (1907) (= *Anacamptis* Rich. × *Gymnadenia* R. Br.); ×*Cupressocyparis* Dallim. (1838) (= *Chamaecyparis* Spach × *Cupressus* L.); ×*Seleniphyllum* G. D. Rowley (1962) (= *Epiphyllum* Haw. × *Selenicereus* (A. Berger) Britton & Rose).

Ex. 2. ×*Amarcrinum* Coutts (1925) is correct for *Amaryllis* L. × *Crinum* L., not "×*Crindonna*". The latter formula was proposed by Ragionieri (1921) for the same nothogenus, but was formed from the generic name adopted for one parent *(Crinum)* and a synonym (*Belladonna* Sweet) of the generic name adopted for the other *(Amaryllis)*. Being contrary to Art. H.6, it is not validly published under Art. 32.1(b).

Ex. 3. The name ×*Leucadenia* Schltr. (1919) is correct for *Leucorchis* E. Mey. × *Gymnadenia* R. Br., but if the generic name *Pseudorchis* Ség. is adopted instead of *Leucorchis*, ×*Pseudadenia* P. F. Hunt (1971) is correct.

Ex. 4. Boivin (1967) published ×*Maltea* for what he considered to be the intergeneric hybrid *Phippsia* (Trin.) R. Br. × *Puccinellia* Parl. As this is not a condensed formula, the name cannot be used for that intergeneric hybrid, for which the correct name is ×*Pucciphippsia* Tzvelev (1971). Boivin did, however, provide a Latin description and designate a type; consequently, *Maltea* B. Boivin is a validly published generic name and is correct if its type is treated as belonging to a separate genus, not to a nothogenus.

H.6.3. The nothogeneric name of an intergeneric hybrid derived from four or more genera is formed from the name of a person to which is added the termination *-ara*; no such name may exceed eight syllables. Such a name is regarded as a condensed formula.

Ex. 5. ×*Potinara* Charlesworth & Co. (1922) (= *Brassavola* R. Br. × *Cattleya* Lindl. × *Laelia* Lindl. × *Sophronitis* Lindl.).

H.6.4. The nothogeneric name of a trigeneric hybrid is either *(a)* a condensed formula in which the three names adopted for the parental genera are combined into a single word not exceeding eight syllables, using the whole or first part of one, followed by the whole or any part of another, followed by the whole or last part of the third (but not the whole of all three) and, optionally, one or two connecting vowels, or *(b)* a name formed like that of a nothogenus derived from four or more genera, i.e., from a personal name to which is added the termination *-ara*.

Ex. 6. ×*Sophrolaeliocattleya* Hurst (1898) (= *Cattleya* Lindl. × *Laelia* Lindl. × *Sophronitis* Lindl.); ×*Vascostylis* Takakura (1964) (= *Ascocentrum* Schltr. ex J. J. Sm. × *Rhynchostylis* Blume × *Vanda* W. Jones ex R. Br.); ×*Rodrettiopsis* Moir (1976) (= *Comparettia* Poepp. & Endl. × *Ionopsis* Kunth × *Rodriguezia* Ruiz & Pav.); ×*Wilsonara* Charlesworth & Co. (1916) (= *Cochlioda* Lindl. × *Odontoglossum* Kunth × *Oncidium* Sw.).

Recommendation H.6A

H.6A.1. When a nothogeneric name is formed from the name of a person by adding the termination *-ara*, that person should preferably be a collector, grower, or student of the group.

Article H.7

H.7.1. The name of a nothotaxon which is a hybrid between subdivisions of a genus is a combination of an epithet, which is a condensed formula formed in the same way as a nothogeneric name (Art. H.6.2), with the name of the genus.

Ex. 1. *Ptilostemon* nothosect. *Platon* Greuter (in Boissiera 22: 159. 1973), comprising hybrids between *Ptilostemon* sect. *Platyrhaphium* Greuter and *Ptilostemon* Cass. sect. *Ptilostemon*; *Ptilostemon* nothosect. *Plinia* Greuter (in Boissiera 22: 158. 1973), comprising hybrids between *Ptilostemon* sect. *Platyrhaphium* and *P.* sect. *Cassinia* Greuter.

Article H.8

H.8.1. When the name or the epithet in the name of a nothotaxon is a condensed formula (Art. H.6 and H.7), the parental names used in its formation must be those which are correct for the particular circumscription, position, and rank accepted for the parental taxa.

Ex. 1. If the genus *Triticum* L. is interpreted on taxonomic grounds as including *Triticum* (s. str.) and *Agropyron* Gaertn., and the genus *Hordeum* L. as including *Hordeum* (s. str.) and *Elymus* L., then hybrids between *Agropyron* and *Elymus* as well as between *Triticum* (s. str.) and *Hordeum* (s. str.) are placed in the same nothogenus, ×*Tritordeum* Asch. & Graebn. (1902). If, however, *Agropyron* is separated generically from *Triticum,* hybrids between *Agropyron* and *Hordeum* (s. str. or s. lat.) are placed in the nothogenus ×*Agrohordeum* A. Camus (1927). Similarly, if *Elymus* is separated generically from *Hordeum,* hybrids between *Elymus* and *Triticum* (s. str. or s. lat.) are placed in the nothogenus ×*Elymotriticum* P. Fourn. (1935). If both *Agropyron* and *Elymus* are given generic rank, hybrids between them are placed in the nothogenus ×*Agroelymus* A. Camus (1927); ×*Tritordeum* is then restricted to hybrids between *Hordeum* (s. str.) and *Triticum* (s. str.), and hybrids between *Elymus* and *Hordeum* are placed in ×*Elyhordeum* Mansf. ex Tsitsin & Petrova (1955), a substitute name for ×*Hordelymus* Bachteev & Darevsk. (1950) non *Hordelymus* (K. Jess.) K. Jess. (1885).

H.8.2. Names ending in *-ara* for nothogenera, which are equivalent to condensed formulae (Art. H.6.3-4), are applicable only to plants which are accepted taxonomically as derived from the parents named.

Ex. 2. If *Euanthe* Schltr. is recognized as a distinct genus, hybrids simultaneously involving its only species, *E. sanderiana* (Rchb.) Schltr., and the three genera *Arachnis* Blume, *Renanthera* Lour., and *Vanda* W. Jones ex R. Br. must be placed in ×*Cogniauxara* Garay & H. R. Sweet (1966); if, on the other hand, *E. sanderiana* is included in *Vanda,* the same hybrids are placed in ×*Holttumara* anon. (1958) *(Arachnis* × *Renanthera* × *Vanda).*

Article H.9

H.9.1. In order to be validly published, the name of a nothogenus or of a nothotaxon with the rank of subdivision of a genus (Art. H.6 and H.7) must be effectively published (see Art. 29-31) with a statement of the names of the parent genera or subdivisions of genera, but no description or diagnosis is necessary, whether in Latin or in any other language.

Ex. 1. Validly published names: ×*Philageria* Mast. (1872), published with a statement of parentage, *Lapageria* Ruiz & Pav. × *Philesia* Comm. ex Juss.; *Eryngium* nothosect. *Alpestria* Burdet & Miège, pro sect. (in Candollea 23: 116. 1968), published with a statement of its parentage, *E.* sect. *Alpina* H. Wolff × *E.* sect. *Campestria* H. Wolff; ×*Agrohordeum* A. Camus (1927) (= *Agropyron* Gaertn. × *Hordeum* L.), of which ×*Hordeopyron* Simonet (1935, *"Hordeopyrum"*) is a later synonym.

Note 1. Since the names of nothogenera and nothotaxa with the rank of a subdivision of a genus are condensed formulae or treated as such, they do not have types.

Ex. 2. The name ×*Ericalluna* Krüssm. (1960) was published for plants (×*E. bealei* Krüssm.) which were thought to be the product of the cross *Calluna vulgaris* (L.) Hull × *Erica cinerea* L. If it is considered that these are not hybrids, but are variants of *E. cinerea,* the name ×*Ericalluna* Krüssm. remains available for use if and when known or postulated plants of *Calluna* Salisb. × *Erica* L. should appear.

Ex. 3. ×*Arabidobrassica* Gleba & Fr. Hoffm. (in Naturwissenschaften 66: 548. 1979), a nothogeneric name which was validly published with a statement of parentage for the result of somatic hybridization by protoplast fusion of *Arabidopsis thaliana* (L.) Heynh. with *Brassica campestris* L., is also available for intergeneric hybrids resulting from normal crosses between *Arabidopsis* Heynh. and *Brassica* L., should any be produced.

Note 2. However, names published merely in anticipation of the existence of a hybrid are not validly published under Art. 34.1(b).

Article H.10

H.10.1. Names of nothotaxa at the rank of species or below must conform with the provisions *(a)* in the body of the *Code* applicable to the same ranks and *(b)* in Art. H.3. Infringements of Art. H.3.1. are treated as errors to be corrected.

H.10.2. Taxa previously published as species or infraspecific taxa which are later considered to be nothotaxa may be indicated as such, without change of rank, in conformity with Art. 3 and 4 and by the application of Art. 50 (which also operates in the reverse direction).

H.10.3. The following are considered to be formulae and not true epithets: designations consisting of the epithets of the names of the parents combined in unaltered form by a hyphen, or with only the termination of one epithet changed, or consisting of the specific epithet of the name of one parent combined with the generic name of the other (with or without change of termination).

Ex. 1. The designation *Potentilla "atrosanguinea-pedata"* published by Maund (in Bot. Gard. 5: No. 385, t. 97. 1833) is considered to be a formula meaning *Potentilla atrosanguinea* Lodd. ex D. Don × *P. pedata* Nestl.

Ex. 2. *Verbascum "nigro-lychnitis"* (Schiede, Pl. Hybr.: 40. 1825) is considered to be a formula, *Verbascum lychnitis* L. × *V. nigrum* L.; the correct binary name for this hybrid is *Verbascum* ×*schiedeanum* W. D. J. Koch (1844).

Ex. 3. The following names include true epithets: *Acaena* ×*anserovina* Orchard (1969) (from *A. anserinifolia* (J. R. Forst. & G. Forst.) Druce and *A. ovina* A. Cunn.); *Micromeria* ×*benthamineolens* Svent. (1969) (from *M. benthamii* Webb & Berthel. and *M. pineolens* Svent.).

Note 1. Since the name of a nothotaxon at the rank of species or below has a type, statements of parentage play a secondary part in determining the application of the name.

Ex. 4. *Quercus* ×*deamii* Trel. (in Mem. Natl. Acad. Sci. 20: 14. 1924) when described was considered as the cross *Q. alba* L. × *Q. muehlenbergii* Engelm. However, progeny grown from acorns from the tree from which the type originated led Bartlett to conclude that the parents were in fact *Q. macrocarpa* Michx. and *Q. muehlenbergii*. If this conclusion is accepted, the name *Q.* ×*deamii* applies to *Q. macrocarpa* × *Q. muehlenbergii*, and not to *Q. alba* × *Q. muehlenbergii*.

Recommendation H.10A

H.10A.1. In forming epithets for names of nothotaxa at the rank of species and below, authors should avoid combining parts of the epithets of the names of the parents.

Recommendation H.10B

H.10B.1. When contemplating the publication of new names for hybrids between named infraspecific taxa, authors should carefully consider whether they are really needed, bearing in mind that formulae, though more cumbersome, are more informative.

Article H.11

H.11.1. The name of a nothospecies of which the postulated or known parent species belong to different genera is a combination of a nothospecific (collective) epithet with a nothogeneric name.

Ex. 1. ×*Heucherella tiarelloides* (Lemoine & E. Lemoine) H. R. Wehrh. is considered to have originated form the cross between a garden hybrid of *Heuchera* L. and *Tiarella cordifolia* L. (see Stearn in Bot. Mag. 165: ad t. 31. 1948). Its original name, *Heuchera* ×*tiarelloides* Lemoine & E. Lemoine (1912), is therefore incorrect.

Ex. 2. When *Orchis fuchsii* Druce was renamed *Dactylorhiza fuchsii* (Druce) Soó the name for its hybrid with *Coeloglossum viride* (L.) Hartm., ×*Orchicoeloglossum mixtum* Asch. & Graebn. (1907), became the basis of the necessary new combination ×*Dactyloglossum mixtum* (Asch. & Graebn.) Rauschert (1969).

H.11.2. The final epithet in the name of an infraspecific nothotaxon, of which the postulated or known parental taxa are assigned to different taxa at a higher rank, may be placed subordinate to the name of a nothotaxon at that higher rank (see Art. 24.1), e.g to a nothospecific name (but see Rec. H.10B).

Ex. 3. Mentha ×*piperita* L. nothosubsp. *piperita* (= *M. aquatica* L. × *M. spicata* L. subsp. *spicata); Mentha* ×*piperita* nothosubsp. *pyramidalis* (Ten.) Harley (in Kew Bull. 37: 604. 1983) (= *M. aquatica* L. × *M. spicata* subsp. *tomentosa* (Briq.) Harley).

Article H.12

H.12.1. Subordinate taxa within nothotaxa of specific or infraspecific rank may be recognized without an obligation to specify parent taxa at the subordinate rank. In this case non-hybrid infraspecific categories of the appropriate rank are used.

Ex. 1. Mentha ×*piperita* f. *hirsuta* Sole; *Populus* ×*canadensis* var. *serotina* (R. Hartig) Rehder and *P.* ×*canadensis* var. *marilandica* (Poir.) Rehder (see also Art. H.4 Note 1).

Note 1. As there is no statement of parentage at the rank concerned there is no control of circumscription at this rank by parentage (compare Art. H.4).

Note 2. It is not feasible to treat subdivisions of nothospecies by the methods of both Art. H.10 and H.12.1 at the same rank.

H.12.2. Names published at the rank of nothomorph[1] are treated as having been published as names of varieties (see Art. 50).

[1] Previous editions of the *Code* (1978, Art. H.10, and the corresponding article in earlier editions) permitted only one rank under provisions equivalent to H.12. That rank was equivalent to variety and the category was termed "nothomorph".

APPENDIX IIA

NOMINA FAMILIARUM ALGARUM, FUNGORUM ET PTERIDOPHYTORUM CONSERVANDA ET REJICIENDA

In the following lists the ***nomina conservanda*** have been inserted in the left column, in ***bold-face italics***. Synonyms and earlier homonyms *(nomina rejicienda)* are listed in the right column.

(H) homonym (Art. 14.10; see also Art. 53), only the earliest being listed.

(=) taxonomic synonym, based on a type different from that of the conserved name to be rejected only in favour of the conserved name (Art. 14.6 and 14.7).

One name listed as conserved has no corresponding *nomen rejiciendum* because its conservation is no longer necessary (Art. 52.3; see Art. 14.13).

A. ALGAE

Acrochaetiaceae Fritsch ex W. R. Taylor, Mar. Alg. N.E. N. Amer., ed. 2: 209, 210. 1957 [*Rhodoph.*].
Typus: *Acrochaetium* Nägeli

(=) *Rhodochortaceae* Nasr in Bull. Fac. Sci. Egypt. Univ. 26: 902. 1947.
Typus: *Rhodochorton* Nägeli

Bangiaceae Engl., Syllabus: 16. Apr 1892 [*Rhodoph.*].
Typus: *Bangia* Lyngb.

(=) *Porphyraceae* Kütz., Phycol. General.: 382. 14-16 Sep 1843.
Typus: *Porphyra* C. Agardh *(nom. cons.)*.

Chromulinaceae Engl. in Engler & Prantl, Nat. Pflanzenfam. 1(2): 570. Dec 1897 [*Chrysoph.*].
Typus: *Chromulina* Cienk.

(=) *Chrysomonadaceae* F. Stein, Organism. Infusionsthiere 3(1): x, 152. 1878.
Typus: *Chrysomonas* F. Stein

Cladophoraceae Wille in Warming, Haandb. Syst. Bot., ed. 2: 30. 1884 [*Chloroph.*].
Typus: *Cladophora* Kütz. *(nom. cons.).*

(=) *Pithophoraceae* Wittr. in Nova Acta Regiae Soc. Sci. Upsal., ser. 3, vol. extra ord. (19): 47. 1877.
Typus: *Pithophora* Wittr.

(=) *Confervaceae* Dumort., Comment. Bot.: 71, 96. Nov (sero) - Dec (prim.) 1822.
Typus: *Conferva* L.

Euglenaceae Dujard., Hist. Nat. Zoophyt.: 347. 1841 [*Euglenoph.*].
Typus: *Euglena* Ehrenb.

(=) *Astasiaceae* Ehrenb., Symb. Phys., Zool.: [32]. 1831.
Typus: *Astasia* Ehrenb. *(nom. rej.).*

Eupodiscaceae Ralfs in Pritchard, Hist. Infus., ed. 4: 758, 842. 1861 [*Bacillarioph.*].
Typus: *Eupodiscus* Bailey *(nom. cons.).*

(H) *Eupodiscaceae* Kütz., Sp. Alg.: 134. 23-24 Jul 1849.
Typus: *Eupodiscus* Ehrenb. *(nom. rej.).*

Isochrysidaceae Bourrelly in Rev. Algol., Mém. Hors-Sér. 1: 227, 228. Oct-Dec 1957 [*Prymnesioph./Haptoph.*].
Typus: *Isochrysis* Parke

(=) *Ruttneraceae* Geitler in Int. Rev. Gesamten Hydrobiol. Hydrogr. 43: 108. 1943.
Typus: *Ruttnera* Geitler

Lomentariaceae J. Agardh, Spec. Gen. Ord. Alg. 3(1): 606, 630. Apr-Aug 1876 [*Rhodoph.*].
Typus: *Lomentaria* Lyngb.

(=) *Chondrosiphonaceae* Kütz., Phycol. General.: 438. 14-16 Sep 1843.
Typus: *Chondrosiphon* Kütz.

Nemastomataceae F. Schmitz in Engler, Syllabus: 22. Apr 1892 [*Rhodoph.*].
Typus: *Nemastoma* J. Agardh *(nom. cons.).*

(=) *Gymnophlaeaceae* Kütz., Phycol. General.: 389, 390. 14-16 Sep 1843.
Typus: *Gymnophlaea* Kütz.

Ochromonadaceae Lemmerm. in Forschungsber. Biol. Stat. Plön 7: 105. 1899 [*Chrysoph.*].
Typus: *Ochromonas* Vysotskij

(=) *Dendromonadaceae* F. Stein, Organism. Infusionsthiere 3(1): x, 153. 1878.
Typus: *Dendromonas* F. Stein

(=) *Spumellaceae* Kent, Man. Infus.: 231. 1880.
Typus: *Spumella* Cienk.

Oscillatoriaceae Engl., Syllabus, ed. 2: 6. Mai 1898 [*Cyanoph.*].
Typus: *Oscillatoria* Vaucher ex Gomont

(=) *Lyngbyaceae* (Gomont) Hansg. in Nuova Notarisia 3: 1. 5 Jan 1892 (trib. *Lyngbyeae* Gomont in Ann. Sci. Nat., Bot., ser. 7, 15: 290. 1 Jan 1892).
Typus: *Lyngbya* C. Agardh ex Gomont

Plocamiaceae Kütz., Phycol. General.: 442, 449. 14-16 Sep 1843 [*Rhodoph.*].
Typus: *Plocamium* J. V. Lamour. *(nom. cons.).*

(=) *Thamnophoraceae* Decne. in Ann. Sci. Nat., Bot., ser. 2, 17: 359, 364. Jun 1842.
Typus: *Thamnophora* C. Agardh

Polyidaceae Kylin, Gatt. Rhodophyc.: 142, 166. 1956 [*Rhodoph.*].
Typus: *Polyides* C. Agardh

(=) *Spongiocarpaceae* Grev., Alg. Brit.: 68. Mar 1830.
Typus: *Spongiocarpus* Grev.

Retortamonadaceae Wenrich in Trans. Amer. Microscop. Soc. 51: 233. Oct 1932 [*Trichomonadoph.*].
Typus: *Retortamonas* Grassi

(=) *Embadomonadaceae* A. G. Alexeev in Compt.-Rend. Séances Mém. Soc. Biol. 80: 358. 31 Mar 1917.
Typus: *Embadomonas* D. L. Mackinnon

(=) *Chilomastigaceae* Wenyon, Protozool.: 268, 287, 620. 1926.
Typus: *Chilomastix* A. G. Alexeev

Rhodomelaceae Aresch. in Nova Acta Regiae Soc. Sci. Upsal. 13: 260. 1847 [*Rhodoph.*].
Typus: *Rhodomela* C. Agardh *(nom. cons.)*.

(=) *Rytiphlaeaceae* Decne. in Arch. Mus. Hist. Nat. 2: 161, 171. 1841.
Typus: *Rytiphlaea* C. Agardh

(=) *Heterocladiaceae* Decne. in Ann. Sci. Nat., Bot., ser. 2, 17: 359, 364. Jun 1842.
Typus: *Heterocladia* Decne.

(=) *Polyphacaceae* Decne. in Ann. Sci. Nat., Bot., ser. 2, 17: 359, 363. Jun 1842.
Typus: *Polyphacus* C. Agardh

(=) *Polysiphoniaceae* Kütz., Phycol. General.: 413, 416. 14-16 Sep 1843.
Typus: *Polysiphonia* Grev. *(nom. cons.)*.

(=) *Amansiaceae* Kütz., Phycol. General.: 442, 447. 14-16 Sep 1843.
Typus: *Amansia* J. V. Lamour.

Siphonocladaceae F. Schmitz in Ber. Sitzungen Naturf. Ges. Halle 1878: 20. 1879 [*Chloroph.*].
Typus: *Siphonocladus* F. Schmitz

Stigonemataceae (Bornet & Flahault) Borzì in Nuova Notarisia 3: 43. 5 Apr 1892 (subtrib. *Stigonematinae* Bornet & Flahault in Ann. Sci. Nat., Bot., ser. 7, 5: 53. 1 Jan 1886. [*Cyanoph.*].
Typus: *Stigonema* C. Agardh ex Bornet & Flahault

(=) *Sirosiphonaceae* A. B. Frank in Leunis, Syn. Pflanzenk., ed. 3, 3: 215, 217. Aug 1886.
Typus: *Sirosiphon* Kütz. ex A. B. Frank

Tetrasporaceae Wittr. in Bih. Kongl. Svenska Vetensk.-Akad. Handl. 1(1): 28. 1872 [*Chloroph.*].
Typus: *Tetraspora* Link ex Desv.

(=) *Palmellaceae* Decne. in Ann. Sci. Nat., Bot., ser. 2, 17: 327, 333. Jun 1842.
Typus: *Palmella* Lyngb.

Trentepohliaceae Hansg., Prodr. Algenfl. Böhmen 1: 37, 85. Jan-Apr 1886 [*Chloroph.*].
Typus: *Trentepohlia* Mart.

(=) *Byssaceae* Adans., Fam. Pl. 2: 1. Jul-Aug 1763.
Typus: *Byssus* L.

Vacuolariaceae Luther in Bih. Kongl. Svenska Vetensk. Akad. Handl. 24 (sect. 3, 13): 19. 1899 [*Raphidoph.*].
Typus: *Vacuolaria* Cienk.

(=) *Coelomonadaceae* Buetschli in Bronn, Kl. Ordn. Thier-Reichs 1: 819. 1884.
Typus: *Coelomonas* F. Stein

B. FUNGI

Corticiaceae Herter in Warnstorf & al., Krypt.-Fl. Brandenburg 6: 70. 30 Jan 1910.
Typus: *Corticium* Pers.

(=) *Cyphellaceae* Lotsy, Vortr. Bot. Stammesgesch. 1: 695, 696. Jan-Mar 1907.
Typus: *Cyphella* Fr. : Fr.

(=) *Peniophoraceae* Lotsy, Vortr. Bot. Stammesgesch. 1: 687, 689. Jan-Mar 1907.
Typus: *Peniophora* Cooke

(=) *Vuilleminiaceae* Maire ex Lotsy, Vortr. Bot. Stammesgesch. 1: 678. Jan-Mar 1907.
Typus: *Vuilleminia* Maire

Cortinariaceae R. Heim ex Pouzar in Česká Mykol. 37: 174. 28 Jul 1983.
Typus: *Cortinarius* (Pers.) Gray

(=) *Crepidotaceae* (S. Imai) Singer in Lilloa 22: 584. 1951 (trib. *Crepidoteae* S. Imai in J. Fac. Agric. Hokkaido Univ. 43: 238. Aug 1938).
Typus: *Crepidotus* (Fr.) Staude

(=) *Galeropsidaceae* Singer in Bol. Soc. Argent. Bot. 10: 61. 15 Aug 1962.
Typus: *Galeropsis* Velen.

(=) *Thaxterogasteraceae* Singer in Bol. Soc. Argent. Bot. 10: 63. 1962.
Typus: *Thaxterogaster* Singer

(=) *Hebelomataceae* Locq., Fl. Mycol. 3: 146. 1977.
Typus: *Hebeloma* (Fr.) P. Kumm.

(=) *Inocybaceae* Jülich in Biblioth. Mycol. 85: 374. 1 Feb 1982.
Typus: *Inocybe* (Fr.) Fr.

(=) *Verrucosporaceae* Jülich in Biblioth. Mycol. 85: 393. 1 Feb 1982.
Typus: *Verrucospora* E. Horak

Gnomoniaceae G. Winter in Rabenh. Krypt.-Fl., ed. 2, 1(2): 570. Mai 1886.
Typus: *Gnomonia* Ces. & De Not.

(=) *Obryzaceae* Körber, Syst. Lich. Germ: 427. Nov-Dec 1855.
Typus: *Obryzum* Wallr.

Mycosphaerellaceae Lindau in Engler & Prantl, Nat. Pflanzenfam. 1(1): 421. Jun 1897.
Typus: *Mycosphaerella* Johanson

(=) *Ascosporaceae* Bonord., Handb. Mykol.: 62. 1851 (sero).
Typus: *Ascospora* Fr.

Physciaceae Zahlbr. in Engler, Syllabus, ed. 2: 46. Mai 1898.
Typus: *Physcia* (Schreb.) Michx.

(=) *Pyxinaceae* (Fr.) Stizenb. in Ber. Tätigk. St. Gallischen Naturwiss. Ges. 1862: 156. 1862 (trib. *Pyxineae* Fr., Syst. Orb. Veg.: 266. Dec 1825).
Typus: *Pyxine* Fr.

Rhytismataceae Chevall., Fl. Gén. Env. Paris 1: 439. 5 Aug 1826.
Typus: *Rhytisma* Fr. : Fr.

(=) *Xylomataceae* Fr., Scleromyceti Sveciae 2: p. post titulum. 1820.
Typus: *Xyloma* Pers. : Fr.

Taphrinaceae Gäum. & C. W. Dodge, Compar. Morph. Fungi: 161. 1928.
Typus: *Taphrina* Fr. : Fr.

(=) *Exoascaceae* G. Winter in Rabenh. Krypt.-Fl. 1(2): 3. Mar 1884.
Typus: *Exoascus* Fuckel

Trapeliaceae M. Choisy ex Hertel in Vortr. Gesamtgeb. Bot., ser. 2, 4: 181. Jun 1970.
Typus: *Trapelia* M. Choisy

(=) *Saccomorphaceae* Elenkin in Ber. Biol. Süsswasser-Stat. Kaiserl. Naturf.-Ges. St. Petersburg 3: 193. 1912.
Typus: *Saccomorpha* Elenkin

Tricholomataceae R. Heim ex Pouzar in Česká Mykol. 37: 175. 28 Jul 1983.
Typus: *Tricholoma* (Fr.) Staude *(nom. cons.)*.

(=) *Hydnangiaceae* Gäum. & C. W. Dodge, Compar. Morph. Fungi: 485. 1928.
Typus: *Hydnangium* Wallr.

(=) *Physalacriaceae* Corner in Beih. Nova Hedwigia 33: 10. Jan 1970.
Typus: *Physalacria* Peck

(=) *Amparoinaceae* Singer in Rev. Mycol. 40: 58. 1976.
Typus: *Amparoina* Singer

(=) *Dermolomataceae* (Bon) Bon in Doc. Mycol. 9(35): 43. 1979.
Typus: *Dermoloma* (J. E. Lange) Herink

(=) *Macrocystidiaceae* Kühner in Bull. Mens. Soc. Linn. Lyon, Soc. Bot. 48: 172. Mar 1979.
Typus: *Macrocystidia* Joss.

(=) *Rhodotaceae* Kühner in Bull. Mens. Soc. Linn. Lyon, Soc. Bot. 49: 235. Apr 1980.
Typus: *Rhodotus* Maire

(=) *Pleurotaceae* Kühner in Bull. Mens. Soc. Linn. Lyon, Soc. Bot. 49: 184. Mar 1980.
Typus: *Pleurotus* (Fr. : Fr.) P. Kumm. *(nom. cons.)*.

(=) *Marasmiaceae* Kühner in Bull. Mens. Soc. Linn. Lyon, Soc. Bot. 49: 76. Feb 1980.
Typus: *Marasmius* Fr. *(nom. cons.).*

(=) *Hygrophoropsidaceae* Kühner in Bull. Mens. Soc. Linn. Lyon, Soc. Bot. 49: 414. Sep 1980.
Typus: *Hygrophoropsis* (J. Schröt.) Maire ex Martin-Sans

(=) *Biannulariaceae* Jülich in Biblioth. Mycol. 85: 356. 1 Feb 1982.
Typus: *Biannularia* Beck

(=) *Cyphellopsidaceae* Jülich in Biblioth. Mycol. 85: 362. 1 Feb 1982.
Typus: *Cyphellopsis* Donk

(=) *Fayodiaceae* Jülich in Biblioth. Mycol. 85: 367. 1 Feb 1982.
Typus: *Fayodia* Kühner

(=) *Laccariaceae* Jülich in Biblioth. Mycol. 85: 374. 1 Feb 1982.
Typus: *Laccaria* Berk. & Broome

(=) *Lentinaceae* Jülich in Biblioth. Mycol. 85: 376. 1 Feb 1982.
Typus: *Lentinus* Fr. : Fr.

(=) *Leucopaxillaceae* (Singer) Jülich in Biblioth. Mycol. 85: 376. 1 Feb 1982.
Typus: *Leucopaxillus* Boursier

(=) *Lyophyllaceae* Jülich in Biblioth. Mycol. 85: 378. 1 Feb 1982.
Typus: *Lyophyllum* P. Karst.

(=) *Nyctalidaceae* Jülich in Biblioth. Mycol. 85: 381. 1 Feb 1982.
Typus: *Nyctalis* Fr.

(=) *Panellaceae* Jülich in Biblioth. Mycol. 85: 382. 1 Feb 1982.
Typus: *Panellus* P. Karst.

(=) *Resupinataceae* (Singer) Jülich in Biblioth. Mycol. 85: 388. 1 Feb 1982.
Typus: *Resupinatus* (Nees) Gray

(=) *Squamanitaceae* Jülich in Biblioth. Mycol. 85: 390. 1 Feb 1982.
Typus: *Squamanita* Imbach

(=) *Termitomycetaceae* Jülich in Biblioth. Mycol. 85: 391. 1 Feb 1982.
Typus: *Termitomyces* R. Heim

(=) *Xerulaceae* Jülich in Biblioth. Mycol. 85: 394. 1 Feb 1982.
Typus: *Xerula* Maire

D. PTERIDOPHYTA

Adiantaceae Newman, Hist. Brit. Ferns: 5. 1-5 Feb 1840.
Typus: *Adiantum* L.

(=) *Parkeriaceae* Hook., Exot. Fl. 2: ad t. 147. Mar 1825.
Typus: *Parkeria* Hook.

Dicksoniaceae (C. Presl) Bower, Origin Land Fl.: 591. 1908 (*Dicksonieae* C. Presl [Tent. Pterid.] in Abh. Königl. Böhm. Ges. Wiss., ser. 4, 5: 133. 1836 ([seors. impr.] ante 2 Dec)).
Typus: *Dicksonia* L'Hér.

(=) *Thyrsopteridaceae* C. Presl, Gefäßbündel Farrn: 22, 38. 1847.
Typus: *Thyrsopteris* Kunze

Dryopteridaceae Ching in Acta Phytotax. Sin. 10: 1. Jan 1965.
Typus: *Dryopteris* Adans. *(nom. cons.)*.

(=) *Peranemataceae* (C. Presl) Ching in Sunyatsenia 5: 246. 30 Oct 1940 *(nom. cons.)*.

Peranemataceae (C. Presl) Ching in Sunyatsenia 5: 246. 30 Oct 1940 (trib. *Peranemateae* C. Presl [Tent. Pterid.] in Abh. Königl. Böhm. Ges. Wiss., ser. 4, 5: 64. 1836 ([seors. impr.] ante 2 Dec)).
Typus: *Peranema* D. Don

(H) *Peranemataceae* Buetschli in Bronn, Kl. Ordn. Thier-Reichs 1: 824. 1884, nom. illeg. [*Algae, Euglenoph.*].
Typus: *Peranema* Dujard., non D. Don

APPENDIX IIB

NOMINA FAMILIARUM BRYOPHYTORUM ET SPERMATOPHYTORUM CONSERVANDA

The names of families printed in ***bold-face italics*** are to be retained in all cases, with priority over unlisted synonyms (Art. 14.5) and homonyms (Art. 14.10).

When two listed names compete, the earlier must be retained unless the contrary is indicated or one of the competing names is listed in Art. 18.5. For any family including the type of an alternative family name, one or the other of these alternative names is to be used.

For purposes of this list the starting point for *Spermatophyta* is Jussieu's *Genera plantarum* (4 Aug 1789); earlier usage of listed names has been disregarded, irrespective of the fact that for family names the starting point is the same as for all other names of *Spermatophyta*, 1 May 1753. A decision of the Tokyo Congress authorizes, for the time being, the maintenance of authorship and dates of publication of names as listed in this Appendix (Art. 14 Note 1, footnote).

C. BRYOPHYTA

Bryoxiphiaceae Besch. in J. Bot. (Morot) 6: 183. 1892 [*Musci*].
Typus: *Bryoxiphium* Mitt. *(nom. cons.)*.

Ditrichaceae Limpr., Laubm. Deutschl. 1: 482. Oct 1887 [*Musci*].
Typus: *Ditrichum* Hampe *(nom. cons.)*.

Entodontaceae Kindb., Gen. Eur. N.-Amer. Bryin.: 7. 1897 [*Musci*].
Typus: *Entodon* Müll. Hal.

Ephemeraceae Schimp., Coroll. Bryol. Eur.: 3. 1856 [*Musci*].
Typus: *Ephemerum* Hampe *(nom. cons.)*.

Eustichiaceae Broth. in Engler & Prantl, Nat. Pflanzenfam., ed. 2, 10: 420. Mai-Jun 1924 [*Musci*].
Typus: *Eustichia* Mitt.

Fossombroniaceae Hazsl., Magyar Birodalom Moh-Flórája: 20, 36. Mai-Dec 1885 [*Hepat.*].
Typus: *Fossombronia* Raddi

Lejeuneaceae Casares-Gil, Fl. Ibér. Brióf. 1: 703. Dec 1919 [*Hepat.*].
Typus: *Lejeunea* Lib. *(nom. cons.).*

Porellaceae Cavers in New Phytol. 9: 292. 1910 [*Hepat.*].
Typus: *Porella* L.

Pottiaceae Schimp., Coroll. Bryol. Eur.: 24. 1856 [*Musci*].
Typus: *Pottia* (Ehrh. ex Rchb.) Fürnr.

Sematophyllaceae Broth. in Engler & Prantl, Nat. Pflanzenfam. 1(3): 1098. 10 Nov 1908 [*Musci*].
Typus: *Sematophyllum* Mitt.

E. SPERMATOPHYTA

Abietaceae Bercht. & J. Presl, Přir. Rostlin 1: 262. 1820.
Typus: *Abies* Mill.
Note: If this family is united with *Pinaceae,* the name *Abietaceae* is rejected in favour of *Pinaceae.*

Acanthaceae Juss., Gen. Pl.: 102. 4 Aug 1789.
Typus: *Acanthus* L.

Aceraceae Juss., Gen. Pl.: 250. 4 Aug 1789.
Typus: *Acer* L.

Achariaceae Harms in Engler & Prantl, Nat. Pflanzenfam., Nachtr. 1: 256. Oct 1897.
Typus: *Acharia* Thunb.

Achatocarpaceae Heimerl in Engler & Prantl, Nat. Pflanzenfam., ed. 2, 16c: 174. Jan-Apr 1934.
Typus: *Achatocarpus* Triana

Actinidiaceae Hutch., Fam. Fl. Pl. 1: 177. Feb-Mar 1926.
Typus: *Actinidia* Lindl.
Note: If this family is united with *Sarauiaceae,* the name *Actinidiaceae* must be used.

Adoxaceae Trautv., Estestv. Istorija Gub. Kievsk. Učebn. Okr.: 35. 1853.
Typus: *Adoxa* L.

Aextoxicaceae Engl. & Gilg in Engler, Syllabus, ed. 8: 250. Jan-Feb 1920.
Typus: *Aextoxicon* Ruiz & Pav.

Agavaceae Endl., Ench. Bot.: 105. 15-21 Aug 1841.
Typus: *Agave* L.

Aizoaceae F. Rudolphi, Syst. Orb. Veg.: 53. 1830.
Typus: *Aizoon* L.

Akaniaceae Stapf in Bull. Misc. Inform. 1912: 380. 13 Dec 1912.
Typus: *Akania* Hook. f.

Alangiaceae DC., Prodr. 3: 203. Mar (med.) 1828.
Typus: *Alangium* Lam. *(nom. cons.).*

Alismataceae Vent., Tabl. Règne Vég. 2: 157. 5 Mai 1799.
Typus: *Alisma* L.

Alliaceae J. Agardh, Theoria Syst. Pl.: 32. Apr-Sep 1858.
Typus: *Allium* L.

Alsinaceae (DC.) Bartl. in Bartling & Wendland, Beitr. Bot. 2: 159. Dec 1825 (trib. *Alsineae* DC., Prodr. 1: 388. Jan 1824).
Typus: *Alsine* L. [= *Stellaria* L.].

Alstroemeriaceae Dumort., Anal. Fam. Pl.: 57, 58. 1829.
Typus: *Alstroemeria* L.

Altingiaceae Lindl., Veg. Kingd.: 253. Jan-Mai 1846.
Typus: *Altingia* Noronha

Amaranthaceae Juss., Gen. Pl.: 87. 4 Aug 1789.
Typus: *Amaranthus* L.

Amaryllidaceae J. St.-Hil., Expos. Fam. Nat. 1: 134. Feb-Apr 1805.
Typus: *Amaryllis* L.

Amborellaceae Pichon in Bull. Mus. Natl. Hist. Nat., ser. 2, 20: 384. 25 Oct 1948.
Typus: *Amborella* Baill.

Ambrosiaceae Link, Handbuch 1: 816. Jan-Aug 1829.
Typus: *Ambrosia* L.

Amygdalaceae (Juss.) D. Don, Prodr. Fl. Nepal.: 239. 26 Jan - 1 Feb 1825 ([unranked] *Amygdaleae* Juss., Gen. Pl.: 340. 4 Aug 1789).
Typus: *Amygdalus* L.

Anacardiaceae Lindl., Intr. Nat. Syst. Bot.: 127. Sep 1830.
Typus: *Anacardium* L.

Ancistrocladaceae Planch. ex Walp. in Ann. Bot. Syst. 2: 175. 15-16 Dec 1851.
Typus: *Ancistrocladus* Wall. *(nom. cons.)*.

Annonaceae Juss., Gen. Pl.: 283. 4 Aug 1789.
Typus: *Annona* L.

Apiaceae Lindl., Intr. Nat. Syst. Bot., ed. 2: 21. Jul 1836; nom. alt.: *Umbelliferae*.
Typus: *Apium* L.

Apocynaceae Juss., Gen. Pl.: 143. 4 Aug 1789.
Typus: *Apocynum* L.

Aponogetonaceae J. Agardh, Theoria Syst. Pl.: 44. Apr-Sep 1858.
Typus: *Aponogeton* L. f. *(nom. cons.)*.

Apostasiaceae Lindl., Nix. Pl.: 22. 17 Sep 1833.
Typus: *Apostasia* Blume

Aquifoliaceae Bartl., Ord. Nat. Pl.: 228, 376. Sep 1830.
Typus: *Aquifolium* Mill., nom. illeg. (*Ilex* L.).

Araceae Juss., Gen. Pl.: 23. 4 Aug 1789.
Typus: *Arum* L.

Araliaceae Juss., Gen. Pl.: 217. 4 Aug 1789.
Typus: *Aralia* L.

Araucariaceae Henkel & W. Hochst., Syn. Nadelhölz.: xvii, 1. Jan 1865.
Typus: *Araucaria* Juss.

Arecaceae Schultz Sch., Nat. Syst. Pflanzenr.: 317. 1832; nom. alt.: *Palmae*.
Typus: *Areca* L.

Aristolochiaceae Juss., Gen. Pl.: 72. 4 Aug 1789.
Typus: *Aristolochia* L.

Asclepiadaceae R. Br., Asclepiadeae: 12, 19. 3 Apr 1810.
Typus: *Asclepias* L.

Asparagaceae Juss., Gen. Pl.: 40. 4 Aug 1789.
Typus: *Asparagus* L.

Asteraceae Dumort., Comment. Bot.: 55. Nov (sero) - Dec (prim.) 1822; nom. alt.: *Compositae*.
Typus: *Aster* L.

Asteranthaceae Knuth in Engler, Pflanzenr. 105: 1. 22 Aug 1939.
Typus: *Asteranthos* Desf.

Austrobaileyaceae (Croizat) Croizat in Cact. Succ. J. (Los Angeles) 15: 64. Mai 1943 (subfam. *Austrobaileyoideae* Croizat in J. Arnold Arbor. 21: 404. 24 Jul 1940).
Typus: *Austrobaileya* C. T. White

Avicenniaceae Endl., Ench. Bot.: 314. 15-21 Aug 1841.
Typus: *Avicennia* L.

Balanitaceae Endl., Ench. Bot.: 547. 15-21 Aug 1841.
Typus: *Balanites* Delile *(nom. cons.)*.

Balanopaceae Benth. & Hook. f., Gen. Pl. 3: v, 341. 7 Feb 1880.
Typus: *Balanops* Baill.

Balanophoraceae Rich. in Mém. Mus. Hist. Nat. 8: 429. Nov 1822.
Typus: *Balanophora* J. R. Forst. & G. Forst.

Balsaminaceae A. Rich. in Bory, Dict. Class. Hist. Nat. 2: 173. 31 Dec 1822.
Typus: *Balsamina* Mill. [= *Impatiens* L.].

Barbeyaceae Rendle in Oliver, Fl. Trop. Afr. 6(2): 14. Mar 1916.
Typus: *Barbeya* Schweinf.

Barringtoniaceae (DC.) F. Rudolphi, Syst. Orb. Veg.: 56. 1830 (trib. *Barringtonieae* DC., Prodr. 3: 288. Mar 1828).
Typus: *Barringtonia* J. R. Forst. & G. Forst. *(nom. cons.).*

Basellaceae Moq., Chenop. Monogr. Enum.: x. Mai 1840.
Typus: *Basella* L.

Bataceae Mart. ex Meisn., Pl. Vasc. Gen., Tab. Diagn.: 349; Comm.: 260. 13-15 Feb 1842.
Typus: *Batis* P. Browne

Begoniaceae C. Agardh, Aphor. Bot.: 200. 13 Jun 1824.
Typus: *Begonia* L.

Berberidaceae Juss., Gen. Pl.: 286. 4 Aug 1789.
Typus: *Berberis* L.

Betulaceae Gray, Nat. Arr. Brit. Pl. 2: 222, 243. 1 Nov 1821.
Typus: *Betula* L.
Note: If this family is united with *Corylaceae,* the name *Betulaceae* must be used.

Bignoniaceae Juss., Gen. Pl.: 137. 4 Aug 1789.
Typus: *Bignonia* L.

Bixaceae Link, Handbuch 2: 371. Jan-Aug 1831.
Typus: *Bixa* L.

Bombacaceae Kunth, Malvac., Büttner., Tiliac.: 5. 20 Apr 1822.
Typus: *Bombax* L.

Boraginaceae Juss., Gen. Pl.: 128. 4 Aug 1789.
Typus: *Borago* L.

Brassicaceae Burnett, Outl. Bot.: 854, 1123. Jun 1835; nom. alt.: *Cruciferae.*
Typus: *Brassica* L.

Bretschneideraceae Engl. & Gilg in Engler, Syllabus, ed. 9 & 10: 218. Nov-Dec 1924.
Typus: *Bretschneidera* Hemsl.

Bromeliaceae Juss., Gen. Pl.: 49. 4 Aug 1789.
Typus: *Bromelia* L.

Brunelliaceae Engl. in Engler & Prantl, Nat. Pflanzenfam., Nachtr. 1: 182. Aug 1897.
Typus: *Brunellia* Ruiz & Pav.

Bruniaceae R. Br. ex DC., Prodr. 2: 43. Nov (med.) 1825.
Typus: *Brunia* Lam. *(nom. cons.).*

Brunoniaceae Dumort., Anal. Fam. Pl.: 19, 21. 1829.
Typus: *Brunonia* Sm. ex R. Br.

Buddlejaceae K. Wilh., Samenpflanzen: 90. Oct 1910.
Typus: *Buddleja* L.

Burmanniaceae Blume, Enum. Pl. Javae: 27. Oct-Dec 1827.
Typus: *Burmannia* L.

Burseraceae Kunth in Ann. Sci. Nat. (Paris) 2: 346. Jun 1824.
Typus: *Bursera* Jacq. ex L. *(nom. cons.).*

Butomaceae Rich. in Mém. Mus. Hist. Nat. 1: 364, 366, 372. 1815-1816.
Typus: *Butomus* L.

Buxaceae Dumort., Comment. Bot.: 54. Nov (sero) - Dec (prim.) 1822.
Typus: *Buxus* L.

Byblidaceae (Engl. & Gilg) Domin in Acta Bot. Bohem. 1: 3. 1922 (subfam. *Byblidoideae* Engl. & Gilg in Engler, Syllabus, ed. 7: 329. Oct 1912 - Mar 1913).
Typus: *Byblis* Salisb.

Byttneriaceae R. Br. in Flinders, Voy. Terra Austr. 2: 540. 19 Jul 1814.
Typus: *Byttneria* Loefl. *(nom. cons.).*
Note: If this family is united with *Sterculiaceae,* the name *Byttneriaceae* is rejected in favour of *Sterculiaceae.*

Cabombaceae A. Rich., Nouv. Elém. Bot., ed. 4: 420. 1828.
Typus: *Cabomba* Aubl.

Cactaceae Juss., Gen. Pl.: 310. 4 Aug 1789.
Typus: *Cactus* L. *(nom. rej.) (Mammillaria* Haw., *nom. cons.).*

Caesalpiniaceae R. Br. in Flinders, Voy. Terra Austr. 2: 551. 19 Jul 1814.
Typus: *Caesalpinia* L.

Callitrichaceae Link, Enum. Hort. Berol. Alt. 1: 7. 16 Mar - 30 Jun 1821.
Typus: *Calliriche* L.

Calycanthaceae Lindl. in Bot. Reg. 5: ad t. 404. 1 Oct 1819.
Typus: *Calycanthus* L. *(nom. cons.).*

Calyceraceae R. Br. ex Rich. in Mém. Mus. Hist. Nat. 6: 74. Nov 1820.
Typus: *Calycera* Cav. *(nom. cons.).*

Campanulaceae Juss., Gen. Pl.: 163. 4 Aug 1789.
Typus: *Campanula* L.

Canellaceae Mart., Nov. Gen. Sp. Pl. 3: 168, 170. Sep 1832.
Typus: *Canella* P. Browne *(nom. cons.).*

Cannabaceae Endl., Gen. Pl.: 286. Oct 1837.
Typus: *Cannabis* L.

Cannaceae Juss., Gen. Pl.: 62. 4 Aug 1789.
Typus: *Canna* L.

Capparaceae Juss., Gen. Pl.: 242. 4 Aug 1789.
Typus: *Capparis* L.

Caprifoliaceae Juss., Gen. Pl.: 210. 4 Aug 1789.
Typus: *Caprifolium* Mill. (*Lonicera* L.).

Cardiopteridaceae Blume, Rumhia 3: 205. Jun 1847.
Typus: *Cardiopteris* Wall. ex Royle

Caricaceae Dumort., Anal. Fam. Pl.: 37, 42. 1829.
Typus: *Carica* L.

Cartonemataceae Pichon in Notul. Syst. (Paris) 12: 219. Feb 1946.
Typus: *Cartonema* R. Br.

Caryocaraceae Szyszył. in Engler & Prantl, Nat. Pflanzenfam. 3(6): 153. Mai 1893.
Typus: *Caryocar* L.

Caryophyllaceae Juss., Gen. Pl.: 299. 4 Aug 1789.
Typus: *Caryophyllus* Mill., non L. (*Dianthus* L.).

Cassythaceae Bartl. ex Lindl., Nix. Pl.: 15. 17 Sep 1833.
Typus: *Cassytha* L.

Casuarinaceae R. Br. in Flinders, Voy. Terra Austr. 2: 571. 19 Jul 1814.
Typus: *Casuarina* L.

Celastraceae R. Br. in Flinders, Voy. Terra Austr. 2: 554. 19 Jul 1814.
Typus: *Celastrus* L.
Note: If this family is united with *Hippocrateaceae,* the name *Celastraceae* must be used.

Centrolepidaceae Endl., Gen. Pl.: 119. Dec 1836.
Typus: *Centrolepis* Labill.

Cephalotaceae Dumort., Anal. Fam. Pl.: 59, 61. 1829.
Typus: *Cephalotus* Labill. *(nom. cons.).*

Cephalotaxaceae Neger, Nadelhölzer: 23, 30. Oct-Nov 1907.
Typus: *Cephalotaxus* Siebold & Zucc. ex Endl.

Ceratophyllaceae Gray, Nat. Arr. Brit. Pl. 2: 395, 554. 1 Nov 1821.
Typus: *Ceratophyllum* L.

Cercidiphyllaceae Engl., Syllabus, ed. 6: 132. Jun-Dec 1909.
Typus: *Cercidiphyllum* Siebold & Zucc.

Chenopodiaceae Vent., Tabl. Règne Vég. 2: 253. 5 Mai 1799.
Typus: *Chenopodium* L.

Chloranthaceae R. Br. ex Lindl., Coll. Bot.: ad t. 17. 21 Oct 1821.
Typus: *Chloranthus* Sw.

Chrysobalanaceae R. Br. in Tuckey, Narr. Exped. Zaire: 433. 5 Mar 1818.
Typus: *Chrysobalanus* L.

Cichoriaceae Juss., Gen. Pl.: 168. 4 Aug 1789.
Typus: *Cichorium* L.

Circaeasteraceae Hutch., Fam. Fl. Pl. 1: 98. Feb-Mar 1926.
Typus: *Circaeaster* Maxim.

Cistaceae Juss., Gen. Pl.: 294. 4 Aug 1789.
Typus: *Cistus* L.

Clethraceae Klotzsch in Linnaea 24: 12. Mai 1851.
Typus: *Clethra* L.

Clusiaceae Lind., Intr. Nat. Syst. Bot., ed. 2: 74. Jul 1836; nom. alt.: *Guttiferae.*
Typus: *Clusia* L.

Cneoraceae Link, Handbuch 2: 440. Jan-Aug 1831.
Typus: *Cneorum* L.

Cochlospermaceae Planch. in London J. Bot. 6: 305. Jun-Jul 1847.
Typus: *Cochlospermum* Kunth *(nom. cons.).*

Colchicaceae DC. in Lamarck & Candolle, Fl. Franç., ed. 3, 3: 192. 17 Sep 1805.
Typus: *Colchicum* L.

Columelliaceae D. Don in Edinburgh New Philos. J. 6: 46, 49. 1828.
Typus: *Columellia* Ruiz & Pav. *(nom. cons.).*

Combretaceae R. Br., Prodr.: 351. 27 Mar 1810.
Typus: *Combretum* Loefl. *(nom. cons.).*

Commelinaceae R. Br., Prodr.: 268. 27 Mar 1810.
Typus: *Commelina* L.

Compositae Giseke, Prael. Ord. Nat. Pl.: 538. Apr 1792; nom. alt.: *Asteraceae.*
Typus: *Aster* L.

Connaraceae R. Br. in Tuckey, Narr. Exped. Zaire: 431. 5 Mar 1818.
Typus: *Connarus* L.

Convolvulaceae Juss., Gen. Pl.: 132. 4 Aug 1789.
Typus: *Convolvulus* L.

Cordiaceae R. Br. ex Dumort., Anal. Fam. Pl.: 20, 25. 1829.
Typus: *Cordia* L.

Coriariaceae DC., Prodr. 1: 739. Jan 1824.
Typus: *Coriaria* L.

Cornaceae (Dumort.) Dumort., Anal. Fam. Pl.: 33, 34. 1829 (trib. *Corneae* Dumort., Fl. Belg.: 83. 1827-1830).
Typus: *Cornus* L.

Corsiaceae Becc. in Malesia 1: 238. Sep 1878.
Typus: *Corsia* Becc.

Corylaceae Mirb., Elém. Physiol. Vég. Bot. 2: 906. 1815.
Typus: *Corylus* L.
Note: If this family is united with *Betulaceae,* the name *Corylaceae* is rejected in favour of *Betulaceae*.

Corynocarpaceae Engl. in Engler & Prantl, Nat. Pflanzenfam., Nachtr. 1: 215. Oct 1897.
Typus: *Corynocarpus* J. R. Forst. & G. Forst.

Crassulaceae DC. in Lamarck & Candolle, Fl. Franç., ed. 3, 4(1): 382. 17 Sep 1805.
Typus: *Crassula* L.

Crossosomataceae Engl. in Engler & Prantl, Nat. Pflanzenfam., Nachtr. 1: 185. Aug 1897.
Typus: *Crossosoma* Nutt.

Cruciferae Juss., Gen. Pl.: 237. 4 Aug 1789; nom. alt.: *Brassicaceae.*
Typus: *Brassica* L.

Crypteroniaceae A. DC., Prodr. 16(2): 677. Jul 1868.
Typus: *Crypteronia* Blume

Cucurbitaceae Juss., Gen. Pl.: 393. 4 Aug 1789.
Typus: *Cucurbita* L.

Cunoniaceae R. Br. in Flinders, Voy. Terra Austr. 2: 548. 19 Jul 1814.
Typus: *Cunonia* L. *(nom. cons.)*.

Cupressaceae Rich. ex Bartl., Ord. Nat. Pl.: 90, 95. Sep 1830.
Typus: *Cupressus* L.

Cuscutaceae (Dumort.) Dumort., Anal. Fam. Pl.: 20, 25. 1829 (trib. *Cuscuteae* Dumort., Fl. Belg.: 50. 1827-1830).
Typus: *Cuscuta* L.

Cyanastraceae Engl. in Bot. Jahrb. Syst. 28: 357. 22 Mai 1900.
Typus: *Cyanastrum* Oliv.

Cycadaceae Pers., Syn. Pl. 2: 630. Sep 1807.
Typus: *Cycas* L.

Cyclanthaceae Poit. ex A. Rich. in Bory, Dict. Class. Hist. Nat. 5: 222. 15 Mai 1824.
Typus: *Cyclanthus* Poit. ex A. Rich.

Cymodoceaceae N. Taylor in N. Amer. Fl. 17: 31. 30 Jun 1909.
Typus: *Cymodocea* K. D. Koenig *(nom. cons.)*.

Cynomoriaceae (C. Agardh) Endl. ex Lindl., Nix. Pl.: 23. 17 Sep 1833 ([unranked] *Cynomorieae* C. Agardh, Aphor. Bot.: 203. 13 Jun 1824).
Typus: *Cynomorium* L.

Cyperaceae Juss., Gen. Pl.: 26. 4 Aug 1789.
Typus: *Cyperus* L.

Cyrillaceae Endl., Ench. Bot.: 578. 15-21 Aug 1841.
Typus: *Cyrilla* Garden ex L.

Daphniphyllaceae Müll. Arg. in Candolle, Prodr. 16(1): 1. Nov 1869.
Typus: *Daphniphyllum* Blume

Datiscaceae R. Br. ex Lindl., Intr. Nat. Syst. Bot.: 109. Sep 1830.
Typus: *Datisca* L.

Degeneriaceae I. W. Bailey & A. C. Sm. in J. Arnold Arbor. 23: 357. 15 Jul 1942.
Typus: *Degeneria* I. W. Bailey & A. C. Sm.

Desfontainiaceae Endl., Ench. Bot.: 336. 15-21 Aug 1841.
Typus: *Desfontainia* Ruiz & Pav.

Dialypetalanthaceae Rizzini & Occhioni in Lilloa 17: 253. 30 Dec 1948.
Typus: *Dialypetalanthus* Kuhlm.

Diapensiaceae (Link) Lindl., Intr. Nat. Syst. Bot., ed. 2: 233. Jul 1836 (subfam. *Diapensioideae* Link, Handbuch 1: 595. Jan-Aug 1829).
Typus: *Diapensia* L.

Dichapetalaceae Baill. in Martius, Fl. Bras. 12(1): 365. 1 Apr 1886.
Typus: *Dichapetalum* Thouars

Dichondraceae Dumort., Anal. Fam. Pl.: 20, 24. 1829.
Typus: *Dichondra* J. R. Forst. & G. Forst.

Diclidantheraceae J. Agardh, Theoria Syst. Pl.: 195. Apr-Sep 1858.
Typus: *Diclidanthera* Mart.

Didiereaceae Drake in Bull. Mus. Hist. Nat. (Paris) 9: 36. 17 Feb 1903.
Typus: *Didierea* Baill.

Dilleniaceae Salisb., Parad. Lond. 2: ad t. 73. Jun 1807.
Typus: *Dillenia* L.

Dioncophyllaceae (Gilg) Airy Shaw in Kew Bull. 1951: 333. 26 Jan 1952 (subtrib. *Dioncophyllinae* Gilg in Engler & Prantl, Nat. Pflanzenfam., ed. 2, 21: 390, 420. Dec 1925).
Typus: *Dioncophyllum* Baill.

Dioscoreaceae R. Br., Prodr.: 294. 27 Mar 1810.
Typus: *Dioscorea* L.

Dipentodontaceae Merr. in Brittonia 4: 69. 16 Dec 1941.
Typus: *Dipentodon* Dunn

Dipsacaceae Juss., Gen. Pl.: 194. 4 Aug 1789.
Typus: *Dipsacus* L.

Dipterocarpaceae Blume, Bijdr. Fl. Ned. Ind.: 222. 20 Sep - 7 Dec 1825.
Typus: *Dipterocarpus* C. F. Gaertn.

Dodonaeaceae Link, Handbuch 2: 441. Jan-Aug 1831.
Typus: *Dodonaea* Mill.

Donatiaceae (Engl.) B. Chandler in Notes Roy. Bot. Gard. Edinburgh 5: 44. Nov 1911 (trib. *Donatieae* Engl. in Engler & Prantl, Nat. Pflanzenfam. 3(2a): 46. Dec 1890).
Typus: *Donatia* J. R. Forst. & G. Forst. *(nom. cons.)*.

Dracaenaceae Salisb., Gen. Pl.: 73. Apr-Mai 1866.
Typus: *Dracaena* Vand. ex L.

Droseraceae Salisb., Parad. Lond. 2: ad t. 95. 1808.
Typus: *Drosera* L.

Dysphaniaceae Pax in Bot. Jahrb. Syst. 61: 230. 15 Jun 1927.
Typus: *Dysphania* R. Br.

Ebenaceae Gürke in Engler & Prantl, Nat. Pflanzenfam. 4(1): 153. Dec 1891.
Typus: *Ebenus* Kuntze, non L. (*Maba* J. R. Forst. & G. Forst.).

Ehretiaceae Mart. ex Lindl., Intr. Nat. Syst. Bot.: 242. Sep 1830.
Typus: *Ehretia* P. Browne

Elaeagnaceae Juss., Gen. Pl.: 74. 4 Aug 1789.
Typus: *Elaeagnus* L.

Elaeocarpaceae Juss. ex DC., Prodr. 1: 519. Jan 1824.
Typus: *Elaeocarpus* L.

Elatinaceae Dumort., Anal. Fam. Pl.: 44, 49. 1829.
Typus: *Elatine* L.

Empetraceae Gray, Nat. Arr. Brit. Pl. 2: 732. 1 Nov 1821.
Typus: *Empetrum* L.

Epacridaceae R. Br., Prodr.: 535. 27 Mar 1810.
Typus: *Epacris* Cav. *(nom. cons.)*.

Ephedraceae Dumort., Anal. Fam. Pl.: 11, 12. 1829.
Typus: *Ephedra* L.

Ericaceae Juss., Gen. Pl.: 159. 4 Aug 1789.
Typus: *Erica* L.

Eriocaulaceae P. Beauv. ex Desv. in Ann. Sci. Nat. (Paris) 13: 47. Jan 1828.
Typus: *Eriocaulon* L.

Erythropalaceae (Hassk.) Sleumer in Engler & Prantl, Nat. Pflanzenfam., ed. 2, 20b: 401. Jan-Feb 1942 (subfam. *Erythropaloideae* Hassk., Pl. Jav. Rar.: 193. Jan-Feb 1848).
Typus: *Erythropalum* Blume

Erythroxylaceae Kunth in Humboldt & al., Nov. Gen. Sp. 5, ed. 4°: 175 [& ed. f°: 135]. 25 Feb 1822.
Typus: *Erythroxylum* P. Browne

Escalloniaceae R. Br. ex Dumort., Anal. Fam. Pl.: 35, 37. 1829.
Typus: *Escallonia* Mutis ex L. f.

Eucommiaceae Engl., Syllabus, ed. 6: 145. Jun-Dec 1909.
Typus: *Eucommia* Oliv.

Eucryphiaceae Endl., Ench. Bot.: 528. 15-21 Aug 1841.
Typus: *Eucryphia* Cav.

Euphorbiaceae Juss., Gen. Pl.: 384. 4 Aug 1789.
Typus: *Euphorbia* L.

Eupomatiaceae Endl., Ench. Bot.: 425. 15-21 Aug 1841.
Typus: *Eupomatia* R. Br.

Eupteleaceae K. Wilh., Samenpflanzen: 17. Oct 1910.
Typus: *Euptelea* Siebold & Zucc.

Fabaceae Lindl., Intr. Nat. Syst. Bot., ed. 2: 148. Jul 1836; nom. alt.: *Leguminosae, Papilionaceae.*
Typus: *Faba* Mill. [= *Vicia* L.].

Fagaceae Dumort., Anal. Fam. Pl.: 11, 12. 1829.
Typus: *Fagus* L.

Flacourtiaceae Rich. ex DC., Prodr. 1: 255. Jan 1824.
Typus: *Flacourtia* Comm. ex L'Hér.
Note: If this family is united with *Samydaceae,* the name *Flacourtiaceae* must be used.

Flagellariaceae Dumort., Anal. Fam. Pl.: 59, 60. 1829.
Typus: *Flagellaria* L.

Fouquieriaceae DC., Prodr. 3: 349. Mar (med.) 1828.
Typus: *Fouquieria* Kunth

Francoaceae A. Juss. in Ann. Sci. Nat. (Paris) 25: 9. Jan 1832.
Typus: *Francoa* Cav.

Frankeniaceae A. St.-Hil. ex Gray, Nat. Arr. Brit. Pl. 2: 623, 633. 1 Nov 1821.
Typus: *Frankenia* L.

Fumariaceae DC., Syst. Nat. 2: 65, 105. Mai 1821.
Typus: *Fumaria* L.

Garryaceae Lindl. in Edwards's Bot. Reg. 20: ad t. 1686. 1 Jul 1834.
Typus: *Garrya* Douglas ex Lindl.

Geissolomataceae Endl., Ench. Bot.: 214. 15-21 Aug 1841.
Typus: *Geissoloma* Lindl. ex Kunth

Gentianaceae Juss., Gen. Pl.: 141. 4 Aug 1789.
Typus: *Gentiana* L.

Geosiridaceae Jonker in Receuil Trav. Bot. Néerl. 36 [Meded. Bot. Mus. Herb. Rijks Univ. Utrecht 60]: 477. [seors. impr.] 18 Mai 1939.
Typus: *Geosiris* Baill.

Geraniaceae Juss., Gen. Pl.: 268. 4 Aug 1789.
Typus: *Geranium* L.

Gesneriaceae Dumort., Comment. Bot.: 57. Nov (sero) - Dec (prim.) 1822.
Typus: *Gesneria* L.

Ginkgoaceae Engl. in Engler & Prantl, Nat. Pflanzenfam., Nachtr. 1: 19. Jul 1897.
Typus: *Ginkgo* L.

Globulariaceae DC. in Lamarck & Candolle, Fl. Franç., ed. 3, 3: 427. 17 Sep 1805.
Typus: *Globularia* L.

Gnetaceae Lindl. in Edwards's Bot. Reg. 20: ad t. 1686. 1 Jul 1834.
Typus: *Gnetum* L.

Gomortegaceae Reiche in Ber. Deutsch. Bot. Ges. 14: 232. 19 Aug 1896.
Typus: *Gomortega* Ruiz & Pav.

Gonystylaceae Gilg in Engler & Prantl, Nat. Pflanzenfam., Nachtr. 1: 231. Oct 1897.
Typus: *Gonystylus* Teijsm. & Binn.

Goodeniaceae R. Br., Prodr.: 573. 27 Mar 1810.
Typus: *Goodenia* Sm.

Gramineae Juss., Gen. Pl.: 28. 4 Aug 1789; nom. alt.: *Poaceae.*
Typus: *Poa* L.

Greyiaceae Hutch., Fam. Fl. Pl. 1: 202. Feb-Mar 1926.
Typus: *Greyia* Hook. & Harv.

Grossulariaceae DC. in Lamarck & Candolle, Fl. Franç., ed. 3, 4(2): 405. 17 Sep 1805.
Typus: *Grossularia* Mill.

Grubbiaceae Endl., Gen. Pl.: xiv. Jan 1839.
Typus: *Grubbia* P. J. Bergius

Gunneraceae Meisn., Pl. Vasc. Gen., Tab. Diagn.: 345, 346; Comm.: 257. 13-15 Feb 1842.
Typus: *Gunnera* L.

Guttiferae Juss., Gen. Pl.: 255. 4 Aug 1789; nom. alt.: *Clusiaceae.*
Typus: *Clusia* L.

Gyrostemonaceae Endl., Ench. Bot.: 509. 15-21 Aug 1841.
Typus: *Gyrostemon* Desf.

Haemodoraceae R. Br., Prodr.: 299. 27 Mar 1810.
Typus: *Haemodorum* Sm.

Haloragaceae R. Br. in Flinders, Voy. Terra Austr. 2: 549. 19 Jul 1814.
Typus: *Haloragis* J. R. Forst. & G. Forst.

Hamamelidaceae R. Br. in Abel, Narr. Journey China: 374. Aug 1818.
Typus: *Hamamelis* L.

Heliotropiaceae Schrad. in Commentat. Soc. Regiae Sci. Gott. Recent. 4: 192. Dec 1819.
Typus: *Heliotropium* L.

Hernandiaceae Blume, Bijdr. Fl. Ned. Ind.: 550. 24 Jan 1826.
Typus: *Hernandia* L.

Heteropyxidaceae Engl. & Gilg in Engler, Syllabus, ed. 8: 281. Jan-Feb 1920.
Typus: *Heteropyxis* Harv. *(nom. cons.).*

Himantandraceae Diels in Bot. Jahrb. Syst. 55: 126. 27 Nov 1917.
Typus: *Himantandra* F. Muell. ex Diels [= *Galbulimima* F. M. Bailey].

Hippocastanaceae DC., Prodr. 1: 597. Jan 1824.
Typus: *Hippocastanum* Mill. (*Aesculus* L.).

Hippocrateaceae Juss. in Ann. Mus. Natl. Hist. Nat. 18: 486. Jul-Aug 1811.
Typus: *Hippocratea* L.
Note: If this family is united with *Celastraceae,* the name *Hippocrateaceae* is rejected in favour of *Celastraceae.*

Hippuridaceae Link, Enum. Hort. Berol. Alt. 1: 5. 16 Mar - 30 Jun 1821.
Typus: *Hippuris* L.

Hoplestigmataceae Gilg in Engler, Syllabus, ed. 9 & 10: 322. Nov-Dec 1924.
Typus: *Hoplestigma* Pierre

Humbertiaceae Pichon in Notul. Syst. (Paris) 13: 23. Jul-Sep 1947.
Typus: *Humbertia* Lam.

Humiriaceae A. Juss. in Saint-Hilaire, Fl. Bras. Merid. 2: 87. 10 Oct 1829.
Typus: *Humiria* Aubl. *(nom. cons.).*

Hydnoraceae C. Agardh, Aphor. Bot.: 88. 21 Dec 1821.
Typus: *Hydnora* Thunb.

Hydrangeaceae Dumort., Anal. Fam. Pl.: 36, 38. 1829.
Typus: *Hydrangea* L.

Hydrocharitaceae Juss., Gen. Pl.: 67. 4 Aug 1789.
Typus: *Hydrocharis* L.

Hydrocotylaceae (Drude) Hyl. [Nomenkl. Stud. Nord. Gefässpfl.] in Uppsala Univ. Årsskr. 1945(7): 20. Jun-Dec 1945 (subfam. *Hydrocotyloideae* Drude in Engler & Prantl, Nat. Pflanzenfam. 3(8): 114, 116. Dec 1897).
Typus: *Hydrocotyle* L.

Hydrophyllaceae R. Br. in Bot. Reg. 3: ad t. 242. 1 Dec 1817.
Typus: *Hydrophyllum* L.

Hydrostachyaceae Engl., Syllabus, ed. 2: 125. Mai 1898.
Typus: *Hydrostachys* Thouars

Hypericaceae Juss., Gen. Pl.: 254. 4 Aug 1789.
Typus: *Hypericum* L.

Hypoxidaceae R. Br. in Flinders, Voy. Terra Austr. 2: 576. 19 Jul 1814.
Typus: *Hypoxis* L.

Icacinaceae (Benth.) Miers in Ann. Mag. Nat. Hist., ser. 2, 8: 174. Sep 1851 (trib. *Icacineae* Benth. in Trans. Linn. Soc. London 18: 679. 1841).
Typus: *Icacina* A. Juss.

Illecebraceae R. Br., Prodr.: 413. 27 Mar 1810.
Typus: *Illecebrum* L.

Illiciaceae (DC.) A. C. Sm. in Sargentia 7: 8. 28 Nov 1947 (trib. *Illicieae* DC., Prodr. 1: 77. Jan 1824).
Typus: *Illicium* L.

Iridaceae Juss., Gen. Pl.: 57. 4 Aug 1789.
Typus: *Iris* L.

Irvingiaceae (Engl.) Exell & Mendonça, Consp. Fl. Angol. 1: 279, 395. 20 Aug 1951 (subfam. *Irvingioideae* Engl., Veg. Erde 9(3,1): 765. Feb-Sep 1915).
Typus: *Irvingia* Hook. f.

Iteaceae J. Agardh, Theoria Syst. Pl.: 151. Apr-Sep 1858.
Typus: *Itea* L.

Ixonanthaceae (Benth. & Hook. f.) Exell & Mendonça in Bol. Soc. Brot., ser. 2, 25: 105. 1951 (trib. *Ixonantheae* Benth. & Hook. f., Gen. Pl. 1: 242, 245. 7 Aug 1862).
Typus: *Ixonanthes* Jack

Juglandaceae A. Rich. ex Kunth in Ann. Sci. Nat. (Paris) 2: 343. Jun 1824.
Typus: *Juglans* L.

Julianiaceae Hemsl. in J. Bot. 44: 379. Oct 1906.
Typus: *Juliania* Schltdl., non La Llave & Lex. (*Amphipterygium* Standl.).

Juncaceae Juss., Gen. Pl.: 43. 4 Aug 1789.
Typus: *Juncus* L.

Juncaginaceae Rich., Démonstr. Bot.: ix. Mai 1808.
Typus: *Juncago* Ség. (*Triglochin* L.).
Note: If this family is united with *Potamogetonaceae,* the name *Juncaginaceae* is rejected in favour of *Potamogetonaceae.*

Koeberliniaceae Engl. in Engler & Prantl, Nat. Pflanzenfam. 3(6): 319. 14 Mai 1895.
Typus: *Koeberlinia* Zucc.

Krameriaceae Dumort., Anal. Fam. Pl.: 20, 23. 1829.
Typus: *Krameria* L.

Labiatae Juss., Gen. Pl.: 110. 4 Aug 1789; nom. alt.: *Lamiaceae.*
Typus: *Lamium* L.

Lacistemataceae Mart., Nov. Gen. Sp. Pl. 1: 158. Jan-Mar 1826.
Typus: *Lacistema* Sw.

Lactoridaceae Engl. in Engler & Prantl, Nat. Pflanzenfam. 3(2): 19. Feb 1888.
Typus: *Lactoris* Phil.

Lamiaceae Lindl., Intr. Nat. Syst. Bot., ed. 2: 275. Jul 1836; nom. alt.: *Labiatae.*
Typus: *Lamium* L.

Lardizabalaceae Decne. in Arch. Mus. Hist. Nat. 1: 185. 1839.
Typus: *Lardizabala* Ruiz & Pav.

Lauraceae Juss., Gen. Pl.: 80. 4 Aug 1789.
Typus: *Laurus* L.

Lecythidaceae Poit. in Mém. Mus. Hist. Nat. 13: 143. 1825.
Typus: *Lecythis* Loefl.

Leeaceae (DC.) Dumort., Anal. Fam. Pl.: 21, 27. 1829 (trib. *Leeeae* DC., Prodr. 1: 635. Jan 1824).
Typus: *Leea* D. Royen ex L. *(nom. cons.).*

Leguminosae Juss., Gen. Pl.: 345. 4 Aug 1789; nom. alt.: *Fabaceae.*
Typus: *Faba* Mill. [= *Vicia* L.].

Leitneriaceae Benth. & Hook. f., Gen. Pl. 3: vi, 396. 7 Feb 1880.
Typus: *Leitneria* Chapm.

Lemnaceae Gray, Nat. Arr. Brit. Pl. 2: 729. 1 Nov 1821.
Typus: *Lemna* L.

Lennoaceae Solms in Abh. Naturf. Ges. Halle 11: 174. 7 Jan 1870.
Typus: *Lennoa* La Llave & Lex.

Lentibulariaceae Rich. in Poiteau & Turpin, Fl. Paris. 1, ed. 4°: 26 [& ed. f°: 23]. 1808.
Typus: *Lentibularia* Ség. (*Utricularia* L.).

Lepidobotryaceae J. Léonard in Bull. Jard. Bot. Etat 20: 38. Jun 1950.
Typus: *Lepidobotrys* Engl.

Lilaeaceae Dumort., Anal. Fam. Pl.: 62, 65. 1829.
Typus: *Lilaea* Bonpl.

Liliaceae Juss., Gen. Pl.: 48. 4 Aug 1789.
Typus: *Lilium* L.

Limnanthaceae R. Br. in London Edinburgh Philos. Mag. & J. Sci. 3: 70. 1833.
Typus: *Limnanthes* R. Br. *(nom. cons.).*

Limoniaceae Ser., Fl. Pharm.: 456. 1851.
Typus: *Limonium* Mill. *(nom. cons.).*

Linaceae DC. ex Gray, Nat. Arr. Brit. Pl. 2: 622, 639. 1 Nov 1821.
Typus: *Linum* L.

Lissocarpaceae Gilg in Engler, Syllabus, ed. 9 & 10: 324. Nov-Dec 1924.
Typus: *Lissocarpa* Benth.

Loasaceae Dumort., Comment. Bot.: 58. Nov-Dec 1822.
Typus: *Loasa* Adans.

Lobeliaceae R. Br. [Observ. Compositae] in Trans. Linn. Soc. London 12: 133. [seors. impr.] Aug 1817.
Typus: *Lobelia* L.

Loganiaceae R. Br. ex Mart., Nov. Gen. Sp. Pl. 2: 133. Jan-Jun 1827.
Typus: *Logania* R. Br. *(nom. cons.).*

Loranthaceae Juss. in Ann. Mus. Natl. Hist. Nat. 12: 292. 1808.
Typus: *Loranthus* Jacq. *(nom. cons.).*

Lowiaceae Ridl., Fl. Malay Penins. 4: 291. 1 Dec 1924.
Typus: *Lowia* Scort. [= *Orchidantha* N. E. Br.].

Lythraceae J. St.-Hil., Expos. Fam. Nat. 2: 175. Feb-Apr 1805.
Typus: *Lythrum* L.

Magnoliaceae Juss., Gen. Pl.: 280. 4 Aug 1789.
Typus: *Magnolia* L.

Malaceae Small in Britton & Small, Fl. S.E. U.S.: 529. 22 Jul 1903.
Typus: *Malus* Mill.

Malesherbiaceae D. Don in Edinburgh New Philos. J. 2: 321. 1827.
Typus: *Malesherbia* Ruiz & Pav.

Malpighiaceae Juss., Gen. Pl.: 252. 4 Aug 1789.
Typus: *Malpighia* L.

Malvaceae Juss., Gen. Pl.: 271. 4 Aug 1789.
Typus: *Malva* L.

Marantaceae Petersen in Engler & Prantl, Nat. Pflanzenfam. 2(6): 33. Oct 1888.
Typus: *Maranta* L.

Marcgraviaceae Choisy in Candolle, Prodr. 1: 565. Jan 1824.
Typus: *Marcgravia* L.

Martyniaceae Stapf in Engler & Prantl, Nat. Pflanzenfam. 4(3b): 265. 12 Mar 1895.
Typus: *Martynia* L.

Mayacaceae Kunth in Abh. Königl. Akad. Wiss. Berlin, Phys. Abh. 1840: 93. 1842.
Typus: *Mayaca* Aubl.

Medusagynaceae Engl. & Gilg in Engler, Syllabus, ed. 9 & 10: 280. Nov-Dec 1924.
Typus: *Medusagyne* Baker

Medusandraceae Brenan in Kew Bull. 1952: 228. 25 Jul 1952.
Typus: *Medusandra* Brenan

Melanthiaceae Batsch, Tab. Affin. Regni Veg.: 133. 2 Mai 1802.
Typus: *Melanthium* L.

Melastomataceae Juss., Gen. Pl.: 328. 4 Aug 1789.
Typus: *Melastoma* L.

Meliaceae Juss., Gen. Pl.: 263. 4 Aug 1789.
Typus: *Melia* L.

Melianthaceae Link, Handbuch 2: 322. Jan-Aug 1831.
Typus: *Melianthus* L.

Menispermaceae Juss., Gen. Pl.: 284. 4 Aug 1789.
Typus: *Menispermum* L.

Menyanthaceae (Dumort.) Dumort., Anal. Fam. Pl.: 20, 25. 1829 (trib. *Menyantheae* Dumort., Fl. Belg.: 52. 1827-1830).
Typus: *Menyanthes* L.

Mesembryanthemaceae Fenzl in Ann. Wiener Mus. Naturgesch. 1: 349. 1836.
Typus: *Mesembryanthemum* L.

Mimosaceae R. Br. in Flinders, Voy. Terra Austr. 2: 551. 19 Jul 1814.
Typus: *Mimosa* L.

Misodendraceae J. Agardh, Theoria Syst. Pl.: 236. Apr-Sep 1858.
Typus: *Misodendrum* Banks ex DC.

Mitrastemonaceae Makino in Bot. Mag. (Tokyo) 25: 252. Dec 1911.
Typus: *Mitrastemon* Makino

Molluginaceae Hutch., Fam. Fl. Pl. 1: 128. Feb-Mar 1926.
Typus: *Mollugo* L.

Monimiaceae Juss. in Ann. Mus. Natl. Hist. Nat. 14: 133. 1809.
Typus: *Monimia* Thouars

Monotropaceae Nutt., Gen. N. Amer. Pl. 1: 272. 14 Jul 1818.
Typus: *Monotropa* L.
Note: If this family is united with *Pyrolaceae,* the name *Monotropaceae* is rejected in favour of *Pyrolaceae*.

Montiniaceae Nakai, Chosakuronbun Mokuroku [Ord. Fam. Trib. Nov.]: 243. 20 Jul 1943.
Typus: *Montinia* Thunb.

Moraceae Link, Handbuch 2: 444. Jan-Aug 1831.
Typus: *Morus* L.

Moringaceae R. Br. ex Dumort., Anal. Fam. Pl.: 43, 48. 1829.
Typus: *Moringa* Adans.

Musaceae Juss., Gen. Pl.: 61. 4 Aug 1789.
Typus: *Musa* L.

Myoporaceae R. Br., Prodr.: 514. 27 Mar 1810.
Typus: *Myoporum* G. Forst.

Myricaceae Blume, Fl. Javae 17-18: 3. 17 Oct 1829.
Typus: *Myrica* L.

Myristicaceae R. Br., Prodr.: 399. 27 Mar 1810.
Typus: *Myristica* Gronov. *(nom. cons.).*

Myrothamnaceae Nied. in Engler & Prantl, Nat. Pflanzenfam. 3(2a): 103. Mar 1891.
Typus: *Myrothamnus* Welw.

Myrsinaceae R. Br., Prodr.: 532. 27 Mar 1810.
Typus: *Myrsine* L.

Myrtaceae Juss., Gen. Pl.: 322. 4 Aug 1789.
Typus: *Myrtus* L.

Najadaceae Juss., Gen. Pl.: 18. 4 Aug 1789.
Typus: *Najas* L.

Nelumbonaceae (DC.) Dumort., Anal. Fam. Pl.: 53. 1829 (trib. *Nelumboneae* DC., Syst. Nat. 2: 43. Mai 1821).
Typus: *Nelumbo* Adans.

Nepenthaceae Dumort., Anal. Fam. Pl.: 14, 16. 1829.
Typus: *Nepenthes* L.

Neuradaceae Link, Handbuch 2: 97. Jan-Aug 1831.
Typus: *Neurada* L.

Nolanaceae Dumort., Anal. Fam. Pl.: 20, 24. 1829.
Typus: *Nolana* L. ex L. f.

Nyctaginaceae Juss., Gen. Pl.: 90. 4 Aug 1789.
Typus: *Nyctago* Juss., nom. illeg. (*Mirabilis* L.).

Nymphaeaceae Salisb. in Ann. Bot. (Koenig & Sims) 2: 70. Jun 1805.
Typus: *Nymphaea* L. *(nom. cons.)*.

Nyssaceae Juss. ex Dumort., Anal. Fam. Pl.: 13. 1829.
Typus: *Nyssa* L.

Ochnaceae DC. in Ann. Mus. Natl. Hist. Nat. 17: 410. Jul-Aug 1811.
Typus: *Ochna* L.

Octoknemaceae Engl. in Bot. Jahrb. Syst. 43: 177. 23 Feb 1909.
Typus: *Octoknema* Pierre

Olacaceae Mirb. ex DC., Prodr. 1: 531. Jan 1824.
Typus: *Olax* L.

Oleaceae Hoffmanns. & Link, Fl. Portug. 1: 385. 1813-1820.
Typus: *Olea* L.

Oliniaceae Harv. & Sond., Fl. Cap. 2: ix. 15-31 Oct 1862.
Typus: *Olinia* Thunb. *(nom. cons.)*.

Onagraceae Juss., Gen. Pl.: 317. 4 Aug 1789.
Typus: *Onagra* Mill. (*Oenothera* L.).

Opiliaceae (Benth.) Valeton, Crit. Overz. Olacin.: 136. 7 Jul 1886 (trib. *Opilieae* Benth. in Trans. Linn. Soc. London 18: 679. 1841).
Typus: *Opilia* Roxb.

Orchidaceae Juss., Gen. Pl.: 64. 4 Aug 1789.
Typus: *Orchis* L.

Orobanchaceae Vent., Tabl. Règne Vég. 2: 292. 5 Mai 1799.
Typus: *Orobanche* L.

Oxalidaceae R. Br. in Tuckey, Narr. Exped. Zaire: 433. 5 Mar 1818.
Typus: *Oxalis* L.

Paeoniaceae F. Rudolphi, Syst. Orb. Veg.: 61. 1830.
Typus: *Paeonia* L.

Palmae Juss., Gen. Pl.: 37. 4 Aug 1789; nom. alt.: *Arecaceae*.
Typus: *Areca* L.

Pandaceae Engl. & Gilg in Engler, Syllabus, ed. 7: 223. Oct 1912 - Mar 1913.
Typus: *Panda* Pierre

Pandanaceae R. Br., Prodr.: 340. 27 Mar 1810.
Typus: *Pandanus* Parkinson

Papaveraceae Juss., Gen. Pl.: 235. 4 Aug 1789.
Typus: *Papaver* L.

Papilionaceae Giseke, Prael. Ord. Nat. Pl.: 415. Apr 1792; nom. alt.: *Fabaceae.*
Typus: *Faba* Mill. [= *Vicia* L.].

Parnassiaceae Gray, Nat. Arr. Brit. Pl. 2: 623, 670. 1 Nov 1821.
Typus: *Parnassia* L.

Passifloraceae Juss. ex Kunth in Humboldt & al., Nov. Gen. Sp. 2, ed. 4°: 126 [& ed. f°: 100]. 8 Dec 1817.
Typus: *Passiflora* L.

Pedaliaceae R. Br., Prodr.: 519. 27 Mar 1810.
Typus: *Pedalium* D. Royen ex L.

Penaeaceae Sweet ex Guill. in Bory, Dict. Class. Hist. Nat. 13: 171. 1 Mar 1828.
Typus: *Penaea* L.

Pentaphragmataceae J. Agardh, Theoria Syst. Pl.: 95. Apr-Sep 1858.
Typus: *Pentaphragma* Wall. ex G. Don

Pentaphylacaceae Engl. in Engler & Prantl, Nat. Pflanzenfam., Nachtr. 1: 214. Oct 1897.
Typus: *Pentaphylax* Gardner & Champ.

Penthoraceae Rydb. ex Britton, Man. Fl. N. States: 475. Oct 1901.
Typus: *Penthorum* L.

Peridiscaceae Kuhlm. in Arq. Serv. Florest. 3: 4. 1950.
Typus: *Peridiscus* Benth.

Periplocaceae Schltr. in Schumann & Lauterbach, Nachtr. Fl. Schutzgeb. Südsee: 351. Nov 1905.
Typus: *Periploca* L.

Petermanniaceae Hutch., Fam. Fl. Pl. 2: 113. 1934.
Typus: *Petermannia* F. Muell. *(nom. cons.).*

Petrosaviaceae Hutch., Fam. Fl. Pl. 2: 36. 1934.
Typus: *Petrosavia* Becc.

Philesiaceae Dumort., Anal. Fam. Pl.: 53, 54, 97. 1829.
Typus: *Philesia* Comm. ex Juss.

Philydraceae Link, Enum. Hort. Berol. Alt. 1: 5. 16 Mar - 30 Jun 1821.
Typus: *Philydrum* Banks ex Gaertn.

Phrymaceae Schauer in Candolle, Prodr. 11: 520. 25 Nov 1847.
Typus: *Phryma* L.

Phytolaccaceae R. Br. in Tuckey, Narr. Exped. Zaire: 454. 5 Mar 1818.
Typus: *Phytolacca* L.

Picrodendraceae Small ex Britton & Millsp., Bahama Fl.: 102. 26 Jun 1920.
Typus: *Picrodendron* Griseb. *(nom. cons.).*

Pinaceae Lindl., Intr. Nat. Syst. Bot., ed. 2: 313. Jul 1836.
Typus: *Pinus* L.
Note: If this family is united with *Abietaceae,* the name *Pinaceae* must be used.

Piperaceae C. Agardh, Aphor. Bot.: 201. 13 Jun 1824.
Typus: *Piper* L.

Pittosporaceae R. Br. in Flinders, Voy. Terra Austr. 2: 542. 19 Jul 1814.
Typus: *Pittosporum* Banks ex Sol. *(nom. cons.).*

Plantaginaceae Juss., Gen. Pl.: 89. 4 Aug 1789.
Typus: *Plantago* L.

Platanaceae Lestiboudois ex Dumort., Anal. Fam. Pl.: 11, 12. 1829.
Typus: *Platanus* L.

Plumbaginaceae Juss., Gen. Pl.: 92. 4 Aug 1789.
Typus: *Plumbago* L.

Poaceae (R. Br.) Barnhart in Bull. Torrey Bot. Club 22: 7. 15 Jan 1895 (trib. *Poeae* R. Br. in Flinders, Voy. Terra Austr. 2: 583. 19 Jul 1814); nom. alt.: *Gramineae.*
Typus: *Poa* L.

Podocarpaceae Endl., Syn. Conif.: 203. Mai-Jun 1847.
Typus: *Podocarpus* L'Hér. ex Pers. *(nom. cons.).*

Podophyllaceae DC., Syst. Nat. 2: 31. Mai 1821.
Typus: *Podophyllum* L.

Podostemaceae Rich. ex C. Agardh, Aphor. Bot.: 125. 19 Jun 1822.
Typus: *Podostemum* Michx.

Polemoniaceae Juss., Gen. Pl.: 136. 4 Aug 1789.
Typus: *Polemonium* L.

Polygalaceae R. Br. in Flinders, Voy. Terra Austr. 2: 542. 19 Jul 1814.
Typus: *Polygala* L.

Polygonaceae Juss., Gen. Pl.: 82. 4 Aug 1789.
Typus: *Polygonum* L.

Pontederiaceae Kunth in Humboldt & al., Nov. Gen. Sp. 1, ed. 4°: 265. Mai 1816.
Typus: *Pontederia* L.

Portulacaceae Juss., Gen. Pl.: 312. 4 Aug 1789.
Typus: *Portulaca* L.

Posidoniaceae Hutch., Fam. Fl. Pl. 2: 41. 1934.
Typus: *Posidonia* K. D. Koenig *(nom. cons.).*

Potamogetonaceae Dumort., Anal. Fam. Pl.: 59, 61. 1829.
Typus: *Potamogeton* L.
Note: If this family is united with *Juncaginaceae,* the name *Potamogetonaceae* must be used.

Primulaceae Vent., Tabl. Règne Vég. 2: 285. 5 Mai 1799.
Typus: *Primula* L.

Proteaceae Juss., Gen. Pl.: 78. 4 Aug 1789.
Typus: *Protea* L. *(nom. cons.).*

Pterostemonaceae Small in N. Amer. Fl. 22(1-6): 183. 18 Dec 1905.
Typus: *Pterostemon* Schauer

Punicaceae Horan., Prim. Lin. Syst. Nat.: 81. Jul-Dec 1834.
Typus: *Punica* L.

Pyrolaceae Dumort., Anal. Fam. Pl.: 43, 47, 80. 1829.
Typus: *Pyrola* L.
Note: If this family is united with *Monotropaceae,* the name *Pyrolaceae* must be used.

Quiinaceae Choisy ex Engl. in Martius, Fl. Bras. 12(1): 475-476. 1 Apr 1888.
Typus: *Quiina* Aubl.

Rafflesiaceae Dumort., Anal. Fam. Pl.: 13, 14. 1829.
Typus: *Rafflesia* R. Br.

Ranunculaceae Juss., Gen. Pl.: 231. 4 Aug 1789.
Typus: *Ranunculus* L.

Rapateaceae Dumort., Anal. Fam. Pl.: 60, 62. 1829.
Typus: *Rapatea* Aubl.

Resedaceae DC. ex Gray, Nat. Arr. Brit. Pl. 2: 622, 665. 1 Nov 1821.
Typus: *Reseda* L.

Restionaceae R. Br., Prodr.: 243. 27 Mar 1810.
Typus: *Restio* Rottb. *(nom. cons.).*

Rhamnaceae Juss., Gen. Pl.: 376. 4 Aug 1789.
Typus: *Rhamnus* L.

Rhizophoraceae R. Br. in Flinders, Voy. Terra Austr. 2: 549. 19 Jul 1814.
Typus: *Rhizophora* L.

Rhoipteleaceae Hand.-Mazz. in Repert. Spec. Nov. Regni Veg. 30: 75. 15 Feb 1932.
Typus: *Rhoiptelea* Diels & Hand.-Mazz.

Roridulaceae Engl. & Gilg in Engler, Syllabus, ed. 9 & 10: 226. Nov-Dec 1924.
Typus: *Roridula* Burm. f. ex L.

Rosaceae Juss., Gen. Pl.: 334. 4 Aug 1789.
Typus: *Rosa* L.

Rubiaceae Juss., Gen. Pl.: 196. 4 Aug 1789.
Typus: *Rubia* L.

Ruppiaceae Horan. ex Hutch., Fam. Fl. Pl. 2: 48. 1934.
Typus: *Ruppia* L.

Ruscaceae Spreng. ex Hutch., Fam. Fl. Pl. 2: 109. 1934.
Typus: *Ruscus* L.

Rutaceae Juss., Gen. Pl.: 296. 4 Aug 1789.
Typus: *Ruta* L.

Sabiaceae Blume, Mus. Bot. 1: 368. 1851.
Typus: *Sabia* Colebr.

Salicaceae Mirb., Elém. Physiol. Vég. Bot. 2: 905. 1815.
Typus: *Salix* L.

Salvadoraceae Lindl., Intr. Nat. Syst. Bot., ed. 2: 269. Jul 1836.
Typus: *Salvadora* L.

Samydaceae Vent. in Mém. Cl. Sci. Math. Inst. Natl. France 1807(2): 149. 1808.
Typus: *Samyda* Jacq. *(nom. cons.)*.
Note: If this family is united with *Flacourtiaceae,* the name *Samydaceae* is rejected in favour of *Flacourtiaceae*.

Santalaceae R. Br., Prodr.: 350. 27 Mar 1810.
Typus: *Santalum* L.

Sapindaceae Juss., Gen. Pl.: 246. 4 Aug 1789.
Typus: *Sapindus* L.

Sapotaceae Juss., Gen. Pl.: 151. 4 Aug 1789.
Typus: *Sapota* Mill. *(Achras* L., *nom. rej.)* [= *Manilkara* Adans. *(nom. cons.)*].

Sarcolaenaceae Caruel in Atti Reale Accad. Lincei Mem. Cl. Sci. Fis., ser. 3, 10: 226, 248. 1881.
Typus: *Sarcolaena* Thouars

Sarcospermataceae H. J. Lam in Bull. Jard. Bot. Buitenzorg, ser. 3, 7: 248. Feb 1925.
Typus: *Sarcosperma* Hook. f.

Sargentodoxaceae Stapf ex Hutch., Fam. Fl. Pl. 1: 100. Feb-Mar 1926.
Typus: *Sargentodoxa* Rehder & E. H. Wilson

Sarraceniaceae Dumort., Anal. Fam. Pl.: 53. 1829.
Typus: *Sarracenia* L.

Saurauiaceae J. Agardh, Theoria Syst. Pl.: 110. Apr-Sep 1858.
Typus: *Saurauia* Willd. *(nom. cons.)*.
Note: If this family is united with *Actinidiaceae,* the name *Saurauiaceae* is rejected in favour of *Actinidiaceae*.

Saururaceae Rich. ex E. Mey., Houttuynia: 20. 5 Dec 1827.
Typus: *Saururus* L.

Saxifragaceae Juss., Gen. Pl.: 308. 4 Aug 1789.
Typus: *Saxifraga* L.

Scheuchzeriaceae F. Rudolphi, Syst. Orb. Veg.: 28. 1830.
Typus: *Scheuchzeria* L.

Schisandraceae Blume, Fl. Javae 32-33: 3. 25 Jun 1830.
Typus: *Schisandra* Michx. *(nom. cons.)*.

Scrophulariaceae Juss., Gen. Pl.: 117. 4 Aug 1789.
Typus: *Scrophularia* L.

Scyphostegiaceae Hutch., Fam. Fl. Pl. 1: 229. Feb-Mar 1926.
Typus: *Scyphostegia* Stapf

Scytopetalaceae Engl. in Engler & Prantl, Nat. Pflanzenfam., Nachtr. 1: 242. Oct 1897.
Typus: *Scytopetalum* Pierre ex Engl.

Selaginaceae Choisy [Mém. Sélag.: 19] in Mém. Soc. Phys. Genève 2: 89. 1823.
Typus: *Selago* L.

Simaroubaceae DC. in Ann. Mus. Natl. Hist. Nat. 17: 422. Jul-Aug 1811.
Typus: *Simarouba* Aubl. *(nom. cons.)*.

Siphonodontaceae (Croizat) Gagnep. & Tardieu in Notul. Syst. (Paris) 14: 102. Jul-Sep 1951 (subfam. *Siphonodontoideae* Croizat in Lilloa 13: 41. 29 Dec 1947).
Typus: *Siphonodon* Griff.

Smilacaceae Vent., Tabl. Règne Vég. 2: 146. 5 Mai 1799.
Typus: *Smilax* L.

Solanaceae Juss., Gen. Pl.: 124. 4 Aug 1789.
Typus: *Solanum* L.

Sonneratiaceae Engl. & Gilg in Engler, Syllabus, ed. 9 & 10: 299. Nov-Dec 1924.
Typus: *Sonneratia* L. f. *(nom. cons.)*.

Sparganiaceae F. Rudolphi, Syst. Orb. Veg.: 27. 1830.
Typus: *Sparganium* L.

Sphenocleaceae Mart. ex DC., Prodr. 7(2): 548. Dec 1839.
Typus: *Sphenoclea* Gaertn. *(nom. cons.)*.

Stachyuraceae J. Agardh, Theoria Syst. Pl.: 152. Apr-Sep 1858.
Typus: *Stachyurus* Siebold & Zucc.

Stackhousiaceae R. Br. in Flinders, Voy. Terra Austr. 2: 555. 19 Jul 1814.
Typus: *Stackhousia* Sm.

Staphyleaceae (DC.) Lindl., Syn. Brit. Fl.: 75. Feb 1829 (trib. *Staphyleeae* DC., Prodr. 2: 2. Nov 1825).
Typus: *Staphylea* L.

Stemonaceae Engl. in Engler & Prantl, Nat. Pflanzenfam. 2(5): 8. 26 Mar 1887.
Typus: *Stemona* Lour.

Stenomeridaceae J. Agardh, Theoria Syst. Pl.: 66. Apr-Sep 1858.
Typus: *Stenomeris* Planch.

Sterculiaceae (DC.) Bartl., Ord. Nat. Pl.: 225, 340. Sep 1830 (trib. *Sterculieae* DC., Prodr. 1: 481. Jan 1824).
Typus: *Sterculia* L.
Note: If this family is united with *Byttneriaceae,* the name *Sterculiaceae* must be used.

Stilbaceae Kunth, Handb. Bot.: 393. 1831.
Typus: *Stilbe* P. J. Bergius

Strasburgeriaceae Engl. & Gilg in Engler, Syllabus, ed. 9 & 10: 282. Nov-Dec 1924.
Typus: *Strasburgeria* Baill.

Strelitziaceae (K. Schum.) Hutch., Fam. Fl. Pl. 2: 72. 1934 (subfam. *Strelitzioideae* K. Schum. in Engler, Pflanzenr. 1: 13. 4 Oct 1900).
Typus: *Strelitzia* Aiton

Stylidiaceae R. Br., Prodr.: 565. 27 Mar 1810.
Typus: *Stylidium* Sw. ex Willd. *(nom. cons.)*.

Styracaceae Dumort., Anal. Fam. Pl.: 28, 29. 1829.
Typus: *Styrax* L.
Note: If this family is united with *Symplocaceae,* the name *Styracaceae* must be used.

Surianaceae Arn. in Wight & Arnott, Prodr. Fl. Ind. Orient.: 360. 10 Oct 1834.
Typus: *Suriana* L.

Symplocaceae Desf. in Mém. Mus. Hist. Nat. 6: 9. 1820.
Typus: *Symplocos* Jacq.
Note: If this family is united with *Styracaceae,* the name *Symplocaceae* is rejected in favour of *Styracaceae*.

Taccaceae Dumort., Anal. Fam. Pl.: 57, 58. 1829.
Typus: *Tacca* J. R. Forst. & G. Forst. *(nom. cons.)*.

Tamaricaceae Link, Enum. Hort. Berol. Alt. 1: 291. 16 Mar - 30 Jun 1821.
Typus: *Tamarix* L.

Taxaceae Gray, Nat. Arr. Brit. Pl. 2: 222, 226. 1 Nov 1821.
Typus: *Taxus* L.

Taxodiaceae Warm., Haandb. Syst. Bot., ed. 2: 163. Jan-Sep 1884.
Typus: *Taxodium* Rich.

Tecophilaeaceae Leyb. in Bonplandia 10: 370. 1862.
Typus: *Tecophilaea* Bertero ex Colla

Tetracentraceae A. C. Sm. in J. Arnold Arbor. 26: 135. 16 Apr 1945.
Typus: *Tetracentron* Oliv.

Tetragoniaceae Nakai in J. Jap. Bot. 18: 103. Mar 1942.
Typus: *Tetragonia* L.

Theaceae D. Don, Prodr. Fl. Nepal.: 224. 26 Jan - 1 Feb 1825.
Typus: *Thea* L. [= *Camellia* L.].

Theligonaceae Dumort., Anal. Fam. Pl.: 15, 17. 1829.
Typus: *Theligonum* L.

Theophrastaceae Link, Handbuch 1: 440. Jan-Aug 1829.
Typus: *Theophrasta* L.

Thismiaceae J. Agardh, Theoria Syst. Pl.: 99. Apr-Sep 1858.
Typus: *Thismia* Griff.

Thurniaceae Engl., Syllabus, ed. 5: 94. Jul 1907.
Typus: *Thurnia* Hook. f.

Thymelaeaceae Juss., Gen. Pl.: 76. 4 Aug 1789.
Typus: *Thymelaea* Mill. *(nom. cons.)*.

Tiliaceae Juss., Gen. Pl.: 289. 4 Aug 1789.
Typus: *Tilia* L.

Tovariaceae Pax in Engler & Prantl, Nat. Pflanzenfam. 3(2): 207. Mar 1891.
Typus: *Tovaria* Ruiz & Pav. *(nom. cons.)*.

Trapaceae Dumort., Anal. Fam. Pl.: 36, 39. 1829.
Typus: *Trapa* L.

Tremandraceae R. Br. ex DC., Prodr. 1: 343. Jan 1824.
Typus: *Tremandra* R. Br. ex DC.

Trichopodaceae Hutch., Fam. Fl. Pl. 2: 143. 1934.
Typus: *Trichopus* Gaertn.

Trigoniaceae Endl., Ench. Bot.: 570. 15-21 Aug 1841.
Typus: *Trigonia* Aubl.

Trilliaceae Lindl., Veg. Kingd.: 218. Jan-Mai 1846.
Typus: *Trillium* L.

Trimeniaceae (Perkins & Gilg) Gibbs, Fl. Arfak Mts.: 135. Jul 1917 (trib. *Trimenieae* Perkins & Gilg in Engler, Pflanzenr. 4: 12, 21. 21 Jun 1901).
Typus: *Trimenia* Seem.

Triuridaceae Gardner in Trans. Linn. Soc. London 19: 160. Jun-Dec 1843.
Typus: *Triuris* Miers

Trochodendraceae Prantl in Engler & Prantl, Nat. Pflanzenfam. 3(2): 21. Feb 1888.
Typus: *Trochodendron* Siebold & Zucc.

Tropaeolaceae Juss. ex DC., Prodr. 1: 683. Jan 1824.
Typus: *Tropaeolum* L.

Turneraceae Kunth ex DC., Prodr. 3: 345. Mar (med.) 1828.
Typus: *Turnera* L.

Typhaceae Juss., Gen. Pl.: 25. 4 Aug 1789.
Typus: *Typha* L.

Ulmaceae Mirb., Elém. Physiol. Vég. Bot. 2: 905. 1815.
Typus: *Ulmus* L.

Umbelliferae Juss., Gen. Pl.: 218. 4 Aug 1789; nom. alt.: *Apiaceae*.
Typus: *Apium* L.

Urticaceae Juss., Gen. Pl.: 400. 4 Aug 1789.
Typus: *Urtica* L.

Uvulariaceae A. Gray ex Kunth, Enum. Pl. 4: 199. 17-19 Jul 1843.
Typus: *Uvularia* L.

Vacciniaceae DC. ex Gray, Nat. Arr. Brit. Pl. 2: 394, 404. 1 Nov 1821.
Typus: *Vaccinium* L.

Valerianaceae Batsch, Tab. Affin. Regni Veg.: 227. 2 Mai 1802.
Typus: *Valeriana* L.

Velloziaceae Endl., Ench. Bot.: 101. 15-21 Aug 1841.
Typus: *Vellozia* Vand.

Verbenaceae J. St.-Hil., Expos. Fam. Nat. 1: 245. Feb-Apr 1805.
Typus: *Verbena* L.

Violaceae Batsch, Tab. Affin. Regni Veg.: 57. 2 Mai 1802.
Typus: *Viola* L.

Vitaceae Juss., Gen. Pl.: 267. 4 Aug 1789.
Typus: *Vitis* L.

Vochysiaceae A. St.-Hil. in Mém. Mus. Hist. Nat. 6: 265. 1820.
Typus: *Vochysia* Aubl.

Welwitschiaceae Markgr. in Engler & Prantl, Nat. Pflanzenfam., ed. 2, 13: 419. 1926.
Typus: *Welwitschia* Hook. f. *(nom. cons.)*.

Winteraceae R. Br. ex Lindl., Intr. Nat. Syst. Bot.: 26. Sep 1830.
Typus: *Wintera* Murray, nom. illeg. *(Drimys* J. R. Forst. & G. Forst., *nom. cons.)*.

Xanthorrhoeaceae Dumort., Anal. Fam. Pl.: 60, 62, 103. 1829.
Typus: *Xanthorrhoea* Sm.

Xyridaceae C. Agardh, Aphor. Bot.: 158. 23 Mai 1823.
Typus: *Xyris* L.

Zannichelliaceae Dumort., Anal. Fam. Pl.: 59, 61. 1829.
Typus: *Zannichellia* L.

Zingiberaceae Lindl., Key Bot.: 69. 15-30 Sep 1835.
Typus: *Zingiber* Boehm. *(nom. cons.)*.

Zosteraceae Dumort., Anal. Fam. Pl.: 65, 66. 1829.
Typus: *Zostera* L.

Zygophyllaceae R. Br. in Flinders, Voy. Terra Austr. 2: 545. 19 Jul 1814.
Typus: *Zygophyllum* L.

APPENDIX IIIA

NOMINA GENERICA CONSERVANDA ET REJICIENDA

In the following lists the ***nomina conservanda*** have been inserted in the left column, in ***bold-face italics***. Synonyms and earlier homonyms *(nomina rejicienda)* are listed in the right column. The numbering of entries of *Spermatophyta* follows Dalla Torre & Harms, *Genera siphonogamarum* (1900-1907).

orth. cons. orthographia conservanda, spelling to be conserved (Art. 14.11).

typ. cons. typus conservandus, type to be conserved (Art. 14.9; see also Art. 14.3 and 10.4); as by Art. 14.8, listed types of conserved names may not be changed even if they are not explicitly designated as *typ. cons.*

typ. des. typi designatio, designation of type (Art. 10.5); used with names that became nomenclatural synonyms by type designation.

vide see; usually followed by a reference to the author and place of publication of first type designation (Art. 10.5); also uses (as *etiam vide,* see also) for cross-reference to another relevant entry.

(H) homonym (Art. 14.10; see also Art. 53), only the earliest being listed.

(≡) nomenclatural synonym, based on the same nomenclatural type as the conserved name (Art. 14.4), usually only the earliest legitimate one being listed (but more than one in some cases in which homotypy results from type designation).

Nomenclatural synonyms of rejected names, when they exist, are cited instead of the type. Nomenclatural synonyms that are part of a type entry are placed in parentheses (round brackets).

(=) taxonomic synonym, based on a type different from that of the conserved name to be rejected only in favour of the conserved name (Art. 14.6 and 14.7).

Taxonomic synonyms following after a type citation (supposedly correct names and their basionym, if any) are placed in [square] brackets. They are given merely for convenience and have no status for nomenclatural purposes.

Some names listed as conserved have no corresponding *nomina rejicienda* because they were conserved solely to maintain a particular type, because evidence after their conservation may have indicated that conservation was unnecessary (see Art. 14.13), or because they were conserved to eliminate doubt about their legitimacy.

A. ALGAE

A1. BACILLARIOPHYCEAE (INCL. FOSS.)

Actinella F. W. Lewis in Proc. Acad. Nat. Sci. Philadelphia 1863: 343. 1864. Typus: *A. punctata* F. W. Lewis	(H)	*Actinella* Pers., Syn. Pl. 2: 469. Sep 1807 [*Comp.*]. ≡ *Actinea* Juss. 1803.
Arachnoidiscus H. Deane pat. ex Shadbolt in Trans. Roy. Microscop. Soc. London 3: 49. 1852. Typus: *A. japonicus* Shadbolt ex A. Pritch. (Hist. Infus., ed. 3: 319. 1852).	(H)	*Arachnodiscus* Bailey ex Ehrenb. in Ber. Bekanntm. Verh. Königl. Preuss. Akad. Wiss. Berlin 1849: 63. 1849 [*Bacillarioph.*]. ≡ *Hemiptychus* Ehrenb. 1848 *(nom. rej.* sub *Arachnoidiscus).*
	(=)	*Hemiptychus* Ehrenb. in Ber. Bekanntm. Verh. Königl. Preuss. Akad. Wiss. Berlin 1848: 7. 1848. Typus: *H. ornatus* Ehrenb.
Aulacodiscus Ehrenb. in Ber. Bekanntm. Verh. Königl. Preuss. Akad. Wiss. Berlin 1844: 73. 1844. Typus: *A. crux* Ehrenb.	(=)	*Tripodiscus* Ehrenb., Lebende Thierart. Kreidebild.: 50. 1840. Typus: *T. germanicus* Ehrenb. (*T. argus* Ehrenb., nom. altern.).
	(=)	*Pentapodiscus* Ehrenb. in Ber. Bekanntm. Verh. Königl. Preuss. Akad. Wiss. Berlin 1843: 165. 1843. Typus: *P. germanicus* Ehrenb.
	(=)	*Tetrapodiscus* Ehrenb. in Ber. Bekanntm. Verh. Königl. Preuss. Akad. Wiss. Berlin 1843: 165. 1843. Typus: *T. germanicus* Ehrenb.
Auricula Castrac. in Atti Accad. Pontif. Sci. Nuovi Lincei 26: 407. 1873. Typus: *A. amphitritis* Castrac.	(H)	*Auricula* Hill, Brit. Herb.: 98. 31 Mar 1756 [*Primul.*]. Typus: non designatus.
Brebissonia Grunow in Verh. K.K. Zool.-Bot. Ges. Wien 10: 512. 1860. Typus: *B. boeckii* (Ehrenb.) O'Meara (in Proc. Roy. Irish Acad., ser. 2, 2(Sci.): 338. Oct 1875) (*Cocconema boeckii* Ehrenb.).	(H)	*Brebissonia* Spach, Hist. Nat. Vég. 4: 401. 11 Apr 1835 [*Onagr.*]. Typus (vide Spach in Ann. Sci. Nat., Bot., ser. 2, 4: 175. 1835): *B. microphylla* (Kunth) Spach (*Fuchsia microphylla* Kunth).
Cerataulina H. Perag. ex F. Schütt in Engler & Prantl, Nat. Pflanzenfam. 1(1b): 95. Dec (sero) 1896. Typus: *C. bergonii* (H. Perag.) F. Schütt (*Cerataulus bergonii* H. Perag.) [= *C. pelagica* (Cleve) Hendey (*Zygoceras pelagicum* Cleve)].	(=)	*Syringidium* Ehrenb. in Ber. Bekanntm. Verh. Königl. Preuss. Akad. Wiss. Berlin 1845: 357. 1845. Typus: *S. bicorne* Ehrenb.

Coscinodiscus Ehrenb. in Abh. Königl. Akad. Wiss. Berlin, Phys. Abh., 1838: 128. 1839.
Typus: *C. argus* Ehrenb. *(typ. cons.).*

Cyclotella (Kütz.) Bréb., Consid. Diatom.: 19. 1838, (*Frustulia* subg. *Cyclotella* Kütz. in Linnaea 8: 535. 1834).
Typus: *C. tecta* Håk. & R. Ross. (in Taxon 33: 529. 1984) *(typ. cons.)* [= *C. distinguenda* Hust.].

Cymatopleura W. Sm. in Ann. Mag. Nat. Hist., ser. 2, 7: 12. Jan 1851.
Typus: *C. solea* (Bréb.) W. Sm. (*Cymbella solea* Bréb.) [= *C. librile* (Ehrenb.) Pant. (*Navicula librile* Ehrenb.)].

(=) *Sphinctocystis* Hassall, Hist. Brit. Freshwater Alg. 1: 436. Jul-Dec 1845.
Typus: *S. librile* (Ehrenb.) Hassall (*Navicula librile* Ehrenb.).

Cymbella C. Agardh, Consp. Diatom.: 1. 4 Dec 1830.
Typus: *C. cymbiformis* C. Agardh *(typ. cons.).*

Diatoma Bory, Dict. Class. Hist. Nat. 5: 461. 15 Mai 1824.
Typus: *D. vulgaris* Bory *(typ. cons.).*

(H) *Diatoma* Lour., Fl. Cochinch.: 290, 295. Sep 1790 [*Rhizophor.*].
Typus: *D. brachiata* Lour.

Diatomella Grev. in Ann. Mag. Nat. Hist., ser. 2, 15: 259. Apr 1855.
Typus: *D. balfouriana* Grev.

(=) *Disiphonia* Ehrenb., Mikrogeologie: 260. 1854.
Typus: *D. australis* Ehrenb.

Didymosphenia Mart. Schmidt in Schmidt, Atlas Diatom.-Kunde: t. 214, f. 1-12. Mar 1899.
Typus: *D. geminata* (Lyngb.) Mart. Schmidt (*Echinella geminata* Lyngb.) (etiam vide *Gomphonema* [*Bacillarioph.*].

(≡) *Dendrella* Bory, Dict. Class. Hist. Nat. 5: 393. 15 Mai 1824 (typ. des. in Regnum Veg. 3: 70. 1952).

(=) *Diomphala* Ehrenb. in Ber. Bekanntm. Verh. Königl. Preuss. Akad. Wiss. Berlin 1842: 336. 1843.
Typus: *D. clava-herculis* Ehrenb.

Eupodiscus Bailey in Smithsonian Contr. Knowl. 2(8): 39. 1851.
Typus: *E. radiatus* Bailey *(typ. cons.).*

(H) *Eupodiscus* Ehrenb. in Ber. Bekanntm. Verh. Königl. Preuss. Akad. Wiss. Berlin 1844: 73. 1844 [*Bacillarioph.*].
≡ *Tripodiscus* Ehrenb. 1840 *(nom. rej.* sub *Aulacodismus).*

Frustulia Rabenh., Süssw.-Diatom.: 50. Mar-Mai 1853.
Typus: *F. saxonica* Rabenh. *(typ. cons.).*

(H) *Frustulia* C. Agardh, Syst. Alg.: xiii, 1. Mai-Sep 1824 [*Bacillarioph.*].
Typus: non designatus.

Gomphonema Ehrenb. in Abh. Königl. Akad. Wiss. Berlin, Phys. Kl., 1831: 87. 1832.
Typus: *G. acuminatum* Ehrenb. *(typ. cons.)*.

(H) *Gomphonema* C. Agardh, Syst. Alg.: xvi, 11. Mai-Sep 1824 [*Bacillarioph.*].
≡ *Didymosphenia* Mart. Schmidt 1899 *(nom. cons.)*.

Gyrosigma Hassall, Hist. Brit. Freshwater Alg. 1: 435. Jul-Dec 1845.
Typus: *G. hippocampus* Hassall, nom. illeg. (*Navicula hippocampus* Ehrenb., nom. illeg., *Frustulia attenuata* Kütz., *Gyrosigma attenuatum* (Kütz.) Rabenh.).

(=) *Scalptrum* Corda in Alman. Carlsbad 5: 193. 1835.
Typus: *S. striatum* Corda

Hantzschia Grunow in Monthly Microscop. J. 18: 174. 1 Oct 1877.
Typus: *H. amphioxys* (Ehrenb.) Grunow (*Eunotia amphioxys* Ehrenb.).

(H) *Hantzschia* Auersw. in Hedwigia 2: 60. 1862 [*Fungi*].
Typus: *H. phycomyces* Auersw.

Hemiaulus Heib., Krit. Overs. Danske Diatom.: 45. 4 Jun 1863.
Typus: *H. proteus* Heib.

(H) *Hemiaulus* Ehrenb. in Ber. Bekanntm. Verh. Königl. Preuss. Akad. Wiss. Berlin 1844: 199. 1844 [*Bacillarioph.: Biddulph.*].
Typus: *H. antarcticus* Ehrenb.

Licmophora C. Agardh in Flora 10: 628. 28 Oct 1827.
Typus: *L. argentescens* C. Agardh

(=) *Styllaria* Drap. ex Bory, Dict. Class. Hist. Nat. 2: 129. 31 Dec 1822.
Typus: *S. paradoxa* (Lyngb.) Bory (Hist. Nat. Zooph.: 709. 1827) (*Echinella paradoxa* Lyngb.).

(=) *Exilaria* Grev., Scott. Crypt. Fl.: ad t. 289. Apr 1827.
Typus: *E. flabellata* Grev.

Melosira C. Agardh, Syst. Alg.: xiv, 8. Mai-Sep 1824 *('Meloseira') (orth. cons.)*.
Typus: *M. nummuloides* C. Agardh

(=) *Lysigonium* Link in Nees, Horae Phys. Berol.: 4. 1-8 Feb 1820.
Typus (vide Regnum Veg. 3: 71. 1952): *Conferva moniliformis* O. F. Müll.

Nitzschia Hassall, Hist. Brit. Freshwater Alg. 1: 435. Jul-Dec 1845.
Typus: *N. elongata* Hassall, nom. illeg. (*Bacillaria sigmoidea* Nitzsch, *N. sigmoidea* (Nitzsch) W. Sm.).

(≡) *Sigmatella* Kütz., Alg. Aq. Dulc. Germ.: No. 2. Jan-Feb 1833.

(=) *Homoeocladia* C. Agardh in Flora 10: 629. 28 Oct 1827.
Typus: *H. martiana* C. Agardh

Pantocsekia Grunow ex Pant., Beitr. Foss. Bacill. Ung. 1: 47. 1886.
Typus: *P. clivosa* Grunow ex Pant.

(H) *Pantocsekia* Griseb. ex Pant. in Oesterr. Bot. Z. 23: 267. Sep 1873 [*Convolvul.*].
Typus: *P. illyrica* Griseb. ex Pant.

Peronia Bréb. & Arn. ex Kitton in Quart. J. Microscop. Sci., ser. 2, 8: 16. 1868.
Typus: *P. erinacea* Bréb. & Arn. ex Kitton, nom. illeg. (*Gomphonema fibula* Bréb. ex Kütz., *P. fibula* (Bréb. ex Kütz.) R. Ross).

(H) *Peronia* Redouté, Liliac.: ad t. 342. 15 Nov 1811 [*Marant.*].
Typus: *P. stricta* F. Delaroche

Pinnularia Ehrenb. in Ber. Bekanntm. Verh. Königl. Preuss. Akad. Wiss. Berlin 1843: 45. 1843.
Typus: *P. viridis* (Nitzsch) Ehrenb. (*Bacillaria viridis* Nitzsch) *(typ. cons.)*.

(H) *Pinnularia* Lindl. & Hutton, Foss. Fl. Gr. Brit. 2: [81], t. 111. 1834 [Foss.].
Typus: *P. capillacea* Lindl. & Hutton

(=) *Stauroptera* Ehrenb. in Ber. Bekanntm. Verh. Königl. Preuss. Akad. Wiss. Berlin 1843: 45. 1843.
Typus: *S. semicruciata* Ehrenb.

Pleurosigma W. Sm. in Ann. Mag. Nat. Hist., ser. 2, 9: 2. Jan 1852.
Typus: *P. angulatum* (E. J. Quekett) W. Sm. (*Navicula angulata* E. J. Quekett).

(=) *Scalptrum* Corda in Alman. Carlsbad 5: 193. 1835.
Typus: *S. striatum* Corda

(=) *Gyrosigma* Hassall, Hist. Brit. Freshwater Alg. 1: 435. Jul-Dec 1845 *(nom. cons.)*.

(=) *Endosigma* Bréb. in Orbigny, Dict. Univ. Hist. Nat. 11: 418, 419. 1848.
Typus: non designatus.

Podocystis Bailey in Smithsonian Contr. Knowl. 7(3): 11. 1854.
Typus: *P. americana* Bailey

(H) *Podocystis* Fr., Summa Veg. Scand.: 512. 1849 [*Fungi*].
Typus (vide Laundon in Mycol. Pap. 99: 14. 1965): *P. capraearum* (DC.) Fr. (*Uredo capraearum* DC.).

(=) *Euphyllodium* Shadbolt in Trans. Roy. Microscop. Soc. London, ser. 2, 2: 14. 1854.
Typus: *E. spathulatum* Shadbolt

Rhabdonema Kütz., Kieselschal. Bacill.: 126. 7-9 Nov 1844.
Typus: *R. minutum* Kütz.

(=) *Tessella* Ehrenb., Zus. Erkenntn. Organis.: 23. 1836.
Typus: *T. catena* Ehrenb.

Rhizosolenia Brightw. in Quart. J. Microscop. Sci. 6: 94. Apr 1858.
Typus: *R. styliformis* Brightw.

(H) *Rhizosolenia* Ehrenb. in Abh. Königl. Akad. Wiss. Berlin, Phys. Abh., 1841: 402. 1843 [*Bacillarioph.*].
Typus: *R. americana* Ehrenb.

Rhopalodia O. Müll. in Bot. Jahrb. Syst. 22: 57. 19 Nov 1895.
Typus: *R. gibba* (Ehrenb.) O. Müll. (*Navicula gibba* Ehrenb.).

(=) *Pyxidicula* Ehrenb. in Abh. Königl. Akad. Wiss. Berlin, Phys. Abh., 1833: 295. 1834.
Typus: *P. operculata* (C. Agardh) Ehrenb. (*Frustulia operculata* C. Agardh).

Tetracyclus Ralfs in Ann. Mag. Nat. Hist. 12: 105. Aug 1843.
Typus: *T. lacustris* Ralfs

(=) *Biblarium* Ehrenb. in Ber. Bekanntm. Verh. Königl. Preuss. Akad. Wiss. Berlin 1843: 47. 1843.
Typus: *B. glans* (Ehrenb.) Ehrenb. (*Navicula glans* Ehrenb.).

A2. BODONOPHYCEAE

Karotomorpha B. V. Travis in Trans. Amer. Microscop. Soc. 53: 277. Jul 1934.
Typus: *K. bufonis* (Dobell) B. V. Travis (*Monocercomonas bufonis* Dobell).

(≡) *Tetramastix* A. G. Alexeev in Compt.-Rend. Séances Mém. Soc. Biol. 79: 1076. 2 Dec 1916.

A3. CHLOROPHYCEAE

Acetabularia J. V. Lamour. in Nouv. Bull. Sci. Soc. Philom. Paris 3: 185. Dec 1812.
Typus: *A. acetabulum* (L.) P. C. Silva (in Univ. Calif. Publ. Bot. 25: 255. 1952) (*Madrepora acetabulum* L.).

(≡) *Acetabulum* Boehm. in Ludwig, Def. Gen. Pl., ed. 3: 504. 1760.

Anadyomene J. V. Lamour. in Nouv. Bull. Sci. Soc. Philom. Paris 3: 187. Dec 1812 (*'Anadyomena'*) (*orth. cons.*).
Typus: *A. flabellata* J. V. Lamour. [= *A. stellata* (Wulfen) C. Agardh (*Ulva stellata* Wulfen)].

Aphanochaete A. Braun, Betracht. Erschein. Verjüng. Natur: 196. 1850.
Typus: *A. repens* A. Braun [= *A. confervicola* (Nägeli) Rabenh. (*Herposteiron confervicola* Nägeli)].

(=) *Herposteiron* Nägeli in Kützing, Sp. Alg.: 424. 23-24 Jul 1849.
Typus: *H. confervicola* Nägeli

Bambusina Kütz., Sp. Alg.: 188. 23-24 Jul 1849.
Typus: *B. brebissonii* Kütz., nom. illeg. (*Didymoprium borreri* Ralfs, *B. borreri* (Ralfs) Cleve).

(=) *Gymnozyga* Ehrenb. ex Kütz., Sp. Alg.: 188. 23-24 Jul 1849.
Typus: *G. moniliformis* Ehrenb. ex Kütz.

Chaetomorpha Kütz., Phycol. Germ.: 203. 14-16 Aug 1845.
Typus: *C. melagonium* (Weber & Mohr) Kütz. (*Conferva melagonium* Weber & Mohr).

(=) *Chloronitum* Gaillon in Cuvier, Dict. Sci. Nat. 53: 389. 1828.
Typus (vide Silva in Univ. Calif. Publ. Bot. 25: 270. 1952): *C. aereum* (Dillwyn) Gaillon (*Conferva aerea* Dillwyn).

(=) *Spongopsis* Kütz., Phycol. General.: 261. 14-16 Sep 1843.
Typus: *S. mediterranea* Kütz.

Chlamydomonas Ehrenb. in Abh. Königl. Akad. Wiss. Berlin, Phys. Abh., 1833: 288. 1834 *('Chlamidomonas') (orth. cons.).*
Typus: *C. pulvisculus* (O. F. Müll.) Ehrenb. (*Monas pulvisculus* O. F. Müll.).

(=) *Protococcus* C. Agardh, Syst. Alg.: xvii, 13. Mai-Sep 1824.
Typus (vide Drouet & Daily in Butler Univ. Bot. Stud. 12: 167. 1956): *P. nivalis* (F. A. Bauer) C. Agardh (*Uredo nivalis* F. A. Bauer).

(=) *Sphaerella* Sommerf. in Mag. Naturvidensk. 4: 252. 1824.
Typus (vide Hazen in Mem. Torrey Bot. Club 6: 238. 1899): *S. nivalis* (Bauer) Sommerf. (*Uredo nivalis* Bauer).

Chlorococcum Menegh. in Mem. Reale Accad. Sci. Torino, ser. 2, 5: 24. 1842.
Typus: *C. infusionum* (Schrank) Menegh. (*Lepra infusionum* Schrank) *(typ. cons.).*

(H) *Chlorococcum* Fr., Syst. Orb. Veg.: 356. Dec 1825 [*Chloroph.: Chlamydomonad.*].
≡ *Protococcus* C. Agardh 1824 *(nom. rej.* sub *Chlamydomonas).*
≡ *Sphaerella* Sommerf. 1824 *(nom. rej.* sub *Chlamydomonas).*

Chloromonas Gobi in Bot. Zap. 15: 232, 255. 1899-1900.
Typus: *C. reticulata* (Gorozh.) Gobi (*Chlamydomonas reticulata* Gorozh.).

(H) *Chloromonas* Kent, Man. Infus.: 369, 401. 1881 [*Euglenoph.: Euglen.*].
≡ *Cryptoglena* Ehrenb. 1832.

(=) *Tetradonta* Korshikov in Russk. Arh. Protistol. 4: 183, 195. 1925.
Typus: *T. variabilis* Korshikov

(=) *Platychloris* Pascher, Süsswasserflora 4: 138, 331. Jan-Mar 1927.
Typus: *P. minima* Pascher (*Chlamydomonas minima* Pascher, non P. A. Dang.).

Cladophora Kütz., Phycol. General.: 262. 14-16 Sep 1843.
Typus: *C. oligoclona* (Kütz.) Kütz. (*Conferva oligoclona* Kütz.).

(=) *Conferva* L., Sp. Pl.: 1164. 1 Mai 1753.
Typus (vide Bonnem. in J. Phys. Chim. Hist. Nat. Arts 94: 198. 1822): *C. rupestris* L.

(=) *Annulina* Link in Nees, Horae Phys. Berol.: 4. 1-8 Feb 1820.
Typus (vide Silva in Univ. Calif. Publ. Bot. 25: 270. 1952): *A. glomerata* (L.) Nees (Horae Phys. Berol.: [index]. 1820) (*Conferva glomerata* L.).

Cladophoropsis Børgesen in Overs. Kongel. Danske Vidensk. Selsk. Forh. Medlemmers Arbeider 1905: 288. 10 Jun 1905.
Typus: *C. membranacea* (C. Agardh) Børgesen (*Conferva membranacea* C. Agardh).

(=) *Spongocladia* Aresch. in Öfvers. Förh. Kongl. Svenska Vetensk.-Akad. 10: 202. 1853.
Typus: *S. vaucheriiformis* Aresch.

Coleochaete Bréb. in Ann. Sci. Nat., Bot., ser. 3, 1: 29. Jan 1844.
Typus: *C. scutata* Bréb.

(=) *Phyllactidium* Kütz., Phycol. General.: 294. 14-16 Sep 1843.
Typus (vide Meneghini in Atti Riunione Sci. Ital. 6: 457. 1845): *P. pulchellum* Kütz.

Enteromorpha Link in Nees, Horae Phys. Berol.: 5. 1-8 Feb 1820.
Typus: *E. intestinalis* (L.) Nees (Horae Phys. Berol.: [index]. 1820.) (*Ulva intestinalis* L.).

(≡) *Splaknon* Adans., Fam. Pl. 2: 13, 607. Jul-Aug 1763 (typ. des.: Silva in Univ. Calif. Publ. Bot. 25: 294. 1952).

Gloeococcus A. Braun, Betracht. Erschein. Verjüng. Natur: 169. 1850.
Typus: *G. minor* A. Braun *(typ. cons.)*.

(H) *Gloiococcus* Shuttlew. in Biblioth. Universelle Genève, ser. 2, 25: 405. Feb 1840 [*Algae* incertae sedis].
Typus: *G. grevillei* (C. Agardh) Shuttlew. (*Haematococcus grevillei* C. Agardh).

Gongrosira Kütz., Phycol. General.: 281. 14-16 Sep 1843.
Typus: *G. sclerococcus* Kütz., nom. illeg. (*Stereococcus viridis* Kütz., *G. viridis* (Kütz.) De Toni).

(≡) *Stereococcus* Kütz. in Linnaea 8: 379. 1833.

Haematococcus Flot. in Nov. Actorum Acad. Caes. Leop.-Carol. Nat. Cur. 20: 413. 1844.
Typus: *H. pluvialis* Flot. *(typ. cons.)* [= *H. lacustris* (Gir.-Chantr.) Rostaf. (*Volvox lacustris* Gir.-Chantr.)].

(H) *Haematococcus* C. Agardh, Icon. Alg. Eur.: ad t. 22. 1830 [*Cyanoph.: Chroococc.*].
Typus (vide Morren in Nouv. Mém. Acad. Roy. Sci. Bruxelles 14(7): 9. 1841): *H. sanguineus* (C. Agardh) C. Agardh (*Palmella sanguinea* C. Agardh).

(=) *Disceraea* Morren & C. Morren in Nouv. Mém. Acad. Roy. Sci. Bruxelles 14(5): 37. 1841.
Typus: *D. purpurea* Morren & C. Morren

Halimeda J. V. Lamour. in Nouv. Bull. Sci. Soc. Philom. Paris 3: 186. Dec 1812 *('Halimedea') (orth. cons.)*.
Typus: *H. tuna* (J. Ellis & Sol.) J. V. Lamour. (Hist. Polyp. Corall.: 309. 1816) (*Corallina tuna* J. Ellis & Sol.).

(≡) *Sertularia* Boehm. in Ludwig, Def. Gen. Pl., ed. 3: 504. 1760 (typ. des.: Silva in Univ. Calif. Publ. Bot. 25: 294. 1952).

Hydrodictyon Roth, Bemerk. Crypt. Wassergew.: 48. Feb-Aug 1797.
Typus: *H. reticulatum* (L.) Bory (Dict. Class. Hist. Nat. 6: 506. 9 Oct 1824) (*Conferva reticulata* L.).

(≡) *Reticula* Adans., Fam. Pl. 2: 3, 598. Jul-Aug 1763.

Microspora Thur. in Ann. Sci. Nat., Bot., ser. 3, 14: 221. Oct 1850.
Typus: *M. floccosa* (Vaucher) Thur. (*Prolifera floccosa* Vaucher).

(H) *Microspora* Hassall in Ann. Mag. Nat. Hist. 11: 363. Mai 1843 [*Chloroph.: Cladophor.*].
Typus: non designatus.

Mougeotia C. Agardh, Syst. Alg.: xxvi, 83. Mai-Sep 1824.
Typus: *M. genuflexa* (Roth) C. Agardh (*Conferva genuflexa* Roth).

(H) *Mougeotia* Kunth in Humboldt & al., Nov. Gen. Sp. 5: ed. f°: 253; ed. 4°: 326. 1823 [*Stercul.*].
Typus: non designatus.

(≡) *Serpentinaria* Gray, Nat. Arr. Brit. Pl. 1: 299. 1 Nov 1821 (typ. des.: Silva in Univ. Calif. Publ. Bot. 25: 252. 1952).

(=) *Agardhia* Gray, Nat. Arr. Brit. Pl. 1: 279, 299. 1 Nov 1821.
Typus: *A. caerulescens* (Sm.) Gray (*Conferva caerulescens* Sm.).

Prasiola Menegh. in Nuovi Saggi Imp. Regia Accad. Sci. Padova 4: 360. 1838.
Typus: *P. crispa* (Lightf.) Kütz. (Phycol. General.: 295. 1843) (*Ulva crispa* Lightf.).

(=) *Humida* Gray, Nat. Arr. Brit. Pl. 1: 278, 281. 1 Nov 1821.
Typus (vide Drouet in Acad. Nat. Sci. Philadelphia Monogr. 15: 312. 1968): *H. muralis* (Dillwyn) Gray (*Conferva muralis* Dillwyn).

Schizogonium Kütz., Phycol. General.: 245. 14-16 Sep 1843.
Typus: *S. murale* (Dillwyn) Kütz. (*Conferva muralis* Dillwyn).

(≡) *Humida* Gray, Nat. Arr. Brit. Pl. 1: 278, 281. 1 Nov 1821.

Sirogonium Kütz., Phycol. General.: 278. 14-16 Sep 1843.
Typus: *S. sticticum* (Sm.) Kütz. (*Conferva stictica* Sm.).

(≡) *Choaspis* Gray, Nat. Arr. Brit. Pl. 1: 279, 299. 1 Nov 1821

Sphaerozosma Ralfs, Brit. Desmid: 65. 1 Jan 1848.
Typus: *S. vertebratum* Ralfs

(H) *Sphaerozosma* Corda, Icon. Fung. 5: 27. Jun 1842 [*Fungi*].
≡ *Sphaerosoma* Klotzsch 1839.

Spirogyra Link in Nees, Horae Phys. Berol.: 5. 1-8 Feb 1820.
Typus: *S. porticalis* (O. F. Müll.) Dumort. (Comment. Bot.: 99. Nov (sero) - Dec (prim.) 1822) (*Conferva porticalis* O. F. Müll.).

(=) *Conjugata* Vaucher, Hist. Conferv. Eau Douce: 3, 37. Mar 1803.
Typus (vide Bonnemaison in J. Phys. Chim. Hist. Nat. Arts 94: 195. 1822): *C. princeps* Vaucher

Stigeoclonium Kütz., Phycol. General.: 253. 14-16 Sep 1843 *('Stygeoclonium') (orth. cons.)*.
Typus: *S. tenue* (C. Agardh) Kütz. (*Draparnaldia tenuis* C. Agardh).

(=) *Myxonema* Fr., Syst. Orb. Veg.: 343. Dec 1825.
Typus (vide Hazen in Mem. Torrey Bot. Club 11: 193. 1902): *M. lubricum* (Dillwyn) Fr. (Fl. Scan.: 329. 1835) (*Conferva lubrica* Dillwyn).

Struvea Sond. in Bot. Zeitung (Berlin) 3: 49. 24 Jan 1845.
Typus: *S. plumosa* Sond.

(H) *Struvea* Rchb., Deut. Bot. Herb.-Buch, Syn.: 222, 236. Jul 1841 [*Tax.*].
≡ *Torreya* Arn. 1838 *(nom. cons.)* (17).

Trentepohlia Mart., Fl. Crypt. Erlang.: 351. Jun 1817.
Typus: *T. aurea* (L.) Mart. (*Byssus aurea* L.).

(H) *Trentepohlia* Roth, Catal. Bot. 2: 73. 1800 [*Cruc.*].
Typus: non designatus.

(=) *Byssus* L., Sp. Pl.: 1168. 1 Mai 1753.
Typus (vide Fries, Stirp. Agri Femsion.: 42. 1825): *B. jolithus* L.

Ulva L., Sp. Pl.: 1163. 1 Mai 1753.
Typus: *U. lactuca* L. *(typ. cons.)*.

Urospora Aresch. in Nova Acta Regiae Soc. Sci. Upsal., ser. 3, 6(2): 15. 1866.
Typus: *U. mirabilis* Aresch.

(=) *Hormiscia* Fr., Fl. Scan.: 326. 1835.
Typus (vide Silva in Univ. Calif. Publ. Bot. 25: 270. 1952): *H. penicilliformis* (Roth) Fr. (*Conferva penicilliformis* Roth).

(=) *Codiolum* A. Braun, Alg. Unicell.: 19. Apr-Oct 1855.
Typus: *C. gregarium* A. Braun

Zygnema C. Agardh in Liljeblad, Utkast Sv. Fl., ed. 3: 492, 595. 1816.
Typus: *Z. cruciatum* (Vaucher) C. Agardh (*Conjugata cruciata* Vaucher).

(=) *Lucernaria* Roussel, Fl. Calvados, ed. 2: 20, 84. 1806.
Typus: *L. pellucida* Roussel

Zygogonium Kütz., Phycol. General.: 280. 14-16 Sep 1843.
Typus: *Z. ericetorum* (Roth) Kütz. (*Conferva ericetorum* Roth).

(≡) *Leda* Bory, Dict. Class. Hist. Nat. 1: 595. 27 Mai 1822 (typ. des.: Silva in Univ. Calif. Publ. Bot. 25: 253. 1952).

A4. CHRYSOPHYCEAE

Anthophysa Bory, Dict. Class. Hist. Nat. 1: 427, 597. 27 Mai 1822 *('Anthophysis')* *(orth. cons.)*.
Typus: *A. muelleri* Bory, nom. illeg. (*Volvox vegetans* O. F. Müll., *A. vegetans* (O. F. Müll.) F. Stein).

Hydrurus C. Agardh, Syst. Alg.: xviii, 24. Mai-Sep 1824.
Typus: *H. vaucheri* C. Agardh, nom. illeg. (*Conferva foetida* Vill., *H. foetidus* (Vill.) Trevis.).

(≡) *Carrodorus* Gray, Nat. Arr. Brit. Pl. 1: 318, 350. 1 Nov 1821.

(=) *Cluzella* Bory, Dict. Class. Hist. Nat. 3: 14. 6 Sep 1823.
Typus: *C. myosurus* (Ducluz.) Bory (Dict. Class. Hist. Nat. 4: 234. 27 Dec 1823) (*Batrachospermum myosurus* Ducluz.).

A5. CYANOPHYCEAE

Anabaena Bory ex Bornet & Flahault in Ann. Sci. Nat., Bot., ser. 7, 7: 180, 224. 1 Jan 1886 .
Typus: *A. oscillarioides* Bory ex Bornet & Flahault

(H) *Anabaena* A. Juss., Euphorb. Gen.: 46. 21 Feb 1824 [*Euphorb.*].
Typus: *A. tamnoides* A. Juss.

Aphanothece Nägeli in Neue Denkschr. Allg. Schweiz. Ges. Gesammten Naturwiss. 10(7): 59. 1849.
Typus: *A. microscopica* Nägeli

(=) *Coccochloris* Spreng., Mant. Prim. Fl. Hal.: 14. 4 Jul 1807.
Typus: *C. stagnina* Spreng.

Gloeocapsa Kütz., Phycol. General.: 173. 14-16 Sep 1843.
Typus: *G. atrata* Kütz., nom. illeg. (*Microcystis atra* Kütz.).

(=) *Bichatia* Turpin in Mém. Mus. Hist. Nat. 16: 163. 1828.
Typus: *B. vesiculinosa* Turpin

Homoeothrix (Thur. ex Bornet & Flahault) Kirchn. in Engler & Prantl, Nat. Pflanzenfam. 1(1a): 85, 87. Aug 1898 (*Calothrix* sect. *Homoeothrix* Thur. ex Bornet & Flahault in Ann. Sci. Nat., Bot., ser. 7, 3: 345, 347. 1 Jan 1886).
Typus: *Calothrix juliana* Bornet & Flahault (*H. juliana* (Bornet & Flahault) Kirchn.).

(=) *Amphithrix* Bornet & Flahault in Ann. Sci. Nat., Bot., ser. 7, 3: 340, 343. 1 Jan 1886.
Typus (vide Geitler in Engler & Prantl, Nat. Pflanzenfam., ed. 2, 1b: 175. 1942): *A. janthina* Bornet & Flahault

(=) *Tapinothrix* Sauv. in Bull. Soc. Bot. France 39: cxxiii. 1892.
Typus: *T. bornetii* Sauv.

Lyngbya C. Agardh ex Gomont in Ann. Sci. Nat., Bot., ser. 7, 16: 95, 118. 1 Jan 1892.
Typus: *L. confervoides* C. Agardh ex Gomont

(H) *Lyngbyea* Sommerf., Suppl. Fl. Lapp.: 189. 1826 [*Bacillarioph.*].
Typus: non designatus.

Microchaete Thur. ex Bornet & Flahault in Ann. Sci. Nat., Bot., ser. 7, 5: 82, 83. 1 Jan 1886.
Typus: *M. grisea* Thur. ex Bornet & Flahault

(H) *Microchaete* Benth., Pl. Hartw.: 209. Nov 1845 [*Comp.*].
Typus (vide Pfeiffer, Nomencl. Bot. 2: 304. 1874): *M. pulchella* (Kunth) Benth. (*Cacalia pulchella* Kunth).

Microcystis Lemmerm., Krypt.-Fl. Brandenburg 3: 45, 72. 4 Mar 1907.
Typus: *M. aeruginosa* (Kütz.) Lemmerm. (*Micraloa aeruginosa* Kütz.) *(typ. cons.)*.

(H) *Microcystis* Kütz. in Linnaea 8: 372. 1833 [*Euglenoph.: Euglen.*].
Typus (vide Drouet & Daily in Butler Univ. Bot. Stud. 12: 152. 1956): *M. noltei* (C. Agardh) Kütz. (*Haematococcus noltei* C. Agardh).

Nodularia Mert. ex Bornet & Flahault in Ann. Sci. Nat., Bot., ser. 7, 7: 180, 243. 1 Jan 1886.
Typus: *N. spumigena* Mert. ex Bornet & Flahault

(H) *Nodularia* Link ex Lyngb., Tent. Hydrophytol. Dan.: xxx, 99. Apr-Aug 1819 [*Rhodoph.: Leman.*].
≡ *Lemanea* Bory 1808 *(nom. cons.)*.

Rivularia C. Agardh ex Bornet & Flahault in Ann. Sci. Nat., Bot., ser. 7, 3: 341; 4: 345. 1 Jan 1886.
Typus: *R. atra* Roth ex Bornet & Flahault

(H) *Rivularia* Roth, Catal. Bot. 1: 212. Jan-Feb 1797 [*Chloroph.: Chaetophor.*]
Typus (vide Hazen in Mem. Torrey Bot. Club 11: 210. 1902): *R. cornu-damae* Roth

Trichodesmium Ehrenb. ex Gomont in Ann. Sci. Nat., Bot., ser. 7, 16: 96, 193. 1 Jan 1892.
Typus: *T. erythraeum* Ehrenb. ex Gomont

(H) *Trichodesmium* Chevall., Fl. Gén. Env. Paris 1: 382. 5 Aug 1826 [*Fungi*].
≡ *Graphiola* Poit. 1824.

A6. DINOPHYCEAE

Abedinium Loebl. & A. R. Loebl. in Stud. Trop. Oceanogr. 3: 1, 14. Jun 1966.
Typus: *A. dasypus* (Cachon & Cachon-Enj.) Loebl. & A. R. Loebl. (*Leptophyllus dasypus* Cachon & Cachon-Enj.).

(≡) *Leptophyllus* Cachon & Cachon-Enj. in Bull. Inst. Océanogr. 62(1292): 7. Feb 1964.

Amphilothus Kof. ex Poche in Arch. Protistenk. 30: 264. 12 Sep 1913.
Typus: *A. elegans* (F. Schütt) Er. Lindem. (in Engler & Prantl, Nat. Pflanzenfam., ed. 2, 2: 69. 1928) (*Amphitholus elegans* F. Schütt).

(≡) *Amphitholus* F. Schütt in Ergebn. Plankt.-Exped. Humboldt-Stiftung IV.M.a.A: 34. Jan 1895.

Dinamoebidium Pascher in Arch. Protistenk. 37: 31. 30 Aug 1916.
Typus: *D. varians* (Pascher) Pascher (*Dinamoeba varians* Pascher).

(≡) *Dinamoeba* Pascher in Arch. Protistenk. 36: 118. 8 Jan 1916.

Dogelodinium Loebl. & A. R. Loebl. in Stud. Trop. Oceanogr. 3: 1, 27. Jun 1966.
Typus: *D. ovoides* (Cachon) Loebl. & A. R. Loebl. (*Collinella ovoides* Cachon).

(≡) *Collinella* Cachon in Ann. Sci. Nat., Zool., ser. 12, 6: 49. Apr-Jun 1964.

Gyrodinium Kof. & Swezy in Mem. Univ. Calif. 5: 273. 28 Jun 1921.
Typus: *G. spirale* (Bergh) Kof. & Swezy (*Gymnodinium spirale* Bergh).

(≡) *Spirodinium* F. Schütt in Engler & Prantl, Nat. Pflanzenfam. 1(1b): 3, 5. 1896.

Keppenodinium Loebl. & A. R. Loebl. in Stud. Trop. Oceanogr. 3: 1, 38. Jun 1966.
Typus: *K. mycetoides* (Cachon) Loebl. & A. R. Loebl. (*Hollandella mycetoides* Cachon).

(≡) *Hollandella* Cachon in Ann. Sci. Nat., Zool., ser. 12, 6: 53. Apr-Jun 1964.

Latifascia Loebl. & A. R. Loebl. in Stud. Trop. Oceanogr. 3: 1, 38. Jun 1966.
Typus: *L. inaequalis* (Kof. & Skogsb.) Loebl. & A. R. Loebl. (*Heteroschisma inaequale* Kof. & Skogsb.).

(≡) *Heteroschisma* Kof. & Skogsb. in Mem. Mus. Comp. Zool. Harvard Coll. 51: 36. Dec 1928.

Sphaeripara Poche in Arch. Naturgesch. 77, Suppl. 1: 80. Sep 1911.
Typus: *S. catenata* (Neresh.) Loebl. & A. R. Loebl. (in Stud. Trop. Oceanogr. 3: 56. Jun 1966) (*Lohmannia catenata* Nehresh.).

(≡) *Lohmannia* Neresh. in Biol. Zentralbl. 23: 757. 1 Nov 1903.

A7. EUGLENOPHYCEAE

Anisonema Dujard., Hist. Nat. Zoophyt.: 327, 344. 1841.
Typus: *A. acinus* Dujard.

(H) *Anisonema* A. Juss., Euphorb. Gen.: 19. 21 Feb 1824 [*Euphorb.*].
Typus: *A. reticulatum* (Poir.) A. Juss. (*Phyllanthus reticulatus* Poir.).

Astasia Dujard., Hist. Nat. Zoophyt.: 353, 356. 1841.
Typus: *A. limpida* Dujard. *(typ. cons.)*.

(H) *Astasia* Ehrenb. in Ann. Phys. Chem. 94: 508. 1830 [*Euglenoph.: Euglen.*].
Typus (vide Silva in Taxon 9: 20. 1960): *A. haematodes* Ehrenb.

Lepocinclis Perty in Mitth. Naturf. Ges. Bern 1849: 28. 15 Feb 1849.
Typus: *L. globulus* Perty

(=) *Crumenula* Dujard. in Ann. Sci. Nat., Zool., ser. 2, 5: 204, 205. Apr 1836.
Typus: *C. texta* Dujard.

Phacus Dujard., Hist. Nat. Zoophyt.: 327, 334. 1841.
Typus: *P. longicauda* (Ehrenb.) Dujard. (*Euglena longicauda* Ehrenb.) *(typ. cons.)*.

(H) *Phacus* Nitzsch in Ersch & Gruber, Allg. Encycl. Wiss. Künste, Sect. 1, 16: 69. 1827 [*Euglenoph.: Euglen.*].
≡ *Virgulina* Bory 1823.

A8. PHAEOPHYCEAE

Agarum Dumort., Comment. Bot.: 102. Nov (sero) - Dec (prim.) 1822.
Typus: *A. clathratum* Dumort. (*Fucus agarum* S. G. Gmel.).

(H) *Agarum* Link in Neues J. Bot. 3(1,2): 7. Apr 1809 [*Rhodoph.: Delesser.*].
Typus: *A. rubens* (L.) Link (*Fucus rubens* L.).

Alaria Grev., Alg. Brit.: xxxix, 25. Mar 1830.
Typus: *A. esculenta* (L.) Grev. (*Fucus esculentus* L.).

(≡) *Musaefolia* Stackh. in Mém. Soc. Imp. Naturalistes Moscou 2: 53, 66. 1809.

Ascophyllum Stackh. in Mém. Soc. Imp. Naturalistes Moscou 2: 54, 66. 1809 *('Ascophylla') (orth. cons.)*.
Typus: *A. laevigatum* Stackh., nom. illeg. (*Fucus nodosus* L., *A. nodosum* (L.) Le Jol.).

(≡) *Nodularius* Roussel, Fl. Calvados, ed. 2, 93. 1806 (typ. des.: Silva in Univ. Calif. Publ. Bot. 25: 299. 1952).

Carpomitra Kütz., Phycol. General.: 343. 14-16 Sep 1843.
Typus: *C. cabrerae* (Clemente) Kütz. (*Fucus cabrerae* Clemente).

(≡) *Dichotomocladia* Trevis. in Atti Riunione Sci. Ital. 4: 333. 15 Aug 1843.
(=) *Chytraphora* Suhr in Flora 17: 721. 14 Dec 1834.
Typus: *C. filiformis* Suhr

Chordaria C. Agardh, Syn. Alg. Scand.: xii. Mai-Dec 1817.
Typus: *C. flagelliformis* (O. F. Müll.) C. Agardh (*Fucus flagelliformis* O. F. Müll.).

(H) *Chordaria* Link in Neues J. Bot. 3(1,2): 8. Apr 1809 [*Phaeoph.: Chord.*].
≡ *Chorda* Stackh. 1797.

Cystophora J. Agardh in Linnaea 15: 3. Apr 1841.
Typus: *C. retroflexa* (Labill.) J. Agardh (*Fucus retroflexus* Labill.).

(=) *Blossevillea* Decne. in Bull. Acad. Roy. Sci. Bruxelles 7(1): 410. 1840 (*'Blosvillea'*).
Typus (vide Silva in Univ. Calif. Publ. Bot. 25: 279. 1952): *B. torulosa* (R. Br. ex Turner) Decne. (in Ann. Sci. Nat., Bot., ser. 2, 17: 331. Jun 1842) (*Fucus torulosus* R. Br. ex Turner).

Cystoseira C. Agardh, Spec. Alg. 1: 50. Jan-Apr 1820.
Typus: *C. concatenata* (L.) C. Agardh (*Fucus concatenatus* L.) [= *C. foeniculacea* (L.) Grev. (*Fucus foeniculaceus* L.)].

(=) *Gongolaria* Boehm. in Ludwig, Def. Gen. Pl., ed. 3: 503. 1760.
Typus: *Fucus abies-marina* S. G. Gmel.
(=) *Baccifer* Roussel, Fl. Calvados, ed. 2: 94. 1806.
Typus: *Fucus baccatus* S. G. Gmel.
(=) *Abrotanifolia* Stackh. in Mém. Soc. Imp. Naturalistes Moscou 2: 56, 81. 1809.
Typus (vide Papenfuss in Hydrobiologia 2: 184. 1950): *A. loeflingii* Stackh. (*Fucus abrotanifolius* L.).
(=) *Ericaria* Stackh. in Mém. Soc. Imp. Naturalistes Moscou 2: 56, 80. 1809.
Typus (vide Papenfuss in Hydrobiologia 2: 185. 1950): *Fucus ericoides* L.

Desmarestia J. V. Lamour. in Ann. Mus. Natl. Hist. Nat. 20: 43. 1813.
Typus: *D. aculeata* (L.) J. V. Lamour. (*Fucus aculeatus* L.).

(≡) *Hippurina* Stackh. in Mém. Soc. Imp. Naturalistes Moscou 2: 59, 89. 1809 (typ. des.: Silva in Univ. Calif. Publ. Bot. 25: 257. 1952).
(=) *Herbacea* Stackh. in Mém. Soc. Imp. Naturalistes Moscou 2: 58, 89. 1809.
Typus: *H. ligulata* Stackh. (*Fucus ligulatus* Lightf., non S. G. Gmel.).
(=) *Hyalina* Stackh. in Mém. Soc. Imp. Naturalistes Moscou 2: 58, 88. 1809.
Typus: *H. mutabilis* Stackh., nom. illeg. (*Fucus viridis* O. F. Müll.).

Desmotrichum Kütz., Phycol. Germ.: 244. 14-16 Aug 1845.
Typus: *D. balticum* Kütz.

(H) *Desmotrichum* Blume, Bijdr.: 329. 20 Sep - 7 Dec 1825 [*Orchid.*].
≡ *Flickingeria* A. D. Hawkes 1961.

(=) *Diplostromium* Kütz., Phycol. General.: 298. 14-16 Sep 1843.
Typus (vide Silva in Univ. Calif. Publ. Bot. 25: 257. 1952): *D. tenuissimum* (C. Agardh) Kütz. (*Zonaria tenuissima* C. Agardh).

Dictyopteris J. V. Lamour. in Nouv. Bull. Sci. Soc. Philom. Paris 1: 332. Mai 1809.
Typus: *D. polypodioides* (DC.) J. V. Lamour. (*Fucus polypodioides* Desf., non S. G. Gmel., *Ulva polypodioides* DC.).

(≡) *Granularius* Roussel, Fl. Calvados, ed. 2: 90. 1806 (typ. des.: Silva in Regnum Veg. 101: 745. 1979).

(=) *Neurocarpus* F. Weber & D. Mohr in Beitr. Naturk. 1: 300. 15 Nov 1805-1806.
Typus: *N. membranaceus* (Stackh.) F. Weber & D. Mohr (*Polypodoidea membranacea* Stackh.; *Fucus membranaceus* Stackh., non Burm. f.).

Dictyosiphon Grev., Alg. Brit.: xliii, 55. Mar 1830.
Typus: *D. foeniculaceus* (Huds.) Grev. (*Conferva foeniculacea* Huds.) (etiam vide *Scytosiphon* [*Phaeoph.*].

Dictyota J. V. Lamour. in J. Bot. (Desvaux) 2: 38. 1-8 Apr 1809.
Typus: *D. dichotoma* (Huds.) J. V. Lamour. (*Ulva dichotoma* Huds.) *(typ. cons.)*.

Ectocarpus Lyngb., Tent. Hydrophytol. Dan.: xxxi, 130. Apr-Aug 1819.
Typus: *E. siliculosus* (Dillwyn) Lyngb. (*Conferva siliculosa* Dillwyn).

(=) *Colophermum* Raf., Précis Découv. Somiol.: 49. Jun-Dec 1814.
Typus: *C. floccosum* Raf.

Elachista Duby, Bot. Gall.: 972. Mai 1830 *('Elachistea') (orth. cons.)*.
Typus: *E. scutellata* Duby, nom. illeg. (*Conferva scutulata* Sm., *E. scutulata* (Sm.) Aresch.).

(=) *Opospermum* Raf., Précis Découv. Somiol.: 48. Jun-Dec 1814.
Typus: *O. nigrum* Raf.

Halidrys Lyngb., Tent. Hydrophytol. Dan.: xxix, 37. Apr-Aug 1819.
Typus: *H. siliquosa* (L.) Lyngb. (*Fucus siliquosus* L.) *(typ. cons.)*.

(H) *Halidrys* Stackh. in Mém. Soc. Imp. Naturalistes Moscou 2: 53, 62. 1809 (typ. des.: Papenfuss in Hydrobiologia 2: 186. 1950) [*Phaeoph.: Fuc.*].
≡ *Fucus* L. 1753 (typ. des.: De Toni in Flora 74: 173. 1891).

(≡) *Siliquarius* Roussel, Fl. Calvados, ed. 2: 94. 1806.

Hesperophycus Setch. & N. L. Gardner in Univ. Calif. Publ. Bot. 4: 127. 26 Aug 1910.
Typus: *H. californicus* P. C. Silva (in Taxon 39: 5. 22 Feb 1990) *(typ. cons.)*.

Himanthalia Lyngb., Tent. Hydrophytol. Dan.: xxix, 36. Apr-Aug 1819.
Typus: *H. lorea* (L.) Lyngb. (*Fucus loreus* L.) [= *H. elongata* (L.) Gray (*Fucus elongatus* L.)].

(≡) *Funicularius* Roussel, Fl. Calvados, ed. 2: 91. 1806.

(=) *Lorea* Stackh. in Mém. Soc. Imp. Naturalistes Moscou 2: 60, 94. 1809.
Typus: *L. elongata* (L.) Stackh. (*Fucus elongatus* L.).

Hormosira (Endl.) Menegh. in Nuovi Saggi Imp. Regia Accad. Sci. Padova 4: 368. 1838 (*Cystoseira* sect. *Hormosira* Endl., Gen. Pl.: 10. Aug 1836).
Typus: *Fucus moniliformis* Labill., non Esper [= *H. banksii* (Turner) Decne. (*Fucus banksii* Turner)].

(≡) *Moniliformia* J. V. Lamour. in Bory, Dict. Class. Hist. Nat. 7: 71. 5 Mar 1825.

Laminaria J. V. Lamour. in Ann. Mus. Natl. Hist. Nat. 20: 40. 1813.
Typus: *L. digitata* (Huds.) J. V. Lamour. (*Fucus digitatus* Huds.).

(=) *Saccharina* Stackh. in Mém. Soc. Imp. Naturalistes Moscou 2: 53, 65. 1809.
Typus (vide Silva in Univ. Calif. Publ. Bot. 25: 259. 1952): *S. plana* Stackh. (*Fucus saccharinus* L.).

Leptonematella P. C. Silva in Taxon 8: 63. 12 Mar 1959.
Typus: *L. fasciculata* (Reinke) P. C. Silva (*Leptonema fasciculatum* Reinke).

Padina Adans., Fam. Pl. 2: 13, 586. Jul-Aug 1763.
Typus: *P. pavonica* (L.) J. V. Lamour. (Hist. Polyp. Corall.: 304. 1816) (*Fucus pavonicus* L.).

Petalonia Derbès & Solier in Ann. Sci. Nat., Bot., ser. 3, 14: 265. Nov 1850.
Typus: *P. debilis* (C. Agardh) Derbès & Solier (*Laminaria debilis* C. Agardh).

(=) *Fasciata* Gray, Nat. Arr. Brit. Pl. 1: 383. 1 Nov 1821.
Typus (vide Silva in Univ. Calif. Publ. Bot. 25: 299. 1952): *F. attenuata* Gray, nom. illeg. (*Fucus fascia* O. F. Müll.).

Saccorhiza Bach. Pyl., Fl. Terre-Neuve: 23. 22 Jan 1830.
Typus: *Laminaria bulbosa* (Huds.) J. Agardh (Spec. Alg. 1: 138. 1848) (*Fucus bulbosus* Huds.) [= *S. polyschides* (Lightf.) Batters (*Fucus polyschides* Lightf.)].

(=) *Polyschidea* Stackh. in Mém. Soc. Imp. Naturalistes Moscou 2: 53, 65. 1809.
Typus (vide Papenfuss in Hydrobiologia 2: 189. 1950): *Fucus polyschides* Lightf.

Sargassum C. Agardh, Spec. Alg. 1: 1. Jan-Apr 1820.
Typus: *S. bacciferum* (Turner) C. Agardh (*Fucus bacciferus* Turner).

(=) *Acinaria* Donati, Essai Hist. Nat. Mer Adriat.: 26, 33. Jan-Mar 1758.
Typus: *Sargassum donatii* (Zanardini) Kütz. (*S. vulgare* var. *donatii* Zanardini).

Scytosiphon C. Agardh, Spec. Alg. 1: 160. Jan-Apr 1820.
Typus: *S. lomentaria* (Lyngb.) Link (Handbuch 3: 232. 1833) (*Chorda lomentaria* Lyngb.) *(typ. cons.)* [= *S. simplicissimus* (Clemente) Cremades (*Ulva simplicissima* Clemente)].

(H) *Scytosiphon* C. Agardh, Disp. Alg. Suec.: 24. 11 Dec 1811 (typ. des.: Silva in Regnum Veg. 8: 205. 1956) [*Phaeoph.: Dictyosiphon.*].
≡ *Dictyosiphon* Grev. 1830 *(nom. cons.)*.

Spermatochnus Kütz., Phycol. General.: 334. 14-16 Sep 1843.
Typus: *S. paradoxus* (Roth) Kütz. (*Conferva paradoxa* Roth) *(typ. cons.)*.

Stilophora J. Agardh in Linnaea 15: 6. Apr 1841.
Typus: *S. rhizodes* (Turner) J. Agardh (*Fucus rhizodes* Turner) *(typ. cons.)*.

(H) *Stilophora* C. Agardh in Flora 10: 642. 7 Nov 1827 [*Phaeoph.: Punctar.*].
≡ *Hydroclathrus* Bory 1825.

Zonaria C. Agardh, Syn. Alg. Scand.: xx. Mai-Dec 1817.
Typus: *Z. flava* C. Agardh *(typ. cons.)* [= *Z. tournefortii* (Lamour.) Mont. (*Fucus tournefortii* Lamour.)].

(H) *Zonaria* Drap. ex F. Weber & D. Mohr in Beitr. Naturk. 1: 247-253. 15 Nov 1805-1806 [*Phaeoph.: Dictyot.*].
≡ *Padina* Adans. 1763 *(nom. cons.)*.

A9. RHODOPHYCEAE

Areschougia Harv. in Trans. Roy. Irish Acad. 22 (Sci.): 554. 1855.
Typus: *A. laurencia* (Hook. f. & Harv.) Harv. (*Thamnocarpus laurencia* Hook. f. & Harv.).

(H) *Areschougia* Menegh. in Giorn. Bot. Ital. 1(1,1): 293. Mai-Jun 1844 [*Phaeoph.: Elachist.*].
Typus (vide Silva in Univ. Calif. Publ. Bot. 25: 283. 1952): *A. stellaris* (Aresch.) Menegh. (*Elachista stellaris* Aresch.).

Audouinella Bory, Dict. Class. Hist. Nat. 3: 340. 6 Sep 1823 *('Auduinella') (orth. cons.)*.
Typus: *A. miniata* Bory [= *A. hermannii* (Roth) Duby (*Conferva hermannii* Roth)].

Bostrychia Mont. in Sagra, Hist. Phys. Cuba, Bot. Pl. Cell.: 39. 1842 (sero).
Typus: *B. scorpioides* (Huds.) Mont. (in Orbigny, Dict. Univ. Hist. Nat. 2: 661. 1842) (*Fucus scorpioides* Huds.).

(H) *Bostrychia* Fr., Syst. Mycol. 1: lii. 1 Jan 1821 [*Fungi*].
Typus: *B. chrysosperma* (Pers. : Fr.) Fr. (*Sphaeria chrysosperma* Pers. : Fr.).

(≡) *Amphibia* Stackh. in Mém. Soc. Imp. Naturalistes Moscou 2: 58, 89. 1809.

Botryocladia (J. Agardh) Kylin in Acta Univ. Lund., ser. 2, sect. 2, 27(11): 17. Mai-Oct 1931 (*Chrysymenia* sect. *Botryocladia* J. Agardh, Spec. Gen. Ord. Alg. 2: 214. Jan-Jun 1851).
Typus: *Chrysymenia uvaria* J. Agardh, nom. illeg. (*B. uvaria* Kylin, nom illeg., *Fucus botryoides* Wulfen, *B. botryoides* (Wulfen) Feldmann).

(=) *Myriophylla* Holmes in Ann. Bot. (London) 8: 340. Sep 1894.
Typus: *M. beckeriana* Holmes

Calliblepharis Kütz., Phycol. General.: 403. 14-16 Sep 1843.
Typus: *C. ciliata* (Huds.) Kütz. (*Fucus ciliatus* Huds.).

(≡) *Ciliaria* Stackh. in Mém. Soc. Imp. Naturalistes Moscou 2: 54, 70. 1809 (typ. des.: Papenfuss in Hydrobiologia 2: 191. 1950).

Catenella Grev., Alg. Brit.: lxiii, 166. Mar 1830.
Typus: *C. opuntia* (Gooden. & Woodw.) Grev. (*Fucus opuntia* Gooden. & Woodw.) [= *C. caespitosa* (With.) L. M. Irvine (*Ulva caespitosa* With.)].

(=) *Clavatula* Stackh. in Mém. Soc. Imp. Naturalistes Moscou 2: 95, 97. 1809.
Typus: *C. caespitosa* Stackh. (*Fucus caespitosus* Stackh., non Forssk.).

Ceramium Roth, Catal. Bot. 1: 146. Jan-Feb 1797.
Typus: *C. virgatum* Roth *(typ. cons.)* [= *Ceramium rubrum* (Huds.) C. Agardh (*Conferva rubra* Huds.)] .

(H) *Ceramion* Adans., Fam. Pl. 2: 13, 535. Jul-Aug 1763 [*Rhodoph.: Gracilar.*].
≡ *Ceramianthemum* Donati ex Léman 1817 *(nom. rej.* sub *Gracilaria).*

Chondria C. Agardh, Syn. Alg. Scand.: xviii. Mai-Dec 1817.
Typus: *C. tenuissima* (With.) C. Agardh (*Fucus tenuissimus* With.).

(=) *Dasyphylla* Stackh., Nereis Brit., ed. 2: ix, xi. 1816 (ante Aug).
Typus (vide Papenfuss in Hydrobiologia 2: 192. 1950): *D. woodwardii* Stackh. (*Fucus dasyphyllus* Woodw.).

Chylocladia Grev. in Hooker, Brit. Fl., ed. 4, 2(1): 256, 297. 1833.
Typus: *C. kaliformis* (With.) Grev. (*Fucus kaliformis* With.) [= *C. verticillata* (Lightf.) Bliding (*Fucus verticillatus* Lightf.)].

(≡) *Kaliformis* Stackh. in Mém. Soc. Imp. Naturalistes Moscou 2: 56, 78. 1809 (typ. des.: Papenfuss in Hydrobiologia 2: 198. 1950).

Corynomorpha J. Agardh in Acta Univ. Lund. 8 (sect. 3, 6): 3. 1872.
Typus: *C. prismatica* (J. Agardh) J. Agardh (*Dumontia prismatica* J. Agardh).

(≡) *Prismatoma* (J. Agardh) Harv., Index Gen. Alg.: 11. Jul-Aug 1860 (*Acrotilus* subg. *Prismatoma* J. Agardh, Spec. Gen. Ord. Alg. 2: 193. Jan-Jun 1851).

Cryptopleura Kütz., Phycol. General.: 444. 14-16 Sep 1843.
Typus: *C. lacerata* (S. G. Gmel.) Kütz. (*Fucus laceratus* S. G. Gmel.) [= *C. ramosa* (Huds.) Kylin ex Newton (*Ulva ramosa* Huds.)].

(H) *Cryptopleura* Nutt. in Trans. Amer. Philos. Soc., ser. 2, 7: 431. 2 Apr 1841 [*Comp.*].
Typus: *C. californica* Nutt.

(≡) *Papyracea* Stackh. in Mém. Soc. Imp. Naturalistes Moscou 2: 56, 76. 1809 (typ. des.: Papenfuss in Index Nom. Gen.: No. 00816. 1955).

Dasya C. Agardh, Syst. Alg.: xxxiv, 211. Mai-Sep 1824 *('Dasia') (orth. cons.)*.
Typus: *D. pedicellata* (C. Agardh) C. Agardh (*Sphaerococcus pedicellatus* C. Agardh) [= *D. baillouviana* (S. G. Gmel.) Mont. (*Fucus baillouviana* S. G. Gmel.)].

(=) *Baillouviana* Adans., Fam. Pl. 2: 13, 523. Jul-Aug 1763.
Typus: *Fucus baillouviana* S. G. Gmel.

Delesseria J. V. Lamour. in Ann. Mus. Natl. Hist. Nat. 20: 122. 1813.
Typus: *D. sanguinea* (Huds.) J. V. Lamour. (*Fucus sanguineus* Huds.).

(≡) *Hydrolapatha* Stackh. in Mém. Soc. Imp. Naturalistes Moscou 2: 54, 67. 1809 (typ. des.: Papenfuss in Hydrobiologia 2: 196. 1950).

Dudresnaya P. Crouan & H. Crouan in Ann. Sci. Nat., Bot., ser. 2, 3: 98. Feb 1835.
Typus: *D. coccinea* (C. Agardh) P. Crouan & H. Crouan (*Mesogloia coccinea* C. Agardh) *(typ. cons.)* [= *D. verticillata* (With.) Le Jol. (*Ulva verticillata* With.)].

(H) *Dudresnaya* Bonnem. in J. Phys. Chim. Hist. Nat. Arts 94: 180. Apr 1822 [*Phaeoph.: Chordar.*].
Typus: *Alcyonidium vermiculatum* (Sm.) J. V. Lamour. (*Rivularia vermiculata* Sm.).

(=) *Borrichius* Gray, Nat. Arr. Brit. Pl. 1: 317, 330. 1 Nov 1821.
Typus: *B. gelatinosus* Gray, nom. illeg. (*Ulva verticillata* With.).

Erythrotrichia Aresch. in Nova Acta Regiae Soc. Sci. Upsal., ser. 2, 14: 435. 1850.
Typus: *E. ceramicola* (Lyngb.) Aresch. (*Conferva ceramicola* Lyngb.) [= *E. carnea* (Dillwyn) J. Agardh (*Conferva carnea* Dillwyn)].

(≡) *Goniotrichum* Kütz., Phycol. General.: 244. 14-16 Sep 1843.

(=) *Porphyrostromium* Trevis., Sagg. Algh. Coccot.: 100. 1848.
Typus: *P. boryi* Trevis., nom. illeg. (*Porphyra boryana* Mont.).

Furcellaria J. V. Lamour. in Ann. Mus. Natl. Hist. Nat. 20: 45. 1813.
Typus: *F. lumbricalis* (Huds.) J. V. Lamour. (*Fucus lumbricalis* Huds.).

(=) *Fastigiaria* Stackh. in Mém. Soc. Imp. Naturalistes Moscou 2: 59, 90. 1809.
Typus (vide Papenfuss in Hydrobiologia 2: 194. 1950): *F. linnaei* Stackh., nom. illeg. (*Fucus fastigiatus* L.).

Gastroclonium Kütz., Phycol. General.: 441. 14-16 Sep 1843.
Typus: *G. ovale* Kütz., nom. illeg. (*Fucus ovalis* Huds., nom. illeg., *Fucus ovatus* Huds., *G. ovatum* (Huds.) Papenf.).

(=) *Sedoidea* Stackh. in Mém. Soc. Imp. Naturalistes Moscou 2: 57, 83. 1809.
Typus (vide Papenfuss in Hydrobiologia 2: 202. 1950): *Fucus sedoides* Gooden. & Woodw., nom. illeg. (*Fucus vermicularis* S. G. Gmel.).

Gelidium J. V. Lamour. in Ann. Mus. Natl. Hist. Nat. 20: 128. 1813.
Typus: *G. corneum* (Huds.) J. V. Lamour. (*Fucus corneus* Huds.).

(≡) *Cornea* Stackh. in Mém. Soc. Imp. Naturalistes Moscou 2: 57, 83. 1809 (typ. des.: Papenfuss in Hydrobiologia 2: 191-192. 1950).

Gracilaria Grev., Alg. Brit.: liv, 121. Mar 1830.
Typus: *G. compressa* (C. Agardh) Grev. (*Sphaerococcus compressus* C. Agardh) *(typ. cons.)*.

(=) *Ceramianthemum* Donati ex Léman in Cuvier, Dict. Sci. Nat., 7: 421. 1817.
Typus (vide Ardissone in Mem. Soc. Crittog. Ital. 1: 240-241. 1883): *Fucus bursa-pastoris* S. G. Gmel.

(=) *Plocaria* Nees, Horae Phys. Berol.: 42. 1-8 Feb 1820.
Typus: *P. candida* Nees

Grateloupia C. Agardh, Spec. Alg. 1: 221. Oct 1822.
Typus: *G. filicina* (J. V. Lamour.) C. Agardh (*Delesseria filicina* J. V. Lamour.).

(H) *Grateloupia* Bonnem. in J. Phys. Chim. Hist. Nat. Arts 94: 189. Apr 1822 [*Rhodoph.: Ceram.*].
Typus: *Conferva arbuscula* Dillwyn

Griffithsia C. Agardh, Syn. Alg. Scand.: xxviii. Mai-Dec 1817 *('Griffitsia') (orth. cons.)*.
Typus: *G. corallina* C. Agardh, nom. illeg. (*Conferva corallina* Murray, nom. illeg., *Conferva corallinoides* L., *G. corallinoides* (L.) Trevis.).

Halymenia C. Agardh, Syn. Alg. Scand.: xix, 35. Mai-Dec 1817.
Typus: *H. floresia* (Clemente) C. Agardh (*Fucus floresius* Clemente) *(typ. cons.)*.

Helminthocladia J. Agardh, Spec. Gen. Ord. Alg. 2: 412. Jan-Jun 1852.
Typus: *H. purpurea* (Harv.) J. Agardh (*Mesogloia purpurea* Harv.) [= *H. calvadosii* (J. V. Lamour. ex Turpin) Setchell (*Dumontia calvadosii* J. V. Lamour. ex Turpin)].

(H) *Helminthocladia* Harv., Gen. S. Afr. Pl.: 396. Aug-Dec 1838 [*Phaeoph.: Chordar.*].
≡ *Mesogloia* C. Agardh 1817.

Helminthora J. Agardh, Spec. Gen. Ord. Alg. 2: 415. Jan-Jun 1852.
Typus: *H. divaricata* (C. Agardh) J. Agardh (*Mesogloia divaricata* C. Agardh).

(H) *Helminthora* Fr., Syst. Orb. Veg.: 341. Dec 1825 [*Rhodoph.: Helminthoclad.*].
Typus: *H. multifida* (F. Weber & D. Mohr) Fr. (Fl. Scan.: 311. 1835) (*Rivularia multifida* F. Weber & D. Mohr).

Heterosiphonia Mont., Prodr. Gen. Phyc.: 4. 1 Aug - 10 Sep 1842.
Typus: *H. berkeleyi* Mont.

(=) *Ellisius* Gray, Nat. Arr. Brit. Pl. 1: 317, 333. 1 Nov 1821.
Typus (vide Silva in Univ. Calif. Publ. Bot. 25: 290. 1952): *E. coccineus* (Huds.) Gray (*Conferva coccinea* Huds.).

Hildenbrandia Nardo in Isis (Oken) 1834: 676. 1834 *('Hildbrandtia') (orth. cons.).*
Typus: *H. prototypus* Nardo

Iridaea Bory, Dict. Class. Hist. Nat. 9: 15. 25 Feb 1826 *(Iridaea, 'Iridea') (orth. cons.).*
Typus: *I. cordata* (Turner) Bory (*Fucus cordatus* Turner) *(typ. cons.).*

(H) *Iridea* Stackh., Nereis Brit., ed. 2: ix, xii. 1816 (ante Aug) [*Phaeoph.: Desmarest.*].
≡ *Hyalina* Stackh. 1809 *(nom. rej.* sub *Desmarestia).*

Laurencia J. V. Lamour. in Ann. Mus. Natl. Hist. Nat. 20: 130. 1813.
Typus: *L. obtusa* (Huds.) J. V. Lamour. (*Fucus obtusus* Huds.).

(=) *Osmundea* Stackh. in Mém. Soc. Imp. Naturalistes Moscou 2: 56, 79. 1809.
Typus (vide Silva in Univ. Calif. Publ. Bot. 25: 292. 1952): *O. expansa* Stackh., nom. illeg. (*Fucus osmunda* S. G. Gmel.).

Lemanea Bory in Ann. Mus. Natl. Hist. Nat. 12: 178. 1808.
Typus: *L. corallina* Bory, nom. illeg. (*Conferva fluviatilis* L., *L. fluviatilis* (L.) C. Agardh).

(≡) *Apona* Adans., Fam. Pl. 2: 2, 519. Jul-Aug 1763.

Lenormandia Sond. in Bot. Zeitung (Berlin) 3: 54. 24 Jan 1845.
Typus: *L. spectabilis* Sond.

(H) *Lenormandia* Delise in Desmazières, Pl. Crypt. N. France: No. 1144. 1841 [*Fungi*].
Typus: *L. jungermanniae* Delise

Lithothamnion Heydr. in Ber. Deutsch. Bot. Ges. 15: 412. 7 Sep 1897.
Typus: *L. muelleri* Lenorm. ex Rozanov (in Mém. Soc. Sci. Nat. Cherbourg 12: 101. 1866) *(typ. cons.).*

(H) *Lithothamniun* Phil. in Arch. Naturgesch. 3(1): 387. 1837 [*Rhodoph.: Corallin.*].
Typus (vide Lemoine in Ann. Inst. Océanogr. 2(2): 66. 1911): *L. byssoides* (Lam.) Phil. (*Nullipora byssoides* Lam.).

Martensia Hering in Ann. Mag. Nat. Hist. 8: 92. Oct 1841.
Typus: *M. elegans* Hering

(H) *Martensia* Giseke, Prael. Ord. Nat. Pl.: 202, 207, 249. Apr 1792 [*Zingiber.*].
Typus: *M. aquatica* (Retz.) Giseke (*Heritiera aquatica* Retz.).

Nemastoma J. Agardh, Alg. Mar. Medit.: 89. 9 Apr 1842 *('Nemostoma') (orth. cons.).*
Typus: *N. dichotomum* J. Agardh

Neurocaulon Zanardini ex Kütz., Sp. Alg.: 744. 23-24 Jul 1849.
Typus: *N. foliosum* (Menegh.) Zanardini ex Kütz. (*Iridaea foliosa* Menegh.) *(typ. cons.).*

Nitophyllum Grev., Alg. Brit.: xlvii, 77. Mar 1830.
Typus: *N. punctatum* (Stackh.) Grev. (*Ulva punctata* Stackh.).

(=) *Scutarius* Roussel, Fl. Calvados, ed. 2: 91. 1806.
Typus (vide Silva in Univ. Calif. Publ. Bot. 25: 268. 1952): *Fucus ocellatus* J. V. Lamour.

Odonthalia Lyngb., Tent. Hydrophytol. Dan.: xxix, 9. Apr-Aug 1819.
Typus: *O. dentata* (L.) Lyngb. (*Fucus dentatus* L.).

(≡) *Fimbriaria* Stackh. in Mém. Soc. Imp. Naturalistes Moscou 2: 95, 96. 1809 (typ. des.: Silva, Univ. Calif. Publ. Bot. 25: 269. 1952).

Phacelocarpus Endl. & Diesing in Bot. Zeitung (Berlin) 3: 289. 25 Apr 1845.
Typus: *P. tortuosus* Endl. & Diesing

(=) *Ctenodus* Kütz., Phycol. General.: 407. 14-16 Sep 1843.
Typus: *C. labillardierei* (Mert. ex Turner) Kütz. (*Fucus labillardierei* Mert. ex Turner).

Phyllophora Grev., Alg. Brit.: lvi, 135. Mar 1830.
Typus: *P. crispa* (Huds.) P. S. Dixon (in Bot. Not. 117: 63. 31 Mar 1964) (*Fucus crispus* Huds.) *(typ. cons.)*.

(≡) *Epiphylla* Stackh., Nereis Brit., ed. 2: x, xii. 1816 (ante Aug).
(=) *Membranifolia* Stackh. in Mém. Soc. Imp. Naturalistes Moscou 2: 55, 75. 1809.
Typus (vide Papenfuss in Hydrobiologia 2: 198. 1950): *M. lobata* Stackh., nom. illeg. (*Fucus membranifolius* Gooden. & Woodw., nom. illeg., *Fucus pseudoceranoides* S. G. Gmel.).

Phymatolithon Foslie in Kongel. Norske Vidensk. Selsk. Skr. (Trondheim) 1898(2): 4. 14 Oct 1898.
Typus: *P. polymorphum* Foslie, nom. illeg. (*Millepora polymorpha* L., nom. illeg., *Millepora calcarea* Pall., *P. calcareum* (Pall.) W. H. Adey & D. L. McKibbin).

(≡) *Apora* Gunnerus in Kongel. Norske Vidensk. Selsk. Skr. (Copenhagen) 4: 72. 1768.

Pleonosporium Nägeli in Sitzungsber. Bayer. Akad. Wiss. München 1861(2): 326, 339. 1862.
Typus: *P. borreri* (Sm.) Nägeli (*Conferva borreri* Sm.).

Plocamium J. V. Lamour. in Ann. Mus. Natl. Hist. Nat. 20: 137. 1813.
Typus: *P. vulgare* J. V. Lamour., nom. illeg. (*Fucus cartilagineus* L., *P. cartilagineum* (L.) P. S. Dixon).

(≡) *Nereidea* Stackh. in Mém. Soc. Imp. Naturalistes Moscou 2: 58, 86. 1809 (typ. des.: Silva in Univ. Calif. Publ. Bot. 25: 264. 1952).

Plumaria F. Schmitz in Nuova Notarisia 7: 5. Jan 1896.
Typus: *P. elegans* (Bonnem.) F. Schmitz (*Ptilota elegans* Bonnem.).

(H) *Plumaria* Heist. ex Fabr., Enum.: 207. 1759 [*Cyper.*].
≡ *Eriophorum* L. 1753.

Polyneura (J. Agardh) Kylin in Acta Univ. Lund., ser. 2, sect. 2, 20(6): 33. 1924 (*Nitophyllum* subg. *Polyneura* J. Agardh, Spec. Gen. Ord. Alg. 3(3): 51. 1898).
Typus: *Nitophyllum hilliae* (Grev.) Grev. (*Delesseria hilliae* Grev., *P. hilliae* (Grev.) Kylin).

(H) *Polyneura* J. Agardh in Acta Univ. Lund. 35 (sect. 2, 4): 60. 1899 [*Rhodoph.: Kallymen.*].
Typus: *P. californica* J. Agardh

Polysiphonia Grev., Scott. Crypt. Fl.: ad t. 90. 1 Dec 1823.
Typus: *P. urceolata* (Dillwyn) Grev. (*Conferva urceolata* Dillwyn) *(typ. cons.)*.

(=) *Grammita* Bonnem. in J. Phys. Chim. Hist. Nat. Arts 94: 186. Apr 1822.
Typus: *Conferva fucoides* Huds.

(=) *Vertebrata* Gray, Nat. Arr. Brit. Pl. 1: 317, 338. 1 Nov 1821.
Typus: *V. fastigiata* Gray, nom. illeg. (*Conferva polymorpha* L.).

(=) *Gratelupella* Bory, Dict. Class. Hist. Nat. 3: 340. 6 Sep 1823.
Typus (vide Bory, Dict. Class. Hist. Nat. 7: 481. 1825): *Ceramium brachygonium* Lyngb.

Porphyra C. Agardh, Syst. Alg.: xxxii, 190. Mai-Sep 1824.
Typus: *P. purpurea* (Roth) C. Agardh (*Ulva purpurea* Roth).

(H) *Porphyra* Lour., Fl. Cochinch.: 63, 69. Sep 1790 [*Verben.*].
Typus: *P. dichotoma* Lour.

(=) *Phyllona* Hill, Hist. Pl., ed. 2: 79. 1773.
Typus: non designatus.

Porphyridium Nägeli in Neue Denkschr. Allg. Schweiz. Ges. Gesammten Naturwiss. 10(7): 71, 138. 1849.
Typus: *P. cruentum* (Gray) Nägeli (*Olivia cruenta* Gray).

(=) *Chaos* Bory ex Desm., Cat. Pl. Omises Botanogr. Belgique: 1. Mar 1823.
Typus: *C. sanguinarius* Bory ex Desm., nom. illeg. (*Phytoconis purpurea* Bory).

(=) *Sarcoderma* Ehrenb. in Ann. Phys. Chem. 94: 504. 1830.
Typus: *S. sanguineum* Ehrenb.

Prionitis J. Agardh, Spec. Gen. Ord. Alg. 2: 185. Jan-Jun 1851.
Typus: *P. ligulata* J. Agardh [= *P. lanceolata* (Harv.) Harv. (*Gelidium lanceolatum* Harv.)].

(H) *Prionitis* Adans., Fam. Pl. 2: 499, 594. Jul-Aug 1763 [*Umbell.*].
≡ *Falcaria* Fabr. 1759 *(nom. cons.)* (6018).

Ptilota C. Agardh, Syn. Alg. Scand.: xix, 39. Mai-Dec 1817.
Typus: *P. plumosa* (Huds.) C. Agardh (*Fucus plumosus* Huds.).

Rhodomela C. Agardh, Spec. Alg. 1: 368. Oct 1822.
Typus: *R. subfusca* (Woodw.) C. Agardh (*Fucus subfuscus* Woodw.).

(=) *Fuscaria* Stackh. in Mém. Soc. Imp. Naturalistes Moscou 2: 59, 93. 1809.
Typus: *F. variabilis* Stackh., nom. illeg. (*Fucus variabilis* With., nom. illeg., *Fucus confervoides* Huds.).

Rhodophyllis Kütz. in Bot. Zeitung (Berlin) 5: 23. 8 Jan 1847.
Typus: *R. bifida* Kütz., nom. illeg. (*Fucus bifudus* Turner, non S. G. Gmel., *Bifida divaricata* Stackh., *R. divaricata* (Stackh.) Papenf.).

(≡) *Bifida* Stackh. in Mém. Soc. Imp. Naturalistes Moscou 2: 95, 97. 1809 (typ. des.: Silva in Univ. Calif. Publ. Bot. 25: 264. 1952).

(=) *Inochorion* Kütz., Phycol. General.: 443. 14-16 Sep 1843.
Typus: *I. dichotomum* Kütz.

Rhodymenia Grev., Alg. Brit.: xlviii, 84. Mar 1830.
Typus: *R. palmetta* (J. V. Lamour.) Grev. (*Delesseria palmetta* J. V. Lamour.) [= *R. pseudopalmata* (J. V. Lamour.) P. C. Silva (*Fucus pseudopalmatus* J. V. Lamour.)].

Schizymenia J. Agardh, Spec. Gen. Ord. Alg. 2: 158, 169. Jan-Jun 1851.
Typus: *S. dubyi* (Chauv.) J. Agardh (*Halymenia dubyi* Chauv.) *(typ. cons.)*.

Suhria J. Agardh ex Endl., Gen. Pl., Suppl. 3: 41. Oct 1843.
Typus: *S. vittata* (L.) J. Agardh ex Endl. (*Fucus vittatus* L.).

(=) *Chaetangium* Kütz., Phycol. General.: 392. 14-16 Sep 1843.
Typus: *C. ornatum* (L.) Kütz. (*Fucus ornatus* L.).

Vidalia J. Agardh, Spec. Gen. Ord. Alg. 2: 1117. Jan-Aug 1863.
Typus: *V. spiralis* (J. V. Lamour.) J. Agardh (*Delesseria spiralis* J. V. Lamour.).

(=) *Volubilaria* J. V. Lamour. ex Bory, Dict. Class. Hist. Nat. 16: 630. 30 Oct 1830.
Typus: *V. mediterranea* J. V. Lamour. ex Bory, nom. illeg. (*Fucus volubilis* L.).

(=) *Spirhymenia* Decne. in Arch. Mus. Hist. Nat. (Paris) 2: 177. 1841.
Typus: *S. serrata* (Suhr) Decne. (in Ann. Sci. Nat., Bot., ser. 2, 17: 358. Jun 1842) (*Carpophyllum serratum* Suhr, *'denticulatum'* lapsu).

(=) *Epineuron* Harv. in London J. Bot. 4: 532. 1845.
Typus (vide Silva in Univ. Calif. Publ. Bot. 25: 293. 1952): *E. colensoi* Hook. f. & Harv. (in London J. Bot. 4: 532. 1845).

A10. TRICHOMONADOPHYCEAE

Chilomastix A. G. Alexeev in Arch. Zool. Exp. Gén. 46: xi. 26 Dec 1910.
Typus: *C. caulleryi* (A. G. Alexeev) A. G. Alexeev (*Macrostoma caulleryi* A. G. Alexeev).

(≡) *Macrostoma* A. G. Alexeev in Compt.-Rend. Séances Mém. Soc. Biol. 67: 200. 17 Jul 1909.

A11. XANTHOPHYCEAE

Botrydiopsis Borzì in Boll. Soc. Ital. Microscop. 1: 69. 1889.
Typus: *B. arhiza* Borzì

(H) *Botrydiopsis* Trevis., Nomencl. Alg.: 70. 1845 [Plantae incertae sedis].
Typus: *B. vulgaris* (Bréb.) Trevis. (*Botrydina vulgaris* Bréb.).

Centritractus Lemmerm. in Ber. Deutsch. Bot. Ges. 18: 274. 24 Jul 1900.
Typus: *C. belonophorus* (Schmidle) Lemmerm. (*Schroederia belonophora* Schmidle).

Monodus Chodat in Beitr. Kryptogamenfl. Schweiz 4(2): 185. 1913.
Typus: *M. acuminata* (Gerneck) Chodat (*Chlorella acuminata* Gerneck) *(typ. cons.)*.

Ophiocytium Nägeli in Neue Denkschr. Allg. Schweiz. Ges. Gesammten Naturwiss. 10(7): 87. 1849.
Typus: *O. apiculatum* Nägeli [= *O. cochleare* (Eichw.) A. Braun (*Spirogyra cochlearis* Eichw.).

(=) *Spirodiscus* Ehrenb. in Abh. Königl. Akad. Wiss. Berlin, Phys. Kl., 1831: 68. 1832.
Typus: *S. fulvus* Ehrenb.

Tetraedriella Pascher in Arch. Protistenk. 69: 423. 5 Feb 1930.
Typus: *T. acuta* Pascher

(=) *Polyedrium* Nägeli in Neue Denkschr. Allg. Schweiz. Ges. Gesammten Naturwiss. 10(7): 83. 1849.
Typus: *P. tetraëdricum* Nägeli

B. FUNGI

Agaricus L., Sp. Pl.: 1171. 1 Mai 1753.
Typus: *A. campestris* L. : Fr. *(typ. cons.)*.

Aleurodiscus Rabenh. ex J. Schröt. in Cohn, Krypt.-Fl. Schlesien 3(1): 429. 2 Jun 1888.
Typus: *A. amorphus* (Pers. : Fr.) J. Schröt. (*Peziza amorpha* Pers. : Fr.).

(=) *Cyphella* Fr., Syst. Mycol. 2: 201. 1822 : Fr., ibid.
Typus: *C. digitalis* (Alb. & Schwein.) Fr.

Alternaria Nees, Syst. Pilze: 72. 1816-1817.
Typus: *A. tenuis* Nees (*Torula alternata* Fr. : Fr., *A. alternata* (Fr. : Fr.) Keissl.).

Amanita Pers., Tent. Disp. Meth. Fung.: 65. 14 Oct - 31 Dec 1797.
Typus: *A. muscaria* (L. : Fr.) Pers. (*Agaricus muscarius* L. : Fr.).

(H) *Amanita* Boehm., Defin. Gen. Pl.: 490. 1760 (typ. des.: Earle in Bull. New York Bot. Gard. 5: 382. 1909; Donk in Beih. Nova Hedwigia 5: 20. 1962) [*Fungi*].
≡ *Agaricus* L. 1753 *(nom. cons.)*.

Amanitopsis Roze in Bull. Soc. Bot. France 23: 50, 51. 1876 (post 11 Feb).
Typus: *A. vaginata* (Bull. : Fr.) Roze (*Agaricus vaginatus* Bull. : Fr.).

(≡) *Vaginarius* Roussel, Fl. Calvados, ed. 2: 59. 1806 (typ. des.: Donk in Beih. Nova Hedwigia 5: 292. 1962).

(=) *Vaginata* Gray, Nat. Arr. Brit. Pl. 1: 601. 1 Nov 1821.
Typus: *V. livida* (Pers.) Gray (*Amanita livida* Pers.).

Amphisphaeria Ces. & De Not. in Comment. Soc. Crittog. Ital. 1: 223. Jan 1863.
Typus: *A. umbrina* (Fr.) De Not. (*Sphaeria umbrina* Fr.) *(typ. cons.)*.

Anema Nyl. ex Forssell, Beitr. Gloeolich.: 40, 91. Mai-Jun 1885.
Typus: *A. decipiens* (A. Massal.) Forssell (*Omphalaria decipiens* A. Massal.).

(≡) *Omphalaria* A. Massal., Framm. Lichenogr.: 13. 1855.
Typus: *O. decipiens* A. Massal.

Anisomeridium (Müll. Arg.) M. Choisy, Icon. Lich. Univ.: 3. Jan 1928 (*Arthopyrenia* sect. *Anisomeridium* Müll. Arg. in Flora 66: 290. 21 Jun 1883).
Typus: *Arthopyrenia xylogena* Müll. Arg. [nomen sub *Anisomeridium* deest].

(=) *Microthelia* Körb., Syst. Lich. Germ.: 372. Jan-Mar 1855.
Typus (vide Fries, Gen. Heterolich. Eur.: 111. 1861): *M. micula* Körb., nom. illeg. (*Verrucaria biformis* Borrer, *Anisomeridium biforme* (Borrer) R. C. Harris).

Anzia Stizenb. in Flora 44: 393. 7 Jul 1861.
Typus: *A. colpodes* (Ach.) Stizenb. (*Lichen colpodes* Ach.).

(=) *Chondrospora* A. Massal. in Atti Reale Ist. Veneto Sci. Lett. Arti, ser. 3, 5: 248. 1860.
Typus: *C. semiteres* (Mont. & Bosch) A. Massal. (*Parmelia semiteres* Mont. & Bosch).

Aposphaeria Sacc. in Michelia 2: 4. 25 Apr 1880.
Typus: *A. pulviscula* (Sacc.) Sacc. (*Phoma pulviscula* Sacc.).

(H) *Aposphaeria* Berk., Outl. Brit. Fungol.: 315. Aug-Dec 1860 [*Fungi*].
Typus: *A. complanata* (Tode : Fr.) Berk. (*Sphaeria complanata* Tode : Fr.).

Arthonia Ach. in Neues J. Bot. 1(3): 3. Jan 1806.
Typus: *A. radiata* (Pers.) Ach. (*Opegrapha radiata* Pers.).

(=) *Coniocarpon* DC. in Lamarck & Candolle, Fl. Franç., ed. 3, 2: 323. 17 Sep 1805.
Typus: *C. cinnabarinum* DC.

Aschersonia Mont. in Ann. Sci. Nat., Bot., ser. 3, 10: 121. Aug 1848.
Typus: *A. taitensis* Mont.

(H) *Aschersonia* Endl., Gen. Pl., Suppl. 2: 103. Mar-Jun 1842 [*Fungi*].
Typus: *A. crustacea* (Jungh.) Endl. (*Laschia crustacea* Jungh.).

Aspicilia A. Massal., Ric. Auton. Lich. Crost.: 36. Jun-Dec 1852.
Typus: [specimen] *"Urceolaria cinerea ß alba"*, Schaerer, Lich. Helv. Exsicc., ed. 2, 6: No. 127 (VER) *(typ. cons.)* [= *A. cinerea* (L.) Körb. (*Lichen cinereus* L.)].

(=) *Circinaria* Link in Neues J. Bot. 3(1-2): 5. Apr 1809.
Typus: *Urceolaria hoffmannii* (Ach.) Ach., nom. illeg. (*Verrucaria contorta* Hoffm.).

(=) *Sagedia* Ach. in Kongl. Vetensk. Acad. Nya Handl. 30: 164. Jul-Sep 1809.
Typus (vide Laundon & Hawksworth in Taxon 37: 478. 1988): *S. zonata* Ach.

(=) *Sphaerothallia* T. Nees in Nova Acta Phys.-Med. Acad. Caes. Leop.-Carol. Nat. Cur. 15(2): 360. 1831.
Typus (vide Choisy in Bull. Soc. Bot. France 76: 525. 1929): *Lecanora esculenta* (Pall.) Eversm. (*Lichen esculentus* Pall.).

(=) *Chlorangium* Link in Bot. Zeitung (Berlin) 7: 731. 12 Oct 1849.
Typus: *C. jussufii* (Link) Link (*Placodium jussufii* Link).

Boletus L., Sp. Pl.: 1176. 1 Mai 1753.
Typus: *B. edulis* Bull. : Fr. *(typ. cons.)*.

Buellia De Not. in Giorn. Bot. Ital. 2(1,1): 195. 1846.
Typus: *B. disciformis* (Fr.) Mudd (Man. Brit. Lich.: 216. 1861) (*Lecidia parasema* var. *disciformis* Fr.).

(=) *Gassicurtia* Fée, Essai Crypt. Ecorc.: xlvi. 4 Dec 1824.
Typus: *G. coccinea* Fée (Essai Crypt. Ecorc.: 100. Mai-Oct 1825).

Caloplaca Th. Fr., Lich. Arct.: 218. Mai-Dec 1860.
Typus: *C. cerina* (Ehrh. ex Hedw.) Th. Fr. (*Lichen cerinus* Ehrh. ex Hedw.).

(=) *Gasparrinia* Tornab., Lichenogr. Sicul.: 27. 1849.
Typus: *G. murorum* (Hoffm.) Tornab. (*Lichen murorum* Hoffm.).

(=) *Pyrenodesmia* A. Massal. in Atti Reale Ist. Veneto Sci. Lett. Arti, ser. 2, 3 (Appunt. 4): 119. 1853.
Typus: *P. chalybaea* (Fr.) A. Massal. (*Parmelia chalybaea* Fr.).

(=) *Xanthocarpia* A. Massal. & De Not. in Massalongo, Alc. Gen. Lich.: 11. Mai-Aug 1853.
Typus: *X. ochracea* (Schaer.) A. Massal. & De Not. (*Lecidea ochracea* Schaer.).

Calvatia Fr., Summa Veg. Scand.: 442. 1849.
Typus: *C. craniiformis* (Schwein.) Fr. (*Bovista craniiformis* Schwein.).

(=) *Langermannia* Rostk. in Sturm, Deutschl. Fl., sect. 3, 5: 23. 3-9 Nov 1839.
Typus: *L. gigantea* (Batsch : Pers.) Rostk. (*Lycoperdon giganteum* Batsch : Pers.).

(=) *Hippoperdon* Mont. in Ann. Sci. Nat., Bot., ser. 2, 17: 121. Feb 1842.
Typus: *H. crucibulum* Mont.

Candida Berkhout, Schimmelgesl. Monilia: 41. 1923.
Typus: *C. vulgaris* Berkhout

(=) *Syringospora* Quinq. in Arch. Physiol. Norm. Pathol. 1: 293. 1868.
Typus: *S. robinii* Quinq., nom. illeg. (*Oidium albicans* C. P. Robin).

(=) *Parendomyces* Queyrat & Laroche in Bull. & Mém. Soc. Méd. Hôp. Paris, ser. 3, 28: 136. 1909.
Typus: *P. albus* Queyrat & Laroche

(=) *Parasaccharomyces* Beurm. & Gougerot in Tribune Méd. (Paris) 42: 502. 7 Aug 1909.
Typus: non designatus.

(=) *Pseudomonilia* A. Geiger in Centralbl. Bakteriol., 2. Abth. 27: 134. 1 Jun 1910.
Typus: *P. albomarginata* A. Geiger

Cetraria Ach., Methodus: 292. Jan-Apr 1803.
Typus: *C. islandica* (L.) Ach. (*Lichen islandicus* L.).

(≡) *Platyphyllum* Vent., Tabl. Règne Vég. 2: 34. 5 Mai 1799.

Ceuthospora Grev., Scott. Crypt. Fl.: ad t. 253-254. Sep 1826.
Typus: *C. lauri* (Grev.) Grev. (*Cryptosphaeria lauri* Grev.).

(H) *Ceuthospora* Fr., Syst. Orb. Veg.: 119. Dec 1825 [*Fungi*].
Typus: *C. phaeocomes* (Rebent. : Fr.) Fr. (*Sphaeria phaeocomes* Rebent. : Fr.).

Chlorociboria Seaver ex C. S. Ramamurthi & al. in Mycologia 49: 857. 28 Mar 1958.
Typus: *C. aeruginosa* (Pers. : Fr.) Seaver ex C. S. Ramamurthi & al. (*Peziza aeruginosa* Pers. : Fr.) *(typ. cons.)*.

Chondropsis Nyl. ex Cromb. in J. Linn. Soc., Bot. 17: 397. 1879.
Typus: *C. semiviridis* (Nyl.) Cromb. (*Parmeliopsis semiviridis* Nyl.).

(H) *Chondropsis* Raf., Fl. Tellur. 3: 29, 97. Nov-Dec 1837 [*Gentian.*].
Typus: *C. trinervis* (L.) Raf. (*Chironia trinervis* L.).

Chrysothrix Mont. in Ann. Sci. Nat., Bot., ser. 3, 18: 312. Nov 1852.
Typus: *C. noli-tangere* Mont., nom. illeg. (*Peribotryon pavonii* Fr. : Fr., *C. pavonii* (Fr. : Fr.) J. R. Laundon).

(≡) *Peribotryon* Fr., Syst. Mycol. 3: 287. 1832.

(=) *Pulveraria* Ach., Methodus: 1. Jan-Apr 1803.
Typus (vide Laundon in Taxon 30: 663. 1981): *P. chlorina* (Ach.) Ach. (*Lichen chlorinus* Ach.).

Cistella Quél., Enchir. Fung.: 319. Jan-Jun 1886.
Typus: *C. dentata* (Pers. : Fr.) Quél. (*Peziza dentata* Pers. : Fr.).

(H) *Cistella* Blume, Bijdr.: 293. 20 Sep - 7 Dec 1825 [*Orchid.*].
Typus: *C. cernua* (Willd.) Blume (*Malaxis cernua* Willd.).

Cladonia P. Browne, Civ. Nat. Hist. Jamaica: 81. 10 Mar 1756.
Typus: *C. subulata* (L.) F. H. Wigg. (Prim. Fl. Holsat.: 90. 29 Mar 1780) (*Lichen subulatus* L.).

Clavaria L., Sp. Pl.: 1182. 1 Mai 1753.
Typus: *C. fragilis* Holmsk. (Beata Ruris 1: 7. 1790) : Fr. (Syst. Mycol. 1: 484. 1 Jan 1821).

Collema F. H. Wigg., Prim. Fl. Holsat.: 89. 29 Mar 1780.
Typus: *C. lactuca* (Weber) F. H. Wigg. (*Lichen lactuca* Weber).

(=) *Gabura* Adans., Fam. Pl. 2: 6, 560. Jul-Aug 1763.
Typus: *Lichen fascicularis* L.

(=) *Kolman* Adans., Fam. Pl. 2: 7, 542. Jul-Aug 1763.
Typus: *Lichen nigrescens* Huds.

Collybia (Fr.) Staude, Schwämme Mitteldeutschl.: xxviii, 119. 1857 (*Agaricus* "trib." *Collybia* Fr., Syst. Mycol. 1: 9, 129. 1 Jan 1821).
Typus: *Agaricus tuberosus* Bull. : Fr. (*C. tuberosa* (Bull. : Fr.) P. Kumm.) .

(=) *Gymnopus* Pers. ex Gray, Nat. Arr. Brit. Pl. 1: 604. 1 Nov 1821.
Typus: *G. fusipes* (Bull. : Fr.) Gray (*Agaricus fusipes* Bull. : Fr.).

Coniothyrium Corda, Icon. Fung. 4: 38. Sep 1840.
Typus: *C. palmarum* Corda

(=) *Clisosporium* Fr., Novit. Fl. Suec.: 80. 18 Dec 1819 : Fr., Syst. Mycol. 1: xlvii. 1 Jan 1821.
Typus: *C. lignorum* Fr. : Fr.

Conocybe Fayod in Ann. Sci. Nat., Bot., ser. 7, 9: 357. Jun 1889.
Typus: *C. tenera* (Schaeff. : Fr.) Fayod (*Agaricus tener* Schaeff. : Fr.).

(=) *Raddetes* P. Karst. in Hedwigia 26: 112. Mai-Jun 1887.
Typus: *R. turkestanicus* P. Karst.

(=) *Pholiotina* Fayod in Ann. Sci. Nat., Bot., ser. 7, 9: 359. Jun 1889.
Typus: *P. blattaria* (Fr. : Fr.) Fayod (*Agaricus blattarius* Fr. : Fr.).

(=) *Pholiotella* Speg. in Bol. Acad. Nac. Ci. 11: 412. 1889.
Typus: *P. blattariopsis* Speg.

Cordyceps Fr., Observ. Mycol. 2 (revis.): 316. 1824.
Typus: *C. militaris* (L. : Fr.) Fr. (*Clavaria militaris* L. : Fr.).

Cortinarius (Pers.) Gray, Nat. Arr. Brit. Pl. 1: 627. 1 Nov 1821 *('Cortinaria') (orth. cons.)* (*Agaricus* sect. *Cortinarius* Pers., Syn. Meth. Fung.: 276. 31 Dec 1801).
Typus: *Agaricus violaceus* L. : Fr. (*C. violaceus* (L. : Fr.) Gray).

Craterellus Pers., Mycol. Eur. 2: 4. Jan-Jul 1825 *('Cratarellus') (orth. cons.)*.
Typus: *C. cornucopioides* (L. : Fr.) Pers. (*Peziza cornucopioides* L. : Fr.) (etiam vide *Pezicula* [*Fungi*]).

(H) *Craterella* Pers. in Neues Mag. Bot. 1: 112. Apr-Aug 1794 [*Fungi*].
Typus: *C. pallida* Pers.

(≡) *Trombetta* Adans., Fam. Pl. 2: 6, 613. Jul-Aug 1763 (typ. des.: Kuntze, Revis. Gen. Pl. 2: 873. 1891).

Crocynia (Ach.) A. Massal. in Atti Reale Ist. Veneto Sci. Lett. Arti, ser. 3, 5: 251. 1860 (*Lecidea* sect. *Crocynia* Ach., Lichenogr. Universalis: 217. 1810).
Typus: *Lecidea gossypina* (Sw.) Ach. (*Lichen gossypinus* Sw., *C. gossypina* (Sw.) A. Massal.).

(≡) *Symplocia* A. Massal., Neagen. Lich.: 4. Nov-Dec 1854.

Cryptococcus Vuill. in Rev. Gen. Sci. Pures Appl. 12: 741. 1901.
Typus: *C. neoformans* (San Felice) Vuill. (*Saccharomyces neoformans* San Felice) (*typ. cons.:* No. 72042 (BPI) e cult. No. CBS 132).

(H) *Cryptococcus* Kütz. in Linnaea 8: 365. 1833 [*Fungi*].
Typus: *C. mollis* Kütz.

Cryptosphaeria Ces. & De Not. in Comment. Soc. Crittog. Ital. 4: 231. 1853.
Typus: *C. millepunctata* Grev. (Fl. Edin.: 360. 18 Mar 1824) *(typ. cons.)*.

(H) *Cryptosphaeria* Grev., Scott. Crypt. Fl.: ad t. 13. Sep 1822 [*Fungi*].
Typus: *C. taxi* (Sowerby : Fr.) Grev. (*Sphaeria taxi* Sowerby : Fr.).

Cryptothecia Stirt. in Proc. Roy. Philos. Soc. Glasgow 10: 164. 1876.
Typus: *C. subnidulans* Stirt.

(=) *Myriostigma* Kremp., Lich. Foliicol.: 22. 1874.
Typus: *M. candidum* Kremp.

Cylindrocarpon Wollenw. in Phytopathology 3: 225. 2 Oct 1913.
Typus: *C. cylindroides* Wollenw.

(=) *Fusidium* Link in Ges. Naturf. Freunde Berlin Mag. Neuesten Entdeck. Gesammten Naturk. 3: 8. Jan-Mar 1809 : Fr., Syst. Mycol. 1: xl. 1 Jan 1821.
Typus: *F. candidum* Link : Fr.

Daldinia Ces. & De Not. in Comment. Soc. Crittog. Ital. 1: 197. Jan 1863.
Typus: *D. concentrica* (Bolton : Fr.) Ces. & De Not. (*Sphaeria concentrica* Bolton : Fr.).

(≡) *Peripherostoma* Gray, Nat. Arr. Brit. Pl. 1: 513. 1 Nov 1821 (per typ. des.).

(≡) *Stromatosphaeria* Grev., Fl. Edin.: lxxiii, 355. 18 Mar 1824 (per typ. des.).

Debaryomyces Lodder & Kreger in Kreger, Yeasts, ed. 3: 130, 145. 1984.
Typus: *D. hansenii* (Zopf) Lodder & Kreger (*Saccharomyces hansenii* Zopf).

(H) *Debaryomyces* Klöcker in Compt.-Rend. Trav. Carlsberg Lab. 7: 273. 1909 [*Fungi*].
Typus: *D. globosus* Klöcker

(≡) *Debaryozyma* Van der Walt & Johannsen in Persoonia 10: 147. 28 Dec 1978.

Dothiora Fr., Summa Veg. Scand.: 418. 1849.
Typus: *D. pyrenophora* (Fr. : Fr.) Fr. (*Dothidea pyrenophora* Fr. : Fr.).

(H) *Dothiora* Fr., Fl. Scan.: 347. 1837 [*Fungi*].
Typus: *Variolaria melogramma* Bull. : Fr.

Drechslera S. Ito in Proc. Imp. Acad. Japan 6: 355. 1930.
Typus: *D. tritici-vulgaris* (Y. Nisik.) S. Ito ex S. Hughes (*Helminthosporium tritici-vulgaris* Y. Nisik.).

(=) *Angiopoma* Lév. in Ann. Sci. Nat., Bot., ser. 2, 16: 235. Oct 1841.
Typus: *A. campanulatum* Lév.

Encoelia (Fr.) P. Karst. in Bidrag Kännedom Finlands Natur Folk 19: 18, 217. 1871 (*Peziza* "trib." *Encoelia* Fr., Syst. Mycol. 2: 74. 1822).
Typus: *Peziza furfuracea* Roth : Fr. (*E. furfuracea* (Roth : Fr.) P. Karst.).

(≡) *Phibalis* Wallr., Fl. Crypt. Germ. 2: 445. Feb-Mar 1833.

Epidermophyton Sabour. in Arch. Méd. Exp. Anat. Pathol. 19: 754. Nov 1907.
Typus: *E. inguinale* Sabour.

(H) *Epidermidophyton* E. Lang in Vierteljahresschr. Dermatol. Syph. 11: 263. 1879 [*Fungi*].
Typus: non designatus.

Eutypella (Nitschke) Sacc. in Atti Soc. Veneto-Trentino Sci. Nat. Padova 4: 80. Oct 1875 (*Valsa* subg. *Eutypella* Nitschke, Pyrenomyc. Germ.: 163. Jan 1870).
Typus: *Valsa sorbi* (Alb. & Schwein. : Fr.) Fr. (*Sphaeria prunastri* var. *sorbi* Alb. & Schwein. : Fr., *E. sorbi* (Alb. & Schwein. : Fr.) Sacc.).

(=) *Scoptria* Nitschke, Pyrenomyc. Germ.: 83. Jan 1867.
Typus: *S. isariphora* Nitschke

Gautieria Vittad., Monogr. Tuberac.: 25. 1831.
Typus: *G. morchelliformis* Vittad.

(H) *Gautiera* Raf., Med. Fl. 1: 202. 11 Jan 1828 [*Eric.*].
≡ *Gaultheria* L. 1753.

Gloeophyllum P. Karst. in Bidrag Kännedom Finlands Natur Folk 37: x, 79. 1882 (*'Gleophyllum'*) (*orth. cons.*).
Typus: *G. sepiarium* (Wulfen : Fr.) P. Karst. (*Agaricus sepiarius* Wulfen : Fr.).

(≡) *Serda* Adans., Fam. Pl. 2: 11, 604. Jul-Aug 1763 (typ. des.: Donk in Persoonia 1: 279. 1960).

(≡) *Sesia* Adans., Fam. Pl. 2: 10, 604. Jul-Aug 1763 (typ. des.: Donk in Persoonia 1: 280. 1960).

(=) *Ceratophora* Humb., Fl. Friberg.: 112. 1793.
Typus: *C. fribergensis* Humb.

Guignardia Viala & Ravaz in Bull. Soc. Mycol. France 8: 63. 22 Mai 1892.
Typus: *G. bidwellii* (Ellis) Viala & Ravaz (*Sphaeria bidwellii* Ellis) *(typ. cons.)*.

Gyalideopsis Vězda in Folia Geobot. Phytotax. 7: 204. 7 Aug 1972.
Typus: *G. peruviana* Vězda

(=) *Diploschistella* Vain. in Ann. Univ. Fenn. Åbo., A, 2(3): 26. 1926.
Typus: *D. urceolata* Vain.

Gymnoderma Nyl. in Flora 43: 546. 21 Sep 1860.
Typus: *G. coccocarpum* Nyl.

(H) *Gymnoderma* Humb., Fl. Friberg.: 109. 1793 [*Fungi*].
Typus: *G. sinuatum* Humb.

Gyrodon Opat. in Arch. Naturgesch. 2(1): 5. 1836.
Typus: *G. sistotremoides* Opat., nom. illeg. (*Boletus sistotremoides* Fr., non Alb. & Schwein., *Boletus sistotrema* Fr. : Fr., *G. sistotrema* (Fr. : Fr.) P. Karst. [= *G. lividus* (Bull. : Fr.) P. Karst. (*Boletus lividus* Bull. : Fr.)].

(=) *Anastomaria* Raf., Ann. Nat.: 16. Mar-Jul 1820.
Typus: *A. campanulata* Raf.

Gyromitra Fr., Summa Veg. Scand.: 346. 1849.
Typus: *G. esculenta* (Pers. : Fr.) Fr. (*Helvella esculenta* Pers. : Fr.).

(=) *Gyrocephalus* Pers. in Mém. Soc. Linn. Paris 3: 77. 1824.
Typus: *G. aginnensis* Pers., nom. illeg. (*Helvella sinuosa* Brond.).

Helminthosporium Link in Ges. Naturf. Freunde Berlin Mag. Neuesten Entdeck. Gesammten Naturk. 3: 10. Jan-Mar 1809 *('Helmisporium') (orth. cons.)*.
Typus: *H. velutinum* Link : Fr.

Hexagonia Fr., Epicr. Syst. Mycol.: 496. 1838 *('Hexagona') (orth. cons.)*.
Typus: *H. hirta* (Beauvérie : Fr.) Fr. (*Favolus hirtus* Beauvérie : Fr.) *(typ. cons.)*.

(H) *Hexagonia* Pollini, Hort. Veron. Pl.: 35. 1816 [*Fungi*].
Typus: *H. mori* Pollini

Hirneola Fr. in Kongl. Vetensk. Acad. Handl. 1848: 144. 1848.
Typus: *H. nigra* Fr., nom. illeg. (*Peziza nigricans* Sw. : Fr., *H. nigricans* (Sw. : Fr.) P. W. Graff).

(H) *Hirneola* Fr., Syst. Orb. Veg.: 93. Dec 1825 [*Fungi*].
≡ *Mycobonia* Patouillard 1894 *(nom. cons.)*.

(=) *Laschia* Fr. in Linnaea 5: 533. Oct 1830 : Fr., Syst. Mycol. 3 (index): 107. 1832.
Typus: *L. delicata* Fr. : Fr.

Hyaloscypha Boud. in Bull. Soc. Mycol. France 1: 118. Mai 1885.
Typus: *H. vitreola* (P. Karst.) Boud. (*Peziza vitreola* P. Karst.) *(typ. cons.)*.

Hydnum L., Sp. Pl.: 1178. 1 Mai 1753.
Typus: *H. repandum* L. : Fr. *(typ. cons.)*.

Hymenochaete Lév. in Ann. Sci. Nat., Bot., ser. 3, 5: 150. 1846.
Typus: *H. rubiginosa* (Dicks. : Fr.) Lév. (*Helvella rubiginosa* Dicks. : Fr.).

(H) *Hymenochaeta* P. Beauv. ex T. Lestib., Essai Cypér.: 43. 29 Mar 1819 [*Cyper.*].
Typus: non designatus.

Hyphodontia J. Erikss. in Symb. Bot. Upsal. 16(1): 101. 1958.
Typus: *H. pallidula* (Bres.) J. Erikss. (*Gonatobotrys pallidula* Bres.) *(typ. cons.)*.

(=) *Grandinia* Fr., Epicr. Syst. Mycol.: 527. 1838.
Typus (vide Clements & Shear, Gen. Fung., ed. 2: 346. 1931): *Thelephora granulosa* Pers. : Fr.

(=) *Lyomices* P. Karst. in Rev. Mycol. (Toulouse) 3(9): 23. 1881.
Typus: *L. serus* (Pers.) P. Karst. (*Hydnum serum* Pers.).

(=) *Kneiffiella* P. Karst. in Bidrag Kännedom Finlands Natur Folk 48: 371. Jan-Sep 1889.
Typus: *K. barba-jovis* (Bull. : Fr.) P. Karst. (*Hydnum barba-jovis* Bull. : Fr.).

(=) *Chaetoporellus* Bondartsev & Singer in Mycologia 36: 67. 1 Feb 1944.
Typus: *C. latitans* (Bourdot & Galzin) Bondartsev & Singer (*Poria latitans* Bourdot & Galzin).

Hypholoma (Fr.) P. Kumm., Führ. Pilzk.: 21, 72. Jul-Aug 1871 (*Agaricus* "trib." *Hypholoma* Fr., Syst. Mycol. 1: 11, 287. 1 Jan 1821).
Typus: *Agaricus fascicularis* Fr. : Fr. (*H. fasciculare* (Fr. : Fr.) P. Kumm.) *(typ. cons.)*.

Hypoderma De Not. in Giorn. Bot. Ital. 2(1,2): 13. 1847.
Typus: *H. rubi* (Pers. : Fr.) DC. ex Chevall. (*Hysterium rubi* Pers. : Fr.) *(typ. cons.)*.

(H) *Hypoderma* DC. in Lamarck & Candolle, Fl. Franç., ed. 3, 2: 304. 17 Sep 1805 (typ. des.: Cannon & Minter in Taxon 32: 580. 1983) [*Fungi*].
≡ *Lophodermium* Chevall. 1826 *(nom. cons.)*.

Hypoxylon Bull., Hist. Champ. France: 168. 1791.
Typus: *H. coccineum* Bull. [= *H. fragiforme* (Pers. : Fr.) Kickx (*Sphaeria fragiformis* Pers. : Fr.)].

(H) *Hypoxylon* Adans., Fam. Pl. 2: 9, 616. Jul-Aug 1763 [*Fungi*].
Typus (vide Donk in Regnum Veg. 34: 16. 1964): *Xylaria polymorpha* (Pers. : Fr.) Grev. (*Sphaeria polymorpha* Pers. : Fr.).

(=) *Sphaeria* Haller, Hist. Stirp. Helv. 3: 120. 25 Mar 1768 : Fr., Syst. Mycol. 1: lii. 1 Jan 1821.
Typus (vide Donk in Regnum Veg. 34: 16. 1964): *Sphaeria fragiformis* Pers. : Fr.

Karstenia Fr. in Acta Soc. Fauna Fl. Fenn. 2(6): 166. Jan-Jun 1885.
Typus: *K. sorbina* (P. Karst.) Fr. (*Propolis sorbina* P. Karst.).

(H) *Karstenia* Göpp. in Nova Acta Phys.-Med. Acad. Caes. Leop.-Carol. Nat. Cur. 17, Suppl.: 451. 18 Nov 1836 [Foss.].
Typus: non designatus.

Lachnocladium Lév., Considér. Mycol.: 108. 1846.
Typus: *L. brasiliense* (Lév.) Pat. (*Eriocladus brasiliensis* Lév.).

(≡) *Eriocladus* Lév. in Ann. Sci. Nat., Bot., ser. 3, 5: 158. 1846.

Lactarius Pers., Tent. Disp. Meth. Fung.: 63. 14 Oct - 31 Dec 1797 (*'Lactaria'*) (*orth. cons.*).
Typus: *L. piperatus* (L. : Fr.) Pers. (*Agaricus piperatus* L. : Fr.).

Laetinaevia Nannf. in Nova Acta Regiae Soc. Sci. Upsal., ser. 4, 8(2): 190. 1932 (post 5 Feb).
Typus: *L. lapponica* (Nannf.) Nannf. (*Naevia lapponica* Nannf.).

(=) *Myridium* Clem., Gen. Fung.: 67. Jul-Oct 1909.
Typus: *M. myriosporum* (W. Phillips & Harkn.) Clem. (*Orbilia myriospora* W. Phillips & Harkn.).

Laetiporus Murrill in Bull. Torrey Bot. Club 31: 607. 26 Nov 1904.
Typus: *L. speciosus* Murrill, nom. illeg. (*Polyporus sulphureus* Bull. : Fr., *L. sulphureus* (Bull. : Fr.) Murrill).

(=) *Cladoporus* (Pers.) Chevall., Fl. Gén. Env. Paris 1: 260. 656. 5 Aug 1826 (*Polyporus* [unranked] *Cladoporus* Pers., Mycol. Eur. 2: 122. Jan-Jul 1825).
Typus: *C. fulvus* Chevall., nom. illeg. (*Polyporus ramosus* Bull.).

Lecanactis Körb., Syst. Lich. Germ.: 275. Jan 1855.
Typus: *L. abietina* (Ach.) Körb. (*Lichen abietinus* Ach.) (*typ. cons.*).

(H) *Lecanactis* Eschw., Syst. Lich.: 14, 25. 1824 [*Fungi*].
Typus: *L. lobata* Eschw.

(=) *Pyrenotea* Fr., Syst. Mycol. 1: xxiii. 1 Jan 1821 : Fr., ibid.
Typus: *P. incrustans* (Ach.) Fr. (in Kongl. Vetensk. Acad. Handl. 1821: 332. 1821) (*Cyphelium incrustans* Ach.).

Lepiota (Pers.) Gray, Nat. Arr. Brit. Pl. 1: 601. 1 Nov 1821 (*Agaricus* sect. *Lepiota* Pers., Tent. Disp. Meth. Fung.: 68. 14 Oct - 31 Dec 1797).
Typus: *Agaricus colubrinus* Pers., non Bull. (*Agaricus clypeolarius* Bull. : Fr., *L. clypeolaria* (Bull. : Fr.) P. Kumm.) *(typ. cons.)*.

Lepraria Ach., Methodus: 3. Jan-Apr 1803.
Typus: *L. incana* (L.) Ach. (*Byssus incana* L.).

(≡) *Pulina* Adans., Fam. Pl. 2: 3, 595. Jul-Aug 1763 (typ. des.: Laundon in Taxon 12: 37. 1963).

(≡) *Conia* Vent., Tabl. Règne Vég. 2: 32. 5 Mai 1799 (typ. des.: Laundon in Taxon 12: 37. 1963).

Leptoglossum P. Karst. in Bidrag Kännedom Finlands Natur Folk 32: xvii, 242. Jul-Dec 1879.
Typus: *L. muscigenum* (Bull. : Fr.) P. Karst. (*Agaricus muscigenus* Bull. : Fr.).

(=) *Boehmia* Raddi, Sp. Nov. Fung. Firenze: 15. 1806.
Typus: *B. muscoides* Raddi

Leptorhaphis Körb., Syst. Lich. Germ.: 371. Jan-Mar 1855.
Typus: *L. oxyspora* (Nyl.) Körb. (*Verrucaria oxyspora* Nyl.).

(=) *Endophis* Norman, Conat. Praem. Gen. Lich.: 28. 1852.
Typus: non designatus.

Leptosphaeria Ces. & De Not. in Comment. Soc. Crittog. Ital. 1: 234. Jan 1863.
Typus: *L. doliolum* (Pers. : Fr.) Ces. & De Not. (*Sphaeria doliolum* Pers. : Fr.).

(≡) *Bilimbiospora* Auersw. in Rabenhorst, Fungi Europaei, ed. 2: No. 261 [in sched. corr.]. 1861.

(=) *Nodulosphaeria* Rabenh., Herb. Mycol., ed. 2: No. 725. 1858 *(nom. cons.)*.

Letharia (Th. Fr.) Zahlbr. in Hedwigia 31: 36. Jan-Apr 1892 (*Evernia* [unranked] *Letharia* Th. Fr., Lichenogr. Scand. 1: 32. 1871).
Typus: *Evernia vulpina* (L.) Ach. (*Lichen vulpinus* L., *Letharia vulpina* (L.) Hue).

(≡) *Chlorea* Nyl. in Mém. Soc. Sci. Nat. Cherbourg 3: 170. Jun 1855.

Lichina C. Agardh, Syn. Alg. Scand.: xii, 9. Mai-Dec 1817.
Typus: *L. pygmaea* (Lightf.) C. Agardh (*Fucus pygmaeus* Lightf.).

(=) *Pygmaea* Stackh. in Mém. Soc. Imp. Naturalistes Moscou 2: 60, 95. 1809.
Typus: *Fucus lichenoides* J. F. Gmel.

Lopadium Körb., Syst. Lich. Germ.: 210. Jan 1855.
Typus: *L. pezizoideum* (Ach.) Körb. (*Lecidea pezizoidea* Ach.).

(=) *Brigantiaea* Trevis., Spighe e Paglie: 7. Jul 1853.
Typus: *B. tricolor* (Mont.) Trevis. (*Biatora tricolor* Mont.).

Lophiostoma Ces. & De Not. in Comment. Soc. Crittog. Ital. 1: 219. Jan 1863.
Typus: *L. macrostomum* (Tode : Fr.) Ces. & De Not. (*Sphaeria macrostoma* Tode : Fr.).

(≡) *Platysphaera* Dumort., Comment. Bot.: 87. Nov (sero) - Dec (prim.) 1822.

Lophodermium Chevall., Fl. Gén. Env. Paris 1: 435. 5 Aug 1826.
Typus: *L. arundinaceum* (Schrad. : Fr.) Chevall. (*Hysterium arundinaceum* Schrad. : Fr.) (etiam vide *Hypoderma* [*Fungi*]).

Marasmius Fr., Fl. Scan.: 339. 1836.
Typus: *M. rotula* (Scop. : Fr.) Fr. (*Agaricus rotula* Scop. : Fr.).

(=) *Micromphale* Gray, Nat. Arr. Brit. Pl. 1: 621. 1 Nov 1821.
Typus: *M. venosa* (Pers.) Gray (*Agaricus venosus* Pers.).

Melanogaster Corda in Sturm, Deutschl. Fl., sect. 3, 3: 1. 1831.
Typus: *M. tuberiformis* Corda

(=) *Bullardia* Jungh. in Linnaea 5: 408. Jul 1830.
Typus: *B. inquinans* Jungh.

Melanoleuca Pat., Cat. Pl. Cell. Tunisie: 22. Jan-Apr 1897.
Typus: *M. vulgaris* (Pat.) Pat. (*Melaleuca vulgaris* Pat., *Agaricus melaleucus* Pers. : Fr., *Melanoleuca melaleuca* (Pers. : Fr.) Murrill).

(≡) *Psammospora* Fayod in Ann. Reale Accad. Agric. Torino 35: 91. 1893.

Melanospora Corda, Icon. Fung. 1: 24. Aug 1837.
Typus: *M. zamiae* Corda

(=) *Ceratostoma* Fr., Observ. Mycol. 2: 337. Apr-Mai 1818.
Typus (vide Fries, Summa Veg. Scand.: 396. 1849): *C. chioneum* (Fr. : Fr.) Fr. (*Sphaeria chionea* Fr. : Fr.).

(=) *Megathecium* Link in Abh. Königl. Akad. Wiss. Berlin, Phys. Kl. 1824: 176. 1826 (typ. des.: Cannon & Hawksworth in Taxon 32: 476. 1983).
≡ *Ceratostoma* Fr. 1818.

Micarea Fr., Syst. Orb. Veg.: 256. Dec 1825.
Typus: *M. prasina* Fr. *(typ. cons.)*.

(H) *Micarea* Fr., Sched. Crit. Lich. Suec. Exsicc.: No. 97. 1825 (ante 7 Mai) [*Fungi*].
Typus: *Biatorea fuliginea* Fr., nom. illeg. (*Lecidea fuliginea* Ach., nom. illeg., *Lecidea icmalea* Ach.).

Mollisia (Fr.) P. Karst. in Bidrag Kännedom Finlands Natur Folk 19: 15, 189. 1871 (*Peziza* "trib." *Mollisia* Fr., Syst. Mycol. 2: 137. 1822).
Typus: *Peziza cinerea* Batsch : Fr. (*M. cinerea* (Batsch : Fr.) P. Karst.).

(=) *Tapesia* (Pers. : Fr.) Fuckel in Jahrb. Nassauischen Vereins Naturk. 23-24: 300. 19 Feb - 24 Nov 1870 (*Peziza* [unranked] *Tapesia* Pers., Mycol. Eur. 1: 270. 1 Jan - 14 Apr 1822 : Fr., Syst. Mycol. 2: 105. 1822).
Typus (vide Clements & Shear, Gen. Fungi: 325. 1931): *Peziza fusca* Pers. : Fr. (*T. fusca* (Pers. : Fr.) Fuckel).

Monilia Bonord., Handb. Mykol.: 76. 1851 (sero).
Typus: *M. cinerea* Bonord.

(H) *Monilia* Link in Ges. Naturf. Freunde Berlin Mag. Neuesten Entdeck. Gesammten Naturk. 3: 16. Jan-Mar 1809 : Fr., Syst. Mycol. 1: xlvi. 1 Jan 1821 [*Fungi*].
Typus: *M. antennata* (Pers. : Fr.) Pers.

Mucor Fresen., Beitr. Mykol.: 7. Aug 1850.
Typus: [icon] *'Mucor mucedo'*, Fresenius, Beitr. Mykol.: t. 1, f. 1-12. Aug 1850 *(typ. cons.)* [= *M. murorum* Naumov].

(H) *Mucor* L., Sp. Pl.: 1185. 1 Mai 1753 : Fr., Syst. Mycol. 3: 317. 1832 [*Fungi*].
Typus (vide Sumstine in Mycologia 2: 127. 1910): *M. mucedo* L. : Fr.

(=) *Hydrophora* Tode, Fungi Mecklenb. Sel. 2: 5. 1791 : Fr., Syst. Mycol. 3: 313. 1832.
Typus (vide Sumstine in Mycologia 2: 132. 1910): *H. stercorea* Tode : Fr.

Mutinus Fr., Summa Veg. Scand.: 434. 1849.
Typus: *M. caninus* (Schaeff. : Pers.) Fr. (*Phallus caninus* Schaeff. : Pers.).

(≡) *Cynophallus* (Fr. : Fr.) Corda, Icon. Fung. 5: 29. Jun 1842 (*Phallus* "trib." *Cynophallus* Fr., Syst. Mycol. 2: 282, 284. 1823 : Fr., ibid.).

(=) *Aedycia* Raf. in Med. Repos., ser. 2, 5: 358. Feb -Apr 1808.
Typus: *A. rubra* Raf.

(=) *Ithyphallus* Gray, Nat. Arr. Brit. Pl. 1: 675. 1 Nov 1821.
Typus: *I. inodorus* (Sowerby) Gray (*Phallus inodorus* Sowerby).

Mycoblastus Norman in Nyt Mag. Naturvidensk. 7: 236. 1853 *('Mykoblastus') (orth. cons.)*.
Typus: *M. sanguinarius* (L.) Norman (*Lichen sanguinarius* L.).

Mycobonia Pat. in Bull. Soc. Mycol. France 10: 76. 30 Jun 1894.
Typus: *M. flava* (Sw. : Fr.) Pat. (*Peziza flava* Sw. : Fr.) (etiam vide *Hirneola* [*Fungi*]).

Mycoporum Flot. ex Nyl. in Mém. Soc. Sci. Nat. Cherbourg 3: 186. Jun 1855.
Typus: *M. elabens* Flot. ex Nyl.

(H) *Mycoporum* G. Mey., Nebenst. Beschaeft. Pflanzenk.: 327. Sep-Dec 1825 [*Fungi*].
Typus: *M. melinostigma* G. Mey.

Nectria (Fr.) Fr., Summa Veg. Scand.: 387. 1849 (*Hypocrea* sect. *Nectria* Fr., Syst. Orb. Veg.: 105. Dec 1825).
Typus: *Sphaeria cinnabarina* Tode : Fr. (*N. cinnabarina* (Tode : Fr.) Fr.) *(typ. cons.)*.

(=) *Ephedrosphaera* Dumort., Comment. Bot.: 90. Nov (sero) - Dec (prim.) 1822.
Typus (vide Cannon & Hawksworth in Taxon 32: 477. 1983): *E. decolorans* (Pers.) Dumort. (*Sphaeria decolorans* Pers.).

(=) *Hydropisphaera* Dumort., Comment. Bot.: 89. Nov (sero) - Dec (prim.) 1822.
Typus: *H. peziza* (Tode : Fr.) Dumort. (*Sphaeria peziza* Tode : Fr.).

Nidularia Fr. in Fries & Nordholm, Symb. Gasteromyc. 1: 2. 22 Mai 1817.
Typus: *N. radicata* Fr.

(H) *Nidularia* Bull., Hist. Champ. France: 163. 1791 [*Fungi*].
Typus: *N. vernicosa* Bull.

Nodulosphaeria Rabenh., Herb. Mycol., ed. 2: No. 725. 1858.
Typus: [specimen] Rabenhorst, Herb. Mycol., ed. 2: No. 725 (S) *(typ. cons.)* [= *N. derasa* (Berk. & Broome) L. Holm (*Sphaeria derasa* Berk. & Broome)] .

Ocellularia G. Mey., Nebenst. Beschaeft. Pflanzenk. 1: 327. Sep-Dec 1825.
Typus: *O. obturata* (Ach.) Spreng. (Syst. Veg. 4(1): 242. 1-7 Jan 1827) (*Thelotrema obturatum* Ach.).

(=) *Ascidium* Fée, Essai Crypt. Ecorc.: xlii. 4 Dec 1824.
Typus: *A. cinchonarum* Fée

Oidium Link in Willdenow, Sp. Pl. 6(1): 121. 1824.
Typus: *O. monilioides* (Nees : Fr.) Link (*Acrosporium monilioides* Nees : Fr.).

(H) *Oidium* Link in Ges. Naturf. Freunde Berlin Mag. Neuesten Entdeck. Gesammten Naturk. 3: 18. Jan-Mar 1809 : Fr., Syst. Mycol. 1: xlv. 1 Jan 1821 [*Fungi*].
Typus: *O. aureum* (Pers. : Fr.) Link (*Trichoderma aureum* (Pers. : Fr.) Pers.).

(≡) *Acrosporium* Nees, Syst. Pilze: 53. 1816-1817 : Fr., Syst. Mycol. 1: xlv. 1 Jan 1821.

Opegrapha Ach. in Kongl. Vetensk. Acad. Nya Handl. 30: 97. 1809.
Typus: *O. vulgata* (Ach.) Ach. (*Lichen vulgatus* Ach.).

(H) *Opegrapha* Humb., Fl. Friberg.: 57. 1793 [*Fungi*].
Typus: *O. vulgaris* Humb., nom. illeg. (*Lichen scriptus* L.).

Panaeolus (Fr.) Quél. in Mém. Soc. Emul. Montbéliard, ser. 2, 5: 151. Aug-Dec 1872 (*Agaricus* subg. *Panaeolus* Fr., Summa Veg. Scand.: 297. 1849).
Typus: *Agaricus papilionaceus* Bull. : Fr. (*P. papilionaceus* (Bull. : Fr.) Quél.).

(≡) *Coprinarius* (Fr.) P. Kumm., Führ. Pilzk.: 20, 68. Jul-Aug 1871.

Panus Fr., Epicr. Syst. Mycol.: 396. 1838.
Typus: *P. conchatus* (Bull. : Fr.) Fr. (*Agaricus conchatus* Bull. : Fr.) *(typ. cons.)*.

Parmelia Ach., Methodus: xxxiii, 153. Jan-Apr 1803.
Typus: *P. saxatilis* (L.) Ach. (*Lichen saxatilis* L.).

(≡) *Lichen* L., Sp. Pl.: 1140. 1 Mai 1753.

Parmeliopsis (Nyl.) Nyl. in Not. Sällsk. Fauna Fl. Fenn. Förh. 8: 121. Jun 1866 (*Parmelia* subg. *Parmeliopsis* Nyl. in Not. Sällsk. Fauna Fl. Fenn. Förh. 5: 130. Jun-Jul 1861).
Typus: *Parmelia ambigua* (Wulfen) Ach. (*Lichen ambiguus* Wulfen, *Parmeliopsis ambigua* (Wulfen) Nyl.) *(typ. cons.)*.

Peccania A. Massal. ex Arnold in Flora 41: 93. 14 Feb 1858.
Typus: *P. coralloides* (A. Massal.) A. Massal. (in Atti Reale Ist. Veneto Sci. Lett. Arti, ser. 3, 5: 335. 1860) (*Corinophoros coralloides* A. Massal.).

(≡) *Corinophoros* A. Massal. in Flora 39: 212. 14 Apr 1856.

Peltigera Willd., Fl. Berol. Prodr.: 347. 1787.
Typus: *P. canina* (L.) Willd. (*Lichen caninus* L.).

(=) *Placodion* P. Browne ex Adans., Fam. Pl. 2: 7, 592. Jul-Aug 1763.
Typus: non designatus.

Peridermium (Link) J. C. Schmidt & Kunze, Deutschl. Schwämme 6: 4. 1817 (*Hypodermium* subg. *Peridermium* Link in Ges. Naturf. Freunde Berlin Mag. Neuesten Entdeck. Gesammten Naturk. 7: 29. 1816).
Typus: *Aecidium elatinum* Alb. & Schwein. (*P. elatinum* (Alb. & Schwein.) J. C. Schmidt & Kunze).

Pertusaria DC. in Lamarck & Candolle, Fl. Franç., ed. 3, 2: 319. 17 Sep 1805.
Typus: *P. communis* DC., nom. illeg. (*Lichen verrucosus* Huds.) [= *P. pertusa* (L.) Tuck. (*Lichen pertusus* L.)].

(=) *Lepra* Scop., Intr. Hist. Nat.: 61. Jan-Apr 1777.
Typus: non designatus.

(=) *Variolaria* Pers. in Ann. Bot. (Usteri) 7: 23. 1794.
Typus: *V. discoidea* Pers.

(=) *Lepronocus* Vent., Tabl. Règne Vég. 2: 32. 5 Mai 1799.
Typus: non designatus.

(=) *Isidium* Ach., Methodus: xxxiii, 136. Jan-Apr 1803.
Typus: *I. corallinum* (L.) Ach. (*Lichen corallinus* L.).

Pezicula Tul. & C. Tul., Select. Fung. Carpol. 3: 182. 1865.
Typus: *P. carpinea* (Pers.) Sacc. (*Peziza carpinea* Pers.).

(H) *Pezicula* Paulet, Tab. Pl. Fung.: 24. 1791 [*Fungi*].
≡ *Craterellus* Pers. 1825 *(nom. cons.)* (typ. des.: Cannon & Hawksworth in Taxon 32: 478. 1983).

Phacidium Fr., Observ. Mycol. 1: 167. 1815.
Typus: *P. lacerum* Fr. : Fr. *(typ. cons.)*.

Phaeocollybia R. Heim, Inocybe: 70. Mai-Jun 1931.
Typus: *P. lugubris* (Fr. : Fr.) R. Heim (*Agaricus lugubris* Fr. : Fr.).

(=) *Quercella* Velen., České Houby: 495. 1921.
Typus: *Q. aurantiaca* Velen.

Phaeotrema Müll. Arg. in Mém. Soc. Phys. Genève 29(8): 10. 1887.
Typus: *P. subfarinosum* (Fée) Müll. Arg. (*Pyrenula subfarinosa* Fée).

(=) *Asteristion* Leight. in Trans. Linn. Soc. London 27: 163. 1870.
Typus: *A. erumpens* Leight.

Phellinus Quél., Enchir. Fung.: 172. Jan-Jun 1886.
Typus: *P. igniarius* (L. : Fr.) Quél. (*Boletus igniarius* L. : Fr.).

(=) *Mison* Adans., Fam. Pl. 2: 10, 578. Jul-Aug 1763 [*Fungi*].
Typus: non designatus.

Phillipsia Berk. in J. Linn. Soc., Bot. 18: 388. 29 Apr 1881.
Typus: *P. domingensis* (Berk.) Berk. (*Peziza domingensis* Berk.).

(H) *Phillipsia* C. Presl in Sternberg, Vers. Fl. Vorwelt 2: 206. 1 Sep - 7 Oct 1838 [Foss.].
Typus: *P. harcourtii* (Witham) C. Presl (*Lepidodendron harcourtii* Witham).

Phlyctis (Wallr.) Flot. in Bot. Zeitung (Berlin) 8: 571. 2 Aug 1850 (*Peltigera* sect. *Phlyctis* Wallr., Fl. Crypt. Germ. 3: 553. 1831).
Typus: *Peltigera agelaea* (Ach.) Wallr. (*Lichen agelaeus* Ach., *P. agelaea* (Ach.) Flot.).

(H) *Phlyctis* Raf., Caratt. Nuov. Gen.: 91. 1810 [*Algae* incertae sedis].
Typus: non designatus.

Pholiota (Fr.) P. Kumm., Führ. Pilzk.: 22, 83. Jul-Aug 1871 (*Agaricus* "trib." *Pholiota* Fr., Syst. Mycol. 1: 240. 1 Jan 1821).
Typus: *Agaricus squarrosus* Batsch : Fr. (*P. squarrosa* (Batsch : Fr.) P. Kumm.).

(≡) *Derminus* (Fr.) Staude, Schwämme Mitteldeutschl.: xxvi, 86. 1857.

Phoma Sacc. in Michelia 2: 4. 25 Apr 1880.
Typus: *P. herbarum* Westend. (in Bull. Acad. Roy. Sci. Belgique 19: 118. 1852) (*typ. cons.*).

(H) *Phoma* Fr., Novit. Fl. Suec.: 80. 18 Dec 1819 : Fr., Syst. Mycol. 1: lii. 1 Jan 1821 [*Fungi*].
Typus: *P. pustula* (Pers. : Fr.) Fr. (*Sphaeria pustula* Pers. : Fr.).

Phomopsis (Sacc.) Bubák in Oesterr. Bot. Z. 55: 78. Feb 1905 (*Phoma* subg. *Phomopsis* Sacc., Syll. Fung. 3: 66. 1884).
Typus: *Phoma lactucae* Sacc. (*P. lactucae* (Sacc.) Bubák).

(H) *Phomopsis* Sacc. & Roum. in Rev. Mycol. (Toulouse) 6: 32. Jan 1884 [*Fungi*].
Typus: *P. brassicae* Sacc. & Roum.

(=) *Myxolibertella* Höhn. in Ann. Mycol. 1: 526. 10 Dec 1903.
Typus (vide Clements & Shear, Gen. Fung.: 359. 1931): *M. aceris* Höhn.

Phyllachora Nitschke ex Fuckel in Jahrb. Nassauischen Vereins Naturk. 23-24: 216. 19 Feb - 24 Nov 1870.
Typus: *P. graminis* (Pers. : Fr.) Fuckel (*Sphaeria graminis* Pers. : Fr.).

(H) *Phyllachora* Nitschke ex Fuckel, Fungi Rhenani: No. 2056. 1867 [*Fungi*].
Typus: *P. agrostis* Fuckel

Phyllosticta Pers., Traité Champ. Comest.: 55, 147. 1818.
Typus: *P. convallariae* Pers.

Physconia Poelt in Nova Hedwigia 9: 30. Mai 1965.
Typus: *P. pulverulacea* Moberg (in Mycotaxon 8: 310. 13 Jan 1979) (*typ. cons.: "Lichen pulverulentus"*, Germania, Lipsia in *Tilia,* 1767, *Schreber* (M)).

Pleospora Rabenh. ex Ces. & De Not. in Comment. Soc. Crittog. Ital. 1: 217. Jan 1863.
Typus: *P. herbarum* (Fr.) Rabenh. ex Ces. & De Not. (*Sphaeria herbarum* Fr.).

(H) *Pleiospora* Harv., Thes. Cap. 1: 51. 1860 [*Legum.*].
Typus: *P. cajanifolia* Harv.

(=) *Clathrospora* Rabenh. in Hedwigia 1: 116. 1857.
Typus: *C. elynae* Rabenh.

Pleurotus (Fr.) P. Kumm., Führ. Pilzk.: 24, 104. Jul-Aug 1871 (*Agaricus* "trib." *Pleurotus* Fr., Syst. Mycol. 1: 178. 1 Jan 1821).
Typus: *Agaricus ostreatus* Jacq. : Fr. (*P. ostreatus* (Jacq. : Fr.) P. Kumm.).

(≡) *Pleuropus* (Pers.) Roussel, Fl. Calvados, ed. 2: 67. 1806 (typ. des.: Donk in Beih. Nova Hedwigia 5: 235. 1962).

(≡) *Crepidopus* Nees ex Gray, Nat. Arr. Brit. Pl. 1: 616. 1 Nov 1821 (per typ. des.).

(=) *Gelona* Adans., Fam. Pl. 2: 11, 561. Jul-Aug 1763.
Typus: non designatus.

(=) *Resupinatus* Gray, Nat. Arr. Brit. Pl. 1: 617. 1 Nov 1821.
Typus: *R. applicatus* (Batsch : Fr.) Gray (*Agaricus applicatus* Batsch : Fr.).

(=) *Pterophyllus* Lév. in Ann. Sci. Nat., Bot., ser. 3, 2: 178. 1844.
Typus: *P. bovei* Lév.

(=) *Hohenbuehelia* Schulzer in Verh. K.K. Zool.-Bot. Ges. Wien 16 (Abh.): 45. Jan-Mai 1866.
Typus: *H. petaloides* (Bull. : Fr.) Schulzer (*Agaricus petaloides* Bull. : Fr.).

Podospora Ces. in Rabenhorst, Klotzschii Herb. Mycol., ed. 2: No. 259 (vel 258). Jan-Jun 1856.
Typus: *P. fimiseda* (Ces. & De Not.) Niessl (in Hedwigia 22: 156. Oct 1883) (*Sordaria fimiseda* Ces. & De Not.) (*typ. cons.*: [specimen] Rabenhorst, Klotzschii Herb. Mycol., ed. 2: No. 259 (S).).

(=) *Schizothecium* Corda, Icon. Fung. 2: 29. Jul 1838.
Typus: *S. fimicola* Corda

Polyblastia A. Massal., Ric. Auton. Lich. Crost.: 147. Jun-Dec 1852.
Typus: *P. cupularis* A. Massal.

(=) *Sporodictyon* A. Massal. in Flora 35: 326. 7 Jun 1852.
Typus: *S. schaererianum* A. Massal.

Porina Müll. Arg. in Flora 66: 320. 11 Jul 1883.
Typus: *P. nucula* Ach. (Syn. Meth. Lich.: 112. 1814).

(H) *Porina* Ach. in Kongl. Vetensk. Acad. Nya Handl. 30: 158. 1809 [*Fungi*].
Typus: *P. pertusa* (L.) Ach. (*Lichen pertusus* L.).

(=) *Ophthalmidium* Eschw., Syst. Lich.: 18. 1824.
Typus: *O. hemisphaericum* Eschw.

(=) *Segestria* Fr., Syst. Orb. Veg.: 263. Dec 1825.
Typus: *S. lectissima* Fr.

Pseudocyphellaria Vain. in Acta Soc. Fauna Fl. Fenn. 7(1): 182. 1-22 Nov 1890.
Typus: *P. aurata* (Ach.) Vain. (*Sticta aurata* Ach.).

(≡) *Crocodia* Link, Handbuch 3: 177. 1833.
(=) *Stictina* Nyl., Syn. Meth. Lich. 1: 333. Apr 1860.
Typus (vide Clements & Shear, Gen. Fung.: 322. 1931): *S. crocata* (L.) Nyl. (*Lichen crocatus* L.).
(=) *Phaeosticta* Trevis., Lichenoth. Veneta: No. 75. Apr 1869.
Typus (vide Choisy in Bull. Mens. Soc. Linn. Lyon, ser. 2, 29: 125. 1960): *P. physciospora* (Nyl.) Trevis. (*Sticta fossulata* subsp. *physciospora* Nyl.).
(=) *Saccardoa* Trevis., Lichenoth. Veneta: No. 75. Apr 1869.
Typus (vide Choisy in Bull. Mens. Soc. Linn. Lyon, ser. 2, 29: 123. 1960): *S. crocata* (L.) Trevis. (*Lichen crocatus* L.).
(=) *Parmostictina* Nyl. in Flora 58: 363. 11 Aug 1875.
Typus: *Sticta hirsuta* Mont.

Pseudographis Nyl. in Mém. Soc. Sci. Nat. Cherbourg 3: 190. Jun 1855.
Typus: *P. elatina* (Ach. : Fr.) Nyl. (*Lichen elatinus* Ach. : Fr.).

(=) *Krempelhuberia* A. Massal., Geneac. Lich.: 34. 1-22 Sep 1854.
Typus: *K. cadubriae* A. Massal.

Psora Hoffm., Deutschl. Fl. 2: 161. Feb-Apr 1796.
Typus: *P. decipiens* (Hedw.) Hoffm. (*Lecidea decipiens* Hedw.).

(H) *Psora* Hill, Veg. Syst. 4: 30. 1762 [*Comp.*].
Typus: non designatus.
(H) *Psora* Hoffm., Descr. Pl. Cl. Crypt. 1: 37. 1789 [*Fungi*].
Typus: *P. caesia* Hoffm.

Pulvinula Boud. in Bull. Soc. Mycol. France 1: 107. Mai 1885.
Typus: *P. convexella* (P. Karst.) Boud. (*Peziza convexella* P. Karst.).

(=) *Pulparia* P. Karst. in Not. Sällsk. Fauna Fl. Fenn. Förh. 8: 205. 1866.
Typus: *P. arctica* P. Karst.

Pycnoporus P. Karst. in Rev. Mycol. (Toulouse) 3(9): 18. 1881.
Typus: *P. cinnabarinus* (Jacq. : Fr.) P. Karst. (*Boletus cinnabarinus* Jacq. : Fr.).

(≡) *Xylometron* Paulet, Prosp. Traité Champ.: 29. 1808.

Pyrenopsis (Nyl.) Nyl., Syn. Meth. Lich. 1(1): 97. 8-14 Aug 1858 (*Synalissa* sect. *Pyrenopsis* Nyl. in Mém. Soc. Sci. Nat. Cherbourg 3: 164. Jun 1855).
Typus: *P. fuscatula* Nyl. *(typ. cons.).*

Pyrenula Ach., Syn. Meth. Lich.: 117. 1814. Typus: *P. nitida* (Weigel) Ach. (*Sphaeria nitida* Weigel) *(typ. cons.)*.

(H) *Pyrenula* Ach. in Kongl. Vetensk. Acad. Nya Handl. 30: 160. 1809 [*Fungi*]. Typus: *P. margacea* (Wahlenb.) Ach. (*Thelotrema margaceum* Wahlenb.).

Pythium Pringsh. in Jahrb. Wiss. Bot. 1: 304. 1858.
Typus: *P. monospermum* Pringsh.

(H) *Pythium* Nees in Nova Acta Phys.-Med. Acad. Caes. Leop.-Carol. Nat. Cur. 11: 515. 1823 [*Fungi*].
Typus: non designatus.

(=) *Artotrogus* Mont. in Gard. Chron. 5: 640. 1845.
Typus: *A. hydnosporus* Mont.

Racodium Fr., Syst. Mycol. 3: 229. 1829.
Typus: *R. rupestre* Pers. : Fr.

(H) *Racodium* Pers. in Neues Mag. Bot. 1: 123. Apr-Aug 1794 : Fr., Syst. Mycol. 1: xlvi. 1 Jan 1821 [*Fungi*].
Typus: *R. cellare* Pers. : Fr.

Ramalina Ach., Lichenogr. Universalis: 122, 598. Apr-Mai 1810.
Typus: *R. fraxinea* (L.) Ach. (*Lichen fraxineus* L.) *(typ. cons.)*.

Ramaria Fr. ex Bonord., Handb. Mykol.: 166. 1851 (sero).
Typus: *R. botrytis* (Pers. : Fr.) Ricken (Vadem. Pilzfr.: 253. Mai-Jun 1918) (*Clavaria botrytis* Pers. : Fr.).

(H) *Ramaria* Holmsk. ex Gray, Nat. Arr. Brit. Pl. 1: 655. 1 Nov 1821 [*Fungi*].
Typus: non designatus.

(≡) *Cladaria* Ritgen, Aufeinanderfolge Org. Gest.: 54. 1828.

Ramularia Unger, Exanth. Pflanzen.: 169. 1833.
Typus: *R. pusilla* Unger

(H) *Ramularia* Roussel, Fl. Calvados, ed. 2: 98. 1806 (typ. des.: Agardh, Spec. Alg. 1: 402. 1823) [*Chloroph.*].
≡ *Ulva* L. 1753 *(nom. cons.)*.

Rhabdospora (Durieu & Mont.) Sacc., Syll. Fung. 3: 578. 15 Dec 1884 (*Septoria* sect. *Rhabdospora* Durieu & Mont. in Durieu, Expl. Sci. Algérie 1: 592. 1849).
Typus: *Septoria oleandri* Durieu & Mont. (*R. oleandri* (Durieu & Mont.) Sacc.).

(=) *Filaspora* Preuss in Linnaea 26: 718. Sep 1855.
Typus: *F. peritheciiformis* Preuss

Rhipidium Cornu in Bull. Soc. Bot. France 18: 58. 1871 (post 24 Mar).
Typus: *R. interruptum* Cornu

(H) *Rhipidium* Wallr., Fl. Crypt. Germ. 2: 742. Feb-Mar 1833 [*Fungi*].
Typus: *R. stipticum* (Bull. : Fr.) Wallr. (*Agaricus stipticus* Bull. : Fr.).

Rhizopus Ehrenb. in Nova Acta Phys.-Med. Acad. Caes. Leop.-Carol. Nat. Cur. 10: 198. 1821.
Typus: *R. nigricans* Ehrenb., nom. illeg. (*Mucor stolonifer* Ehrenb. : Fr., *R. stolonifer* (Ehrenb. : Fr.) Vuill.).

(=) *Ascophora* Tode, Fungi Mecklenb. Sel. 1: 13. 1790 : Fr., Syst. Mycol. 3: 309. 1832.
Typus (vide Kirk in Taxon 35: 374. 1986): *A. mucedo* Tode : Fr.

Robillarda Sacc. in Michelia 2: 8. 25 Apr 1880.
Typus: *R. sessilis* (Sacc.) Sacc. (*Pestalotia sessilis* Sacc.).

(H) *Robillarda* Castagne, Cat. Pl. Marseille: 205. 1845 [*Fungi*].
Typus: *R. glandicola* Castagne

Roccella DC. in Lamarck & Candolle, Fl. Franç., ed. 3, 2: 334. 17 Sep 1805.
Typus: *R. fuciformis* (L.) DC. (*Lichen fuciformis* L.).

(=) *Thamnium* Vent., Tabl. Règne Vég. 2: 35. 5 Mai 1799.
Typus (vide Ahti in Taxon 33: 330. 1984): *T. roccella* (L.) J. St.-Hil. (Expos. Fam. Nat. 1: 21. Feb-Apr 1805) (*Lichen roccella* L.).

Rutstroemia P. Karst. in Bidrag Kännedom Finlands Natur Folk 19: 12, 105. 1871.
Typus: *R. firma* (Pers. : Fr.) P. Karst. (*Peziza firma* Pers. : Fr.) *(typ. cons.)*.

Schaereria Körb., Syst. Lich. Germ.: 232. Jan 1855.
Typus: [specimen] *"Schaereria lugubris"*, Falkenstein, *Krempelhuber* (M) *(typ. cons.)* [= *S. cinereorufa* (Schaer.) Th. Fr. (*Lecidea cinereorufa* Schaer.)].

Sclerotinia Fuckel in Jahrb. Nassauischen Vereins Naturk. 23-24: 330. 19 Feb - 24 Nov 1870.
Typus: *S. libertiana* Fuckel, nom. illeg. (*Peziza sclerotiorum* Lib., *S. sclerotiorum* (Lib.) de Bary) *(typ. cons.)*.

Scutellinia (Cooke) Lambotte, Fl. Mycol. Belge, Suppl. 1: 299. 1887 (*Peziza* subg. *Scutellinia* Cooke, Mycographia: 259. Feb 1879).
Typus: *Peziza scutellata* L. : Fr. (*S. scutellata* (L. : Fr.) Lambotte).

(=) *Patella* F. H. Wigg., Prim. Fl. Holsat.: 106. 29 Mar 1780.
Typus (vide Korf & Schumacher in Taxon 35: 378. 1986): *P. ciliata* F. H. Wigg.

Septobasidium Pat. in J. Bot. (Morot) 6: 63. 16 Feb 1892.
Typus: *S. velutinum* Pat.

(=) *Gausapia* Fr., Syst. Orb. Veg.: 302. Dec 1825.
Typus: *Thelephora pedicellata* Schwein.
(=) *Glenospora* Berk. & Desm. in J. Hort. Soc. London 4: 255. 1849.
Typus: *G. curtisii* Berk. & Desm.
(=) *Campylobasidium* Lagerh. ex F. Ludw., Lehrb. Nied. Krypt.: 474. Jul 1892.
Typus: non designatus.

Septoria Sacc., Syll. Fung. 3: 474. 15 Dec 1884.
Typus: *S. cytisi* Desm. (in Ann. Sci. Nat., Bot., ser. 3, 8: 24. Jul 1847).

(H) *Septaria* Fr., Novit. Fl. Suec.: 78. 18 Dec 1819 : Fr., Syst. Mycol. 1: xl. 1 Jan 1821 [*Fungi*].
Typus: *S. ulmi* Fr. : Fr.

Simocybe P. Karst. in Bidrag Kännedom Finlands Natur Folk 32: xxii, 416. Jul-Dec 1879.
Typus: *S. centunculus* (Fr. : Fr.) P. Karst. (*Agaricus centunculus* Fr. : Fr.) *(typ. cons.)*.

Siphula Fr., Lichenogr. Eur. Reform.: 7, 406. Jun-Jul 1831.
Typus: *S. ceratites* (Wahlenb.) Fr. (*Baeomyces ceratites* Wahlenb.).

(H) *Siphula* Fr., Sched. Crit. Lich. Suec. Exsicc. 1: 3. 1824 [*Fungi*].
≡ *Dufourea* Ach. ex Luyk. 1809 *(nom. rej.* sub *Xanthoria)*.

Sordaria Ces. & De Not. in Comment. Soc. Crittog. Ital. 1: 225. Jan 1863.
Typus: *S. fimicola* (Roberge ex Desm.) Ces. & De Not. (*Sphaeria fimicola* Roberge ex Desm.) *(typ. cons.)*.

Sphaerophorus Pers. in Ann. Bot. (Usteri) 7: 23. 1794.
Typus: *S. coralloides* Pers., nom. illeg. (*Lichen globiferus* L.) *(typ. cons.)* [= *S. globosus* (Huds.) Vain., *Lichen globosus* Huds.].

Sphaeropsis Sacc. in Michelia 2: 105. 25 Apr 1880.
Typus: *S. visci* (Alb. & Schwein. : Fr.) Sacc. (*Sphaeria atrovirens* var. *visci* Alb. & Schwein. : Fr.).

(H) *Sphaeropsis* Lév. in Demidov, Voy. Russie Mér. 2: 112. 1842 [*Fungi*].
Typus: *S. conica* Lév.
(=) *Macroplodia* Westend. in Bull. Acad. Roy. Sci. Belgique, ser. 2, 2: 562. 1857.
Typus: *M. aquifolia* Westend.

Sphaerotheca Lév. in Ann. Sci. Nat., Bot., ser. 3, 15: 138. Mar 1851.
Typus: *S. pannosa* (Wallr. : Fr.) Lév. (*Alphitomorpha pannosa* Wallr. : Fr.).

(H) *Sphaerotheca* Desv. in Mém. Soc. Imp. Naturalistes Moscou 5: 68. 1817 [*Fungi*].
Typus: *S. albescens* Desv., nom. illeg. (*Aecidium thesii* Desv.).

Spongipellis Pat., Hyménomyc. Eur.: 140. Jan-Mar 1887.
Typus: *S. spumeus* (Sowerby : Fr.) Pat. (*Boletus spumeus* Sowerby : Fr.).

(=) *Somion* Adans., Fam. Pl. 2: 5, 606. Jul-Aug 1763.
Typus (vide Donk in Verh. Kon. Ned. Akad. Wetensch., Afd. Natuurk., Tweede Sect. 62: 175. 1974): *Hydnum occarium* Batsch : Fr.

Stagonospora (Sacc.) Sacc., Syll. Fung. 3: 445. 15 Dec 1884 (*Hendersonia* subg. *Stagonospora* Sacc. in Michelia 2: 8. 25 Apr 1880).
Typus: *Hendersonia paludosa* Sacc. & Speg. (*S. paludosa* (Sacc. & Speg.) Sacc.).

(=) *Hendersonia* Berk. in Ann. Mag. Nat. Hist. 6: 430. 1841.
Typus: *H. elegans* Berk.

Staurothele Norman, Conat. Praem. Gen. Lich.: 28. 1852.
Typus: *S. clopima* (Wahlenb.) Th. Fr. (Lich. Arct.: 263. Mai-Dec 1860) (*Verrucaria clopima* Wahlenb.).

(=) *Paraphysorma* A. Massal., Ric. Auton. Lich. Crost.: 116. Jun-Dec 1852.
Typus: *P. protuberans* (Schaer.) A. Massal. (*Parmelia cervina* var. *protuberans* Schaer.).

Stereocaulon Hoffm., Deutschl. Fl. 2: 128. Feb-Apr 1796.
Typus: *S. paschale* (L.) Hoffm. (*Lichen paschalis* L.).

(H) *Stereocaulon* (Schreb.) Schrad., Spic. Fl. Germ.: 113. 16 Mai - 5 Jun 1794 (*Lichen* sect. *Stereocaulon* Schreb., Gen. Pl.: 768. Mai 1791) [*Fungi*].
Typus: *Lichen corallinus* L. (*S. corallinum* (L.) Schrad.).

Stilbella Lindau in Engler & Prantl, Nat. Pflanzenfam. 1(1**): 489. Sep 1900.
Typus: *S. erythrocephala* (Ditmar : Fr.) Lindau (*Stilbum erythrocephalum* Ditmar : Fr.).

(=) *Botryonipha* Preuss in Linnaea 25: 79. Jun 1852.
Typus: *B. alba* Preuss

Telamonia (Fr.) Wünsche, Pilze: 87, 122. 1877 (*Agaricus* "trib." *Telamonia* Fr., Syst. Mycol. 1: 10, 210. 1 Jan 1821).
Typus: *Agaricus torvus* Fr. : Fr. (*T. torva* (Fr. : Fr.) Wünsche).

(≡) *Raphanozon* P. Kumm., Führ. Pilzk.: 22. Jul-Aug 1871.

Thamnolia Ach. ex Schaer., Enum. Crit. Lich. Eur.: 243. Aug-Sep 1850.
Typus: *T. vermicularis* (Sw.) Ach. ex Schaer. (*Lichen vermicularis* Sw.).

(≡) *Cerania* Ach. ex Gray, Nat. Arr. Brit. Pl. 1: 413. 1 Nov 1821.

Thelopsis Nyl. in Mém. Soc. Sci. Nat. Cherbourg 3: 194. Jun 1855.
Typus: *T. rubella* Nyl.

(=) *Sychnogonia* Körb., Syst. Lich. Germ.: 332. Jan-Mar 1855.
Typus: *S. bayrhofferi* Zwackh ex Körb.

Tholurna Norman in Flora 44: 409. 14 Jul 1861.
Typus: *T. dissimilis* (Norman) Norman (*Podocratera dissimilis* Norman).

(≡) *Podocratera* Norman in Förh. Skand. Naturf. Möte 1860: 426. 6-12 Apr 1861.

Tomentella Pers. ex Pat., Hyménomyc. Eur.: 154. Jan-Mar 1887.
Typus: *T. ferruginea* (Pers. : Fr.) Pat. (*Thelephora ferruginea* Pers. : Fr.).

(=) *Caldesiella* Sacc., Fungi Ital.: t. 125. Mai 1877.
Typus: *C. italica* Sacc.

(=) *Odontia* Pers. in Neues Mag. Bot. 1: 110. Apr-Aug 1794.
Typus (vide Banker in Bull. Torrey Bot. Club 29: 448. 1902): *O. ferruginea* Pers.

Tremella Pers. in Neues Mag. Bot. 1: 111. Apr-Aug 1794.
Typus: *T. mesenterica* Schaeff. : Fr. (Syst. Mycol. 2: 210. 1822) *(typ. cons.)*.

Tricholoma (Fr.) Staude, Schwämme Mitteldeutschl.: xxviii, 125. 1857 (*Agaricus* "trib." *Tricholoma* Fr., Syst. Mycol. 1: 9, 36. 1 Jan 1821).
Typus: *Agaricus flavovirens* Alb. & Schwein. : Fr. (*T. flavovirens* (Alb. & Schwein. : Fr.) S. Lundel).

(H) *Tricholoma* Benth. in Candolle, Prodr. 10: 426. 8 Apr 1846 [*Scrophular.*].
Typus: *T. elatinoides* Benth.

Trypethelium Spreng., Anleit. Kennt. Gew. 3: 350. 28 Mar 1804.
Typus: *T. eluteriae* Spreng.

(=) *Bathelium* Ach., Methodus: 111. Jan-Apr 1803.
Typus: *B. mastoideum* Afzel. ex Ach.

Tubercularia Tode, Fungi Mecklenb. Sel. 1: 18. 1790.
Typus: *T. vulgaris* Tode : Fr.

Urocystis Rabenh. ex Fuckel in Jahrb. Nassauischen Vereins Naturk. 23-24: 41. 19 Feb - 24 Nov 1870.
Typus: *U. occulta* (Wallr.) Fuckel (*Erysibe occulta* Wallr.).

(=) *Polycystis* Lév. in Ann. Sci. Nat., Bot., ser. 3, 5: 269. 1846.
Typus: *P. pompholygodes* (Schltdl.) Lév. (*Caeoma pompholygodes* Schltdl.).

(=) *Tuburcinia* Fr., Syst. Mycol. 3: 439. 1832 : Fr., ibid.
Typus: *T. orobanches* (Mérat) Fr. (*Rhizoctonia orobanches* Mérat).

Uromyces (Link) Unger, Exanth. Pflanzen.: 277. 1832 (*Hypodermium* subg. *Uromyces* Link in Ges. Naturf. Freunde Berlin Mag. Neuesten Entdeck. Gesammten Naturk. 7: 28. 1816).
Typus: *Uredo appendiculata* Pers. : Pers. (*Uromyces appendiculatus* (Pers. : Pers.) Unger).

(=) *Coeomurus* Link ex Gray, Nat. Arr. Brit. Pl. 1: 541. 1 Nov 1821.
Typus: *C. phaseolorum* (R. Hedw. ex DC.) Gray (*Puccinia phaseolorum* R. Hedw. ex DC.).

(=) *Pucciniola* L. Marchand in Bijdr. Natuurk. Wetensch. 4: 47. 1829.
Typus: *P. diadelphiae* L. Marchand

Valsa Fr., Summa Veg. Scand.: 410. 1849.
Typus: *V. ambiens* (Pers. : Fr.) Fr. (*Sphaeria ambiens* Pers. : Fr.).

(H) *Valsa* Adans., Fam. Pl. 2: 9, 617. Jul-Aug 1763 [*Fungi*].
Typus (vide Cannon & Hawksworth in Taxon 32: 478. 1983): *Sphaeria disciformis* Hoffm.

Venturia Sacc., Syll. Fung. 1: 586. 13 Jun 1882.
Typus: *V. inaequalis* (Cooke) G. Winter (in Thümen, Mycoth. Univ.: No. 261. 1875) (*Sphaerella inaequalis* Cooke).

(H) *Venturia* De Not. in Giorn. Bot. Ital. 1(1,1): 332. Mai-Jun 1844 [*Fungi*].
Typus: *V. rosae* De Not.

Verrucaria Schrad., Spic. Fl. Germ.: 108. 16 Mai - 5 Jun 1794.
Typus: *V. rupestris* Schrad.

(H) *Verrucaria* Scop., Intr. Hist. Nat.: 61. Jan-Apr 1777 [*Fungi*].
Typus: *Baeomyces roseus* Pers.

Volutella Fr., Syst. Mycol. 3: 458, 466. 1832.
Typus: *V. ciliata* (Alb. & Schwein. : Fr.) Fr. (*Tubercularia ciliata* Alb. & Schwein. : Fr.) *(typ. cons.)*.

Volvariella Speg. in Anales Mus. Nac. Hist. Nat. Buenos Aires 6: 119. 4 Apr 1899.
Typus: *V. argentina* Speg.

(=) *Volvarius* Roussel, Fl. Calvados, ed. 2: 59. 1806.
Typus (vide Earle in Bull. New York Bot. Gard. 5: 395, 449. 1909): *Agaricus volvaceus* Bull. : Fr.

Xanthoria (Fr.) Th. Fr., Lich. Arct.: 166. Mai-Dec 1860 (*Parmelia* [unranked] *Xanthoria* Fr., Syst. Orb. Veg.: 243. Dec 1825).
Typus: *Parmelia parietina* (L.) Ach. (*Lichen parietinus* L., *X. parietina* (L.) Th. Fr.).

(≡) *Blasteniospora* Trevis., Tornab. Blasteniosp.: 2. Feb 1853.

(=) *Dufourea* Ach. ex Luyk., Tent. Hist. Lich.: 93. 21 Dec 1809.
Typus (vide De Notaris in Giorn. Bot. Ital. 2(1,1): 224. 1846): *D. flammea* (L. f.) Ach. (Lichenogr. Universalis: 103, 524. 1810) (*Lichen flammeus* L. f.).

Xerocomus Quél. in Mougeot & Ferry, Fl. Vosges, Champ.: 477. 1887.
Typus: *X. subtomentosus* (L. : Fr.) Quél. (*Boletus subtomentosus* L. : Fr.).

(≡) *Versipellis* Quél., Enchir. Fung.: 157. Jan-Jun 1886.

Xylaria Hill ex Schrank, Baier. Fl. 1: 200. 1789.
Typus: *X. hypoxylon* (L. : Fr.) Grev. (Fl. Edin.: 355. 18 Mar 1824) (*Clavaria hypoxylon* L. : Fr.) *(typ. cons.)*.

C. BRYOPHYTA

C1. HEPATICAE

Acrolejeunea (Spruce) Schiffn. in Engler & Prantl, Nat. Pflanzenfam. 1(3): 119, 128. Sep 1893 (*Lejeunea* subg. *Acrolejeunea* Spruce in Trans. & Proc. Bot. Soc. Edinburgh 15: 74, 115. Apr 1884).
Typus: *Lejeunea torulosa* (Lehm. & Lindenb.) Spruce (*Jungermannia torulosa* Lehm. & Lindenb., *A. torulosa* (Lehm. & Lindenb.) Schiffn.).

(H) *Acro-lejeunea* Steph. in Bot. Gaz. 15: 286. Nov 1890 [*Hepat.*].
Typus: *A. parviloba* Steph.

Adelanthus Mitt. in J. Proc. Linn. Soc., Bot. 7: 243. 5 Apr 1864.
Typus: *A. falcatus* (Hook.) Mitt. (*Jungermannia falcata* Hook.).

(H) *Adelanthus* Endl., Gen. Pl.: 1327. Oct 1840 [*Icacin.*].
Typus: *A. scandens* (Thunb.) Endl. ex Baill. (*Cavanilla scandens* Thunb.).

Asterella P. Beauv. in Cuvier, Dict. Sci. Nat. 3: 257. 30 Jan 1805.
Typus: *A. tenella* (L.) P. Beauv. (*Marchantia tenella* L.) *(typ. cons.)*.

Bazzania Gray, Nat. Arr. Brit. Pl. 1: 704, 775. 1 Nov 1821 *('Bazzanius') (orth. cons.)*.
Typus: *B. trilobata* (L.) Gray (*Jungermannia trilobata* L.).

Calypogeia Raddi, Jungermanniogr. Etrusca: 31. 1818 *('Calypogeja') (orth. cons.)*.
Typus: *C. fissa* (L.) Raddi (*Mnium fissum* L., *nom. cons.*) (etiam vide *Mnium* [*Hepat.*], *nom. cons.*).

Cephaloziella (Spruce) Schiffn. in Engler & Prantl, Nat. Pflanzenfam. 1(3): 98. Sep 1893 (*Cephalozia* subg. *Cephaloziella* Spruce, Cephalozia: 23, 62. Oct-Dec 1882).
Typus: *Cephalozia divaricata* (Sm.) Dumort. (*Jungermannia divaricata* Sm., *C. divaricata* (Sm.) Schiffn.).

(=) *Dichiton* Mont., Syll. Gen. Sp. Crypt. 52. Feb 1856.
Typus: *D. perpusillus* Mont., nom. illeg. (*Jungermannia calyculata* Durieu & Mont., *D. calyculatus* (Durieu & Mont.) Trevis.).

Chiloscyphus Corda in Naturalientausch 12 [Opiz, Beitr. Naturgesch.]: 651. Sep 1829 *('Cheilocyphos') (orth. cons.)*.
Typus: *C. polyanthos* (L.) Corda (*Jungermannia polyanthos* L.).

Conocephalum Hill, Gener. Nat. Hist., ed. 2, 2: 118. 1773 *('Conicephala') (orth. cons.)*.
Typus: *C. conicum* (L.) Dumort. (Comment. Bot.: 115. Nov (sero) - Dec (prim.) 1822) *('Conocephalus conicus')* (*Marchantia conica* L.).

Diplophyllum (Dumort.) Dumort., Recueil Observ. Jungerm.: 15. 1835 (*Jungermannia* sect. *Diplophyllum* Dumort., Syll. Jungerm. Europ.: 44. 1831).
Typus: *Jungermannia albicans* L. (*D. albicans* (L.) Dumort.).

(H) *Diplophyllum* Lehm. in Ges. Naturf. Freunde Berlin Mag. Neuesten Entdeck. Gesammten Naturk. 8: 310. 1818 [*Scrophular.*].
≡ *Oligospermum* D. Y. Hong 1984.

Gymnomitrion Corda in Naturalientausch 12 [Opiz, Beitr. Naturgesch.]: 651. Sep 1829.
Typus: *G. concinnatum* (Lightf.) Corda (*Jungermannia concinnata* Lightf.).

(≡) *Cesius* Gray, Nat. Arr. Brit. Pl. 1: 705. 1 Nov 1821.

Haplomitrium Nees, Naturgesch. Eur. Leberm. 1: 109. 15 Sep - 15 Dec 1833.
Typus: *H. hookeri* (Sm.) Nees (*Jungermannia hookeri* Sm.).

(≡) *Scalius* Gray, Nat. Arr. Brit. Pl. 1: 704. 1 Nov 1821.

Heteroscyphus Schiffn. in Oesterr. Bot. Z. 60: 171. Mai 1910.
Typus: *H. aselliformis* (Reinw. & al.) Schiffn. (*Jungermannia aselliformis* Reinw. & al.).

(≡) *Gamoscyphus* Trevis. in Mem. Reale Ist. Lombardo Sci., Ser. 3, Cl. Sci. Mat. 4: 422. 1877.

Jubula Dumort., Comment. Bot.: 112. Nov (sero) - Dec (prim.) 1822.
Typus: *J. hutchinsiae* (Hook.) Dumort. (*Jungermannia hutchinsiae* Hook.) *(typ. cons.)*.

Lejeunea Lib. in Ann. Gén. Sci. Phys. 6: 372. 1820 *('Lejeunia') (orth. cons.)*.
Typus: *L. serpillifolia* Lib. (non *Jungermannia serpillifolia* Scop. 1772, nec *Jungermannia serpyllifolia* Dicks. 1801) [= *L. cavifolia* (Ehrh.) Lindb. (*Jungermannia cavifolia* Ehrh.)].

Lembidium Mitt. in Hooker, Handb. N. Zeal. Fl.: 754. 1867.
Typus: *L. nutans* (Hook. f. & Taylor) A. Evans (in Trans. Connecticut Acad. Arts 8: 266. 1892) (*Jungermannia nutans* Hook. f. & Taylor).

(H) *Lembidium* Körb., Syst. Lich. Germ.: 358. Jan-Mar 1855 [*Fungi*].
Typus: *L. polycarpum* Körb.

Lepidozia (Dumort.) Dumort., Recueil Observ. Jungerm.: 19. 1835 (*Pleuroschisma* sect. *Lepidozia* Dumort., Syll. Jungerm. Europ.: 69. 1831).
Typus: *Pleuroschisma reptans* (L.) (*Jungermannia reptans* L., *L. reptans* (L.) Dumort.) (etiam vide *Mastigophora* [*Hepat.*]).

Lethocolea Mitt. in Hooker, Handb. N. Zeal. Fl.: 751, 753. 1867.
Typus: *L. drummondii* Mitt., nom. illeg. (*Gymnanthe drummondii* Mitt., nom. illeg., *Podanthe squamata* Taylor, *L. squamata* (Taylor) E. A. Hodgs.).

(≡) *Podanthe* Taylor in London J. Bot. 5: 413. 1846.

Lopholejeunea (Spruce) Schiffn. in Engler & Prantl, Nat. Pflanzenfam. 1(3): 119, 129. Sep 1893 (*Lejeunea* subg. *Lopholejeunea* Spruce in Trans. & Proc. Bot. Soc. Edinburgh 15: 74, 119. Apr 1884).
Typus: *Lejeunea sagraeana* (Mont.) Gottsche & al. (*Phragmicoma sagraeana* Mont., *L. sagraeana* (Mont.) Schiffn.).

(H) *Lopho-lejeunea* Steph. in Bot. Gaz. 15: 285. Nov 1890 [*Hepat.*].
Typus: *L. multilacera* Steph.

Mannia Opiz in Naturalientausch 12 [Opiz, Beitr. Naturgesch.]: 646. Sep 1829.
Typus: *M. michelii* Opiz, nom. illeg. (*Grimaldia dichotoma* Raddi, nom. illeg., *Marchantia androgyna* L., *Mannia androgyna* (L.) A. Evans).

(=) *Cyathophora* Gray, Nat. Arr. Brit. Pl. 1: 678, 683. 1 Nov 1821.
Typus: *C. angustifolia* Gray, nom. illeg. (*Marchantia androgyna* L.).

Marchesinia Gray, Nat. Arr. Brit. Pl. 1: 679 (*'Marchesinius'*), 689, 817 (*'Marchesinus'*). 1 Nov 1821 (*orth. cons.*).
Typus: *M. mackaii* (Hook.) Gray (*Jungermannia mackaii* Hook.).

Mastigophora Nees, Naturgesch. Eur. Leberm. 3: 89. Apr 1838.
Typus: *M. woodsii* (Hook.) Nees (*Jungermannia woodsii* Hook.).

(H) *Mastigophora* Nees, Naturgesch. Eur. Leberm. 1: 95, 101. 15 Sep - 15 Dec 1833 [*Hepat.*].
≡ *Lepidozia* (Dumort.) Dumort. 1835 (*nom. cons.*).

Mylia Gray, Nat. Arr. Brit. Pl. 1: 693. 1 Nov 1821 *('Mylius') (orth. cons.)*.
Typus: *M. taylorii* (Hook.) Gray (*Jungermannia taylorii* Hook.).

Nardia Gray, Nat. Arr. Brit. Pl. 1: 694. 1 Nov 1821 *('Nardius') (orth. cons.)*.
Typus: *N. compressa* (Hook.) Gray (*Jungermannia compressa* Hook.).

Pallavicinia Gray, Nat. Arr. Brit. Pl. 1: 775. 1 Nov 1821 *('Pallavicinius') (orth. cons.)*.
Typus: *P. lyellii* (Hook.) Carruth. (in J. Bot. 3: 302. 1 Oct 1865) (*Jungermannia lyellii* Hook.).

Pellia Raddi, Jungermanniogr. Etrusca: 38. 1818.
Typus: *P. fabroniana* Raddi, nom. illeg. (*Jungermannia epiphylla* L., *P. epiphylla* (L.) Corda).

(≡) *Merkia* Borkh., Tent. Disp. Pl. German.: 156. Apr 1792.

Plagiochasma Lehm. & Lindenb. in Lehmann, Nov. Stirp. Pug. 4: 13. Feb-Mar 1832.
Typus: *P. cordatum* Lehm. & Lindenb.

(=) *Aytonia* J. R. Forst. & G. Forst., Char. Gen. Pl.: 74. 29 Nov 1775.
Typus: *A. rupestris* J. R. Forst. & G. Forst.

Plagiochila (Dumort.) Dumort., Recueil Observ. Jungerm.: 14. 1835 (*Radula* sect. *Plagiochila* Dumort., Syll. Jungerm. Europ.: 42. 1831).
Typus: *Radula asplenioides* (L.) Dumort. (*Jungermannia asplenioides* L., *P. asplenioides* (L.) Dumort.).

(=) *Carpolepidum* P. Beauv., Fl. Oware 1: 21. Jun 1805.
Typus (vide Bonner, Index Hepat. 3: 526. 1963): *C. dichotomum* P. Beauv.

Radula Dumort., Comment. Bot.: 112. Nov (sero) - Dec (prim.) 1822.
Typus: *R. complanata* (L.) Dumort. (*Jungermannia complanata* L.).

(≡) *Martinellius* Gray, Nat. Arr. Brit. Pl. 1: 690. 1 Nov 1821.

Reboulia Raddi in Opusc. Sci. 2: 357. Nov 1818-1819 (prim.) *('Rebouillia') (orth. cons.)*.
Typus: *R. hemisphaerica* (L.) Raddi (*Marchantia hemisphaerica* L.).

Riccardia Gray, Nat. Arr. Brit. Pl. 1: 679, 683. 1 Nov 1821 *('Riccardius') (orth. cons.)*.
Typus: *R. multifida* (L.) Gray (*Jungermannia multifida* L.).

Riccia L., Sp. Pl.: 1138. 1 Mai 1753.
Typus: *R. glauca* L. *(typ. cons.)*.

Saccogyna Dumort., Comment. Bot.: 113. Nov (sero) - Dec (prim.) 1822.
Typus: *S. viticulosa* (L.) Dumort. (*Jungermannia viticulosa* L.).

(≡) *Lippius* Gray, Nat. Arr. Brit. Pl. 1: 679, 706. 1 Nov 1821.

Scapania (Dumort.) Dumort., Recueil Observ. Jungerm.: 14. 1835 (*Radula* sect. *Scapania* Dumort., Syll. Jungerm. Europ.: 38. 1831).
Typus: *Radula undulata* (L.) Dumort. (*Jungermannia undulata* L., *S. undulata* (L.) Dumort.) *(typ. cons.)*.

Solenostoma Mitt. in J. Linn. Soc., Bot. 8: 51. 30 Jun 1864.
Typus: *S. tersum* (Nees) Mitt. (*Jungermannia tersa* Nees).

(=) *Gymnoscyphus* Corda in Sturm, Deutschl. Fl., sect. 2, Heft 26-27: 158. 1-7 Mar 1835.
Typus: *G. repens* Corda

Taxilejeunea (Spruce) Schiffn. in Engler & Prantl, Nat. Pflanzenfam. 1(3): 118, 125. Sep 1893 (*Lejeunea* subg. *Taxilejeunea* Spruce in Trans. & Proc. Bot. Soc. Edinburgh 15: 77, 212. Apr 1884).
Typus: *Lejeunea chimborazensis* Spruce (*T. chimborazensis* (Spruce) Steph.).

(H) *Taxilejeunea* Steph. in Hedwigia 28: 262. Jul-Aug 1889 [*Hepat.*].
Typus: *T. convexa* Steph.

Trachylejeunea (Spruce) Schiffn. in Engler & Prantl, Nat. Pflanzenfam. 1(3): 119, 126. Sep 1893 (*Lejeunea* subg. *Trachylejeunea* Spruce in Trans. & Proc. Bot. Soc. Edinburgh 15: 76, 180. Apr 1884).
Typus: *Lejeunea acanthina* Spruce (*T. acanthina* (Spruce) Schiffn.).

(H) *Trachylejeunea* Steph. in Hedwigia 28: 262. Jul-Aug 1889 [*Hepat.*].
Typus: *T. elegantissima* Steph.

Treubia Goebel in Ann. Jard. Bot. Buitenzorg 9: 1. 1890 (ante 1 Oct).
Typus: *T. insignis* Goebel

Trichocolea Dumort., Comment. Bot.: 113. Nov (sero) - Dec (prim.) 1822 *('Thricholea')* *(orth. cons.)*.
Typus: *T. tomentella* (Ehrh.) Dumort. (*Jungermannia tomentella* Ehrh.)

C2. MUSCI

Acidodontium Schwägr., Sp. Musc. Frond. Suppl. 2(2): 152. Mai 1827.
Typus: *A. kunthii* Schwägr., nom. illeg. (*Bryum megalocarpum* Hook., *A. megalocarpum* (Hook.) Renauld & Cardot).

(≡) *Megalangium* Brid., Bryol. Univ. 2: 28. 1827 (ante 21 Nov).

Aloina Kindb. in Bih. Kongl. Svenska Vetensk.-Akad. Handl. 6(19): 22. 1882.
Typus: *A. aloides* (W. D. J. Koch ex Schultz) Kindb. (*Trichostomum aloides* W. D. J. Koch ex Schultz).

(=) *Aloidella* (De Not.) Venturi in Comment. Fauna Veneto Trentino 1(3): 124. 1 Jan 1868 (*Tortula* sect. *Aloidella* De Not., Musci Ital.: 3, 14. 1862).
Typus: non designatus.

Amblyodon P. Beauv. in Mag. Encycl. 5: 323. 21 Feb 1804 *('Amblyodum') (orth. cons.)*.
Typus: *A. dealbatus* (Hedw.) P. Beauv. (*Meesia dealbata* Hedw.) *(typ. cons.)*.

Amphidium Schimp., Coroll. Bryol. Eur.: 39. 1856.
Typus: *A. lapponicum* (Hedw.) Schimp. (*Anictangium lapponicum* Hedw.).

(H) *Amphidium* Nees in Sturm, Deutschl. Fl., Abt. 2, 5(17): 2. 25 Apr 1819 [*Musci*].
Typus: *A. pulvinatum* Nees

Anacolia Schimp., Syn. Musc. Eur., ed. 2: 513. 1876.
Typus: *A. webbii* (Mont.) Schimp. (*Glyphocarpa webbii* Mont.).

(=) *Glyphocarpa* R. Br. in Trans. Linn. Soc. London 12: 575. Aug 1819.
Typus: *G. capensis* R. Br.

Anoectangium Schwägr., Sp. Musc. Frond. Suppl. 1(1): 33. Jun-Jul 1811.
Typus: *A. compactum* Schwägr. [= *A. aestivum* (Hedw.) Mitt. (*Gymnostomum aestivum* Hedw.)].

(=) *Anictangium* Hedw., Sp. Musc. Frond.: 40. 1 Jan 1801.
Typus: non designatus.

Atractylocarpus Mitt. in J. Linn. Soc., Bot. 12: 13, 71. Jun 1869.
Typus: *A. mexicanus* Mitt. [= *A. flagellaceus* (Müll. Hal.) R. S. Wiliams (*Dicranum flagellaceum* Müll. Hal.)].

Atrichum P. Beauv. in Mag. Encycl. 5: 329. 21 Feb 1804.
Typus: *A. undulatum* (Hedw.) P. Beauv. (*Polytrichum undulatum* Hedw.).

(≡) *Catharinea* Ehrh. ex F. Weber & D. Mohr, Index Mus. Pl. Crypt.: 2. Aug-Dec 1803.

Aulacomnium Schwägr., Sp. Musc. Frond. Suppl.: ad t. 215. Nov-Dec 1827 *('Aulacomnion') (orth. cons.).* Typus: *A. androgynum* (Hedw.) Schwägr. (*Bryum androgynum* Hedw.).	(≡)	*Gymnocephalus* Schwägr., Sp. Musc. Frond. Suppl. 1(2): 87. 1 Jan - 9 Mai 1816.
	(=)	*Arrhenopterum* Hedw., Sp. Musc. Frond.: 198. 1 Jan 1801. Typus: *A. heterostichum* Hedw.
	(=)	*Orthopixis* P. Beauv. in Mag. Encycl. 5: 322. 21 Feb 1804. Typus: non designatus.
Barbula Hedw., Sp. Musc. Frond.: 115. 1 Jan 1801. Typus: *B. unguiculata* Hedw.	(H)	*Barbula* Lour., Fl. Cochinch.: 357, 366. Sep 1790 [*Verben.*]. Typus: *B. sinensis* Lour.
Bartramia Hedw., Sp. Musc. Frond.: 164. 1 Jan 1801. Typus: *B. halleriana* Hedw.	(H)	*Bartramia* L., Sp. Pl.: 389. 1 Mai 1753 [*Til.*]. Typus: *B. indica* L.
Bartramidula Bruch & Schimp. in Bruch & al., Bryol. Europ. 4: 55. Apr 1846. Typus: *B. wilsonii* Bruch & Schimp., nom. illeg. (*Glyphocarpa cernua* Wilson, *B. cernua* (Wilson) Lindb.).	(=)	*Glyphocarpa* R. Br. in Trans. Linn. Soc. London 12: 575. Aug 1819. Typus: *G. capensis* R. Br.
Bryoxiphium Mitt. in J. Linn. Soc., Bot. 12: 24, 580. Jun 1869 *('Bryoziphium') (orth. cons.).* Typus: *B. norvegicum* (Brid.) Mitt. (*Phyllogonium norvegicum* Brid.).	(≡)	*Eustichium* Bruch & Schimp. in Bruch & al., Bryol. Europ. 2: 159. Dec 1849.
Callicostella (Müll. Hal.) Mitt. in J. Proc. Linn. Soc., Suppl. Bot. 1: 66, 136. 21 Feb 1859 (*Hookeria* sect. *Callicostella* Müll. Hal., Syn. Musc. Frond. 2: 216. Jul 1851). Typus: *Hookeria papillata* Mont. (*C. papillata* (Mont.) Mitt.).	(=)	*Schizomitrium* Schimp. in Bruch & al., Bryol. Europ. 5: 59. Jul 1851. Typus (vide Crosby in Taxon 24: 355. 1975): *S. martianum* (Hornsch.) Crosby (*Hookeria martiana* Hornsch.).
Crossidium Jur., Laubm.-Fl. Oesterr.-Ung.: 127. Apr-Jun 1882. Typus: *C. squamigerum* (Viv.) Jur. (*Barbula squamigera* Viv.).	(=)	*Chloronotus* Venturi in Comment. Fauna Veneto Trentino 1(3): 124. 1 Jan 1868. Typus: non designatus.
Cynodontium Bruch & Schimp. in Schimper, Coroll. Bryol. Eur.: 12. 1856. Typus: *C. polycarpum* (Hedw.) Schimp. (*Fissidens polycarpus* Hedw.).	(H)	*Cynodontium* Brid., Muscol. Recent. Suppl. 1: 155. 20 Apr 1806 [*Musci*]. Typus: non designatus.

Daltonia Hook. & Taylor, Muscol. Brit.: 80. 1 Jan 1818.
Typus: *D. splachnoides* Hook. & Taylor *(typ. cons.).*

Distichium Bruch & Schimp. in Bruch & al., Bryol. Europ. 2: 153. Apr 1846.
Typus: *D. capillaceum* (Hedw.) Bruch & Schimp. (*Cynontodium capillaceum* Hedw.).

(=) *Cynontodium* Hedw., Sp. Musc. Frond.: 57. 1 Jan 1801.
Typus: non designatus.

Ditrichum Hampe in Flora 50: 181. 26 Apr 1867.
Typus: *D. homomallum* (Hedw.) Hampe (*Didymodon homomallus* Hedw.) *(typ. cons.)* [= *D. heteromallum* (Hedw.) E. Britton (*Weissia heteromalla* Hedw.)] .

(H) *Ditrichum* Cass. in Bull. Sci. Soc. Philom. Paris 1817: 33. Feb 1817 [*Comp.*].
Typus: *D. macrophyllum* Cass.

(≡) *Diaphanophyllum* Lindb. in Öfvers. Förh. Kongl. Svenska Vetensk.-Akad. 19: 605. 1 Feb - 28 Mai 1863.

(=) *Aschistodon* Mont. in Ann. Sci. Nat., Bot., ser. 3, 4: 109. Aug 1845.
Typus: *A. conicus* Mont.

(=) *Lophiodon* Hook. f. & Wilson in London J. Bot. 3: 543. Sept-Oct 1844.
Typus: *L. strictus* Hook. f. & Wilson

Drepanocladus (Müll. Hal.) G. Roth in Hedwigia 38, Beibl.: (6). 28 Feb 1899 (*Hypnum* subsect. *Drepanocladus* Müll. Hal., Syn. Musc. Frond. 2: 321. Jul 1851).
Typus: *Hypnum aduncum* Hedw. (*D. aduncus* (Hedw.) Warnst.) *(typ. cons.).*

(H) *Drepanocladus* Müll. Hal. in Nuovo Giorn. Bot. Ital., ser. 2, 5: 203. 1898 [*Musci*].
Typus: *D. sinensi-uncinatus* Müll. Hal.

(=) *Drepano-hypnum* Hampe in Linnaea 37: 518. Oct 1872.
Typus: *D. fontinaloides* Hampe

Drummondia Hook. in Drummond, Musc. Amer.: No. 62. 1828.
Typus: *D. clavellata* Hook. [= *D. prorepens* (Hedw.) E. Britton (*Gymnostomum prorepens* Hedw.)].

(=) *Anodontium* Brid., Muscol. Recent. Suppl. 1: 41. 20 Apr 1806.
Typus: *A. prorepens* (Hedw.) Brid. (*Gymnostomum prorepens* Hedw.).

(=) *Leiotheca* Brid., Bryol. Univ. 1: 304, 726. Jan-Mar 1826.
Typus: *L. prorepens* (Hook.) Brid. (*Orthotrichum prorepens* Hook.).

Ephemerella Müll. Hal., Syn. Musc. Frond. 1: 34. Feb 1848.
Typus: *E. pachycarpa* (Schwägr.) Müll. Hal. (*Phascum pachycarpum* Schwägr.).

(=) *Physedium* Brid., Bryol. Univ. 1: 51. Jan-Mar 1826.
Typus: *P. splachnoides* (Hornsch.) Brid. (*Phascum splachnoides* Hornsch.).

Ephemerum Hampe in Flora 20: 285. 14 Mai 1837.
Typus: *E. serratum* (Hedw.) Hampe (*Phascum serratum* Hedw.).

(H) *Ephemeron* Mill., Gard. Dict. Abr., ed. 4: [470]. 28 Jan 1754 [*Commelin.*].
≡ *Tradescantia* L. 1753.

Gymnostomum Nees & Hornsch. in Nees & al., Bryol. Germ. 1: 112, 153. 14 Feb - 15 Apr 1823.
Typus: *G. calcareum* Nees & Hornsch.

(H) *Gymnostomum* Hedw., Sp. Musc. Frond.: 30. 1 Jan 1801 [*Musci*].
Typus: non designatus.

Gyroweisia Schimp., Syn. Musc. Eur., ed. 2: 38. 1876.
Typus: *G. tenuis* (Hedw.) Schimp. (*Gymnostomum tenue* Hedw.).

(=) *Weisiodon* Schimp., Coroll. Bryol. Eur.: 9. 1856.
Typus: *W. reflexus* (Brid.) Schimp. (*Weissia reflexa* Brid.).

Haplohymenium Dozy & Molk., Musc. Frond. Ined. Archip. Ind.: 125. 1846.
Typus: *H. sieboldii* (Dozy & Molk.) Dozy & Molk. (*Leptohymenium sieboldii* Dozy & Molk.).

(H) *Haplohymenium* Schwägr., Sp. Musc. Frond. Suppl.: ad t. 271. Jan 1829 [*Musci*].
Typus: *H. microphyllum* Schwägr. [= *Thuidium haplohymenium* (Harv.) A. Jaeger (*Hypnum haplohymenium* Harv.)].

Hedwigia P. Beauv. in Mag. Encycl. 5: 304. 21 Feb 1804.
Typus: *H. ciliata* (Hedw.) P. Beauv. (*Anictangium ciliatum* Hedw.).

(H) *Hedwigia* Sw., Prodr.: 4, 62. 20 Jun - 29 Jul 1788 [*Burser.*].
Typus: *H. balsamifera* Sw.

Helodium Warnst., Krypt.-Fl. Brandenburg 2: 675, 692. 9 Oct 1905.
Typus: *H. blandowii* (F. Weber & D. Mohr) Warnst. (*Hypnum blandowii* F. Weber & D. Mohr).

(H) *Helodium* Dumort., Fl. Belg.: 77. 1827 [*Umbell.*].
≡ *Helosciadium* W. D. J. Koch 1824.

Holomitrium Brid., Bryol. Univ. 1: 226. Jan-Mar 1826 *('Olomitrium') (orth. cons.)*.
Typus: *H. perichaetiale* (Hook.) Brid. (*Trichostomum perichaetiale* Hook.) *(typ. cons.)*.

Hookeria Sm. in Trans. Linn. Soc. London 9: 275. 23 Nov 1808.
Typus: *H. lucens* (Hedw.) Sm. (*Hypnum lucens* Hedw.).

(H) *Hookera* Salisb., Parad. Lond.: ad t. 98. 1 Mar 1808 [*Lil.*].
≡ *Brodiaea* Sm. 1810 *(nom. cons.)* (1053).

Hygroamblystegium Loeske, Moosfl. Harz.: 298. Jan-Mar 1903.
Typus: *H. irriguum* (Hook. & Wilson) Loeske (*Hypnum irriguum* Hook. & Wilson) [= *H. tenax* (Hedw.) Jenn. (*Hypnum tenax* Hedw.)].

(=) *Drepanophyllaria* Müll. Hal. in Nuovo Giorn. Bot. Ital., ser. 2, 3: 114. 1896.
≡ *Cratoneuron* (Sull.) Spruce 1867.

Hypnum Hedw., Sp. Musc. Frond.: 236. 1 Jan 1801.
Typus: *H. cupressiforme* Hedw.

Lepidopilum (Brid.) Brid., Bryol. Univ. 2: 267. 1827 (ante 21 Nov) (*Pilotrichum* subg. *Lepidopilum* Brid., Muscol. Recent. Suppl. 4: 141. 18 Dec 1818).
Typus: *Pilotrichum scabrisetum* (Schwägr.) Brid. (*Neckera scabriseta* Schwägr., *L. scabrisetum* (Schwägr.) Steere).

(=) *Actinodontium* Schwägr., Sp. Musc. Frond. Suppl. 2(2): 75. Mai 1826.
Typus: *A. ascendens* Schwägr.

Leptodon D. Mohr, Observ. Bot.: 27. 1803 (post 19 Mar).
Typus: *L. smithii* (Hedw.) F. Weber & D. Mohr (Index Mus. Pl. Crypt.: 2. Aug-Dec 1803) (*Hypnum smithii* Hedw.) *(typ. cons.)*.

Leptostomum R. Br. in Trans. Linn. Soc. London 10: 320. 7 Sep 1811.
Typus: *L. inclinans* R. Br.

(=) *Orthopixis* P. Beauv. in Mag. Encycl. 5: 322. 21 Feb 1804.
Typus: non designatus.

Leucoloma Brid., Bryol. Univ. 2: 218. 1827 (ante 21 Nov).
Typus: *L. bifidum* (Brid.) Brid. (*Hypnum bifidum* Brid.).

(≡) *Macrodon* Arn., Disp. Méth. Mousses: 42. 20 Dec 1825 - 6 Feb 1826.

(=) *Sclerodontium* Schwägr., Sp. Musc. Frond. Suppl. 2(1): 124. 1824.
Typus: *S. pallidum* (Hook.) Schwägr. (*Leucodon pallidus* Hook.).

Meesia Hedw., Sp. Musc. Frond.: 173. 1 Jan 1801.
Typus: *M. longiseta* Hedw.

(H) *Meesia* Gaertn., Fruct. Sem. Pl. 1: 344. Dec 1788 [*Ochn.*].
Typus: *M. serrata* Gaertn.

Mittenothamnium Henn. in Hedwigia 41, Beibl.: 225. 15 Dec 1902.
Typus: *M. reptans* (Hedw.) Cardot (in Rev. Bryol. Lichénol. 40: 21. 1913) (*Hypnum reptans* Hedw.) *(typ. cons.)*.

Mniobryum Limpr. in Rabenh. Krypt.-Fl., ed. 2, 4(2): 272. Jan 1892.
Typus: *M. carneum* Limpr., nom. illeg. (*Bryum delicatulum* Hedw., *M. delicatulum* (Hedw.) Dixon).

Mnium Hedw., Sp. Musc. Frond.: 188. 1 Jan 1801.
Typus: *M. hornum* Hedw.

(H) *Mnium* L., Sp. Pl.: 1109. 1 Mai 1753 [*Hepat.*].
≡ *Calypogeia* Raddi 1818 *(nom. cons.)*.

Muelleriella Dusén in Bot. Not. 1905: 304. 1905.
Typus: *M. crassifolia* (Hook. f. & Wilson) Dusén (*Orthotrichum crassifolium* Hook. f. & Wilson).

(H) *Muelleriella* Van Heurck, Treat. Diatom.: 435. Oct-Nov 1896 [*Bacillarioph.*].
Typus: *M. limbata* (Ehrenb.) Van Heurck (*Pyxidicula limbata* Ehrenb.).

Myrinia Schimp., Syn. Musc. Eur.: 482. Mar-Apr 1860.
Typus: *M. pulvinata* (Wahlenb.) Schimp. (*Leskea pulvinata* Wahlenb.).

(H) *Myrinia* Lilja, Fl. Sv. Odl. Vext., Suppl. 1: 25. 1840 [*Onagr.*].
Typus: *M. microphylla* Lilja

Neckera Hedw., Sp. Musc. Frond.: 200. 1 Jan 1801.
Typus: *N. pennata* Hedw. *(typ. cons.)*.

(H) *Neckeria* Scop., Intr. Hist. Nat.: 313. Jan-Apr 1777 [*Papaver.*].
≡ *Capnoides* Mill. 1754 (*nom. rej.* sub 2858).

Orthothecium Schimp. in Bruch & al., Bryol. Europ. 5: 105. Jul 1851.
Typus: *O. rufescens* (Sm.) Schimp. (*Hypnum rufescens* Sm.).

(H) *Orthothecium* Schott & Endl., Melet. Bot.: 31. 1832 [*Stercul.*].
Typus: *O. lhotskyanum* Schott & Endl.

Papillaria (Müll. Hal.) Lorentz, Moosstudien: 165. 1864 (*Neckera* subsect. *Papillaria* Müll. Hal., Syn. Musc. Frond. 2: 134. Sep 1850).
Typus: *Neckera nigrescens* (Sw. ex Hedw.) Schwägr. (*Hypnum nigrescens* Sw. ex Hedw., *P. nigrescens* (Sw. ex Hedw.) A. Jaeger) *(typ. cons.)*.

(H) *Papillaria* J. Kickx f., Fl. Crypt. Louvain: 73, 104. 20 Jun 1835 [*Fungi*].
≡ *Pycnothelia* Dufour 1821.

Pelekium Mitt. in J. Linn. Soc., Bot. 10: 176. 19 Mar 1868.
Typus: *P. velatum* Mitt.

(=) *Lorentzia* Hampe in Flora 50: 75. 26 Feb 1867.
Typus: *L. longirostris* Hampe (in Nuovo Giorn. Bot. Ital. 4: 288. 1872).

Platygyrium Schimp. in Bruch & al., Bryol. Europ. 5: 95. Jul 1851.
Typus: *P. repens* (Brid.) Schimp. (*Pterigynandrum repens* Brid.).

(=) *Pterigynandrum* Hedw., Sp. Musc. Frond.: 80. 1 Jan 1801.
Typus (vide Schimper in Bruch & al., Bryol. Eur. 5: 121. 1851): *P. filiforme* Hedw.

(=) *Pterogonium* Sw. in Monthly Rev. 34: 537. 1 Jun 1801.
Typus (vide Schimper in Bruch & al., Bryol. Eur. 5: 125. 1851): *P. gracile* (Hedw.) Sm. (Engl. Bot. 16: ad t. 1085. 1802) (*Pterigynandrum gracile* Hedw.).

(=) *Leptohymenium* Schwägr., Sp. Musc. Frond. Suppl. 3(1): ad t. 246c. Apr-Dec 1828.
Typus: *L. tenue* (Hook.) Schwägr. (*Neckera tenuis* Hook.).

Pleuridium Rabenh., Deutschl. Krypt.-Fl. 2(3): 79. Jul 1848.
Typus: *P. subulatum* (Hedw.) Rabenh. (*Phascum subulatum* Hedw.).

(H) *Pleuridium* Brid., Muscol. Recent. Suppl. 4: 10. 18 Dec 1818 [*Musci*].
Typus (vide Snider & Margadant in Taxon 22: 693. 1973): *P. globiferum* Brid.

Pleurozium Mitt. in J. Linn. Soc., Bot. 12: 22, 537. Jun 1869.
Typus: *P. schreberi* (Brid.) Mitt. (*Hypnum schreberi* Brid.).

Pterygoneurum Jur., Laubm.-Fl. Oesterr.-Ung.: 95. Apr-Jun 1882 (*'Pterigoneurum'*) (*orth. cons.*).
Typus: *P. cavifolium* Jur., nom. illeg. (*Pottia cavifolia* Fürnr., nom. illeg., *Gymnostomum ovatum* Hedw., *P. ovatum* (Hedw.) Dixon).

(=) *Pharomitrium* Schimp., Syn. Musc. Eur.: 120. Mar-Apr 1860.
Typus: *P. subsessile* (Brid.) Schimp. (*Gymnostomum subsessile* Brid.).

Ptychomitrium Fürnr. in Flora 12(2, Ergänzungsbl.): 19. Jul-Oct 1829 (*'Pthychomitrium'*) (*orth. cons.*).
Typus: *P. polyphyllum* (Sw.) Bruch & Schimp. (in Bruch & al., Bryol. Europ. 3: 82. Dec 1837) (*Dicranum polyphyllum* Sw.).

(=) *Brachysteleum* Rchb., Consp. Regni Veg.: 34. Dec 1828 - Mar 1829.
Typus: *B. crispatum* (Hedw.) Hornsch. (in Martius, Fl. Bras. 1(2): 20. 1840) (*Encalypta crispata* Hedw.).

Rhodobryum (Schimp.) Limpr., Laubm. Deutschl. 2: 444. Dec 1892 (*Bryum* subg. *Rhodobryum* Schimp., Syn. Musc. Eur.: 381. Mar-Apr 1860).
Typus: *Bryum roseum* (Hedw.) Crome (*Mnium roseum* Hedw., *R. roseum* (Hedw.) Limpr.).

(H) *Rhodo-bryum* Hampe in Linnaea 38: 663. Dec 1874 [*Musci*].
Typus: *R. leucocanthum* Hampe

Schistidium Bruch & Schimp. in Bruch & al., Bryol. Europ. 3: 93. Aug 1845.
Typus: *S. maritimum* (Turn.) Bruch & Schimp. (*Grimmia maritima* Turn.) *(typ. cons.)*.

(H) *Schistidium* Brid., Muscol. Recent. Suppl. 4: 20. 18 Dec 1818 [*Musci*].
Typus (vide Mårtensson in Kung. Svenska Vetenskapsakad. Avh. Naturskyddsärenden 14: 106. 1956): *S. pulvinatum* (Hedw.) Brid. (*Gymnostomum pulvinatum* Hedw.).

Timmia Hedw., Sp. Musc. Frond.: 176. 1 Jan 1801.
Typus: *T. megapolitana* Hedw.

(H) *Timmia* J. F. Gmel., Syst. Nat. 2: 524, 538. Sep (sero) - Nov 1791 [*Amaryllid.*].
Typus: non designatus.

Tortella (Müll. Hal.) Limpr. in Rabenh. Krypt.-Fl., ed. 2, 4(1): 599. Oct 1888 (*Barbula* sect. *Tortella* Müll. Hal., Syn. Musc. Frond. 1: 599. Mar 1849).
Typus: *Barbula caespitosa* Schwägr. (*T. caespitosa* (Schwägr.) Limpr.) [= *T. humilis* (Hedw.) Jenn. (*Barbula humilis* Hedw.)].

(=) *Pleurochaete* Lindb. in Öfvers. Förh. Kongl. Svenska Vetensk.-Akad. 21: 253. 21 Aug 1864.
Typus: *P. squarrosa* (Brid.) Lindb. (*Barbula squarrosa* Brid.).

Tortula Hedw., Sp. Musc. Frond.: 122. 1 Jan 1801.
Typus: *T. subulata* Hedw. *(typ. cons.)*.

(H) *Tortula* Roxb. ex Willd., Sp. Pl. 3: 6, 359. 1800 [*Verben.*].
Typus: *T. aspera* Roxb. ex Willd.

Trichostomum Bruch in Flora 12: 396. 7 Jul 1829.
Typus: *T. brachydontium* Bruch

(H) *Trichostomum* Hedw., Sp. Musc. Frond.: 107. 1 Jan 1801 [*Musci*].
Typus: non designatus.

(=) *Plaubelia* Brid., Bryol. Univ. 1: 522. Jan-Mar 1826.
Typus: *P. tortuosa* Brid.

D. PTERIDOPHYTA

Anemia Sw., Syn. Fil.: 6, 155. Mar-Apr 1806.
Typus: *A. phyllitidis* (L.) Sw. (*Osmunda phyllitidis* L.).

(=) *Ornithopteris* Bernh. in Neues J. Bot. 1(2): 40. Oct-Nov 1805.
Typus (vide Reed in Bol. Soc. Brot., ser. 2, 21: 153. 1947): *O. adiantifolia* (L.) Bernh. (*Osmunda adiantifolia* L.).

Angiopteris Hoffm. in Commentat. Soc. Regiae Sci. Gott. 12: 29. 1796.
Typus: *A. evecta* (G. Forst.) Hoffm. (*Polypodium evectum* G. Forst.).

(H) *Angiopteris* Adans., Fam. Pl. 2: 21, 518. Jul-Aug 1763 [*Pteridoph.*].
≡ *Onoclea* L. 1753.

Araiostegia Copel. in Philipp. J. Sci. 34: 240. 1927.
Typus: *A. hymenophylloides* (Blume) Copel. (*Aspidium hymenophylloides* Blume).

(=) *Gymnogrammitis* Griff., Ic. Pl. Asiat. 2: t. 129, f. 1. 1849; Not. Pl. Asiat. 2: 608. 1849.
Typus: *G. dareiformis* (Hook.) Ching ex Tardieu & C. Chr. (in Notul. Syst. (Paris) 6: 2. 1937) (*Polypodium dareiforme* Hook.).

Ceterach Willd., Anleit. Selbststud. Bot.: 578. 1804.
Typus: *C. officinarum* Willd.

(H) *Ceterac* Adans., Fam. Pl. 2: 20, 536. Jul-Aug 1763 [*Pteridoph.*].
Typus: non designatus.

Cheilanthes Sw., Syn. Fil.: 5, 126. Mar-Apr 1806.
Typus: *C. micropteris* Sw.

(=) *Allosorus* Bernh. in Neues J. Bot. 1(2): 5, 36. Oct-Nov 1805.
Typus (vide Pichi Sermolli in Webbia 9: 394. 1953): *A. pusillus* (Willd. ex Bernh.) Bernh. (*Adiantum pusillum* Willd. ex Bernh.).

Coniogramme Fée [Mém. Foug. 5] in Mém. Soc. Mus. Hist. Nat. Strasbourg 5: 167. 1852.
Typus: *C. javanica* (Blume) Fée (*Gymnogramma javanica* Blume).

(=) *Dictyogramme* Fée [Mém. Foug. 5] in Mém. Soc. Mus. Hist. Nat. Strasbourg 4(1): 206. 1850.
Typus: *D. japonica* (Thunb.) Fée ([Mém. Foug. 5] in Mém. Soc. Mus. Hist. Nat. Strasbourg 5: 375. 1852) (*Hemionitis japonica* Thunb.).

Cystodium J. Sm. in Hooker, Gen. Fil.: ad t. 96. 1841.
Typus: *C. sorbifolium* (Sm.) J. Sm. (*Dicksonia sorbifolia* Sm.).

(H) *Cystodium* Fée, Essai Crypt. Ecorc. 2: 13. 1837 [*Fungi*].
≡ *Gassicurtia* Fée 1824.

Cystopteris Bernh. in Neues J. Bot. 1(2): 5, 26. Oct-Nov 1805.
Typus: *C. fragilis* (L.) Bernh. (*Polypodium fragile* L.).

Danaea Sm. in Mém. Acad. Roy. Sci. (Turin) 5: 420. 1793.
Typus: *D. nodosa* (L.) Sm. (*Acrostichum nodosum* L.).

(H) *Danaa* All., Fl. Pedem. 2: 34. Apr-Jul 1785 [*Umbell.*].
Typus: *D. aquilegiifolia* (All.) All. (*Coriandrum aquilegiifolium* All.).

Doryopteris J. Sm. in J. Bot. (Hooker) 3: 404. Mai 1841.
Typus: *D. palmata* (Willd.) J. Sm. (*Pteris palmata* Willd.).

(=) *Cassebeera* Kaulf., Enum. Filic.: 216. 1824.
Typus (vide Fée [Mém. Foug. 5] in Mém. Soc. Mus. Hist. Nat. Strasbourg 5: 119. 1852): *C. triphylla* (Lam.) Kaulf. (*Adiantum triphyllum* Lam.).

Drymoglossum C. Presl [Tent. Pterid.] in Abh. Königl. Böhm. Ges. Wiss., ser. 4, 5: 227. 1836 (ante 2 Dec).
Typus: *D. piloselloides* (L.) C. Presl (*Pteris piloselloides* L.).

(=) *Pteropsis* Desv. in Mém. Soc. Linn. Paris 6(3): 218. Jul 1827.
Typus (vide Pichi Sermolli in Webbia 9: 403. 1953): *Acrostichum heterophyllum* L.

Drynaria (Bory) J. Sm. in J. Bot. (Hooker) 4: 60. Jul 1841 (*Polypodium* subg. *Drynaria* Bory in Ann. Sci. Nat. (Paris) 5: 463. 1825).
Typus: *Polypodium linnaei* Bory, nom. illeg. (*Polypodium quercifolium* L., *D. quercifolia* (L.) J. Sm.).

Dryopteris Adans., Fam. Pl. 2: 20, 551. Jul-Aug 1763.
Typus: *D. filix-mas* (L.) Schott (Gen. Fil.: ad t. 9. 1834) (*Polypodium filix-mas* L.).

(≡) *Filix* Ség., Pl. Veron. 3: 53. Jul-Aug 1754.

Elaphoglossum Schott ex J. Sm. in J. Bot. (Hooker) 4: 148. Aug 1841.
Typus: *E. conforme* (Sw.) J. Sm. (*Acrostichum conforme* Sw.) *(typ. cons.)*.

(=) *Aconiopteris* C. Presl [Tent. Pterid.] in Abh. Königl. Böhm. Ges. Wiss., ser. 4, 5: 236. 1836 (ante 2 Dec).
Typus: *A. subdiaphana* (Hook. & Grev.) C. Presl (*Acrostichum subdiaphanum* Hook. & Grev.).

Gleichenia Sm. in Mém. Acad. Roy. Sci. (Turin) 5: 419. 1793.
Typus: *G. polypodioides* (L.) Sm. (*Onoclea polypodioides* L.).

Lygodium Sw. in J. Bot. (Schrader) 1800(2): 7, 106. Oct-Dec 1801.
Typus: *L. scandens* (L.) Sw. (*Ophioglossum scandens* L.).

(=) *Ugena* Cav., Icon. 6: 73. Jan-Mai 1801.
Typus (vide Pichi Sermolli in Webbia 9: 418. 1953): *U. semihastata* Cav., nom. illeg. (*Ophioglossum flexuosum* L.).

Marsilea L., Sp. Pl.: 1099. 1 Mai 1753.
Typus: *M. quadrifolia* L. *(typ. cons.)*.

Matteuccia Tod. in Giorn. Sci. Nat. Econ. Palermo 1: 235. 1866.
Typus: *M. struthiopteris* (L.) Tod. (*Osmunda struthiopteris* L.).

(≡) *Pteretis* Raf. in Amer. Monthly Mag. & Crit. Rev. 2: 268. Feb 1818.

Pellaea Link, Fil. Spec.: 59. 3-10 Sep 1841.
Typus: *P. atropurpurea* (L.) Link (*Pteris atropurpurea* L.).

Polystichum Roth, Tent. Fl. Germ. 3: 31, 69. Jun-Sep 1799.
Typus: *P. lonchitis* (L.) Roth (*Polypodium lonchitis* L.).

(=) *Hypopeltis* Michx., Fl. Bor.-Amer. 2: 266. 19 Mar 1803.
Typus: *H. lobulata* Bory (in Fouché & al., Exp. Sci. Morée, Bot.: 286. Sep 1832).

Pteridium Gled. ex Scop., Fl. Carniol.: 169. 15 Jun - 21 Jul 1760.
Typus: *P. aquilinum* (L.) Kuhn (in Ascherson & al., Bot. Ost-Afrika: 11. Aug-Sep 1879) (*Pteris aquilina* L.).

Schizaea Sm. in Mém. Acad. Roy. Sci. (Turin) 5: 419. 1793.
Typus: *S. dichotoma* (L.) Sm. (*Acrostichum dichotomum* L.).

(=) *Lophidium* Rich. in Actes Soc. Hist. Nat. Paris 1: 114. 1792.
Typus: *L. latifolium* Rich.

Selaginella P. Beauv., Prodr. Aethéogam.: 101. 10 Jan 1805.
Typus: *S. spinosa* P. Beauv., nom. illeg. (*Lycopodium selaginoides* L., *S. selaginoides* (L.) Link).

(≡) *Selaginoides* Ség., Pl. Veron. 3: 51. Jul-Aug 1754.

(=) *Lycopodioides* Boehm. in Ludwig, Def. Gen. Pl. Ed. 3: 485. 1760.
Typus (vide Rothmaler in Feddes Repert. Spec. Nov. Regni Veg. 54: 69. 1944): *L. denticulata* (L.) Kuntze (Revis. Gen. Pl. 1-2: 824. 5 Nov 1891) (*Lycopodium denticulatum* L.).

(=) *Stachygynandrum* P. Beauv. ex Mirb. in Lamarck & Mirbel, Hist. Nat. Vég. 3: 477. 1802, 4: 312. 1802.
Typus (vide Pichi Sermolli in Webbia 26: 164. 1971): *S. flabellatum* (L.) P. Beauv. (Prodr. Aethéogam.: 113. 10 Jan 1805) (*Lycopodium flabellatum* L.).

Sphenomeris Maxon in J. Wash. Acad. Sci. 3: 144. 1913.
Typus: *S. clavata* (L.) Maxon (*Adiantum clavatum* L.).

(≡) *Stenoloma* Fée [Mém. Foug. 5] in Mém. Soc. Mus. Hist. Nat. Strasbourg 5: 330. 1852 (typ. des.: Morton in Taxon 8: 29. 1959).

Thelypteris Schmidel, Icon. Pl., ed. Keller: 3, 45. 18 Oct 1763.
Typus: *T. palustris* Schott (Gen. Fil.: ad t. 10. 1834) (*Acrostichum thelypteris* L.).

(H) *Thelypteris* Adans., Fam. Pl. 2: 20, 610. Jul-Aug 1763 [*Pteridoph.*].
≡ *Pteris* L. 1753.

E. SPERMATOPHYTA

CYCADACEAE

4 **_Dioon_** Lindl. in Edwards's Bot. Reg. 29 (Misc.): 59. Aug 1843 *('Dion') (orth. cons.)*.
Typus: *D. edule* Lindl.

7 **_Zamia_** L., Sp. Pl., ed. 2: 1659. Jul-Aug 1763.
Typus: *Z. pumila* L.
(≡) *Palma-filix* Adans., Fam. Pl. 2: 21, 587. Jul-Aug 1763 (typ. des.: Florin in Taxon 5: 189. 1956).

TAXACEAE

13 **_Podocarpus_** L'Hér. ex Pers., Syn. Pl. 2: 580. Sep 1807.
Typus: *P. elongatus* (Aiton) L'Hér. ex Pers. (*Taxus elongata* Aiton) *(typ. cons.)*.
(H) *Podocarpus* Labill., Nov. Holl. Pl. 2: 71. Aug 1806 [*Podocarp.*].
≡ *Phyllocladus* Rich. & Mirb. 1825 *(nom. cons.)* (15).
(=) *Nageia* Gaertn., Fruct. Sem. Pl. 1: 191. Dec 1788.
Typus: *N. japonica* Gaertn., nom. illeg. (*Myrica nagi* Thunb.).

15 **_Phyllocladus_** Rich. & Mirb. in Mém. Mus. Hist. Nat. 13: 48. 1825.
Typus: *P. billardierei* Mirb., nom. illeg. (*Podocarpus aspleniifolius* Labill., *Phyllocladus aspleniifolius* (Labill.) Hook. f.) (etiam vide 13).

17 **_Torreya_** Arn. in Ann. Nat. Hist. 1: 130. Apr 1838.
Typus: *T. taxifolia* Arn.
(H) *Torreya* Raf. in Amer. Monthly Mag. & Crit. Rev. 3: 356. Sep 1818 [*Lab.*].
Typus: *T. grandiflora* Raf.

PINACEAE

20 **_Agathis_** Salisb. in Trans. Linn. Soc. London 8: 311. 9 Mar 1807.
Typus: *A. loranthifolia* Salisb., nom. illeg. (*Pinus dammara* Lamb., *A. dammara* (Lamb.) Rich.).

23 **_Cedrus_** Trew, Cedr. Lib. Hist. 1: 6. 12 Mai - 13 Oct 1757.
Typus: *C. libani* A. Rich. (in Bory, Dict. Class. Hist. Nat. 3: 299. 6 Sep 1823) (*Pinus cedrus* L.).
(H) *Cedrus* Duhamel, Traité Arbr. Arbust. 1: xxviii, 139. 1755 [*Cupress.*].
Typus: non designatus.

25 ***Pseudolarix*** Gordon, Pinetum: 292. Jun-Dec 1858.
Typus: [specimen cult. in Anglia] ex Herb. George Gordon (K No. 0003455) (typ. cons.) [= *P. amabilis* (J. Nelson) Rehder (*Larix amabilis* J. Nelson)] .

31 ***Cunninghamia*** R. Br. in Richard, Comm. Bot. Conif. Cycad.: 80, 149. Sep-Nov 1826.
Typus: *C. sinensis* R. Br., nom. illeg. (*Pinus lanceolata* Lamb., *C. lanceolata* (Lamb.) Hook.).

(H) *Cunninghamia* Schreb., Gen. Pl.: 789. Mai 1791 [*Rub.*].
≡ *Malanea* Aubl. 1775.

(≡) *Belis* Salisb. in Trans. Linn. Soc. London 8: 315. 9 Mar 1807.

32 ***Sequoia*** Endl., Syn. Conif.: 197. Mai-Jun 1847.
Typus: *S. sempervirens* (D. Don) Endl. (*Taxodium sempervirens* D. Don).

32a ***Metasequoia*** Hu & W. C. Cheng in Bull. Fan Mem. Inst. Biol., Bot., ser. 2, 1: 154. 15 Mai 1948.
Typus: *M. glyptostroboides* Hu & W. C. Cheng *(typ. cons.)*.

(H) *Metasequoia* Miki in Jap. J. Bot. 11: 261. 1941 (post Mar) [Foss.].
Typus: *M. disticha* (Heer) Miki (*Sequoia disticha* Heer).

GNETACEAE

48 ***Welwitschia*** Hook. f. in Gard. Chron. 1862: 71. 25 Jan 1862.
Typus: *W. mirabilis* Hook. f.

(H) *Welwitschia* Rchb., Handb. Nat. Pfl.-Syst.: 194. 1-7 Oct 1837 [*Polemon.*].
≡ *Eriastrum* Wooton & Standl. 1913.

(≡) *Tumboa* Welw. in Gard. Chron. 1861: 75. Jan 1861.

POTAMOGETONACEAE

57 ***Posidonia*** K. D. Koenig in Ann. Bot. (König & Sims) 2: 95. 1 Jun 1805.
Typus: *P. caulinii* K. D. Koenig, nom. illeg. (*Zostera oceanica* L., *P. oceanica* (L.) Delile).

(=) *Alga* Boehm. in Ludwig, Def. Gen. Pl., ed. 3: 503. 1760.
Typus: non designatus.

60 ***Cymodocea*** K. D. Koenig in Ann. Bot. (König & Sims) 2: 96. 1 Jun 1805.
Typus: *C. aequorea* K. D. Koenig

(=) *Phucagrostis* Cavolini, Phucagr. Theophr. Anth.: xiii. 1792.
Typus: *P. major* Cavolini

APONOGETONACEAE

65	***Aponogeton*** L. f., Suppl. Pl.: 32, 214. Apr 1782. Typus: *A. monostachyos* L. f., nom. illeg. (*Saururus natans* L., *A. natans* (L.) Engl. & Krause).	(H)	*Aponogeton* Hill, Brit. Herb.: 480. Dec 1756 [*Potamogeton.*]. ≡ *Zannichellia* L. 1753.

GRAMINEAE (POACEAE)

124	***Vossia*** Wall. & Griff. in J. Asiat. Soc. Bengal 5: 572. Sep 1836. Typus: *V. procera* Wall. & Griff., nom. illeg. (*Ischaemum cuspidatum* Roxb., *V. cuspidata* (Roxb.) Griff.).	(H)	*Vossia* Adans., Fam. Pl. 2: 243, 619. Jul-Aug 1763 [*Aiz.*]. Typus: non designatus.
127	***Rottboellia*** L. f., Suppl. Pl.: 13, 114. Apr 1782. Typus: *R. exaltata* L. f. 1782, non (L.) Naezen 1779 *(typ. cons.)* [= *R. cochinchinensis* (Lour.) Clayton (*Stegosia cochinchinensis* Lour.)].	(H) (=)	*Rottboelia* Scop., Intr. Hist. Nat.: 233. Jan-Apr 1777 [*Olac.*]. ≡ *Heymassoli* Aubl. 1775. *Manisuris* L., Mant. Pl. 2: 164, 300. Oct 1771. Typus: *M. myurus* L.
134a	***Diectomis*** Kunth in Mém. Mus. Hist. Nat. 2: 69.1815. Typus: *D. fastigiata* (Sw.) P. Beauv. (Ess. Agrostogr.: 132, 160. Dec 1812) (*Andropogon fastigiatus* Sw.) *(typ. cons.)*.	(H)	*Diectomis* P. Beauv., Ess. Agrostogr.: 132. Dec 1812 [*Gram.*]. Typus: *D. fasciculata* P. Beauv.
134b	***Sorghum*** Moench, Methodus: 207. 4 Mai 1794. Typus: *S. bicolor* (L.) Moench (*Holcus bicolor* L.).	(H)	*Sorgum* Adans., Fam. Pl. 2: 38, 606. Jul-Aug 1763 [*Gram.*]. ≡ *Holcus* L. 1753 *(nom. cons.)* (257).
134c	***Chrysopogon*** Trin., Fund. Agrost.: 187. Jan 1820. Typus: *C. gryllus* (L.) Trin. (*Andropogon gryllus* L.).	(≡) (=) (=)	*Pollinia* Spreng., Pl. Min. Cogn. Pug. 2: 10. 1815. *Rhaphis* Lour., Fl. Cochinch.: 538, 552. Sep 1790. Typus: *R. trivialis* Lour. *Centrophorum* Trin., Fund. Agrost.: 106. Jan 1820. Typus: *C. chinense* Trin.
143	***Tragus*** Haller, Hist. Stirp. Helv. 2: 203. 25 Mar 1768. Typus: *T. racemosus* (L.) All. (Fl. Pedem. 2: 241. Apr-Jun 1785) (*Cenchrus racemosus* L.).	(≡)	*Nazia* Adans., Fam. Pl. 2: 31, 581. Jul-Aug 1763.

150 ***Zoysia*** Willd. in Ges. Naturf. Freunde Berlin Neue Schriften 3: 440. 1801 (post 21 Apr).
Typus: *Z. pungens* Willd.

166 ***Echinochloa*** P. Beauv., Ess. Agrostogr.: 53. Dec 1812.
Typus: *E. crusgalli* (L.) P. Beauv. (*Panicum crusgalli* L.).

(≡) *Tema* Adans., Fam. Pl. 2: 496, 610. Jul-Aug 1763.

166a ***Digitaria*** Haller, Hist. Stirp. Helv. 2: 244. 25 Mar 1768.
Typus: *D. sanguinalis* (L.) Scop. (Fl. Carniol., ed. 2, 1: 52. 1771) (*Panicum sanguinale* L.) *(typ. cons.)*.

(H) *Digitaria* Heist. ex Fabr., Enum.: 207. 1759 [*Gram.*].
Typus: non designatus.

169 ***Oplismenus*** P. Beauv., Fl. Oware 2: 14. 6 Aug 1810.
Typus: *O. africanus* P. Beauv.

(=) *Orthopogon* R. Br., Prodr.: 194. 27 Mar 1810.
Typus (vide Hitchcock in U.S.D.A. Bull. 772: 238. 1920): *O. compositus* (L.) R. Br. (*Panicum compositum* L.).

171 ***Setaria*** P. Beauv., Ess. Agrostogr.: 51, 178. Dec 1812.
Typus: *S. viridis* (L.) P. Beauv. (*Panicum viride* L.) *(typ. cons.)*.

(H) *Setaria* Ach. ex Michx., Fl. Bor.-Amer. 2: 331. 19 Mar 1803 [*Fungi*].
Typus: *S. trichodes* Michx.

194 ***Leersia*** Sw., Prodr.: 1, 21. 20 Jun - 29 Jul 1788.
Typus: *L. oryzoides* (L.) Sw. (*Phalaris oryzoides* L.) *(typ. cons.)*.

(≡) *Homalocenchrus* Mieg in Acta Helv. Phys.-Math. 4: 307. 1760.

201 ***Ehrharta*** Thunb. in Kongl. Vetensk. Acad. Handl. 40: 217. Jul-Dec 1779.
Typus: *E. capensis* Thunb.

(=) *Trochera* Rich. in Observ. Mém. Phys. 13: 225. Mar 1779.
Typus: *T. striata* Rich.

206 ***Hierochloë*** R. Br., Prodr.: 208. 27 Mar 1810.
Typus: *H. odorata* (L.) P. Beauv. (Ess. Agrostogr.: 62, 164. Dec 1812) (*Holcus odoratus* L.) *(typ. cons.)*.

(=) *Savastana* Schrank, Baier. Fl. 1: 100, 337. Jun-Dec 1789.
Typus: *S. hirta* Schrank

(=) *Torresia* Ruiz & Pav., Fl. Peruv. Prodr.: 125. Oct (prim.) 1794.
Typus: *T. utriculata* Ruiz & Pav. (Syst. Veg. Fl. Peruv. Chil.: 251. Dec (sero) 1798).

(=) *Disarrenum* Labill., Nov. Holl. Pl. 2: 82. Mar 1807.
Typus: *D. antarcticum* (G. Forst.) Labill. (*Aira antarctica* G. Forst.).

212 ***Piptochaetium*** J. Presl in Presl, Reliq. Haenk. 1: 222. Jan-Jun 1830.
Typus: *P. setifolium* J. Presl

(=) *Podopogon* Raf., Neogenyton: 4. 1825.
Typus (vide Clayton in Taxon 32: 649. 1983): *Stipa avenacea* L.

221 ***Crypsis*** Aiton, Hort. Kew. 1: 48. 7 Aug - 1 Oct 1789.
Typus: *C. aculeata* (L.) Aiton (*Schoenus aculeatus* L.).

228 ***Coleanthus*** Seidel in Roemer & Schultes, Syst. Veg. 2: 11, 276. Nov 1817.
Typus: *C. subtilis* (Tratt.) Seidel (*Schmidtia subtilis* Tratt.).

257 ***Holcus*** L., Sp. Pl.: 1047. 1 Mai 1753.
Typus: *H. lanatus* L. *(typ. cons.)*.

269 ***Corynephorus*** P. Beauv., Ess. Agrostogr.: 90, 159. Dec 1812.
Typus: *C. canescens* (L.) P. Beauv. (*Aira canescens* L.).

(≡) *Weingaertneria* Bernh., Syst. Verz.: 23, 51. 1800.

272 ***Ventenata*** Koeler, Descr. Gram.: 272. 1802.
Typus: *V. avenacea* Koeler, nom. illeg. (*Avena dubia* Leers, *V. dubia* (Leers) Coss.) *(typ. cons.)*.

(H) *Vintenatia* Cav., Icon. 4: 28. Sep-Dec 1797 [*Epacrid.*].
Typus: non designatus.

(=) *Heteranthus* Borkh. [Fl. Grafsch. Catzenelnb. 2] in Andre, Botaniker Compend. Biblioth. 16-18: 71. 1796.
Typus: *H. bromoides* Borkh., nom. illeg. (*Bromus triflorus* L.).

278a ***Loudetia*** Hochst. ex Steud., Syn. Pl. Glumac. 1: 238. 12-13 Apr 1854.
Typus: *L. elegans* Hochst. ex A. Braun (in Flora 24: 713. 7 Dec 1841).

(H) *Loudetia* Hochst. ex A. Braun in Flora 24: 713. 7 Dec 1841 [*Gram.*].
≡ *Tristachya* Nees 1829.

280 ***Danthonia*** DC. in Lamarck & Candolle, Fl. Franç., ed. 3, 3: 32. 17 Sep 1805.
Typus: *D. spicata* (L.) Roem. & Schult. (Syst. Veg. 2: 690. Nov 1817) (*Avena spicata* L.) *(typ. cons.)*.

(=) *Sieglingia* Bernh., Syst. Verz.: 44. 1800.
Typus: *S. decumbens* (L.) Bernh. (*Festuca decumbens* L.).

282 ***Cynodon*** Rich. in Persoon, Syn. Pl. 1: 85. 1 Apr - 15 Jun 1805.
Typus: *C. dactylon* (L.) Pers. (*Panicum dactylon* L.).

(≡) *Dactilon* Vill., Hist. Pl. Dauphiné 2: 69. Feb 1787.

(=) *Capriola* Adans., Fam. Pl. 2: 31, 532. Jul-Aug 1763.
Typus: non designatus.

286 ***Ctenium*** Panz., Ideen Rev. Gräser: 38, 61. 1813.
Typus: *C. carolinianum* Panz., nom. illeg. (*Chloris monostachya* Michx.) [= *Ctenium aromaticum* (Walter) Wood (*Aegilops aromatica* Walter)].

(≡) *Campulosus* Desv. in Nouv. Bull. Sci. Soc. Philom. Paris 2: 189. Dec 1810.

295 ***Bouteloua*** Lag. in Varied. Ci. 2(4): 134. 1805 (*'Botelua') (orth. cons.)*.
Typus: *B. racemosa* Lag.

308 ***Buchloë*** Engelm. in Trans. Acad. Sci. St. Louis 1: 432. Jan-Apr 1859.
Typus: *B. dactyloides* (Nutt.) Engelm. (*Sesleria dactyloides* Nutt.).

312 ***Schmidtia*** Steud. ex J. A. Schmidt, Beitr. Fl. Cap Verd. Ins.: 144. 1 Jan - 13 Feb 1852.
Typus: *S. pappophoroides* Steud. ex J. A. Schmidt

(H) *Schmidtia* Moench, Suppl. Meth.: 217. 2 Mai 1802 [*Comp.*].
Typus: *S. fruticosa* Moench

320 ***Echinaria*** Desf., Fl. Atlant. 2: 385. Feb-Jul 1799.
Typus: *E. capitata* (L.) Desf. (*Cenchrus capitatus* L.).

(H) *Echinaria* Fabr., Enum.: 206. 1759 [*Gram.*].
≡ *Cenchrus* L. 1753.

(≡) *Panicastrella* Moench, Methodus: 205. 4 Mai 1794.

329 ***Cortaderia*** Stapf in Gard. Chron., ser. 3, 22: 378. 27 Nov 1897.
Typus: *C. argentea* Stapf, nom. illeg. (*Gynerium argenteum* Nees, nom. illeg., *Arundo dioica* Spreng. 1825, non Lour. 1790, *Arundo selloana* Schult. & Schult. f., *C. selloana* (Schult. & Schult. f.) Asch. & Graebn.).

(≡) *Moorea* Lem. in Ill. Hort. 2: 14. 2 Feb 1855.

356 ***Diarrhena*** P. Beauv., Ess. Agrostogr.: 142, 160, 162. Dec 1812.
Typus: *D. americana* P. Beauv. (*Festuca diandra* Michx. 1803, non Moench 1794).

357 ***Centotheca*** Desv. in Nouv. Bull. Sci. Soc. Philom. Paris 2: 189. Dec 1810 (*'Centosteca') (orth. cons.)*.
Typus: *C. lappacea* (L.) Desv. (*Cenchrus lappaceus* L.).

358 ***Zeugites*** P. Browne, Civ. Nat. Hist. Jamaica: 341. 10 Mar 1756.
Typus: *Z. americanus* Willd. (Sp. Pl. 4: 204. 1805) (*Apluda zeugites* L.).

374 ***Lamarckia*** Moench, Methodus: 201. 4 Mai 1794 *('Lamarkia') (orth. cons.)*.
Typus: *L. aurea* (L.) Moench (*Cynosurus aureus* L.).

(H) *Lamarckia* Olivi, Zool. Adriat.: 258. Sep-Dec 1792 [*Chloroph.*].
Typus: non designatus.

(≡) *Achyrodes* Boehm. in Ludwig, Def. Gen. Pl., ed. 3: 420. 1760.

381 ***Scolochloa*** Link, Hort. Berol. 1: 136. 1 Oct - 27 Nov 1827.
Typus: *S. festucacea* (Willd.) Link (*Arundo festucacea* Willd.).

(H) *Scolochloa* Mert. & W. D. J. Koch, Deutschl. Fl., ed. 3, 1: 374, 528. Jan-Mai 1823 [*Gram.*].
Typus: *S. arundinacea* (P. Beauv.) Mert. & W. D. J. Koch (*Donax arundinaceus* P. Beauv.).

383 ***Glyceria*** R. Br., Prodr.: 179. 27 Mar 1810.
Typus: *G. fluitans* (L.) R. Br. (*Festuca fluitans* L.).

384 ***Puccinellia*** Parl., Fl. Ital. 1: 366. 1848.
Typus: *P. distans* (Jacq.) Parl. (*Poa distans* Jacq.) *(typ. cons.)*.

(≡) *Atropis* (Trin.) Rupr. ex Griseb. in Ledebour, Fl. Ross. 4: 388. Sep 1852 (*Poa* sect. *Atropis* Trin. in Mém. Acad. Imp. Sci. Saint-Pétersbourg, Sér. 6, Sci. Math., Seconde Pt. Sci. Nat. 6: 68. Mar 1836).

417 ***Phyllostachys*** Siebold & Zucc. in Abh. Math.-Phys. Cl. Königl. Bayer. Akad. Wiss. 3: 745. 1843.
Typus: *P. bambusoides* Siebold & Zucc.

424 ***Bambusa*** Schreb., Gen. Pl.: 236. Apr 1789.
Typus: *B. arundinacea* (Retz.) Willd. (Sp. Pl. 2: 245. Mar 1799) (*Bambos arundinacea* Retz.).

(≡) *Bambos* Retz., Observ. Bot. 5: 24. Sep 1788.

CYPERACEAE

452 ***Lipocarpha*** R. Br. in Tuckey, Narr. Exped. Zaire: 459. 5 Mar 1818.
Typus: *L. argentea* R. Br., nom. illeg. (*Hypaelyptum argenteum* Vahl, nom. illeg., *Scirpus senegalensis* Lam., *L. senegalensis* (Lam.) T. Durand & H. Durand).

(=) *Hypaelyptum* Vahl, Enum. Pl. 2: 283. Oct-Dec 1805.
Typus (vide Panigrahi in Taxon 34: 511. 1985): *H. filiforme* Vahl

454 ***Ascolepis*** Nees ex Steud., Syn. Pl. Glumac. 2: 105. 10-11 Apr 1855.
Typus: *A. eriocauloides* (Steud.) Nees ex Steud. (*Kyllinga eriocauloides* Steud.) *(typ. cons.)*.

459 ***Mariscus*** Vahl, Enum. Pl. 2: 372. Oct-Dec 1805.
Typus: *M. capillaris* (Sw.) Vahl (*Schoenus capillaris* Sw.) *(typ. cons.)*.

(H) *Mariscus* Scop., Meth. Pl.: 22. 25 Mar 1754 [*Cyper.*].
Typus: *Schoenus mariscus* L.

462 ***Kyllinga*** Rottb., Descr. Icon. Rar. Pl.: 12. Jan-Jul 1773.
Typus: *K. nemoralis* (J. R. Forst. & G. Forst.) Dandy ex Hutch. & Dalziel (Fl. W. Trop. Afr. 2: 487. Feb 1936) (*Thyrocephalon nemorale* J. R. Forst. & G. Forst.) *(typ. cons.)*.

(H) *Killinga* Adans., Fam. Pl. 2: 498, 539. Jul-Aug 1763 [*Umbell.*].
≡ *Athamantha* L. 1753.

465 ***Ficinia*** Schrad. in Commentat. Soc. Regiae Sci. Gott. Recent. 7: 143. 1832.
Typus: *F. filiformis* (Lam.) Schrad. (*Schoenus filiformis* Lam.) *(typ. cons.)*.

(=) *Melancranis* Vahl, Enum. Pl. 2: 239. Oct-Dec 1805.
Typus: non designatus.

(=) *Hemichlaena* Schrad. in Gött. Gel. Anz. 1821: 2066. 29 Dec 1821.
Typus (vide Goetghebeur & Arnold in Taxon 33: 114. 1984): *H. capillifolia* Schrad.

466a ***Trichophorum*** Pers., Syn. Pl. 1: 69. 1 Apr - 15 Jun 1805.
Typus: *T. alpinum* (L.) Pers. (*Eriophorum alpinum* L.) *(typ. cons.)*.

468 ***Scirpus*** L., Sp. Pl.: 47. 1 Mai 1753.
Typus: *S. sylvaticus* L. *(typ. cons.)*.

468a ***Blysmus*** Panz. ex Schult., Mant. 2: 41. Jan-Apr 1824.
Typus: *B. compressus* (L.) Panz. ex Link (Hort. Berol. 1: 278. 1 Oct - 27 Nov 1827) (*Schoenus compressus* L.).

(≡) *Nomochloa* P. Beauv. ex T. Lestib., Essai Cypér.: 37. 29 Mar 1819.

468b ***Schoenoplectus*** (Rchb.) Palla in Verh. K.K. Zool.-Bot. Ges. Wien 38 (Sitzungsber.): 49. 1888 (*Scirpus* subg. *Schoenoplectus* Rchb., Icon. Fl. Germ. Helv.: 8: 40. 1846).
Typus: *Scirpus lacustris* L. (*Schoenoplectus lacustris* (L.) Palla).

(=) *Heleophylax* P. Beauv. ex T. Lestib., Essai Cypér.: 41. 29 Mar 1819.
Typus: non designatus.

(=) *Elytrospermum* C. A. Mey. in Mém. Acad. Imp. Sci. St.-Pétersbourg Divers Savans 1: 200. Oct 1831.
Typus: *E. californicum* C. A. Mey.

471 ***Fimbristylis*** Vahl, Enum. Pl. 2: 285. Oct-Dec 1805.
Typus: *F. dichotoma* (L.) Vahl (*Scirpus dichotomus* L.) *(typ. cons.)*.

(=) *Iria* (Rich. ex Pers.) R. Hedw., Gen. Pl.: 360. Jul 1806 (*Cyperus* subg. *Iria* Rich. ex Pers., Syn. Pl. 1: 65. 1 Apr - 15 Jun 1805).
Typus: *Cyperus monostachyos* L.

471a ***Bulbostylis*** Kunth, Enum. Pl. 2: 205. 6 Mai 1837.
Typus: *B. capillaris* (L.) Kunth ex C. B. Clarke (in Hooker, Fl. Brit. India 6: 652. Sep 1893) (*Scirpus capillaris* L.) *(typ. cons.)*.

(H) *Bulbostylis* Steven in Mém. Soc. Imp. Naturalistes Moscou 5: 355. 1817 [*Cyper.*].
Typus: non designatus.

(=) *Stenophyllus* Raf., Neogenyton: 4. 1825.
Typus: *S. cespitosus* Raf. (*Scirpus stenophyllus* Elliott).

492 ***Rhynchospora*** Vahl, Enum. Pl. 2: 229. Oct-Dec 1805 *('Rynchospora')* *(orth. cons.)*.
Typus: *R. alba* (L.) Vahl (*Schoenus albus* L.) *(typ. cons.)*.

(=) *Dichromena* Michx., Fl. Bor.-Amer. 1: 37. 19 Mar 1803.
Typus: *D. leucocephala* Michx.

PALMAE (ARECACEAE)

542 ***Pritchardia*** Seem. & H. Wendl. in Bonplandia 10: 197. 1 Jul 1862.
Typus: *P. pacifica* Seem. & H. Wendl.

(H) *Pritchardia* Unger ex Endl., Gen. Pl., Suppl. 2: 102. Mar-Jun 1842 [Foss.].
Typus: *P. insignis* Unger ex Endl.

543 ***Washingtonia*** H. Wendl. in Bot. Zeitung (Berlin) 37: lxi, 68, 148. 31 Jan 1879.
Typus: *W. filifera* (Linden ex André) H. Wendl. (*Pritchardia filifera* Linden ex André).

565 ***Metroxylon*** Rottb. in Nye Saml. Kongel. Danske Vidensk. Selsk. Skr. 2: 527. 1783.
Typus: *M. sagu* Rottb.

(=) *Sagus* Steck, Sagu: 21. 21 Sep 1757.
Typus (vide Moore in Taxon 11: 165. 1965): *S. genuina* Giseke (Prael. Ord. Nat. Pl.: 93. Apr 1792).

567 ***Pigafetta*** (Blume) Becc. in Malesia 1: 89. 1877 *('Pigafettia')* (*Sagus* sect. *Pigafetta* Blume, Rumphia 2: 154. Jan-Aug 1843) *(orth. cons.)*.
Typus: *Sagus filaris* Giseke (*P. filaris* (Giseke) Becc.) *(typ. cons.)*.

(H) *Pigafetta* Adans., Fam. Pl. 2: 223, 590. Jul-Aug 1763 [*Acanth.*].
≡ *Eranthemum* L. 1753.

575 ***Arenga*** Labill. in Bull. Sci. Soc. Philom. Paris 2: 162. Nov (sero) 1800.
Typus: *A. saccharifera* Labill. [= *A. pinnata* (Wurmb) Merr. (*Saguerus pinnatus* Wurmb)].

(=) *Saguerus* Steck, Sagu: 15. 21 Sep 1757.
Typus: *S. pinnatus* Wurmb (in Verh. Batav. Genootsch. Kunsten 1: 351. 1781).

594	***Chamaedorea*** Willd., Sp. Pl. 4: 638, 800. Apr 1806. Typus: *C. gracilis* Willd., nom. illeg. (*Borassus pinnatifrons* Jacq., *C. pinnatifrons* (Jacq.) Oerst.).	(=) (=)	*Morenia* Ruiz & Pav., Fl. Peruv. Prodr.: 150. Oct (prim.) 1794. Typus: *M. fragrans* Ruiz & Pav. (Syst. Veg. Fl. Peruv. Chil.: 299. Dec 1798). *Nunnezharia* Ruiz & Pav., Fl. Peruv. Prodr.: 147. Oct (prim.) 1794. Typus: *N. fragrans* Ruiz & Pav. (Syst. Veg. Fl. Peruv. Chil.: 294. Dec 1798).
612	***Prestoea*** Hook. f. in Bentham & Hooker, Gen. Pl. 3: 875, 899. 14 Apr 1883. Typus: *P. pubigera* (Griseb. & H. Wendl.) Hook. f. (in Rep. (Annual) Roy. Bot. Gard. Kew 1882: 56. 1884) (*Hyospathe pubigera* Griseb. & H. Wendl.).	(=) (=)	*Martinezia* Ruiz & Pav., Fl. Peruv. Prodr.: 148. Oct (prim.) 1794. Typus: *M. ensiformis* Ruiz & Pav. (Syst. Veg. Fl. Peruv. Chil.: 297. Dec 1798). *Oreodoxa* Willd. in Deutsch. Abh. Königl. Akad. Wiss. Berlin 1801: 251. 1803. Typus: *O. acuminata* Willd.
631	***Euterpe*** Mart., Hist. Nat. Palm. 2: 28. Nov 1823. Typus: *E. oleracea* Mart.	(H) (=) (=)	*Euterpe* Gaertn., Fruct. Sem. Pl. 1: 24. Dec 1788 [*Palm.*]. Typus: *E. pisifera* Gaertn. *Martinezia* Ruiz & Pav., Fl. Peruv. Prodr.: 148. Oct (prim.) 1794. Typus: *M. ensiformis* Ruiz & Pav. (Syst. Veg. Fl. Peruv. Chil.: 297. Dec 1798). *Oreodoxa* Willd. in Deutsch. Abh. Königl. Akad. Wiss. Berlin 1801: 251. 1803. Typus: *O. acuminata* Willd.
639	***Veitchia*** H. Wendl. in Seemann, Fl. Vit.: 270. 31 Jul 1868. Typus: *V. joannis* H. Wendl. *(typ. cons.)*.	(H)	*Veitchia* Lindl. in Gard. Chron. 1861: 265. Mar 1861 [*Pin.*]. Typus: *V. japonica* Lindl.
657	***Orbignya*** Mart. ex Endl., Gen. Pl.: 257. Oct 1837. Typus: *O. phalerata* Mart. (Hist. Nat. Palm. 3: 302. 19 Sep 1845).	(H)	*Orbignya* Bertero in Mercurio Chileno 16: 737. 15 Jul 1829 [*Euphorb.*]. Typus: *O. trifolia* Bertero
660	***Maximiliana*** Mart., Palm. Fam.: 20. 13 Apr 1824. Typus: *M. regia* Mart. (Hist. Nat. Palm. 2: 131. Jan-Mar 1826), non *Maximilianea regia* Mart. 1819 (*M. martiana* H. Karst.) *(typ. cons.)*.	(H)	*Maximilianea* Mart. in Flora 2: 452. 7 Aug 1819 [*Cochlosperm.*]. Typus: *M. regia* Mart.
668	***Astrocaryum*** G. Mey., Prim. Fl. Esseq.: 265. Nov 1818. Typus: *A. aculeatum* G. Mey.	(=)	*Avoira* Giseke, Prael. Ord. Nat. Pl.: 38, 53. Apr 1792. Typus: *A. vulgaris* Giseke

670 ***Desmoncus*** Mart., Palm. Fam.: 20. 13 Apr 1824.
Typus: *D. polyacanthos* Mart. (Hist. Nat. Palm. 2: 85. 1824, serius) *(typ. cons.)*.

CYCLANTHACEAE

678a ***Asplundia*** Harling in Acta Horti Berg. 17: 41. 1954 (post 3 Nov).
Typus: *A. latifolia* (Ruiz & Pav.) Harling (*Carludovica latifolia* Ruiz & Pav.).

(=) *Sarcinanthus* Oersted in Vidensk. Meddel. Dansk Naturhist. Foren. Kjøbenhavn 1857: 196. 1857.
Typus: *S. utilis* Oersted

682 ***Ludovia*** Brongn. in Ann. Sci. Nat., Bot., ser. 4, 15: 361. Jun 1861.
Typus: *L. lancifolia* Brongn.

(H) *Ludovia* Pers., Syn. Pl. 2: 576. Sep 1807 [*Cyclanth.*].
Typus: non designatus.

ARACEAE

690 ***Culcasia*** P. Beauv., Fl. Oware, ed. 4°: 4. 2 Oct 1803.
Typus: *C. scandens* P. Beauv. *(typ. cons.)*.

700 ***Monstera*** Adans., Fam. Pl. 2: 470, 578. Jul-Aug 1763.
Typus: *M. adansonii* Schott (in Wiener Z. Kunst 1830: 1028. 23 Oct 1830) (*Dracontium pertusum* L., non *M. pertusa* (Roxb.) Schott) *(typ. cons.)*.

708 ***Symplocarpus*** Salisb. ex Nutt., Gen. N. Amer. Pl. 1: 105. 14 Jul 1818.
Typus: *S. foetidus* (L.) Nutt. (*Dracontium foetidum* L.).

723 ***Amorphophallus*** Blume ex Decne. in Nouv. Ann. Mus. Hist. Nat. 3: 366. 1834.
Typus: *A. campanulatus* Decne.

(=) *Thomsonia* Wall., Pl. Asiat. Rar. 1: 83. 1 Sep 1830.
Typus: *T. napalensis* Wall.

(=) *Pythion* Mart. in Flora 14: 458. 1831 (med.).
Typus: *Arum campanulatum* Roxb., nom. illeg. (*Dracontium paeoniifolium* Dennst., *Amorphophallus paeoniifolius* (Dennst.) Nicolson).

730 ***Montrichardia*** Crueg. in Bot. Zeitung (Berlin) 12: 25. 13 Jan 1854.
Typus: *M. aculeata* (G. Mey.) Schott (Syn. Aroid.: 72. Mar 1856) (*Caladium aculeatum* G. Mey.).

(=) *Pleurospa* Raf., Fl. Tellur. 4: 8. 1838 (med.).
Typus (vide Nicolson in Regnum Veg. 34: 55. 1964): *P. reticulata* Raf., nom. illeg. (*Arum arborescens* L.).

739 ***Philodendron*** Schott in Wiener Z. Kunst 1829: 780. 6 Aug 1829 *('Philodendrum') (orth. cons.).*
Typus: *P. grandifolium* (Jacq.) Schott (*Arum grandifolium* Jacq.).

747 ***Peltandra*** Raf. in J. Phys. Chim. Hist. Nat. Arts 89: 103. Aug 1819.
Typus: *P. undulata* Raf.

748 ***Zantedeschia*** Spreng., Syst. Veg. 3: 756, 765. Jan-Mar 1826.
Typus: *Z. aethiopica* (L.) Spreng. (*Calla aethiopica* L.) (etiam vide 755).

752 ***Alocasia*** (Schott) G. Don in Sweet, Hort. Brit., ed. 3: 631. 1839 (sero) (*Colocasia* sect. *Alocasia* Schott in Schott & Endlicher, Melet. Bot.: 18. 1832).
Typus: *Colocasia cucullata* (Lour.) Schott (*Arum cucullatum* Lour., *Alocasia cucullata* (Lour.) G. Don) *(typ. cons.).*

(H) *Alocasia* Neck. ex Raf., Fl. Tellur. 3: 64. Nov-Dec 1837 [*Ar.*].
Typus: non designatus.

755 ***Colocasia*** Schott in Schott & Endlicher, Melet. Bot.: 18. 1832.
Typus: *C. antiquorum* Schott (*Arum colocasia* L.) *(typ. cons.).*

(H) *Colocasia* Link, Diss. Bot.: 77. 1795 [*Ar.*].
≡ *Zantedeschia* Spreng. 1826 *(nom. cons.)* (748).

756 ***Hapaline*** Schott, Gen. Aroid.: 44. 1858.
Typus: *H. benthamiana* (Schott) Schott (*Hapale benthamiana* Schott).

(≡) *Hapale* Schott in Oesterr. Bot. Wochenbl. 7: 85. 12 Mar 1857.

779 ***Helicodiceros*** Schott in Klotzsch, App. Gen. Sp. Nov. 1855: 2. Dec 1855 - 1856 (prim.).
Typus: *H. muscivorus* (L. f.) Engl. (in Candolle & Candolle, Monogr. Phan. 2: 605. Sep 1879) (*Arum muscivorum* L. f.).

(≡) *Megotigea* Raf., Fl. Tellur. 3: 64. Nov-Dec 1837.

784 ***Biarum*** Schott in Schott & Endlicher, Melet. Bot.: 17. 1832.
Typus: *B. tenuifolium* (L.) Schott (*Arum tenuifolium* L.).

(≡) *Homaid* Adans., Fam. Pl. 2: 470, 584. Jul-Aug 1763.

787 ***Pinellia*** Ten. in Atti Reale Accad. Sci. Sez. Soc. Reale Borbon. 4: 69. 1839.
Typus: *P. tuberifera* Ten., nom. illeg. (*Arum subulatum* Desf.) [= *P. ternata* (Thunb.) Makino (*Arum ternatum* Thunb.)].

(=) *Atherurus* Blume, Rumphia 1: 135. Apr-Jun 1837.
Typus (vide Nicolson in Taxon 16: 515. 1967): *A. tripartitus* Blume

LEMNACEAE

796 ***Wolffia*** Horkel ex Schleid., Beitr. Bot. 1: 233. 11-13 Jul 1844.
Typus: *W. michelii* Schleid.

(H) *Wolfia* Schreb., Gen. Pl.: 801. Mai 1791 [*Flacourt.*].
Typus: non designatus.

RESTIONACEAE

800 ***Lyginia*** R. Br., Prodr.: 248. 27 Mar 1810.
Typus: *L. barbata* R. Br. *(typ. cons.)*.

804 ***Restio*** Rottb., Descr. Pl. Rar.: 9. 1772.
Typus: *R. triticeus* Rottb.

(H) *Restio* L., Syst. Nat., ed. 12, 2: 735. 15-31 Oct 1767 [*Restion.*].
Typus: *R. dichotomus* L.

808 ***Leptocarpus*** R. Br., Prodr.: 250. 27 Mar 1810.
Typus: *L. aristatus* R. Br. *(typ. cons.)*.

(=) *Schoenodum* Labill., Nov. Holl. Pl. 2: 79. Aug 1806.
Typus (vide Kunth, Enum. Pl. 3: 445. 1841): *S. tenax* Labill.

815 ***Hypolaena*** R. Br., Prodr.: 251. 27 Mar 1810.
Typus: *H. fastigiata* R. Br. *(typ. cons.)*.

(=) *Calorophus* Labill., Nov. Holl. Pl. 2: 78. Aug 1806.
Typus: *C. elongata* Labill.

816 ***Hypodiscus*** Nees in Lindley, Intr. Nat. Syst. Bot., ed. 2: 450. Jul 1836.
Typus: *H. aristatus* (Thunb.) Nees ex Mast. (in J. Linn. Soc., Bot. 10: 252. 23 Mai 1868) (*Restio aristatus* Thunb.).

(=) *Lepidanthus* Nees in Linnaea 5: 665. Oct 1830.
Typus: *L. willdenowia* Nees, nom. illeg. (*Willdenowia striata* Thunb.).

ERIOCAULACEAE

830 ***Paepalanthus*** Kunth, Enum. Pl. 3: 498. 23-29 Mai 1841.
Typus: *P. lamarckii* Kunth *(typ. cons.)*.

(H) *Paepalanthus* Mart. in Nova Acta Phys.-Med. Acad. Caes. Leop.-Carol. Nat. Cur. 17: 13. 5-11 Jul 1835 [*Eriocaul.*].
Typus (vide Moldenke in N. Amer. Fl. 19: 37. 1937): *P. corymbosus* (Bong.) Kunth (Enum. Pl. 3: 509. 23-29 Mai 1849) (*Eriocaulon corymbosum* Bong.).

(=) *Dupathya* Vell., Fl. Flumin.: 35. 7 Sep - 28 Nov 1829.
Typus: non designatus.

BROMELIACEAE

846 ***Cryptanthus*** Otto & A. Dietr. in Allg. Gartenzeitung 4: 297. 1836.
Typus: *C. bromelioides* Otto & A. Dietr.

(H) *Cryptanthus* Osbeck, Dagb. Ostind. Resa: 215. 1757 [*Spermatoph.*].
Typus: *C. chinensis* Osbeck

861 ***Aechmea*** Ruiz & Pav., Fl. Peruv. Prodr.: 47. Oct (prim.) 1794.
Typus: *A. paniculata* Ruiz & Pav. (Fl. Peruv. 3: 37. Aug 1802).

(=) *Hoiriri* Adans., Fam. Pl. 2: 67, 584. Jul-Aug 1763.
Typus: *Bromelia nudicaulis* L.

878 ***Pitcairnia*** L'Hér., Sert. Angl.: 7. Jan (prim.) 1789.
Typus: *P. bromeliifolia* L'Hér.

(=) *Hepetis* Sw., Prodr.: 4, 56. 20 Jun - 29 Jul 1788.
Typus: *H. angustifolia* Sw.

891 ***Vriesea*** Lindl. in Edwards's Bot. Reg. 29: ad t. 10. 7 Feb 1843 *('Vriesia') (orth. cons.)*.
Typus: *V. psittacina* (Hook.) Lindl. (*Tillandsia psittacina* Hook.).

(≡) *Hexalepis* Raf., Fl. Tellur. 4: 24. 1838 (med.).

COMMELINACEAE

894 ***Palisota*** Rchb. ex Endl., Gen. Pl.: 125. Dec 1836.
Typus: *P. ambigua* (P. Beauv.) C. B. Clarke (in Candolle & Candolle, Monogr. Phan. 3: 131. Jun 1881) (*Commelina ambigua* P. Beauv.).

(=) *Duchekia* Kostel., Allg. Med.-Pharm. Fl. 1: 213. Mai 1831.
Typus: *D. hirsuta* (Thunb.) Kostel. (*Dracaena hirsuta* Thunb.).

899a **Murdannia** Royle, Ill. Bot. Himal. Mts.: 403. Mai-Apr 1840.
Typus: *M. scapiflora* (Roxb.) Royle (*Commelina scapiflora* Roxb.).

(=) *Dilasia* Raf., Fl. Tellur. 4: 122. 1838 (med.).
Typus: *D. vaginata* (L.) Raf. (*Commelina vaginata* L.).

(=) *Streptylis* Raf., Fl. Tellur. 4: 122. 1838 (med.).
Typus: *S. bracteolata* (Lam.) Raf. (*Commelina bracteolata* Lam.).

904 **Cyanotis** D. Don, Prodr. Fl. Nepal.: 45. 26 Jan - 1 Feb 1825.
Typus: *C. barbata* D. Don

909 **Dichorisandra** J. C. Mikan, Del. Fl. Faun. Bras.: ad t. 3. 1820 (sero).
Typus: *D. thyrsiflora* J. C. Mikan

910 **Tinantia** Scheidw. in Allg. Gartenzeitung 7: 365. 16 Nov 1839.
Typus: *T. fugax* Scheidw.

(H) *Tinantia* Dumort., Anal. Fam. Pl.: 58. 1829 [*Irid.*].
Typus: non designatus.

(=) *Pogomesia* Raf., Fl. Tellur. 3: 67. Nov-Dec 1837.
Typus: *P. undata* (Humb. & Bonpl. ex Willd.) Raf. (*Tradescantia undata* Humb. & Bonpl. ex Willd.).

PONTEDERIACEAE

921 **Eichhornia** Kunth, Eichhornia: 3. 1842.
Typus: *E. azurea* (Sw.) Kunth (*Pontederia azurea* Sw.) (*typ. cons.*).

(=) *Piaropus* Raf., Fl. Tellur. 2: 81. Jan-Mar 1837.
Typus: non designatus.

923 **Reussia** Endl., Gen. Pl.: 139. Dec 1836.
Typus: *R. triflora* Seub. (in Mart., Fl. Bras. 3(1): 96. 1 Jun 1847).

924 **Heteranthera** Ruiz & Pav., Fl. Peruv. Prodr.: 9. Oct (prim.) 1794.
Typus: *H. reniformis* Ruiz & Pav. (Fl. Peruv. 1: 43. 1798, med.).

JUNCACEAE

937 **Luzula** DC. in Lamarck & Candolle, Fl. Franç., ed. 3, 3: 158. 17 Sep 1805.
Typus: *L. campestris* (L.) DC. (*Juncus campestris* L.) (*typ. cons.*).

(≡) *Juncoides* Ség., Pl. Veron. 3: 88. Jul-Aug 1754.

LILIACEAE

944 ***Narthecium*** Huds., Fl. Angl.: 127. Jan-Jun 1762.
Typus: *N. ossifragum* (L.) Huds. (*Anthericum ossifragum* L.).

(H) *Narthecium* Gérard, Fl. Gallo-Prov.: 142. Mar-Oct 1761 [*Lil.*].
Typus: *Anthericum calyculatum* L.

951 ***Chionographis*** Maxim. in Bull. Acad. Imp. Sci. Saint-Pétersbourg 11: 435. 31 Mai 1867.
Typus: *C. japonica* (Willd.) Maxim. (*Melanthium japonicum* Willd.).

(=) *Siraitos* Raf., Fl. Tellur. 4: 26. 1838 (med.).
Typus: *S. aquaticus* Raf.

952 ***Heloniopsis*** A. Gray in Mem. Amer. Acad. Arts, ser. 2, 6: 416. 1858.
Typus: *H. pauciflora* A. Gray

(=) *Hexonix* Raf., Fl. Tellur. 2: 13. Jan-Mar 1837.
Typus: *H. japonica* (Thunb.) Raf. (*Scilla japonica* Thunb.).

(=) *Kozola* Raf., Fl. Tellur. 2: 25. Jan-Mar 1837.
Typus: *K. japonica* (Thunb.) Raf. (*Scilla japonica* Thunb.).

955 ***Amianthium*** A. Gray in Ann. Lyceum Nat. Hist. New York 4: 121. Nov 1837.
Typus: *A. muscitoxicum* (Walter) A. Gray (*Melanthium muscitoxicum* Walter) *(typ. cons.)*.

(=) *Chrosperma* Raf., Neogenyton: 3. 1825.
Typus: *Melanthium laetum* Aiton

957 ***Stenanthium*** (A. Gray) Kunth, Enum. Pl. 4: 189. 17-19 Jul 1843 (*Veratrum* subg. *Stenanthium* A. Gray in Ann. Lyceum Nat. Hist. New York 4: 119. Nov 1837).
Typus: *Veratrum angustifolium* Pursh (*S. angustifolium* (Pursh) Kunth).

(≡) *Anepsa* Raf., Fl. Tellur. 2: 31. Jan-Mar 1837.

962 ***Schelhammera*** R. Br., Prodr.: 273. 27 Mar 1810.
Typus: *S. undulata* R. Br. *(typ. cons.)*.

(H) *Schelhameria* Heist. ex Fabr., Enum.: 161. 1759 [*Cruc.*].
Typus: non designatus.

967 ***Tricyrtis*** Wall., Tent. Fl. Napal.: 61. Sep-Dec 1826.
Typus: *T. pilosa* Wall.

(=) *Compsoa* D. Don, Prodr. Fl. Nepal.: 50. 26 Jan - 1 Feb 1825.
Typus: *C. maculata* D. Don

968 ***Burchardia*** R. Br., Prodr.: 272. 27 Mar 1810.
Typus: *B. umbellata* R. Br.

(H) *Burcardia* Heist. ex Duhamel, Traité Arbr. Arbust. 1: xxx, 11. 1755 [*Verben.*].
≡ *Callicarpa* L. 1753.

974 ***Anguillaria*** R. Br., Prodr.: 273. 27 Mar 1810.
Typus: *A. dioica* R. Br. *(typ. cons.).*

(H) *Anguillaria* Gaertn., Fruct. Sem. Pl. 1: 372. Dec 1788 (typ. des.: Rickett & Stafleu in Taxon 8: 234. 1959) [*Myrsin.*].
≡ *Heberdenia* Banks ex DC. 1841 *(nom. cons.)* (6288).

975 ***Iphigenia*** Kunth, Enum. Pl. 4: 212. 17-19 Jul 1843.
Typus: *I. indica* (L.) Kunth (*Melanthium indicum* L.).

(=) *Aphoma* Raf., Fl. Tellur. 2: 31. Jan-Mar 1837.
Typus: *A. angustiflora* Raf.

982 ***Paradisea*** Mazzuc., Viaggio Bot. Alpi Giulie: 27. 1811.
Typus: *P. hemeroanthericoides* Mazzuc., nom. illeg. (*Hemerocallis liliastrum* L., *P. liliastrum* (L.) Bertol.).

(≡) *Liliastrum* Fabr., Enum.: 4. 1759.

985 ***Bulbine*** Wolf, Gen. Pl.: 84. 1776.
Typus: *B. frutescens* (L.) Willd. (Enum. Hort. Berol.: 372. Apr 1809) (*Anthericum frutescens* L.).

985a ***Trachyandra*** Kunth, Enum. Pl. 4: 573. 17-19 Jul 1843.
Typus: *T. hispida* (L.) Kunth (*Anthericum hispidum* L.).

(≡) *Obsitila* Raf., Fl. Tellur. 2: 27. Jan-Mar 1837 (typ. des.: Merrill, Ind. Raf.: 92. 1949).
(=) *Lepicaulon* Raf., Fl. Tellur. 2: 27. Jan-Mar 1837.
Typus: *L. squameum* (L. f.) Raf. (*Anthericum squameum* L. f.).

987 ***Simethis*** Kunth, Enum. Pl. 4: 618. 17-19 Jul 1843.
Typus: *S. bicolor* Kunth, nom. illeg. (*Anthericum planifolium* Vand. ex L., *S. planifolia* (L.) Gren. & Godr.).

992 ***Thysanotus*** R. Br., Prodr.: 282. 27 Mar 1810.
Typus: *T. junceus* R. Br., nom. illeg. (*Chlamysporum juncifolium* Salisb., *T. juncifolius* (Salisb.) Willis & Court).

(≡) *Chlamysporum* Salisb., Parad. Lond.: ad t. 103. 1 Apr 1808.

1006 ***Schoenolirion*** Torr. in J. Acad. Nat. Sci. Philadelphia, ser. 2, 3: 103. Dec 1855.
Typus: *S. croceum* (Michx.) A. Gray (in Amer. Naturalist 10: 427. 1876) (*Phalangium croceum* Michx.).

(=) *Amblostima* Raf., Fl. Tellur. 2: 26. Jan-Mar 1837.
Typus: non designatus.
(=) *Oxytria* Raf., Fl. Tellur. 2: 26. Jan-Mar 1837.
Typus: *O. crocea* Raf. (*Phalangium croceum* Nutt. 1818, non Michx. 1803).

1007 ***Chlorogalum*** Kunth, Enum. Pl. 4: 681. 17-19 Jul 1843.
Typus: *C. pomeridianum* (DC.) Kunth (*Scilla pomeridiana* DC.) *(typ. cons.)*.
(≡) *Laothoë* Raf., Fl. Tellur. 3: 53. Nov-Dec 1837.

1011 ***Bowiea*** Harv. ex Hook. f. in Bot. Mag.: ad t. 5619. 1 Jan 1867.
Typus: *B. volubilis* Harv. ex Hook. f.
(H) *Bowiea* Haw. in Philos. Mag. J. 64: 299. 1824 [*Lil.*].
Typus: *B. africana* Haw.

1018 ***Hosta*** Tratt., Arch. Gewächsk. 1: 55. 1812.
Typus: *H. japonica* Tratt.
(H) *Hosta* Jacq., Pl. Hort. Schoenbr. 1: 60. 1797 [*Verben.*].
Typus: non designatus.

1021 ***Blandfordia*** Sm., Exot. Bot. 1: 5. 1 Dec 1804.
Typus: *B. nobilis* Sm.
(H) *Blandfordia* Andrews in Bot. Repos.: ad t. 343. 9 Feb 1804 [*Diapens.*].
Typus: *B. cordata* Andrews

1024 ***Kniphofia*** Moench, Methodus: 631. 4 Mai 1794.
Typus: *K. alooides* Moench, nom. illeg. (*Aloë uvaria* L., *K. uvaria* (L.) Hook.).
(H) *Kniphofia* Scop., Intr. Hist. Nat.: 327. Jan-Apr 1777 [*Combret.*].
Typus: non designatus.

1029 ***Haworthia*** Duval, Pl. Succ. Horto Alencon.: 7. 1809.
Typus: *H. arachnoidea* (L.) Duval (*Aloë pumila* var. *arachnoidea* L.).
(=) *Catevala* Medik., Theodora: 67. 1786.
Typus: non designatus.

1032 ***Laxmannia*** R. Br., Prodr.: 285. 27 Mar 1810.
Typus: *L. gracilis* R. Br. *(typ. cons.)*.
(H) *Laxmannia* J. R. Forst. & G. Forst., Char. Gen. Pl.: 47. 29 Nov 1775 [*Comp.*].
Typus: *L. arborea* J. R. Forst. & G. Forst.

1037 ***Johnsonia*** R. Br., Prodr.: 287. 27 Mar 1810.
Typus: *J. lupulina* R. Br.
(H) *Johnsonia* Mill., Gard. Dict. Abr., ed. 4: [693]. 28 Jan 1754 [*Verben.*].
Typus: non designatus.

1044 ***Baxteria*** R. Br. in London J. Bot. 2: 494. 1843.
Typus: *B. australis* R. Br.
(H) *Baxtera* Rchb., Consp. Regni Veg.: 131. Dec 1828 - Mar 1829 [*Asclepiad.*].
≡ *Harrisonia* Hook. 1826, non R. Br. ex Juss. 1825 *(nom. cons.)* (4117).

1046 ***Agapanthus*** L'Hér., Sert. Angl.: 17. Jan (prim.) 1789.
Typus: *A. umbellatus* L'Hér., nom. illeg. (*Crinum africanum* L., *A. africanus* (L.) Hoffmanns.) (etiam vide 1047).
(≡) *Abumon* Adans., Fam. Pl. 2: 54, 511. Jul-Aug 1763.

1047 ***Tulbaghia*** L., Mant. Pl.: 148, 223. Oct 1771 *('Tulbagia') (orth. cons.)*.
Typus: *T. capensis* L.

(H) *Tulbaghia* Heist., Beschr. Neu. Geschl.: 15. 1755 [*Lil.*].
≡ *Agapanthus* L'Hér. 1789 *(nom. cons.)* (1046).

1050 ***Nothoscordum*** Kunth, Enum. Pl. 4: 457. 17-19 Jul 1843.
Typus: *N. striatum* Kunth, nom. illeg. (*Ornithogalum bivalve* L., *N. bivalve* (L.) Britton) *(typ. cons.)*.

1053 ***Brodiaea*** Sm. in Trans. Linn. Soc. London 10: 2. Feb 1810.
Typus: *B. grandiflora* Sm., nom. illeg. (*Hookera coronaria* Salisb., *B. coronaria* (Salisb.) Jeps.) *(typ. cons.)* (etiam vide *Hookeria* [*Musci*]).

1055 ***Bessera*** Schult. f. in Linnaea 4: 121. Jan 1829.
Typus: *B. elegans* Schult. f.

(H) *Bessera* Schult., Observ. Bot.: 27. 1809 [*Boragin.*].
Typus: *B. azurea* Schult.

1077 ***Lloydia*** Rchb., Fl. Germ. Excurs.: 102. Mar-Apr 1830.
Typus: *L. serotina* (L.) Rchb. (*Anthericum serotinum* L.).

1087 ***Camassia*** Lindl. in Edwards's Bot. Reg.: ad t. 1486. 1 Apr 1832.
Typus: *C. esculenta* Lindl., nom. illeg. (*Phalangium quamash* Pursh, *C. quamash* (Pursh) Greene).

(=) *Cyanotris* Raf. in Amer. Monthly Mag. & Crit. Rev. 3: 356. Sep 1818.
Typus: *C. scilloides* Raf.

1088 ***Eucomis*** L'Hér., Sert. Angl.: 17. Jan (prim.) 1789.
Typus: *E. regia* (L.) L'Hér. (*Fritillaria regia* L.) *(typ. cons.)*.

(≡) *Basilaea* Juss. ex Lam., Encycl. 1: 382. 1 Aug 1785.

1093 ***Bellevalia*** Lapeyr. in J. Phys. Chim. Hist. Nat. Arts 67: 425. Dec 1808.
Typus: *B. operculata* Lapeyr.

(H) *Bellevalia* Scop., Intr. Hist. Nat.: 198. Jan-Apr 1777 [*Verben.*].
≡ *Marurang* Rumph. ex Adans. 1763.

1095a ***Leopoldia*** Parl., Fl. Palerm. 1: 435. 1845.
Typus: *L. comosa* (L.) Parl. (*Hyacinthus comosus* L.).

(H) *Leopoldia* Herb. in Trans. Hort. Soc. London 4: 181. Jan-Feb 1821 [*Amaryllid.*].
Typus: non designatus.

1108 ***Cordyline*** Comm. ex R. Br., Prodr.: 280. 27 Mar 1810.
Typus: *C. cannifolia* R. Br. *(typ. cons.).*

(H) *Cordyline* Adans., Fam. Pl. 2: 54, 543. Jul-Aug 1763 [*Lil.*].
≡ *Sansevieria* Thunb. 1794 *(nom. cons.)* (1110).

(=) *Taetsia* Medik., Theodora: 82. 1786.
Typus: *T. ferrea* Medik., nom. illeg. (*Dracaena ferrea* L., nom. illeg., *Convallaria fruticosa* L.).

1110 ***Sansevieria*** Thunb., Prodr. Pl. Cap. 1: [xii], 65. 1794.
Typus: *S. thyrsiflora* Thunb., nom. illeg. (*Aloë hyacinthoides* L., *S. hyacinthoides* (L.) Druce) (etiam vide 1108).

(≡) *Acyntha* Medik., Theodora: 76. 1786.

(=) *Sanseverinia* Petagna, Inst. Bot. 3: 643. 1787.
Typus: *S. thyrsiflora* Petagna

1111 ***Astelia*** Banks & Sol. ex R. Br., Prodr.: 291. 27 Mar 1810.
Typus: *A. alpina* R. Br.

(=) *Funckia* Willd. in Ges. Naturf. Freunde Berlin Mag. Neuesten Entdeck. Gesammten Naturk. 2: 19. 1808.
Typus: *F. magellanica* Willd., nom. illeg. (*Melanthium pumilum* G. Forst.).

1112 ***Milligania*** Hook. f. in Hooker's J. Bot. Kew Gard. Misc. 5: 296. Oct 1853.
Typus: *M. longifolia* Hook. f. *(typ. cons.).*

(H) *Milligania* Hook. f. in Icon. Pl.: ad t. 299. 6 Jan - 6 Feb 1840 [*Gunner.*].
Typus: *M. cordifolia* Hook. f.

1118 ***Smilacina*** Desf. in Ann. Mus. Natl. Hist. Nat. 9: 51. 1807.
Typus: *S. stellata* (L.) Desf. (*Convallaria stellata* L.) *(typ. cons.).*

(=) *Vagnera* Adans., Fam. Pl. 2: 496, 617. Jul-Aug 1763.
Typus: non designatus.

(=) *Polygonastrum* Moench, Methodus: 637. 4 Mai 1794.
Typus: *P. racemosum* (L.) Moench (*Convallaria racemosa* L.).

1119 ***Maianthemum*** F. H. Wigg., Prim. Fl. Holsat.: 14. 29 Mar 1780.
Typus: *M. convallaria* F. H. Wigg., nom. illeg. (*Convallaria bifolia* L., *M. bifolium* (L.) F. W. Schmidt).

1129 ***Reineckea*** Kunth in Abh. Königl. Akad. Wiss. Berlin 1842: 29. 1844.
Typus: *R. carnea* (Andrews) Kunth (*Sansevieria carnea* Andrews).

1140 ***Ophiopogon*** Ker Gawl. in Bot. Mag.: ad t. 1063. 1 Nov 1807.
Typus: *O. japonicus* (L. f.) Ker Gawl. (*Convallaria japonica* L. f.).

(=) *Mondo* Adans., Fam. Pl. 2: 496, 578. Jul-Aug 1763.
Typus: non designatus.

1146 ***Luzuriaga*** Ruiz & Pav., Fl. Peruv. 3: 65. Aug 1802.
Typus: *L. radicans* Ruiz & Pav.

(=) *Enargea* Banks ex Gaertn., Fruct. Sem. Pl. 1: 283. Dec 1788.
Typus: *E. marginata* Gaertn.

(=) *Callixene* Comm. ex Juss., Gen. Pl.: 41. 4 Aug 1789.
Typus: non designatus.

HAEMODORACEAE

1161 ***Lachnanthes*** Elliott, Sketch Bot. S.-Carolina 1: 47. 26 Sep 1816.
Typus: *L. tinctoria* Elliott

AMARYLLIDACEAE

1166 ***Hessea*** Herb., Amaryllidaceae: 289. Apr (sero) 1837.
Typus: *H. stellaris* (Jacq.) Herb. (*Amaryllis stellaris* Jacq.).

(H) *Hessea* P. J. Bergius ex Schltdl. in Linnaea 1: 252. Apr 1826 [*Amaryllid.*].
≡ *Carpolyza* Salisb. 1807.

1175 ***Nerine*** Herb. in Bot. Mag.: ad t. 2124. 1 Jan 1820.
Typus: *N. sarniensis* (L.) Herb. (*Amaryllis sarniensis* L.) *(typ. cons.)*.

(≡) *Imhofia* Heist., Beschr. Neu. Geschl.: 29. 1755.

1176 ***Amaryllis*** L., Sp. Pl.: 292. 1 Mai 1753.
Typus: *A. belladonna* L. (*typ. cons.*: [specimen] in Herb. Clifford.: 135, *Amaryllis* No. 2 (BM)).

1178 ***Vallota*** Salisb. ex Herb., Appendix: 29. Dec 1821.
Typus: *V. purpurea* Herb., nom. illeg. (*Crinum speciosum* L. f., *V. speciosa* (L. f.) Voss).

(H) *Valota* Adans., Fam. Pl. 2: 495, 617. Jul-Aug 1763 [*Gram.*].
Typus: *V. insularis* (L.) Chase (in Proc. Biol. Soc. Wash. 19: 188. 1906) (*Andropogon insulare* L.).

1181 ***Zephyranthes*** Herb., Appendix: 36. Dec 1821.
Typus: *Z. atamasca* (L.) Herb. (*Amaryllis atamasca* L.) *(typ. cons.)*.

(≡) *Atamosco* Adans., Fam. Pl. 2: 57, 522. Jul-Aug 1763.

1191 ***Cyrtanthus*** Aiton, Hort. Kew. 1: 414. 7 Aug - 1 Oct 1789.
Typus: *C. angustifolius* (L. f.) Aiton (*Crinum angustifolium* L. f.).

(H) *Cyrtanthus* Schreb., Gen. Pl.: 122. Apr 1789 [*Rub.*].
≡ *Posqueria* Aubl. 1775.

1196	***Eucharis*** Planch. & Linden in Linden, Cat. Pl. Exot. 8: 3. 1853. Typus: *E. candida* Planch. & Linden	(=)	*Caliphruria* Herb. in Edwards's Bot. Reg. 30 (Misc.): 87. Dec 1844. Typus: *C. hartwegiana* Herb.
1208	***Hippeastrum*** Herb., Appendix: 31. Dec 1821. Typus: *H. reginae* (L.) Herb. (*Amaryllis reginae* L.) *(typ. cons.)*.	(=)	*Leopoldia* Herb. in Trans. Hort. Soc. London 4: 181. Jan-Feb 1821. Typus: non designatus.
1211	***Urceolina*** Rchb., Consp. Regni Veg.: 61. Dec 1828 - Mar 1829. Typus: *Urceolaria pendula* Herb., nom. illeg. (*Crinum urceolatum* Ruiz & Pav., *U. urceolata* (Ruiz & Pav.) M. L. Green).	(=)	*Leperiza* Herb., Appendix: 41. Dec 1821. Typus: *L. latifolia* (Ruiz & Pav.) Herb. (*Pancratium latifolium* Ruiz & Pav.).
1236	***Lanaria*** Aiton, Hort. Kew. 1: 462. 7 Aug - 1 Oct 1789. Typus: *L. plumosa* Aiton, nom. illeg. (*Hyacinthus lanatus* L., *L. lanata* (L.) Druce).	(H) (=)	*Lanaria* Adans., Fam. Pl. 2: 255, 568. Jul-Aug 1763 [*Caryophyll.*]. ≡ *Gypsophila* L. 1753. *Argolasia* Juss., Gen. Pl.: 60. 4 Aug 1789. Typus: non designatus.

TACCACEAE

1248	***Tacca*** J. R. Forst. & G. Forst., Char. Gen. Pl.: 35. 29 Nov 1775. Typus: *T. pinnatifida* J. R. Forst. & G. Forst.	(=)	*Leontopetaloides* Boehm. in Ludwig, Def. Gen. Pl., ed. 3: 512. 1760. Typus: *Leontice leontopetaloides* L.

DIOSCOREACEAE

1258	***Petermannia*** F. Muell., Fragm. 2: 92. Aug 1860. Typus: *P. cirrosa* F. Muell.	(H)	*Petermannia* Rchb., Deut. Bot. Herb.-Buch, Syn.: 236. Jul 1841 [*Chenopod.*]. ≡ *Cycloloma* Moq. 1840.

IRIDACEAE

1260	***Syringodea*** Hook. f. in Bot. Mag.: ad t. 6072. 1 Dec 1873. Typus: *S. pulchella* Hook. f.	(H)	*Syringodea* D. Don in Edinburgh New Philos. J. 17: 155. Jul 1834 [*Eric.*]. Typus: *S. vestita* (Thunb.) D. Don (*Erica vestita* Thunb.).

1261 ***Romulea*** Maratti, Pl. Romul. Saturn.: 13. 1772.
Typus: *R. bulbocodium* (L.) Sebast. & Mauri (Fl. Roman. Prodr.: 17. 1818) (*Crocus bulbocodium* L.) (*typ. cons.*).

(≡) *Ilmu* Adans., Fam. Pl. 2: 497, 566. Jul-Aug 1763.

1265 ***Moraea*** Mill., Fig. Pl. Gard. Dict.: 159. 27 Jun 1758 (*'Morea'*) (*orth. cons.*).
Typus: *M. vegeta* L. (Sp. Pl., ed. 2: 59. Sep 1762) (*typ. cons.*).

1265a ***Dietes*** Salisb. ex Klatt in Linnaea 34: 583. Feb 1866.
Typus: *D. compressa* (L. f.) Klatt (*Iris compressa* L. f.) [= *D. iridioides* (L.) Klatt (*Moraea iridioides* L.)].

(=) *Naron* Medik. in Hist. & Commentat. Acad. Elect. Sci. Theod.-Palat. 6: 419. Apr-Jun 1790.
Typus: *N. orientale* Medik. nom. illeg. (*Moraea iridioides* L., *N. iridioides* (L.) Moench).

1283 ***Libertia*** Spreng., Syst. Veg. 1: 127. 1824 (sero).
Typus: *L. ixioides* (G. Forst.) Spreng. (*Sisyrinchium ixioides* G. Forst.).

(H) *Libertia* Dumort., Comment. Bot.: 9. Nov (sero) -Dec (prim.) 1822 [*Lil.*].
Typus: *L. recta* Dumort., nom. illeg. (*Hemerocallis caerulea* Andrews).

(=) *Tekel* Adans., Fam. Pl. 2: 497, 610. Jul-Aug 1763.
Typus: non designatus.

1284 ***Bobartia*** L., Sp. Pl.: 54. 1 Mai 1753.
Typus: *B. indica* L. (*typ. cons.*: [specimen sup. laev.] in Herb. Hermann 4: 80 (BM)).

1285 ***Belamcanda*** Adans., Fam. Pl. 2: 60 (*'Belam-canda'*), 524 (*'Belamkanda'*). Jul-Aug 1763 (*orth. cons.*).
Typus: *B. chinensis* (L.) DC. (in Redouté, Liliac. 3: ad t. 121. Jul 1805) (*Ixia chinensis* L.) (*typ. cons.*).

1289 ***Patersonia*** R. Br. in Bot. Mag.: ad t. 1041. 1 Aug 1807.
Typus: *P. sericea* R. Br.

(=) *Genosiris* Labill., Nov. Holl. Pl. 1: 13. Jan 1805.
Typus: *G. fragilis* Labill.

1292 ***Eleutherine*** Herb. in Edwards's Bot. Reg. 29: ad t. 57. 1 Nov 1843.
Typus: *Marica plicata* Ker Gawl., nom. illeg. (*Moraea plicata* Sw., nom. illeg., *Sisyrinchium latifolium* Sw.) [= *E. bulbosa* (Mill.) Urb. (*Sisyrinchium bulbosum* Mill.)].

1302 ***Ixia*** L., Sp. Pl., ed. 2: 51. Sep 1762.
Typus: *I. polystachya* L. *(typ. cons.)*.

(H) *Ixia* L., Sp. Pl.: 36. 1 Mai 1753 [*Irid.*].
Typus (vide Regnum Veg. 46: 272. 1966): *I. africana* L.

1310 ***Babiana*** Ker Gawl. ex Sims in Bot. Mag.: ad t. 539. 1 Nov 1801.
Typus: *B. plicata* Ker Gawl. (in Bot. Mag.: ad t. 576. 1 Aug 1802), nom. illeg. (*Gladiolus fragrans* Jacq., *B. fragrans* (Jacq.) Eckl.).

(=) *Beverna* Adans., Fam. Pl. 2: (20). Jul-Aug 1763.
Typus: non designatus.

1313 ***Micranthus*** (Pers.) Eckl., Topogr. Verz. Pflanzensamml. Ecklon: 43. Oct 1827 (*Gladiolus* subg. *Micranthus* Pers., Syn. Pl. 1: 46. 1 Apr - 15 Jun 1805).
Typus: *Gladiolus alopecuroides* L. (*M. alopecuroides* (L.) Eckl.) *(typ. cons.)*.

(H) *Micranthus* J. C. Wendl., Bot. Beob.: 38. 1798 [*Acanth.*].
≡ *Phaulopsis* Willd. 1800 *(nom. cons.)* (7932).

1315 ***Watsonia*** Mill., Fig. Pl. Gard. Dict.: 184. 22 Dec 1758.
Typus: *W. meriana* (L.) Mill. (Gard. Dict., ed. 8: *Watsonia* No. 1. 16 Apr 1768) (*Antholyza meriana* L.) (etiam vide 5692).

1316 ***Freesia*** Klatt in Linnaea 34: 672. Dec 1866.
Typus: *F. refracta* (Jacq.) Klatt (*Gladiolus refractus* Jacq.) *(typ. cons.)*.

MUSACEAE

1321 ***Heliconia*** L., Mant. Pl.: 147, 211. Oct 1771.
Typus: *H. bihai* (L.) L. (*Musa bihai* L.).

(≡) *Bihai* Mill., Gard. Dict. Abr., ed. 4: [194]. 28 Jan 1754.

ZINGIBERACEAE

1324 ***Zingiber*** Mill., Gard. Dict. Abr., ed. 4: [1545]. 28 Jan 1754 *('Zinziber') (orth. cons.)*.
Typus: *Z. officinale* Roscoe (in Trans. Linn. Soc. London 8: 358. 9 Mar 1807) (*Amomum zingiber* L.) (etiam vide 1344).

1328 ***Alpinia*** Roxb. in Asiat. Res. 11: 350. 1810.
Typus: *A. galanga* (L.) Willd. (Sp. Pl. 1: 12. Jun 1797) (*Maranta galanga* L.) *(typ. cons.)*.

(H) *Alpinia* L., Sp. Pl.: 2. 1 Mai 1753 [*Zingiber.*].
Typus: *A. racemosa* L.

(=) *Albina* Giseke, Prael. Ord. Nat. Pl.: 207, 227, 248. Apr 1792.
Typus: non designatus.

(=) *Buekia* Giseke, Prael. Ord. Nat. Pl.: 204, 216, 239. Apr 1792.
Typus: *B. malaccensis* (K. D. Koenig) Raeusch. (Nomencl. Bot. 1. 1797) (*Costus malaccensis* K. D. Koenig).

(=) *Zerumbet* J. C. Wendl., Sert. Hannov. 4: 3. Apr-Mai 1798.
Typus: *Z. speciosum* J. C. Wendl.

1331 ***Renealmia*** L. f., Suppl. Pl.: 7, 79. Apr 1782.
Typus: *R. exaltata* L. f.

(H) *Renealmia* L., Sp. Pl.: 286. 1 Mai 1753 [*Bromel.*].
Typus (vide Rickett & Stafleu in Taxon 8: 241. 1959): *R. paniculata* L.

1332 ***Riedelia*** Oliv. in Hooker's Icon. Pl. 15: 15. Mar 1883.
Typus: *R. curviflora* Oliv.

(H) *Riedelia* Cham. in Linnaea 7: 240 ('224'). 1832 [*Verben.*].
Typus: *R. lippioides* Cham.

(=) *Nyctophylax* Zipp. in Alg. Konst- Lett.-Bode 1829(1): 298. 8 Mai 1829.
Typus: *N. alba* Zipp.

1337a ***Nicolaia*** Horan., Prodr. Monogr. Scitam.: 32. 1862.
Typus: *N. imperialis* Horan. *(typ. cons.)* [= *N. elatior* (Jack) Horan.].

(=) *Diracodes* Blume, Enum. Pl. Javae 1: 55. Oct-Dec 1827.
Typus: *D. javanica* Blume

1344 ***Amomum*** Roxb., Pl. Coromandel 3: 75. 18 Feb 1820 .
Typus: *A. subulatum* Roxb. *(typ. cons.).*

(H) *Amomum* L., Sp. Pl.: 1. 1 Mai 1753 (typ. des.: Burtt & Smith in Taxon 17: 730. 1968) [*Zingiber.*].
≡ *Zingiber* Mill. 1754 *(nom. cons.)* (1324).

(=) *Etlingera* Giseke, Prael. Ord. Nat. Pl.: 209. Apr 1792.
Typus: *E. littoralis* (K. D. Koenig) Raeusch. (Nomencl. Bot.: 1. 1797) (*Amomum littorale* K. D. Koenig).

(=) *Meistera* Giseke, Prael. Ord. Nat. Pl.: 205. Apr 1792.
Typus: *Amomum koenigii* J. F. Gmel.

(=) *Paludana* Giseke, Prael. Ord. Nat. Pl.: 207. Apr 1792.
Typus: *Amomum globba* J. F. Gmel.

(=) *Wurfbainia* Giseke, Prael. Ord. Nat. Pl.: 206. Apr 1792.
Typus: *W. uliginosa* (K. D. Koenig) Giseke (*Amomum uliginosum* K. D. Koenig).

1351 ***Curcuma*** L., Sp. Pl.: 2. 1 Mai 1753.
Typus: *C. longa* L. *(typ. cons.).*

1360 ***Tapeinochilos*** Miq. in Ann. Mus. Lugduno-Batavum 4: 101. 21 Feb 1869 *('Tapeinocheilos') (orth. cons.).*
Typus: *T. pungens* (Teijsm. & Binn.) Miq. (*Costus pungens* Teijsm. & Binn.).

MARANTACEAE

1368 ***Phrynium*** Willd., Sp. Pl. 1: 1, 17. Jun 1797.
Typus: *P. capitatum* Willd., nom. illeg. (*Pontederia ovata* L., *Phrynium rheedei* Suresh & Nicolson).

(=) *Phyllodes* Lour., Fl. Cochinch.: 1, 13. 1790.
Typus: *P. placentaria* Lour.

BURMANNIACEAE

1386 ***Arachnitis*** Phil. in Bot. Zeitung (Berlin) 22: 217. 15 Jul 1864.
Typus: *A. uniflora* Phil.

(H) *Arachnites* F. W. Schmidt, Fl. Boëm. 1: 74. 9 Apr - 7 Oct 1793 [*Orchid.*].
Typus: non designatus.

ORCHIDACEAE

1393a ***Paphiopedilum*** Pfitzer, Morph. Stud. Orchideenbl.: 11. Jan-Jul 1886.
Typus: *P. insigne* (Wall. ex Lindl.) Pfitzer (*Cypripedium insigne* Wall. ex Lindl.) *(typ. cons.)*.

(≡) *Cordula* Raf., Fl. Tellur. 4: 46. 1838 (med.).
(=) *Stimegas* Raf., Fl. Tellur. 4: 45. 1838 (med.).
Typus: *S. venustum* (Wall. ex Sims) Raf. (*Cypripedium venustum* Wall. ex Sims).

1393b ***Phragmipedium*** Rolfe in Orchid Rev. 4: 330. Nov 1896.
Typus: *P. caudatum* (Lindl.) Rolfe (*Cypripedium caudatum* Lindl.).

(=) *Uropedium* Lindl., Orchid. Linden.: 28. Nov-Dec 1846.
Typus: *U. lindenii* Lindl.

1397 ***Serapias*** L., Sp. Pl.: 949. 1 Mai 1753.
Typus: *S. lingua* L. *(typ. cons.)*.

1399 ***Himantoglossum*** Spreng., Syst. Veg. 3: 675, 694. Jan-Mar 1826.
Typus: *H. hircinum* (L.) Spreng. (*Satyrium hircinum* L.) *(typ. cons.)*.

1403a ***Peristylus*** Blume, Bijdr.: 404. 20 Sep - 7 Dec 1825.
Typus: *P. grandis* Blume

(=) *Glossula* Lindl. in Bot. Reg. 10: ad t. 862. Feb 1825.
Typus: *G. tentaculata* Lindl.

1408 ***Holothrix*** Rich. ex Lindl., Gen. Sp. Orchid. Pl.: 257. Aug 1835.
Typus: *H. parvifolia* Lindl., nom. illeg. (*Orchis hispidula* L. f., *H. hispidula* (L. f.) Durand & Schinz) *(typ. cons.)*.

(=) *Monotris* Lindl. in Edwards's Bot. Reg.: ad t. 1701. 1 Sep 1834.
Typus: *M. secunda* Lindl.
(=) *Scopularia* Lindl. in Edwards's Bot. Reg. 20: ad t. 1701. 1 Sep 1834.
Typus: *S. burchellii* Lindl.
(=) *Saccidium* Lindl., Gen. Sp. Orchid. Pl.: 258. Aug 1835.
Typus: *S. pilosum* Lindl.
(=) *Tryphia* Lindl., Gen. Sp. Orchid. Pl.: 258. Aug 1835.
Typus: *T. secunda* (Thunb.) Lindl. (*Orchis secunda* Thunb.).

1410 ***Platanthera*** Rich., De Orchid. Eur.: 20, 26, 35. Aug-Sep 1817.
Typus: *P. bifolia* (L.) Rich. (*Orchis bifolia* L.).

1430 ***Satyrium*** Sw. in Kongl. Vetensk. Acad. Nya Handl. 21: 214. Jul-Sep 1800.
Typus: *S. bicorne* (L.) Thunb. (Prodr. Pl. Cap.: 6. 1794) (*Orchis bicornis* L.) *(typ. cons.)*.

(H) *Satyrium* L., Sp. Pl.: 944. 1 Mai 1753 [*Orchid.*].
Typus (vide Green in Sprague & al., Nom. Prop. Brit. Bot.: 185. 1929): *S. viride* L.

1449 ***Pterostylis*** R. Br., Prodr.: 326. 27 Mar 1810.
Typus: *P. curta* R. Br. *(typ. cons.)*.

(=) *Diplodium* Sw. in Ges. Naturf. Freunde Berlin Mag. Neuesten Entdeck. Gesammten Naturk. 4: 84. Jul 1810.
Typus: *Disperis alata* Labill. (Nov. Holl. Pl. 2: 59. 1806).

1468 ***Nervilia*** Comm. ex Gaudich., Voy. Uranie, Bot.: 421. 12 Sep 1829.
Typus: *N. aragoana* Gaudich.

(=) *Stellorkis* Thouars in Nouv. Bull. Sci. Soc. Philom. Paris 1: 317. Apr 1809.
Typus (vide Thouars, Hist. Orchid.: ad t. 24. 1822): *Arethusa simplex* L.

1482 ***Epipactis*** Zinn, Cat. Pl. Hort. Gott.: 85. 20 Apr - 21 Mai 1757.
Typus: *E. helleborine* (L.) Crantz (Stirp. Austr. Fasc., ed. 2: 467. Jan-Jul 1769) (*Serapias helleborine* L.) *(typ. cons.)*.

(H) *Epipactis* Ség., Pl. Veron. 3: 253. Jul-Aug 1754 [*Orchid.*].
Typus: *Satyrium repens* L.

(≡) *Helleborine* Mill., Gard. Dict. Abr., ed. 4: [622]. 28 Jan 1754.

1483 ***Limodorum*** Boehm. in Ludwig, Def. Gen. Pl., ed. 3: 358. 1760.
Typus: *L. abortivum* (L.) Sw. (in Nova Acta Regiae Soc. Sci. Upsal. 6: 80. 1799) (*Orchis abortiva* L.).

(H) *Limodorum* L., Sp. Pl.: 950. 1 Mai 1753 [*Orchid.*].
≡ *Calopogon* R. Br. 1813 *(nom. cons.)* (1534).

1488 ***Pelexia*** Poit. ex Lindl. in Bot. Reg.: ad t. 985. 1 Jun 1826.
Typus: *P. spiranthoides* Lindl., nom. illeg. (*Satyrium adnatum* Sw., *P. adnata* (Sw.) Spreng.).

(≡) *Collea* Lindl. in Bot. Reg. 9: ad t. 760. 1 Dec 1823.

1490 ***Spiranthes*** Rich., De Orchid. Eur.: 20, 28, 36. Aug-Sep 1817.
Typus: *S. autumnalis* Rich., nom. illeg. (*Ophrys spiralis* L., *S. spiralis* (L.) Chevall.) *(typ. cons.)*.

(≡) *Orchiastrum* Ség., Pl. Veron. 3: 252. Jul-Dec 1754.

1494 ***Listera*** R. Br. in Aiton, Hort. Kew., ed. 2, 5: 201. Nov 1813.
Typus: *L. ovata* (L.) R. Br. (*Ophrys ovata* L.) *(typ. cons.)*.

(H) *Listera* Adans., Fam. Pl. 2: 321, 572. Jul-Aug 1763 [*Legum.*].
Typus: non designatus.

(=) *Diphryllum* Raf. in Med. Repos., ser. 2, 5: 357. Feb-Apr 1808.
Typus: *D. bifolium* Raf.

1495 ***Neottia*** Guett. in Hist. Acad. Roy. Sci. Mém. Math. Phys. (Paris, 4°) 1750: 374. 1754.
Typus: *N. nidus-avis* (L.) Rich. (De Orchid. Eur.: 59. Aug-Sep 1817) (*Ophrys nidus-avis* L.).

1500 ***Anoectochilus*** Blume, Bijdr.: 411. 20 Sep - 7 Dec 1825 *('Anecochilus') (orth. cons.).*
Typus: *A. setaceus* Blume

1502 ***Zeuxine*** Lindl., Orchid. Scelet.: 9. Jan 1826 *('Zeuxina') (orth. cons.).*
Typus: *Z. sulcata* (Roxb.) Lindl. (Gen. Sp. Orchid. Pl.: 485. Sep 1840) (*Pterygodium sulcatum* Roxb.).

1507 ***Hetaeria*** Blume, Bijdr.: 409. 20 Sep - 7 Dec 1825 *('Etaeria') (orth. cons.).*
Typus: *H. oblongifolia* Blume *(typ. cons.).*

1516 ***Platylepis*** A. Rich. in Mém. Soc. Hist. Nat. Paris 4: 34. Sep 1828.
Typus: *P. goodyeroides* A. Rich., nom. illeg. (*Goodyera occulta* Thouars, *P. occulta* (Thouars) Rchb.).

(=) *Erporkis* Thouars in Nouv. Bull. Sci. Soc. Philom. Paris 1: 317. Apr 1809.
Typus: non designatus.

1533 ***Bletilla*** Rchb. f. in Fl. Serres Jard. Eur. 8: 246. 5 Oct 1853.
Typus: *B. gebinae* (Lindl.) Rchb. f. (*Bletia gebinae* Lindl.) *(typ. cons.).*

(=) *Jimensia* Raf., Fl. Tellur. 4: 38. 1838 (med.).
Typus: *J. nervosa* Raf., nom. illeg. (*Limodorum striatum* Thunb., *J. striata* (Thunb.) Garay & R. E. Schult.).

1534 ***Calopogon*** R. Br. in Aiton, Hort. Kew., ed. 2, 5: 204. Nov 1813.
Typus: *C. pulchellus* R. Br., nom. illeg. (*Limodorum pulchellum* Salisb., nom. illeg., *Limodorum tuberosum* L., *C. tuberosus* (L.) Britton & al.) (etiam vide 1483).

1553 ***Microstylis*** (Nutt.) Eaton, Man. Bot., ed. 3: 115, 347, 353. 23 Mar - 23 Apr 1822 (*Malaxis* sect. *Microstylis* Nutt., Gen. N. Amer. Pl. 2: 196. 14 Jul 1818).
Typus: *Malaxis ophioglossoides* Muhl. ex Willd., nom. illeg. (*Malaxis unifolia* Michx., *Microstylis unifolia* (Michx.) Britton & al.).

(≡) *Achroanthes* Raf. in Amer. Monthly Mag. & Crit. Rev. 4: 195. Jan 1819.

1556 ***Liparis*** Rich., De Orchid. Eur.: 21, 30, 38. Aug-Sep 1817.
Typus: *L. loeselii* (L.) Rich. (*Ophrys loeselii* L.).

(=) *Leptorkis* Thouars in Nouv. Bull. Sci. Soc. Philom. Paris 1: 317. Apr 1809.
Typus: non designatus.

1558	***Oberonia*** Lindl., Gen. Sp. Orchid. Pl.: 15. Apr 1830. Typus: *O. iridifolia* Lindl., nom. illeg. (*Malaxis ensiformis* Sm., *O. ensiformis* (Sm.) Lindl.) *(typ. cons.)*.	(=)	*Iridorkis* Thouars in Nouv. Bull. Sci. Soc. Philom. Paris 1: 319. Apr 1809. Typus: non designatus.
1559	***Calypso*** Salisb., Parad. Lond.: ad t. 89. 1 Dec 1807. Typus: *C. borealis* Salisb., nom. illeg. (*Cypripedium bulbosum* L., *Calypso bulbosa* (L.) Oakes).	(H)	*Calypso* Thouars, Pl. Iles. Afriq. Austral.: t. 6. 16 Apr 1804 [*Hippocrat.*]. Typus: *Epidendrum distichum* Lam.
1565	***Polystachya*** Hook., Exot. Fl. 2: ad t. 103. Mai 1824. Typus: *P. luteola* Hook., nom. illeg. (*Epidendrum minutum* Aubl.) [= *P. extinctoria* Rchb.].	(=)	*Dendrorkis* Thouars in Nouv. Bull. Sci. Soc. Philom. Paris 1: 318. Apr 1809. Typus: non designatus.
1569	***Claderia*** Hook. f., Fl. Brit. India 5: 810. Apr 1890. Typus: *C. viridiflora* Hook. f.	(H)	*Claderia* Raf., Sylva Tellur.: 12. Oct-Dec 1838 [*Rut.*]. Typus: *C. parviflora* Raf.
1587	***Stelis*** Sw. in J. Bot. (Schrader) 1799(2): 239. Apr 1800. Typus: *S. ophioglossoides* (Jacq.) Sw. (*Epidendrum ophioglossoides* Jacq.) *(typ. cons.)*.		
1614	***Epidendrum*** L., Sp. Pl., ed. 2: 1347. Jul-Aug 1763. Typus: *E. nocturnum* Jacq. (Enum. Syst. Pl.: 29. Aug-Sep 1760) *(typ. cons.)*.	(H)	*Epidendrum* L., Sp. Pl.: 952. 1 Mai 1753 [*Orchid.*]. Typus (vide Britton & Wilson in Sci. Surv. Porto Rico & Virgin Islands 5: 203. 1924): *E. nodosum* L.
1617	***Laelia*** Lindl., Gen. Sp. Orchid. Pl.: 96, 115. Jul 1831. Typus: *L. grandiflora* (La Llave & Lex.) Lindl. (*Bletia grandiflora* La Llave & Lex.) *(typ. cons.)*.	(H)	*Laelia* Adans., Fam. Pl. 2: 423, 567. Jul-Aug 1763 [*Cruc.*]. Typus: *L. orientalis* (L.) Desv. (in J. Bot. Agric. 3: 160. 1815 (prim.)) (*Bunias orientalis* L.).
1619	***Brassavola*** R. Br. in Aiton, Hort. Kew., ed. 2, 5: 216. Nov 1813. Typus: *B. cucullata* (L.) R. Br. (*Epidendrum cucullatum* L.).	(H)	*Brassavola* Adans., Fam. Pl. 2: 127, 527. Jul-Aug 1763 [*Comp.*]. ≡ *Helenium* L. 1753.

1631 ***Calanthe*** R. Br. in Bot. Reg.: ad t. 573 ('578'). 1 Oct 1821.
Typus: *C. veratrifolia* Ker Gawl. (in Bot. Reg.: ad t. 720. 1 Jul 1823), nom. illeg. (*Orchis triplicata* Willemet, *C. triplicata* (Willemet) Ames).

(=) *Alismorkis* Thouars in Nouv. Bull. Sci. Soc. Philom. Paris 1: 318. Apr 1809.
Typus: non designatus.

1648 ***Eulophia*** R. Br. in Bot. Reg.: ad t. 573 ('578'). 1 Oct 1821 (*'Eulophus'*) (*orth. cons.*).
Typus: *E. guineensis* Ker Gawl. (*typ. cons.*).

(=) *Graphorkis* Thouars in Nouv. Bull. Sci. Soc. Philom. Paris 1: 318. Apr 1809 (*nom. cons.*) (1648a).
(=) *Lissochilus* R. Br. in Bot. Reg.: ad t. 573 ('578'). 1 Oct 1821.
Typus: *L. speciosus* R. Br.

1648a ***Graphorkis*** Thouars in Nouv. Bull. Sci. Soc. Philom. Paris 1: 318. Apr 1809.
Typus: [specimen] Réunion or Madagascar, *Thouars s.n.* (P) (*typ. cons.*) [= *G. concolor* (Thouars) Kuntze (*Limodorum concolor* Thouars)] (etiam vide 1648).

1694 ***Dendrobium*** Sw. in Nova Acta Regiae Soc. Sci. Upsal., ser. 2, 6: 82. 1799.
Typus: *D. moniliforme* (L.) Sw. (*Epidendrum moniliforme* L.) (*typ. cons.*).

(=) *Callista* Lour., Fl. Cochinch.: 516, 519. Sep 1790.
Typus: *C. amabilis* Lour.
(=) *Ceraia* Lour., Fl. Cochinch.: 518. Sep 1790.
Typus: *C. simplicissima* Lour.

1697 ***Eria*** Lindl. in Bot. Reg.: ad t. 904. 1 Aug 1825.
Typus: *E. stellata* Lindl.

1704 ***Cirrhopetalum*** Lindl., Gen. Sp. Orchid. Pl.: 45, 58. Mai 1830.
Typus: *C. thouarsii* Lindl., nom. illeg. (*Epidendrum umbellatum* G. Forst., *C. umbellatum* (G. Forst.) Frappier ex Cordemoy).

(=) *Ephippium* Blume, Bijdr.: 308. 20 Sep - 7 Dec 1825.
Typus: non designatus.
(=) *Zygoglossum* Reinw. in Syll. Pl. Nov. 2: 4. 1825.
Typus: *Z. umbellatum* Reinw.

1704a ***Genyorchis*** Schltr., Westafr. Kautschuk-Exped.: 280. Dec 1900.
Typus: Cameroon, *Schlechter 12737* (BR) (*typ. cons.*) [= *G. apetala* (Lindl.) J. J. Verm. (*Bulbophyllum apetalum* Lindl.)].

1705 ***Bulbophyllum*** Thouars, Hist. Orchid., Tabl. Esp.: 3. 1822.
Typus: *B. nutans* Thouars (*typ. cons.*).

(≡) *Phyllorkis* Thouars in Nouv. Bull. Sci. Soc. Philom. Paris 1: 319. Apr 1809.

1714 ***Panisea*** (Lindl.) Lindl., Fol. Orchid. 5, *Panisea:* 1. 20 Jan 1854 (*Coelogyne* sect. *Panisea* Lindl., Gen. Sp. Orchid. Pl.: 44. Mai 1830).
Typus: *Coelogyne parviflora* Lindl. (*P. parviflora* (Lindl.) Lindl.).

(≡) *Androgyne* Griff., Not. Pl. Asiat. 3: 279. 1851.

1739 ***Warmingia*** Rchb. f., Otia Bot. Hamburg.: 87. 8 Aug 1881.
Typus: *W. eugenii* Rchb. f.

(H) *Warmingia* Engl. in Martius, Fl. Bras. 12(2): 281. 1 Sep 1874 [*Anacard.*].
Typus: *W. pauciflora* Engl.

1751 ***Brachtia*** Rchb. f. in Linnaea 22: 853. Mai 1850.
Typus: *B. glumacea* Rchb. f.

(H) *Brachtia* Trevis., Sagg. Algh. Coccot.: 57. 1848 [*Chloroph.*].
Typus: *B. crassa* (Naccari) Trevis. (*Palmella crassa* Naccari).

1778 ***Miltonia*** Lindl. in Edwards's Bot. Reg.: ad t. 1976. 1 Aug 1837.
Typus: *M. spectabilis* Lindl.

1779 ***Oncidium*** Sw. in Kongl. Vetensk. Acad. Nya Handl. 21: 239. Jul-Sep 1800.
Typus: *O. altissimum* (Jacq.) Sw. (*Epidendrum altissimum* Jacq.) *(typ. cons.)*.

1822 ***Saccolabium*** Blume, Bijdr.: 292. 20 Sep - 7 Dec 1825.
Typus: *S. pusillum* Blume

(=) *Gastrochilus* D. Don, Prodr. Fl. Nepal.: 32. 26. Jan - 1 Feb 1825.
Typus: *G. calceolaris* D. Don

1824 ***Acampe*** Lindl., Fol. Orchid. 4, *Acampe:* 1. 20 Aug 1853.
Typus: *A. multiflora* (Lindl.) Lindl. (*Vanda multiflora* Lindl.).

(=) *Sarcanthus* Lindl. in Bot. Reg. 10: ad t. 817. 1 Aug 1824.
Typus: *Epidendrum praemorsum* Roxb.

1834 ***Oeonia*** Lindl. in Bot. Reg.: ad t. 817. 1 Aug 1824 *('Aeonia') (orth. cons.)*.
Typus: *O. aubertii* Lindl., nom. illeg. (*Epidendrum volucre* Thouars, *O. volucris* (Thouars) Durand & Schinz).

1834a ***Symphyglossum*** Schltr. in Orchis 13: 8. 15 Feb 1919.
Typus: *S. sanguineum* (Rchb. f.) Schltr. (*Mesospinidium sanguineum* Rchb. f.).

(H) *Symphyoglossum* Turcz. in Bull. Soc. Imp. Naturalistes Moscou 21: 255. 1848 [*Asclepiad.*].
Typus: *S. hastatum* (Bunge) Turcz. (*Asclepias hastata* Bunge).

SAURURACEAE

1857	***Houttuynia*** Thunb. in Kongl. Vetensk. Acad. Nya Handl. 4: 149. Apr-Jun 1783 *('Houtuynia') (orth. cons.).* Typus: *H. cordata* Thunb.	(H)	*Houttuynia* Houtt., Nat. Hist. 2(12): 448. 5 Jul 1780 [*Irid.*]. Typus: *H. capensis* Houtt.

JUGLANDACEAE

1882	***Carya*** Nutt., Gen. N. Amer. Pl. 2: 220. 14 Jul 1818. Typus: *C. tomentosa* (Poir.) Nutt. (*Juglans tomentosa* Poir.) *(typ. cons.).*	(=)	*Hicorius* Raf., Fl. Ludov.: 109. Oct-Dec (prim.) 1817. Typus: non designatus.

BETULACEAE

1885	***Ostrya*** Scop., Fl. Carniol.: 414. 15 Jun - 21 Jul 1760. Typus: *O. carpinifolia* Scop. (*Carpinus ostrya* L.).	(H)	*Ostrya* Hill, Brit. Herb.: 513. Jan 1757 [*Betul.*]. ≡ *Carpinus* L. 1753.

FAGACEAE

1889	***Nothofagus*** Blume, Mus. Bot. 1: 307. 1851 (prim.). Typus: *N. antarctica* (G. Forst.) Oerst. (Bidr. Egefam.: 24. 1871) (*Fagus antarctica* G. Forst.) *(typ. cons.).*	(=)	*Fagaster* Spach, Hist. Nat. Vég. 11: 142. 25 Dec 1841. Typus: *Fagus dombeyi* Mirbel
		(=)	*Calucechinus* Hombr. & Jacquinot in Urville, Voy. Pôle Sud, Bot., Atlas (Dicot.): t. 6. Sep-Dec 1843. Typus: *C. antarctica* Hombr. & Jacquinot
		(=)	*Calusparassus* Hombr. & Jacquinot in Urville, Voy. Pôle Sud, Bot., Atlas (Dicot.): t. 6. Sep-Dec 1843. Typus: *C. forsteri* Hombr. & Jacquinot
1891a	***Castanopsis*** (D. Don) Spach, Hist. Nat. Vég. 11: 142, 185. 25 Dec 1841 (*Quercus* [unranked] *Castanopsis* D. Don, Prodr. Fl. Nepal.: 56. 26 Jan - 1 Feb 1825). Typus: *Quercus armata* Roxb. (*C. armata* (Roxb.) Spach).	(=)	*Balanoplis* Raf., Alsogr. Amer.: 29. 1838. Typus (vide Pichi Sermolli in Taxon 3: 113. 1954): *B. tribuloides* (Sm.) Raf. (*Quercus tribuloides* Sm.).
1893	***Cyclobalanopsis*** Oerst. in Vidensk. Meddel. Dansk Naturhist. Foren. Kjøbenhavn 1866: 77. 5 Jul 1867. Typus: *C. velutina* Oerst. (*Quercus velutina* Lindl. ex Wall., non Lam.) *(typ. cons.).*	(=)	*Perytis* Raf., Alsogr. Amer.: 29. 1838. Typus (vide Pichi Sermolli in Taxon 3: 114. 1954): *P. lamellosa* (Sm.) Raf. (*Quercus lamellosa* Sm.).

ULMACEAE

1901 ***Zelkova*** Spach in Ann. Sci. Nat., Bot., ser. 2, 15: 356. 1 Jun 1841.
Typus: *Z. crenata* Spach, nom. illeg. (*Rhamnus carpinifolius* Pall., *Z. carpinifolia* (Pall.) K. Koch).

1904 ***Aphananthe*** Planch. in Ann. Sci. Nat., Bot., ser. 3, 10: 265. Nov 1848.
Typus: *A. philippinensis* Planch. in Ann. Sci. Nat., Bot., ser. 3, 10: 337. Dec 1848.

(H) *Aphananthe* Link, Enum. Hort. Berol. Alt. 1: 383. 16 Mar - 30 Jun 1821 [*Phytolacc.*].
Typus: *A. celosioides* (Spreng.) Link (*Galenia celosioides* Spreng.).

MORACEAE

1917 ***Trophis*** P. Browne, Civ. Nat. Hist. Jamaica: 357. 10 Mar 1756.
Typus: *T. americana* L. (Syst. Nat., ed. 10: 1289. 7 Jun 1759).

(=) *Bucephalon* L., Sp. Pl.: 1190. 1 Mai 1753.
Typus: *B. racemosum* L.

1918 ***Maclura*** Nutt., Gen. N. Amer. Pl. 2: 233. 14 Jul 1818.
Typus: *M. aurantiaca* Nutt.

(=) *Ioxylon* Raf. in Amer. Monthly Mag. & Crit. Rev. 2: 118. Dec 1817.
Typus: *I. pomiferum* Raf.

1923 ***Broussonetia*** L'Hér. ex Vent., Tabl. Règne Vég. 3: 547. 5 Mai 1799.
Typus: *B. papyrifera* (L.) Vent. (*Morus papyrifera* L.).

(H) *Broussonetia* Ortega, Nov. Pl. Descr. Dec.: 61. 1798 (post 15 Mai) [*Legum.*].
Typus: *B. secundiflora* Ortega

1937 ***Clarisia*** Ruiz & Pav., Fl. Peruv. Prodr.: 128. Oct (prim.) 1794.
Typus: *C. racemosa* Ruiz & Pav. (Syst. Veg. Fl. Peruv. Chil.: 255. Dec (sero) 1798) *(typ. cons.)*.

(H) *Clarisia* Abat in Mem. Acad. Soc. Med. Sevilla 10: 418. 1792 [*Basell.*].
≡ *Anredera* Juss. 1789.

1942 ***Cudrania*** Trécul in Ann. Sci. Nat., Bot., ser. 3, 8: 122. Jul-Dec 1847.
Typus: *C. javanensis* Trécul *(typ. cons.)*.

(=) *Vanieria* Lour., Fl. Cochinch.: 539, 564. Sep 1790.
Typus (vide Merrill in Trans. Amer. Philos. Soc., ser. 2, 24: 134. 1935): *V. cochinchinensis* Lour.

1946 ***Artocarpus*** J. R. Forst. & G. Forst., Char. Gen. Pl.: 51. 29 Nov 1775.
Typus: *A. communis* J. R. Forst. & G. Forst.

(=) *Sitodium* Parkinson, J. Voy. South Seas: 45. Jul 1773.
Typus: *S. altile* Parkinson

1956 ***Antiaris*** Lesch. in Ann. Mus. Natl. Hist. Nat. 16: 478. 1810.
Typus: *A. toxicaria* Lesch.

(=) *Ipo* Pers., Syn. Pl. 2: 566. Sep 1807.
Typus: *I. toxicaria* Pers.

1957	***Brosimum*** Sw., Prodr. 1: 12. 20 Jun - 29 Jul 1788. Typus: *B. alicastrum* Sw. *(typ. cons.)*.	(≡) (=) (=)	*Alicastrum* P. Browne, Civ. Nat. Hist. Jamaica: 372. 10 Mar 1756. *Piratinera* Aubl., Hist. Pl. Guiane: 888. Jun-Dec 1775. Typus: *P. guianensis* Aubl. *Ferolia* Aubl., Hist. Pl. Guiane 2 [Misc.]: 7. Jun-Dec 1775. Typus: *F. guianensis* Aubl.
1971	***Cecropia*** Loefl., Iter Hispan.: 272. Dec 1758. Typus: *C. peltata* L. (Syst. Nat., ed. 10: 1286. 7 Jun 1759).	(≡)	*Coilotapalus* P. Browne, Civ. Nat. Hist. Jamaica: 111. 10 Mar 1756.

URTICACEAE

1980	***Laportea*** Gaudich., Voy. Uranie, Bot.: 498. 6 Mar 1830. Typus: *L. canadensis* (L.) Wedd. (in Ann. Sci. Nat., Bot., ser. 4, 1: 181: Mar 1854) (*Urtica canadensis* L.).	(≡)	*Urticastrum* Heist. ex Fabr., Enum.: 204. 1759.
1984	***Pilea*** Lindl., Collect. Bot.: ad t. 4. 1 Apr 1821. Typus: *P. muscosa* Lindl., nom. illeg. (*Parietaria microphylla* L., *Pilea microphylla* (L.) Liebm.).		
1987	***Pellionia*** Gaudich., Voy. Uranie, Bot.: 494. 6 Mar 1830. Typus: *P. elatostemoides* Gaudich. *(typ. cons.)*.	(=)	*Polychroa* Lour., Fl. Cochinch.: 538, 559. Sep 1790. Typus: *P. repens* Lour.
1988	***Elatostema*** J. R. Forst. & G. Forst., Char. Gen. Pl.: 53. 29 Nov 1775. Typus: *E. sessile* J. R. Forst. & G. Forst. *(typ. cons.)*.		

PROTEACEAE

2023	***Persoonia*** Sm. in Trans. Linn. Soc. London 4: 215. 24 Mai 1798. Typus: *P. lanceolata* Andrews (in Bot. Repos.: ad t. 74. Nov-Dec 1799) *(typ. cons.)*.	(=)	*Linkia* Cav., Icon. 4: 61. 14 Mai 1798. Typus: *L. levis* Cav.

2026 ***Isopogon*** R. Br. ex Knight, Cult. Prot.: 93. Dec 1809.
Typus: *I. anemonifolius* (Salisb.) Knight (*Protea anemonifolia* Salisb.) *(typ. cons.)*.

(=) *Atylus* Salisb., Parad. Lond.: ad t. 67. 1 Jun 1806 -1 Mai 1807.
Typus: non designatus.

2028 ***Sorocephalus*** R. Br. in Trans. Linn. Soc. London 10: 139. Feb 1810.
Typus: *S. imbricatus* (Thunb.) R. Br. (*Protea imbricata* Thunb.) *(typ. cons.)*.

(=) *Soranthe* Salisb. ex Knight, Cult. Prot.: 71. Dec 1809.
Typus: non designatus.

2035 ***Protea*** L., Mant. Pl.: 187, 194, 328. Oct 1771.
Typus: *P. cynaroides* (L.) L. (*Leucadendron cynaroides* L.) *(typ. cons.)*.

(H) *Protea* L., Sp. Pl.: 94. 1 Mai 1753 (typ. des.: Hitchcock in Sprague & al., Nom. Prop. Brit. Bot.: 113. 1929) [*Prot.*].
≡ *Leucadendron* R. Br. 1810 *(nom. cons.)* (2037).

(=) *Lepidocarpus* Adans., Fam. Pl. 2: 284, 569. Jul-Aug 1763.
≡ *Leucadendron* L. 1753 (*nom. rej.* sub 2037).

2036 ***Leucospermum*** R. Br. in Trans. Linn. Soc. London 10: 95. Feb 1810.
Typus: *L. hypophyllum* R. Br., nom. illeg. (*Leucadendron hypophyllocarpodendron* L., *Leucospermum hypophyllocarpodendron* (L.) Druce) *(typ. cons.)*.

2037 ***Leucadendron*** R. Br. in Trans. Linn. Soc. London 10: 50. Feb 1810.
Typus: *L. argenteum* (L.) R. Br. (*Protea argentea* L.) *(typ. cons.)* (etiam vide 2035).

(H) *Leucadendron* L., Sp. Pl.: 91. 1 Mai 1753 [*Prot.*].
Typus (vide Hitchcock in Sprague & al., Nom. Prop. Brit. Bot.: 122. 1929): *L. lepidocarpodendron* L.

2045 ***Grevillea*** R. Br. ex Knight, Cult. Prot.: xvii, 120. Dec 1809 *('Grevillia') (orth. cons.)*.
Typus: *G. aspleniifolia* R. Br. ex Knight.

(=) *Lysanthe* Salisb. ex Knight, Cult. Prot.: 116. Dec 1809.
Typus: non designatus.

(=) *Stylurus* Salisb. ex Knight, Cult. Prot.: 115. Dec 1809.
Typus: non designatus.

2062 ***Telopea*** R. Br. in Trans. Linn. Soc. London 10: 197. Feb 1810.
Typus: *T. speciosissima* (Sm.) R. Br. (*Embothrium speciosissimum* Sm.) *(typ. cons.)*.

(≡) *Hylogyne* Salisb. ex Knight, Cult. Prot.: 126. Dec 1809.

2063 ***Lomatia*** R. Br. in Trans. Linn. Soc. London 10: 199. Feb 1810.
Typus: *L. silaifolia* (Sm.) R. Br. (*Embothrium silaifolium* Sm.) *(typ. cons.)*.

(=) *Tricondylus* Salisb. ex Knight, Cult. Prot.: 121. Dec 1809.
Typus: non designatus.

2064 ***Knightia*** R. Br. in Trans. Linn. Soc. London 10: 193. Feb 1810.
Typus: *K. excelsa* R. Br.

(=) *Rymandra* Salisb. ex Knight, Cult. Prot.: 124. Dec 1809.
Typus: *R. excelsa* Knight

2066 ***Stenocarpus*** R. Br. in Trans. Linn. Soc. London 10: 201. Feb 1810.
Typus: *S. forsteri* R. Br., nom. illeg. (*Embothrium umbelliferum* J. R. Forst. & G. Forst., *S. umbelliferus* (J. R. Forst. & G. Forst.) Druce) *(typ. cons.)*.

(≡) *Cybele* Salisb. ex Knight, Cult. Prot.: 123. Dec 1809.

2068 ***Banksia*** L. f., Suppl. Pl.: 15, 126. Apr 1782.
Typus: *B. serrata* L. f.

(H) *Banksia* J. R. Forst. & G. Forst., Char. Gen. Pl.: 4. 29 Nov 1775 [*Thymel.*].
Typus: non designatus.

2069 ***Dryandra*** R. Br. in Trans. Linn. Soc. London 10: 211. Feb 1810.
Typus: *D. formosa* R. Br. *(typ. cons.)*.

(H) *Dryandra* Thunb., Nov. Gen. Pl.: 60. 18 Jun 1783 [*Euphorb.*].
Typus: *D. cordata* Thunb.

(=) *Josephia* R. Br. ex Knight, Cult. Prot.: 110. Dec 1809.
Typus: non designatus.

LORANTHACEAE

2074 ***Loranthus*** Jacq., Enum. Stirp. Vindob.: 55, 230. Mai 1762.
Typus: *L. europaeus* Jacq.

(H) *Loranthus* L., Sp. Pl.: 331. 1 Mai 1753 [*Loranth.*].
Typus: *L. americanus* L.

(=) *Scurrula* L., Sp. Pl.: 110. 1 Mai 1753.
Typus: *S. parasitica* L.

2074a ***Tapinanthus*** (Blume) Rchb., Deut. Bot. Herb.-Buch [1]: 73. Jul 1841 (*Loranthus* sect. *Tapinanthus* Blume, Fl. Javae (Loranth.): 15. 16 Aug 1830).
Typus: *Loranthus sessilifolius* P. Beauv. (*T. sessilifolius* (P. Beauv.) Tiegh.).

(H) *Tapeinathus* Herb., Amaryllidaceae: 59, 73, 190, 414. Apr (sero) 1837 [*Amaryllid.*].
Typus: *T. humilis* (Cav.) Herb. (*Pancratium humile* Cav.).

2078 ***Struthanthus*** Mart. in Flora 13: 102. 21 Feb 1830.
Typus: *S. syringifolius* (Mart.) Mart. (*Loranthus syringifolius* Mart.) *(typ. cons.)*.

(=) *Spirostylis* C. Presl in Schult. & Schult. f., Syst. Veg. 7: xvii. 1829.
Typus: *S. haenkei* C. Presl ex Schult. & Schult. f. (Syst. Veg. 7: 163. 1829).

2091 ***Arceuthobium*** M. Bieb., Fl. Taur.-Caucas. 3: 629. 1819 (sero) - 1820 (prim.).
Typus: *A. oxycedri* (DC.) M. Bieb. (*Viscum oxycedri* DC.).
(=) *Razoumofskya* Hoffm., Hort. Mosq.: 1. Jun-Dec 1808.
Typus: *R. caucasica* Hoffm.

SANTALACEAE

2097 ***Exocarpos*** Labill., Voy. Rech. Pérouse 1: 155. 22 Feb - 4 Mar 1800.
Typus: *E. cupressiformis* Labill.
(=) *Xylophylla* L., Mant. Pl.: 147, 221. Oct 1771.
Typus: *X. longifolia* L.

2103 ***Scleropyrum*** Arn. in Mag. Zool. Bot. 2: 549. 1838.
Typus: *S. wallichianum* (Wight & Arn.) Arn. (*Sphaerocarya wallichiana* Wight & Arn.).
(=) *Heydia* Dennst. ex Kostel., Allg. Med.-Pharm. Fl. 5: 2005. Jan-Sep 1836.
Typus: *H. horrida* Dennst. ex Kostel.

2109 ***Buckleya*** Torr. in Amer. J. Sci. Arts 45: 170. Jun 1843.
Typus: *B. distichophylla* (Nutt.) Torr. (*Borya distichophylla* Nutt.).
(=) *Nestronia* Raf., New Fl. 3: 12. Jan-Mar 1838.
Typus: *N. umbellula* Raf.

2120 ***Quinchamalium*** Molina, Sag. Stor. Nat. Chili: 151, 350. 12-13 Oct 1782.
Typus: *Q. chilense* Molina

OPILIACEAE

2124 ***Cansjera*** Juss., Gen. Pl.: 448. 4 Aug 1789.
Typus: *C. rheedei* J. F. Gmel. (Syst. Nat. 2: 4, 20. Sep-Nov 1791).
(≡) *Tsjeru-caniram* Adans., Fam. Pl. 2: 80, 614. Jul-Aug 1763.

OLACACEAE

2147 ***Heisteria*** Jacq., Enum. Syst. Pl.: 4, 20. Aug-Sep 1760.
Typus: *H. coccinea* Jacq.
(H) *Heisteria* L., Opera Var.: 242. 1758 [*Polygal.*].
≡ *Muraltia* DC. 1824 *(nom. cons.)* (4278).

BALANOPHORACEAE

2163 ***Helosis*** Rich. in Mém. Mus. Hist. Nat. 8: 416, 432. 1822.
Typus: *H. guyannensis* Rich., nom. illeg. (*Cynomorium cayanense* Sw., *H. cayanensis* (Sw.) Spreng.).

RAFFLESIACEAE

2180 ***Cytinus*** L., Gen. Pl., ed. 6: 576 ('566'). Jun 1764.
Typus: *C. hypocistis* (L.) L. (*Asarum hypocistis* L.).

(≡) *Hypocistis* Mill., Gard. Dict. Abr., ed. 4: [662]. 28 Jan 1754.

POLYGONACEAE

2194 ***Emex*** Campd., Monogr. Rumex: 56. 1819.
Typus: *E. spinosa* (L.) Campd. (*Rumex spinosus* L.) *(typ. cons.)*.

(≡) *Vibo* Medik., Philos. Bot. 1: 178. Apr 1789.

2201 ***Polygonum*** L., Sp. Pl.: 359. 1 Mai 1753.
Typus: *P. aviculare* L. *(typ. cons.)*.

2202 ***Fagopyrum*** Mill., Gard. Dict. Abr., ed. 4: [495]. 28 Jan 1754.
Typus: *F. esculentum* Moench (Methodus: 290. 4 Mai 1794) (*Polygonum fagopyrum* L.) *(typ. cons.)*.

2208 ***Muehlenbeckia*** Meisn., Pl. Vasc. Gen. 1: 316; 2: 227. 18-24 Jul 1841.
Typus: *M. australis* (G. Forst.) Meisn. (*Coccoloba australis* G. Forst.).

(≡) *Calacinum* Raf., Fl. Tellur. 2: 33. Jan-Mar 1837.

(=) *Karkinetron* Raf., Fl. Tellur. 3: 11. Nov-Dec 1837.
Typus: non designatus.

2209 ***Coccoloba*** P. Browne, Civ. Nat. Hist. Jamaica: 209. 10 Mar 1756 *('Coccolobis') (orth. cons.)*.
Typus: *C. uvifera* (L.) L. (*Polygonum uvifera* L.) *(typ. cons.)*.

(=) *Guaiabara* Mill., Gard. Dict. Abr., ed. 4: [590]. 28 Jan 1754.
Typus: non designatus.

CHENOPODIACEAE

2261 ***Suaeda*** Forssk. ex J. F. Gmel., Onomat. Bot. Compl. 8: 797. 1776.
Typus: *S. vera* Forssk. *(typ. cons.)*.

AMARANTHACEAE

2297 ***Chamissoa*** Kunth in Humboldt & al., Nov. Gen. Sp. 2, ed. 4°: 196 [& ed. f°: 158]. Feb 1818.
Typus: *C. altissima* (Jacq.) Kunth (*Achyranthes altissima* Jacq.) *(typ. cons.)*.

(=) *Kokera* Adans., Fam. Pl. 2: 269, 541. Jul-Aug 1763.
Typus: non designatus.

2312 ***Cyathula*** Blume, Bijdr.: 548. 24 Jan 1826.
Typus: *C. prostrata* (L.) Blume (*Achyranthes prostrata* L.).

(H) *Cyathula* Lour., Fl. Cochinch.: 93, 101. Sep 1790 [*Amaranth.*].
Typus: *C. geniculata* Lour.

2314 ***Pupalia*** Juss. in Ann. Mus. Natl. Hist. Nat. 2: 132. 1803.
Typus: *P. lappacea* (L.) Juss. (*Achyranthes lappacea* L.).

(≡) *Pupal* Adans., Fam. Pl. 2: 268, 596. Jul-Aug 1763.

2317 ***Aerva*** Forssk., Fl. Aegypt.-Arab.: 170. 1 Oct 1775.
Typus: *A. tomentosa* Forssk. *(typ. cons.)*.

(=) *Ouret* Adans., Fam. Pl. 2: 268, 596. Jul-Aug 1763.
Typus (vide Rickett & Stafleu in Taxon 8: 268. 1959): *Achyranthes lanata* L.

2339 ***Iresine*** P. Browne, Civ. Nat. Hist. Jamaica: 358. 10 Mar 1756.
Typus: [specimen] *P. Browne,* Herb. Linn. No. 288.5 (LINN) *(typ. cons.)* [= *I. diffusa* L.].

NYCTAGINACEAE

2348 ***Allionia*** L., Syst. Nat., ed. 10: 883, 890, 1361. 7 Jun 1759.
Typus: *A. incarnata* L. *(typ. cons.)* (etiam vide 9192).

(H) *Allionia* Loefl., Iter Hispan.: 181. Dec 1758 [*Nyctagin.*].
Typus: *A. violacea* L. (Syst. Nat., ed. 10: 890. 7 Jun 1759).

2350 ***Bougainvillea*** Comm. ex Juss., Gen. Pl.: 91. 4 Aug 1789 *('Buginvillaea') (orth. cons.)*.
Typus: *B. spectabilis* Willd. (Sp. Pl. 2: 348. 1799) *(typ. cons.)*.

AIZOACEAE

2405 ***Mesembryanthemum*** L., Sp. Pl.: 480. 1 Mai 1753.
Typus: *M. nodiflorum* L. *(typ. cons.)*.

2405a ***Lampranthus*** N. E. Br. in Gard. Chron., ser. 3, 87: 71. 25 Jan 1930.
Typus: *L. multiradiatus* (Jacq.) N. E. Br. (*Mesembryanthemum multiradiatum* Jacq.).

(=) *Oscularia* Schwantes in Möller's Deutsche Gärtn.-Zeitung 42: 187. 21 Mai 1927.
Typus: *O. deltoides* (L.) Schwantes (*Mesembryanthemum deltoides* L.).

PORTULACACEAE

2406 ***Talinum*** Adans., Fam. Pl. 2: 245, 609. Jul-Aug 1763.
Typus: *T. triangulare* (Jacq.) Willd. (Sp. Pl. 2: 862. Dec 1799) (*Portulaca triangularis* Jacq.) *(typ. cons.)*.

2407 ***Calandrinia*** Kunth in Humboldt & al., Nov. Gen. Sp. 6, ed. f°: 62. 14 Apr 1823.
Typus: *C. caulescens* Kunth *(typ. cons.)*.

(=) *Baitaria* Ruiz & Pav., Fl. Peruv. Prodr.: 63. Oct (prim.) 1794.
Typus: *B. acaulis* Ruiz et Pav. (Syst. Veg. Fl. Peruv. Chil.: 111. Dec (sero) 1798).

2412 ***Anacampseros*** L., Opera Var.: 232. 1758.
Typus: *A. telephiastrum* DC. (Cat. Pl. Horti Monsp.: 77. Feb-Mar 1813) (*Portulaca anacampseros* L.).

(H) *Anacampseros* Mill., Gard. Dict. Abr., ed. 4: [73]. 28 Jan 1754 [*Crassul.*].
Typus: non designatus.

CARYOPHYLLACEAE

2432 ***Moenchia*** Ehrh. in Neues Mag. Aerzte 5: 203. 1783 (post 11 Jun).
Typus: *M. quaternella* Ehrh., nom. illeg. (*Sagina erecta* L., *M. erecta* (L.) P. Gaertn. & al.).

2450 ***Spergularia*** (Pers.) J. Presl & C. Presl, Fl. Čech.: 94. 1819 (*Arenaria* subg. *Spergularia* Pers., Syn. Pl. 1: 504. 1 Apr - 15 Jun 1805).
Typus: *Arenaria rubra* L. (*S. rubra* (L.) J. Presl & C. Presl).

(≡) *Tissa* Adans., Fam. Pl. 2: 507, 611. Jul-Aug 1763 (typ. des.: Swart in Regnum Veg. 102: 1764. 1979).

(=) *Buda* Adans., Fam. Pl. 2: 507, 528. Jul-Aug 1763.
Typus: non designatus

2455 ***Polycarpaea*** Lam. in J. Hist. Nat. 2: 3, 5. 1792.
Typus: *P. teneriffae* Lam. *(typ. cons.)*.

(=) *Polia* Lour., Fl Cochinch.: 97, 164. Sep 1790.
Typus: *P. arenaria* Lour.

2467 ***Pollichia*** Aiton, Hort. Kew. 1: 5. 7 Aug - 1 Oct 1789.
Typus: *P. campestris* Aiton

(H) *Polichia* Schrank in Acta Acad. Elect. Mogunt. Sci. Util. Erfurti 3: 35. 1781 [*Lab.*].
Typus: *P. galeobdolon* (L.) Willd. (Fl. Berol. Prodr.: 198. 1787) (*Galeopsis galeobdolon* L.).

2477 ***Siphonychia*** Torr. & A. Gray, Fl. N. Amer. 1: 173. Jul 1838.
Typus: *S. americana* (Nutt.) Torr. & A. Gray (*Herniaria americana* Nutt.).

2490 ***Silene*** L., Sp. Pl.: 416. 1 Mai 1753. Typus: *S. anglica* L.
(=) *Lychnis* L., Sp. Pl.: 436. 1 Mai 1753. Typus (vide Britton & Brown, Ill. Fl. N. U.S., ed. 2, 2: 68. 1913): *L. chalcedonica* L.

NYMPHAEACEAE

2513 ***Nymphaea*** L., Sp. Pl.: 510. 1 Mai 1753. Typus: *N. alba* L. *(typ. cons.)*.

2514 ***Nuphar*** Sm., Fl. Graec. Prodr. 1: 361. Mai-Nov 1809. Typus: *N. lutea* (L.) Sm. (*Nymphaea lutea* L.).
(≡) *Nymphozanthus* Rich., Démonstr. Bot.: 63, 68, 103. Mai 1808.

2515 ***Barclaya*** Wall. in Trans. Linn. Soc. London 15: 442. 11-20 Dec 1827. Typus: *B. longifolia* Wall.
(≡) *Hydrostemma* Wall. in Philos. Mag. Ann. Chem. 1: 454. Jun 1827.

RANUNCULACEAE

2528 ***Eranthis*** Salisb. in Trans. Linn. Soc. London 8: 303. 9 Mar 1807. Typus: *E. hyemalis* (L.) Salisb. (*Helleborus hyemalis* L.).
(≡) *Cammarum* Hill, Brit. Herb.: 47. 23 Feb 1756.

2542 ***Naravelia*** Adans., Fam. Pl. 2: 460, 581. Jul-Aug 1763 *('Naravel') (orth. cons.)*. Typus: *N. zeylanica* (L.) DC. (Syst. Nat. 1: 167. 1-15 Nov 1817) (*Atragene zeylanica* L.).

LARDIZABALACEAE

2551 ***Decaisnea*** Hook. f. & Thomson in Proc. Linn. Soc. Lond. 2: 350. 1 Mai 1855. Typus: *D. insignis* (Griff.) Hook. f. & Thomson (*Slackia insignis* Griff.).
(H) *Decaisnea* Brongn. in Duperrey, Voy. Monde, Phan.: 192. Jan 1834 [*Orchid.*]. Typus: *D. densiflora* Brongn.

BERBERIDACEAE

2566 ***Mahonia*** Nutt., Gen. N. Amer. Pl. 1: 211. 14 Jul 1818. Typus: *M. aquifolium* (Pursh) Nutt. (*Berberis aquifolium* Pursh) *(typ. cons.)*.

MENISPERMACEAE

2568 ***Pericampylus*** Miers in Ann. Mag. Nat. Hist., ser. 2, 7: 36, 40. Jan 1851.
Typus: *P. incanus* (Colebr.) Hook. f. & Thomson (Fl. Ind.: 194. 1-19 Jul 1855) (*Cocculus incanus* Colebr.).

(=) *Pselium* Lour., Fl. Cochinch.: 600, 621. Sep 1790.
Typus: *P. heterophyllum* Lour.

2570 ***Cocculus*** DC., Syst. Nat. 1: 515. 1-15 Nov 1817.
Typus: *C. hirsutus* (L.) W. Theob. (in Mason, Burmah 2: 657. 1883) (*Menispermum hirsutum* L.) *(typ. cons.)*.

(=) *Cebatha* Forssk., Fl. Aegypt.-Arab.: 171. 1 Oct 1775.
Typus: *Cocculus cebatha* DC.

(=) *Leaeba* Forssk., Fl. Aegypt.-Arab.: 172. 1 Oct 1775.
Typus: *L. dubia* J. F. Gmel. (Syst. Nat. 2: 567. Sep-Nov 1791).

(=) *Epibaterium* J. R. Forst. & G. Forst., Char. Gen. Pl.: 54. 29 Nov 1775.
Typus: *E. pendulum* J. R. Forst. & G. Forst.

(=) *Nephroia* Lour., Fl. Cochinch.: 539, 565. Sep 1790.
Typus: *N. sarmentosa* Lour.

(=) *Baumgartia* Moench, Methodus: 650. 4 Mai 1794.
Typus: *B. scandens* Moench

(=) *Androphylax* J. C. Wendl., Bot. Beob.: 37, 38. 1798.
Typus: *A. scandens* J. C. Wendl.

2577 ***Tiliacora*** Colebr. in Trans. Linn. Soc. London 13: 53, 67. 23 Mai - 21 Jun 1821.
Typus: *T. racemosa* Colebr.

(=) *Braunea* Willd., Sp. Pl. 4: 638, 797. Apr 1806.
Typus: *B. menispermoides* Willd.

2583 ***Tinospora*** Miers in Ann. Mag. Nat. Hist., ser. 2, 7: 35, 38. Jan 1851.
Typus: *T. cordifolia* (Willd.) Miers ex Hook. f. & Thomson (Fl. Ind. (2): 184. 1-19 Jul 1855) (*Menispermum cordifolium* Willd.).

(=) *Campylus* Lour., Fl. Cochinch.: 94, 113. Sep 1790.
Typus: *C. sinensis* Lour.

2611 ***Hyperbaena*** Miers ex Benth. in J. Proc. Linn. Soc., Bot. 5, Suppl. 2: 47. 1861.
Typus: *H. domingensis* (DC.) Benth. (*Cocculus domingensis* DC.) *(typ. cons.)*.

(=) *Alina* Adans., Fam. Pl. 2: 84, 512. Jul-Aug 1763.
Typus: non designatus.

MAGNOLIACEAE

2656 ***Schisandra*** Michx., Fl. Bor.-Amer. 2: 218. 19 Mar 1803.
Typus: *S. coccinea* Michx.

(=) *Stellandria* Brickell in Med. Repos. 6: 327. Feb-Mar 1803.
Typus: *S. glabra* Brickell

2658 ***Drimys*** J. R. Forst. & G. Forst., Char. Gen. Pl.: 42. 29 Nov 1775.
Typus: *D. winteri* J. R. Forst. & G. Forst. *(typ. cons.)*.

CALYCANTHACEAE

2663 ***Calycanthus*** L., Syst. Nat., ed. 10: 1053, 1066, 1371. 7 Jun 1759.
Typus: *C. floridus* L.

(=) *Basteria* Mill., Fig. Pl. Gard. Dict.: 40. 30 Dec 1755.
Typus: non designatus.

2663a ***Chimonanthus*** Lindl. in Bot. Reg.: ad t. 404. 1 Oct 1819.
Typus: *C. fragrans* Lindl., nom. illeg. (*Calycanthus praecox* L., *Chimonanthus praecox* (L.) Link).

(≡) *Meratia* Loisel., Herb. Gén. Amat. 3: ad t. 173. Jul 1818.

ANNONACEAE

2679 ***Guatteria*** Ruiz & Pav., Fl. Peruv. Prodr.: 85. Oct (prim.) 1794.
Typus: *G. glauca* Ruiz & Pav. (Syst. Veg. Fl. Peruv. Chil.: 145. Dec (sero) 1798).

(=) *Aberemoa* Aubl., Hist. Pl. Guiane 1: 610. Jun-Dec 1775.
Typus: *A. guianensis* Aubl.

2680 ***Duguetia*** A. St.-Hil., Fl. Bras. Merid. 1, ed. 4°: 35 [& ed. f°: 28]. 23 Feb 1824.
Typus: *D. lanceolata* A. St.-Hil.

2684 ***Cananga*** (DC.) Hook. f. & Thomson, Fl. Ind.: 129. 1-19 Jul 1855 (*Unona* subsect. *Cananga* DC., Syst. Nat. 1: 485. 1-15 Nov 1817).
Typus: *Unona odorata* (Lam.) Dunal (*Uvaria odorata* Lam., *C. odorata* (Lam.) Hook. f. & Thomson).

(H) *Cananga* Aubl., Hist. Pl. Guiane: 607. Jun-Dec 1775 [*Annon.*].
Typus: *C. ouregou* Aubl.

2691a ***Enneastemon*** Exell in J. Bot. 70, Suppl. 1: 209. Feb 1932.
Typus: *E. angolensis* Exell

(=) *Clathrospermum* Planch. ex Hook. f. in Bentham & Hooker, Gen. Pl. 1: 29. 7 Aug 1862.
Typus: *C. vogelii* (Hook. f.) Benth. (*Uvaria vogelii* Hook. f.).

2717 ***Xylopia*** L., Syst. Nat., ed. 10: 1241, 1250, 1378. 7 Jun 1759.
Typus: *X. muricata* L. *(typ. cons.).*

(≡) *Xylopicrum* P. Browne, Civ. Nat. Hist. Jamaica: 250. 10 Mar 1756.

MYRISTICACEAE

2750 ***Myristica*** Gronov., Fl. Orient.: 141. Apr-Nov 1755.
Typus: *M. fragrans* Houtt. (Nat. Hist. 2(3): 333. Dec (sero) 1774).

MONIMIACEAE

2758 ***Trimenia*** Seem., Fl. Vit.: 425. Feb 1873.
Typus: *T. weinmanniifolia* Seem.

(=) *Piptocalyx* Oliv. ex Benth., Fl. Austral. 5: 292. Aug-Oct 1870.
Typus: *P. moorei* Oliv. ex Benth.

2759 ***Peumus*** Molina, Sag. Stor. Nat. Chili: 185, 350. 12-13 Oct 1782.
Typus: *P. boldus* Molina *(typ. cons.).*

(=) *Boldu* Adans., Fam. Pl. 2: 446, 526. Jul-Aug 1763.
Typus: non designatus.

2775 ***Laurelia*** Juss. in Ann. Mus. Natl. Hist. Nat. 14: 134. 1809.
Typus: *L. sempervirens* (Ruiz & Pav.) Tul. (in Arch. Mus. Hist. Nat. 8: 416. 1855) (*Pavonia sempervirens* Ruiz & Pav.) *(typ. cons.).*

LAURACEAE

2782 ***Cinnamomum*** Schaeff., Bot. Exped.: 74. Oct-Dec 1760.
Typus: *C. verum* J. Presl (in Berchtold & Presl, Přir. Rostlin 2(2): 36. 1825) (*Laurus cinnamomum* L.).

(=) *Camphora* Fabr., Enum.: 218. 1759.
Typus: *Laurus camphora* L.

2783 ***Persea*** Mill., Gard. Dict. Abr., ed. 4: [1030]. 28 Jan 1754.
Typus: *P. americana* Mill. (Gard. Dict., ed. 8: *Persea* No. [1]. 16 Apr 1768) (*Laurus persea* L.).

2789 ***Umbellularia*** (Nees) Nutt., N. Amer. Sylva 1: 87. Jul-Dec 1842 (*Oreodaphne* subg. *Umbellularia* Nees, Syst. Laur.: 381, 462. 30 Oct - 5 Nov 1836).
Typus: *Oreodaphne californica* (Hook. & Arn.) Nees (*Tetranthera californica* Hook. & Arn., *U. californica* (Hook. & Arn.) Nutt.).

(≡) *Sciadiodaphne* Rchb., Deut. Bot. Herb.-Buch [1]: 70; [2]: 118. Jul 1841.

2790 ***Nectandra*** Rol. ex Rottb. in Acta Lit. Univ. Hafn. 1: 279. 1778.
Typus: *N. sanguinea* Rol. ex Rottb. *(typ. cons.)*.

(H) *Nectandra* P. J. Bergius, Descr. Pl. Cap.: 131. Sep 1767 [*Thymel.*].
Typus (vide Mansfeld in Bull. Misc. Inform. Kew 1935: 439. 1935): *N. sericea* (L.) P. J. Bergius (*Passerina sericea* L.).

2793 ***Eusideroxylon*** Teijsm. & Binn. in Natuurk. Tijdschr. Ned. Indië 25: 292. 1863.
Typus: *E. zwageri* Teijsm. & Binn.

2797 ***Neolitsea*** (Benth. & Hook. f.) Merr. in Philipp. J. Sci., C, 1, Suppl. 1: 56. 15 Apr 1906 (*Litsea* sect. *Neolitsea* Benth. & Hook. f., Gen. Pl. 3: 161. Feb 1880).
Typus: *Litsea zeylanica* Nees & T. Nees (*N. zeylanica* (Nees & T. Nees) Merr.).

(=) *Bryantea* Raf., Sylva Tellur.: 165. Oct-Dec 1838.
Typus: *B. dealbata* (R. Br.) Raf. (*Tetranthera dealbata* R. Br.).

2798 ***Litsea*** Lam., Encycl. 3: 574. 13 Feb 1792.
Typus: *L. chinensis* Lam.

(=) *Malapoënna* Adans., Fam. Pl. 2: 447, 573. Jul-Aug 1763.
Typus: *Darwinia quinqueflora* Dennst.

2804 ***Bernieria*** Baill. in Bull. Mens. Soc. Linn. Paris: 434. 1884.
Typus: *B. madagascariensis* Baill.

(H) *Berniera* DC., Prodr. 7: 18. Apr (sero) 1838 [*Comp.*].
Typus: *B. nepalensis* DC., nom. illeg. (*Chaptalia maxima* D. Don).

2811a ***Endlicheria*** Nees in Linnaea 8: 37. 1833.
Typus: *E. hirsuta* (Schott) Nees (*Cryptocarya hirsuta* Schott) *(typ. cons.)*.

(H) *Endlichera* C. Presl, Symb. Bot. 1: 73. Jan-Feb 1832 [*Rub.*].
Typus: *E. brasiliensis* C. Presl

2813 ***Cryptocarya*** R. Br., Prodr.: 402. 27 Mar 1810.
Typus: *C. glaucescens* R. Br.

(=) *Ravensara* Sonn., Voy. Indes Orient. 2: 226. 1782.
Typus: *R. aromatica* Sonn.

2821 ***Lindera*** Thunb., Nov. Gen. Pl.: 64. 18 Jun 1783.
Typus: *L. umbellata* Thunb.

(H) *Lindera* Adans., Fam. Pl. 2: 499, 571. Jul-Aug 1763 [*Umbell.*].
Typus: *Chaerophyllum coloratum* L.

(=) *Benzoin* Schaeff., Bot. Exped.: 60. 1 Oct - 24 Dec 1760.
Typus: *B. odoriferum* Nees (in Wallich, Pl. Asiat. Rar. 2: 63. 6 Sep 1831) (*Laurus benzoin* L.).

PAPAVERACEAE

2856 ***Dicentra*** Bernh. in Linnaea 8: 457, 468. 1833 (post Jul).
Typus: *D. cucullaria* (L.) Bernh. (*Fumaria cucullaria* L.).

(≡) *Diclytra* Borkh. in Arch. Bot. (Leipzig) 1(2): 46. Mai-Dec 1797.

(=) *Capnorchis* Mill., Gard. Dict. Abr., ed. 4: [250]. 28 Jan 1754.
Typus: non designatus.

(=) *Bikukulla* Adans., Fam. Pl. 2: (23). Jul-Aug 1763.
Typus: non designatus.

(=) *Dactylicapnos* Wall., Tent. Fl. Napal.: 51. Sep-Dec 1826.
Typus: *D. thalictrifolia* Wall.

2857 ***Adlumia*** Raf. ex DC., Syst. Nat. 2: 111. Mai (sero) 1821.
Typus: *A. cirrhosa* Raf. ex DC., nom. illeg. (*Fumaria fungosa* Aiton, *A. fungosa* (Aiton) Greene ex Britton & al.).

2858 ***Corydalis*** DC. in Lamarck & Candolle, Fl. Franç., ed. 3, 4: 637. 17 Sep 1805.
Typus: *C. bulbosa* (L.) DC., *comb. rej.* (*Fumaria bulbosa* L., *nom. rej.*, *Fumaria bulbosa* var. *solida* L., *C. solida* (L.) Clairv.) (*typ. cons.:* Herb. Linn. No. 881.5 (LINN)).

(H) *Corydalis* Medik., Philos. Bot. 1: 96. Apr 1789 [*Papaver.*].
≡ *Cysticapnos* Mill. 1754 *(nom. rej.)*.

(≡) *Pistolochia* Bernh., Syst. Verz.: 57, 74. 1800.

(=) *Capnoides* Mill., Gard. Dict. Abr., ed. 4: [249]. 28 Jan 1754.
Typus: *C. sempervirens* (L.) Borkh. (in Arch. Bot. (Leipzig) 1(2): 44. Mai-Dec 1797) (*Fumaria sempervirens* L.).

(=) *Cysticapnos* Mill., Gard. Dict. Abr., ed. 4: [427]. 28 Jan 1754.
Typus: *C. vesicaria* (L.) Fedde (in Repert. Spec. Nov. Regni Veg. 19: 287. 20 Feb 1924) (*Fumaria vesicaria* L.).

(=) *Pseudo-fumaria* Medik., Philos. Bot. 1: 110. Apr 1789.
Typus: *P. lutea* (L.) Borkh. (in Arch. Bot. (Leipzig) 1(2): 45. Mai-Dec 1797) (*Fumaria lutea* L.).

CRUCIFERAE (BRASSICACEAE)

2884 ***Coronopus*** Zinn, Cat. Pl. Hort. Gott.: 325. 20 Apr - 21 Mai 1757.
Typus: *C. ruellii* All. (Fl. Pedem. 1: 256. Apr-Jul 1785) (*Cochlearia coronopus* L.).

(H) *Coronopus* Mill., Gard. Dict. Abr., ed. 4: [387]. 28 Jan 1754 [*Plantagin.*].
Typus: non designatus.

2902 ***Bivonaea*** DC. in Mém. Mus. Hist. Nat. 7: 241. 20 Apr 1821.
Typus: *B. lutea* (Biv.) DC. (Syst. Nat. 2: 255. Mai (sero) 1821) (*Thlaspi luteum* Biv.).

(H) *Bivonea* Raf. in Specchio Sci. 1: 156. 1 Mai 1814 [*Euphorb.*].
Typus: *B. stimulosa* (Michx.) Raf. (*Jatropha stimulosa* Michx.).

2908 ***Kernera*** Medik., Pfl.-Gatt.: 77, 95. 22 Apr 1792.
Typus: *K. myagrodes* Medik., nom. illeg. (*Cochlearia saxatilis* L., *K. saxatilis* (L.) Rchb.).

(H) *Kernera* Schrank, Baier. Reise: 50. post 5 Apr 1786 [*Scrophular.*].
Typus: *K. bavarica* Schrank

2923 ***Goldbachia*** DC. in Mém. Mus. Hist. Nat. 7: 242. 20 Apr 1821.
Typus: *G. laevigata* (M. Bieb.) DC. (Syst. Nat. 2: 577. Mai (sero) 1821) (*Raphanus laevigatus* M. Bieb.).

(H) *Goldbachia* Trin. in Sprengel, Neue Entd. 2: 42. Jan 1821 [*Gram.*].
Typus: *G. mikanii* Trin.

2936 ***Carrichtera*** DC. in Mém. Mus. Hist. Nat. 7: 244. 20 Apr 1821.
Typus: *C. annua* (L.) DC. (*Vella annua* L.) *(typ. cons.)*.

(H) *Carrichtera* Adans., Fam. Pl. 2: 421, 533. Jul-Aug 1763 [*Cruc.*].
Typus: *Vella pseudocytisus* L.

2940 ***Schouwia*** DC. in Mém. Mus. Hist. Nat. 7: 244. 20 Apr 1821.
Typus: *S. arabica* DC. (Syst. Nat. 2: 644. Mai (sero) 1821.), nom. illeg. (*Subularia purpurea* Forssk., *Schouwia purpurea* (Forssk.) Schweinf.).

2956 ***Rapistrum*** Crantz, Class. Crucif. Emend.: 105. Jan-Aug 1769.
Typus: *R. hispanicum* (L.) Crantz (*Myagrum hispanicum* L.) *(typ. cons.)*.

(H) *Rapistrum* Scop., Meth. Pl.: 13. 25 Mar 1754 [*Cruc.*].
≡ *Neslia* Desv. 1814 *(nom. cons.)* (2988).

2961 ***Barbarea*** W. T. Aiton, Hort. Kew., ed. 2, 4: 109. Dec 1812.
Typus: *B. vulgaris* W. T. Aiton (*Erysimum barbarea* L.).

(H) *Barbarea* Scop., Fl. Carniol.: 522. 15 Jan - 21 Jul 1760 [*Cruc.*].
Typus: *Dentaria bulbifera* L.

2965 ***Nasturtium*** W. T. Aiton, Hort. Kew., ed. 2, 4: 109. Dec 1812.
Typus: *N. officinale* W. T. Aiton (*Sisymbrium nasturtium-aquaticum* L.) *(typ. cons.)*.

(H) *Nasturtium* Mill., Gard. Dict. Abr., ed. 4: [946]. 28 Jan 1754 [*Cruc.*].
Typus: non designatus.

(≡) *Cardaminum* Moench, Methodus: 262. 4 Mai 1794.

2965a ***Armoracia*** P. Gaertn. & al., Oekon. Fl. Wetterau 2: 426. Mai-Jul 1800.
Typus: *A. rusticana* P. Gaertn. & al. (*Cochlearia armoracia* L.).

(≡) *Raphanis* Moench, Methodus: 267. 4 Mai 1794.

2968 ***Ricotia*** L., Sp. Pl., ed. 2: 912. Jul-Aug 1763.
Typus: *R. aegyptiaca* L., nom. illeg. (*Cardamine lunaria* L., *R. lunaria* (L.) DC.) (etiam vide 7393).

2973 ***Mancoa*** Wedd., Chlor. Andina 2: t. 86d. 10 Oct 1859.
Typus: *M. hispida* Wedd.

(H) *Mancoa* Raf., Fl. Tellur. 3: 56. Nov-Dec 1837 [*Phytolacc.*].
Typus: *M. secunda* (Ruiz & Pav.) Raf. (*Rivina secunda* Ruiz & Pav.).

2986 ***Capsella*** Medik., Pfl.-Gatt.: 85, 99. 22 Apr 1792.
Typus: *C. bursa-pastoris* (L.) Medik. (*Thlaspi bursa-pastoris* L.) *(typ. cons.)*.

(≡) *Bursa-pastoris* Ség., Pl. Veron. 3: 166. Jul-Aug 1754.

2988 ***Neslia*** Desv. in J. Bot. Agric. 3: 162. 1815 (prim.).
Typus: *N. paniculata* (L.) Desv. (*Myagrum paniculatum* L.) (etiam vide 2956).

2989a ***Erophila*** DC. in Mém. Mus. Hist. Nat. 7: 234. 20 Apr 1821.
Typus: *E. verna* (L.) Chevall. (Fl. Gén. Env. Paris 2: 898. 5 Jan 1828) (*Draba verna* L.) *(typ. cons.)*.

(≡) *Gansblum* Adans., Fam. Pl. 2: 420, 561. Jul-Aug 1763.

2997 ***Descurainia*** Webb & Berthel., Hist. Nat. Iles Canaries 3(2,1): 72. Nov 1836.
Typus: *D. sophia* (L.) Webb ex Prantl (in Engler & Prantl, Nat. Pflanzenfam. 3(2): 192. Mar 1891) (*Sisymbrium sophia* L.) *(typ. cons.)*.

(≡) *Sophia* Adans., Fam. Pl. 2: 417, 606. Jul-Aug 1763.

(=) *Hugueninia* Rchb., Fl. Germ. Excurs.: 691. 1832.
Typus: *H. tanacetifolia* (L.) Rchb. (*Sisymbrium tanacetifolium* L.).

2999 ***Arabidopsis*** Heynh. in Holl & Heynhold, Fl. Sachsen 1(2): 538. 1842.
Typus: *A. thaliana* (L.) Heynh. (*Arabis thaliana* L.) *(typ. cons.)*.

3013 ***Lobularia*** Desv. in J. Bot. Agric. 3: 162. 1815 (prim.).
Typus: *L. maritima* (L.) Desv. (*Clypeola maritima* L.).

(≡) *Aduseton* Adans., Fam. Pl. 2: (23), 420 *('Konig')*, 542 *('Konig')*. Jul-Aug 1763.

3022 ***Lepidostemon*** Hook. f. & Thomson in J. Proc. Linn. Soc., Bot. 5: 131, 156. 27 Mar 1861.
Typus: *L. pedunculosus* Hook. f. & Thomson *(typ. cons.)*.

3032 ***Malcolmia*** W. T. Aiton, Hort. Kew., ed. 2, 4: 121. Dec 1812 *('Malcomia') (orth. cons.)*.
Typus: *M. maritima* (L.) W. T. Aiton (*Cheiranthus maritimus* L.) *(typ. cons.)*.

(≡) *Wilckia* Scop., Intr. Hist. Nat.: 317. Jan-Apr 1777.

3038 ***Euclidium*** W. T. Aiton, Hort. Kew., ed. 2, 4: 74. Dec 1812.
Typus: *E. syriacum* (L.) W. T. Aiton (*Anastatica syriaca* L.).

(≡) *Hierochontis* Medik., Pfl.-Gatt.: 51. 22 Apr 1792.
(=) *Soria* Adans., Fam. Pl. 2: 421, 606. Jul-Aug 1763.
Typus: non designatus.

3042 ***Matthiola*** W. T. Aiton, Hort. Kew., ed. 2, 4: 119. Dec 1812 *('Mathiola') (orth. cons.)*.
Typus: *M. incana* (L.) W. T. Aiton (*Cheiranthus incanus* L.) *(typ. cons.)*.

(H) *Matthiola* L., Sp. Pl.: 1192. 1 Mai 1753 [*Rub.*].
Typus: *M. scabra* L.

3050 ***Dontostemon*** Andrz. ex C. A. Mey. in Ledeb., Fl. Altaic. 3: 4, 118. Jul-Dec 1831.
Typus: *D. integrifolius* (L.) C. A. Mey. (*Sisymbrium integrifolium* L.).

3051 ***Chorispora*** R. Br. ex DC. in Mém. Mus. Hist. Nat. 7: 237. 20 Apr 1821.
Typus: *C. tenella* (Pall.) DC. (Syst. Nat. 2: 435. Mai (sero) 1821) (*Raphanus tenellus* Pall.).

(≡) *Chorispermum* W. T. Aiton, Hort. Kew., ed. 2, 4: 129. Dec 1812.

TOVARIACEAE

3081 ***Tovaria*** Ruiz & Pav., Fl. Peruv. Prodr.: 49. Oct (prim.) 1794.
Typus: *T. pendula* Ruiz & Pav. (Fl. Peruv. 3: 73. Aug 1802).

(H) *Tovara* Adans., Fam. Pl. 2: 276, 612. Jul-Aug 1763 [*Polygon.*].
Typus: *T. virginiana* (L.) Raf. (Fl. Tellur. 3: 12. Nov-Dec 1837) (*Polygonum virginianum* L.).

CAPPARACEAE

3087 ***Gynandropsis*** DC., Prodr. 1: 237. Jan (med.) 1824.
Typus: *G. pentaphylla* DC., nom. illeg. (*Cleome pentaphylla* L., nom. illeg., *Cleome gynandra* L., *G. gynandra* (L.) Briq.).

(≡) *Pedicellaria* Schrank in Bot. Mag. (Römer & Usteri) 3(8): 10. Apr 1790.

3103 ***Steriphoma*** Spreng., Syst. Veg. 4(2): 130, 139. Jan-Jun 1827.
Typus: *S. cleomoides* Spreng., nom. illeg. (*Capparis paradoxa* Jacq., *S. paradoxum* (Jacq.) Endl.).

(=) *Hermupoa* Loefl., Iter Hispan.: 307. Dec 1758.
Typus: *H. loeflingiana* DC. (Prodr. 1: 254. Jan (med.) 1824).

3106 ***Boscia*** Lam. ex J. St.-Hil., Expos. Fam. Nat. 2: 3. Feb-Apr 1805.
Typus: *B. senegalensis* (Pers.) Lam. ex Poir. (in Lamarck, Encycl., Suppl. 1: 680. 2 Mai 1811) (*Podoria senegalensis* Pers.) *(typ. cons.)*.

(H) *Boscia* Thunb., Prodr. Pl. Cap.: [x], 32. 1794 [*Rut.*].
Typus: *B. undulata* Thunb.

RESEDACEAE

3122 ***Caylusea*** A. St.-Hil., Deux. Mém. Réséd.: 29. 1837 (sero) - Jan (prim.) 1838.
Typus: *C. canescens* Webb (in Hooker, Niger Fl.: 101. Nov-Dec 1849), non (L.) Walp. 1843 [= *C. hexagyna* (Forssk.) M. L. Green (*Reseda hexagyna* Forssk.)].

3126 ***Oligomeris*** Cambess. in Jacquemont, Voy. Inde 4, Bot.: 23. 1839.
Typus: *O. glaucescens* Cambess.

(=) *Dipetalia* Raf., Fl. Tellur. 3: 73. Nov-Dec 1837.
Typus: *D. capensis* (Burm. f.) Raf. (*Reseda capensis* Burm. f.).

(=) *Ellimia* Nutt. in Torrey & Gray, Fl. N. Amer. 1: 125. Jul 1838.
Typus: *E. ruderalis* Nutt.

SARRACENIACEAE

3131 ***Darlingtonia*** Torr. in Smithsonian Contr. Knowl. 6(4): 4. Apr 1853.
Typus: *D. californica* Torr.

(H) *Darlingtonia* DC. in Ann. Sci. Nat. (Paris) 4: 97. Jan 1825 [*Legum.*].
Typus (vide Pichi Sermolli in Taxon 3: 115. 1954): *D. brachyloba* (Willd.) DC. (*Acacia brachyloba* Willd.).

CRASSULACEAE

3171 ***Rochea*** DC., Pl. Hist. Succ.: ad t. 103. 16 Oct 1802.
Typus: *R. coccinea* (L.) DC. (Pl. Hist. Succ., index) (*Crassula coccinea* L.) *(typ. cons.)*.

(H) *Rochea* Scop., Intr. Hist. Nat.: 296. Jan-Apr 1777 [*Legum.*].
Typus: non designatus.

3172 ***Diamorpha*** Nutt., Gen. N. Amer. Pl. 1: 293. 14 Jul 1818.
Typus: [specimen] N. of Camden, S. Carolina, winter 1816, *Nuttall s.n.* (PH) *(typ. cons.)* [= *D. smallii* Britton].

CEPHALOTACEAE

3176 ***Cephalotus*** Labill., Nov. Holl. Pl. 2: 6. Feb 1806.
Typus: *C. follicularis* Labill.

(H) *Cephalotos* Adans., Fam. Pl. 2: 189, 534. Jul-Aug 1763 [*Lab.*].
Typus: *Thymus cephalotos* L.

SAXIFRAGACEAE

3182 ***Bergenia*** Moench, Methodus: 664. 4 Mai 1794.
Typus: *B. bifolia* Moench, nom. illeg. (*Saxifraga crassifolia* L., *B. crassifolia* (L.) Fritsch).

(H) *Bergena* Adans., Fam. Pl. 2: 345, 525. Jul-Aug 1763 [*Lecythid.*].
≡ *Lecythis* Loefl. 1758.

3185 ***Boykinia*** Nutt. in J. Acad. Nat. Sci. Philadelphia 7: 113. 28 Oct 1834.
Typus: *B. aconitifolia* Nutt.

(H) *Boykiana* Raf., Neogenyton: 2. 1825 [*Lythr.*].
Typus: *B. humilis* (Michx.) Raf. (*Ammannia humilis* Michx.).

3187 ***Suksdorfia*** A. Gray in Proc. Amer. Acad. Arts 15: 41. 1 Oct 1879.
Typus: *S. violacea* A. Gray

(=) *Hemieva* Raf., Fl. Tellur. 2: 70. Jan-Mar 1837.
Typus (vide Rydberg in N. Amer. Fl. 22: 121. 1905): *H. ranunculifolia* (Hook.) Raf. (*Saxifraga ranunculifolia* Hook.).

3196 ***Tolmiea*** Torr. & A. Gray, Fl. N. Amer. 1: 582. Jun 1840.
Typus: *T. menziesii* (Pursh) Torr. & A. Gray (*Tiarella menziesii* Pursh).

(H) *Tolmiea* Hook., Fl. Bor.-Amer. 2: 44. 1834 (sero) [*Eric.*].
Typus: *T. occidentalis* Hook.

(≡) *Leptaxis* Raf., Fl. Tellur. 2: 75. Jan-Mar 1837.

3197 ***Lithophragma*** (Nutt.) Torr. & A. Gray, Fl. N. Amer. 1: 583. Jun 1840 (*Tellima* [unranked] *Lithophragma* Nutt. in J. Acad. Nat. Sci. Philadelphia 7: 26. 28 Oct 1834).
Typus: *Tellima parviflora* Hook. (*L. parviflorum* (Hook.) Torr. & A. Gray) *(typ. cons.)*.

(≡) *Pleurendotria* Raf., Fl. Tellur. 2: 73. Jan-Mar 1837.

3201 ***Vahlia*** Thunb., Nov. Gen. Pl.: 36. 10 Jul 1782.
Typus: *V. capensis* (L. f.) Thunb. (*Russelia capensis* L. f.).

(=) *Bistella* Adans., Fam. Pl. 2: 226, 525. Jul-Aug 1763.
Typus: *B. geminiflora* Delile (in Cailliaud, Voy. Méroé 2: 97. Jan-Jun 1826).

3204 ***Donatia*** J. R. Forst. & G. Forst., Char. Gen. Pl.: 5. 29 Nov 1775.
Typus: *D. fascicularis* J. R. Forst. & G. Forst.

3209 ***Jamesia*** Torr. & A. Gray, Fl. N. Amer. 1: 593. Jun 1840.
Typus: *J. americana* Torr. & A. Gray

(H) *Jamesia* Raf. in Atlantic J. 1: 145. 1832 (sero) [*Legum.*].
Typus: *J. obovata* Raf., nom. illeg. (*Psoralea jamesii* Torr.).

3225 ***Brexia*** Noronha ex Thouars, Gen. Nov. Madagasc.: 20. 17 Nov 1806.
Typus: *B. madagascariensis* (Lam.) Ker Gawl. (in Bot. Reg.: ad t. 730. 1 Aug 1823) (*Venana madagascariensis* Lam.).

(≡) *Venana* Lam., Tabl. Encycl. 2: 99. 6 Nov 1797.

PITTOSPORACEAE

3252 ***Pittosporum*** Banks ex Gaertn., Fruct. Sem. Pl. 1: 286. Dec 1788.
Typus: *P. tenuifolium* Gaertn. *(typ. cons.)*.

(=) *Tobira* Adans., Fam. Pl. 2: 449, 611. Jul-Aug 1763.
Typus: non designatus.

CUNONIACEAE

3269 ***Platylophus*** D. Don in Edinburgh New Philos. J. 9: 92. Apr-Jun 1830.
Typus: *P. trifoliatus* (L. f.) D. Don (*Weinmannia trifoliata* L. f.).

(H) *Platylophus* Cass. in Cuvier, Dict. Sci. Nat. 44: 36. Dec 1826 [*Comp.*].
Typus: *Centaurea nigra* L.

3275 ***Cunonia*** L., Syst. Nat., ed. 10: 1013, 1025, 1368. 7 Jun 1759.
Typus: *C. capensis* L.

(H) *Cunonia* Mill., Fig. Pl. Gard. Dict. 1: 75. 28 Sep 1756 [*Irid.*].
Typus: *C. antholyza* Mill. (Gard. Dict., ed. 8: *Cunonia* No. [1]. 16 Apr 1768) (*Antholyza cunonia* L.).

3276 ***Weinmannia*** L., Syst. Nat., ed. 10: 997, 1005, 1367. 7 Jun 1759.
Typus: *W. pinnata* L.
(≡) *Windmannia* P. Browne, Civ. Nat. Hist. Jamaica: 212. 10 Mar 1756.

3277 ***Pancheria*** Brongn. & Gris in Bull. Soc. Bot. France 9: 74. 1862.
Typus: *P. elegans* Brongn. & Gris
(H) *Panchezia* Montrouz. in Mém. Acad. Roy. Sci. Lyon, Sect. Sci. 10: 223. 1860 [*Rub.*].
Typus: *P. collina* Montrouz.

BRUNIACEAE

3284 ***Thamnea*** Sol. ex Brongn. in Ann. Sci. Nat. (Paris) 8: 386. Aug 1826.
Typus: *T. uniflora* Sol. ex Brongn.
(H) *Thamnia* P. Browne, Civ. Nat. Hist. Jamaica: 245. 10 Mar 1756 [*Flacourt.*].
Typus: *Laetia thamnia* L. (Pl. Jamaic. Pug.: 31. 28 Nov 1759).

3285 ***Tittmannia*** Brongn. in Ann. Sci. Nat. (Paris) 8: 385. Aug 1826.
Typus: *T. lateriflora* Brongn.
(H) *Tittmannia* Rchb., Iconogr. Bot. Exot. 1: 26. Jan-Jun 1824 [*Scrophular.*].
Typus: *T. viscosa* (Hornem.) Rchb. (*Gratiola viscosa* Hornem.).

3286 ***Lonchostoma*** Wikstr. in Kongl. Vetensk. Acad. Handl. 1818: 350. 1818.
Typus: *L. obtusiflorum* Wikstr., nom. illeg. (*Passerina pentandra* Thunb., *L. pentandrum* (Thunb.) Druce).
(=) *Ptyxostoma* Vahl in Skr. Naturhist.-Selsk. 6: 95. 1810.
Typus: non designatus.

3292 ***Brunia*** Lam., Encycl. 1: 474. 1 Aug 1785.
Typus: *B. paleacea* Bergius (Descr. Pl. Cap.: 56. Sep 1767).
(H) *Brunia* L., Sp. Pl.: 199. 1 Mai 1753 [*Brun.*].
Typus: *B. lanuginosa* L.

ROSACEAE

3316 ***Physocarpus*** (Camb.) Raf., New Fl. 3: 73. Jan-Mar 1838 *(‘Physocarpa’)* *(Spiraea* sect. *Physocarpus* Camb. in Ann. Sci. Nat. (Paris) 1: 239, 385. 1824, *‘Physocarpos’) (orth. cons.).*
Typus: *Spiraea opulifolia* L. (*P. opulifolius* (L.) Maxim.).
(=) *Epicostorus* Raf. in Atlantic J. 1: 144. 1832 (sero).
Typus: *E. montanus* Raf., nom. illeg. (*Spiraea monogyna* Torr.).

3323 ***Sorbaria*** (Ser. ex DC.) A. Braun in Ascherson, Fl. Brandenburg 1: 177. Jan 1860 (*Spiraea* sect. *Sorbaria* Ser. ex DC., Prodr. 2: 545. Nov (med.) 1825).
Typus: *Spiraea sorbifolia* L. (*Sorbaria sorbifolia* (L.) A. Braun).
(≡) *Schizonotus* Lindl., Intr. Nat. Syst., Bot.: 81. Sep 1830.

3328 ***Lindleya*** Kunth in Humboldt & al., Nov. Gen. Sp. 6, ed. 4°: 239 [& ed. f°: 188]. 5 Jan 1824.
Typus: *L. mespiloides* Kunth

(H) *Lindleya* Nees in Flora 4: 299. 21 Mai 1821 [*The.*].
≡ *Wikstroemia* Schrad., 5 Mai 1821 (*nom. rej.* sub 5446).

3332 ***Holodiscus*** (C. Koch) Maxim. in Trudy Imp. S.-Peterburgsk. Bot. Sada 6: 253. Jul-Dec 1879 (*Spiraea* "Gruppe" *Holodiscus* K. Koch, Dendrologie 1: 309. Jan 1869).
Typus: *Spiraea discolor* Pursh (*H. discolor* (Pursh) Maxim.) *(typ. cons.)*.

3336a ***Chaenomeles*** Lindl. in Trans. Linn. Soc. London 13: 96, 97. 23 Mai - 21 Jun 1821 *('Choenomeles') (orth. cons.)*.
Typus: *C. japonica* (Thunb.) Lindl. ex Spach (Hist. Nat. Vég. 2: 159. 12 Jul 1834) (*Pyrus japonica* Thunb.).

3338a ***Aronia*** Medik., Philos. Bot. 1: 155. Apr 1789.
Typus: *A. arbutifolia* (L.) Pers. (Syn. Pl. 2: 39. Nov 1806) (*Mespilus arbutifolia* L.).

(H) *Aronia* Mitch., Diss. Princ. Bot.: 28. 1769 [*Ar.*].
≡ *Orontium* L. 1753.

3339 ***Rhaphiolepis*** Lindl. in Bot. Reg.: ad t. 468. 1 Jul 1820 *('Raphiolepis') (orth. cons.)*.
Typus: *R. indica* (L.) Lindl. (*Crataegus indica* L.).

(=) *Opa* Lour., Fl. Cochinch.: 304, 308. Sep 1790.
Typus (vide McVaugh in Taxon 5: 144. 1956): *O. metrosideros* Lour.

3377 ***Aremonia*** Neck. ex Nestl., Monogr. Potentilla: iv, 17. Jun 1816.
Typus: *A. agrimonoides* (L.) DC. (Prodr. 2: 588. Nov (med.) 1825) (*Agrimonia agrimonoides* L.).

(≡) *Agrimonoides* Mill., Gard. Dict. Abr., ed. 4: [42]. 28 Jan 1754.

CONNARACEAE

3424 ***Rourea*** Aubl., Hist. Pl. Guiane: 467. Jun-Dec 1775.
Typus: *R. frutescens* Aubl.

3424a ***Santaloides*** G. Schellenb., Beitr. Anat. Syst. Connar.: 38. 1910.
Typus: *S. minor* (Gaertn.) G. Schellenb. (*Aegiceras minus* Gaertn.) *(typ. cons.)*.

(H) *Santalodes* Kuntze, Revis. Gen. Pl. 1: 155. 5 Nov 1891 [*Connar.*].
Typus: *S. hermanniana* Kuntze, nom. illeg. (*Connarus santaloides* Vahl).
(≡) *Kalawael* Adans., Fam. Pl. 2: 344, 530. Jul-Aug 1763.

LEGUMINOSAE (FABACEAE)

3441 ***Pithecellobium*** Mart. in Flora 20 (2, Beibl.): 114. 21 Oct 1837 *('Pithecollobium') (orth. cons.).*
Typus: *P. unguis-cati* (L.) Benth. (in London J. Bot. 3: 200. 1844) (*Mimosa unguis-cati* L.) *(typ. cons.).*

(=) *Zygia* P. Browne, Civ. Nat. Hist. Jamaica: 279. 10 Mar 1756.
Typus (vide Fawcett & Rendle, Fl. Jamaica 4: 150. 1920): *Z. latifolia* (L.) Fawc. & Rendle (Fl. Jamaica 4: 150. Jul 1920) (*Mimosa latifolia* L.).

3444 ***Calliandra*** Benth. in J. Bot. (Hooker) 2: 138. Apr 1840.
Typus: *C. houstonii* Benth., nom. illeg. (*Mimosa houstoniana* Mill., *C. houstoniana* (Mill.) Standl.) *(typ. cons.).*

3448 ***Schrankia*** Willd., Sp. Pl. 4: 888, 1041. Apr 1806.
Typus: *S. aculeata* Willd., nom. illeg. (*Mimosa quadrivalvis* L., *S. quadrivalvis* (L.) Merr.).

(H) *Schranckia* J. F. Gmel., Syst. Nat. 2: 312, 515. Sep (sero) - Nov 1791 [*Celastr.*].
Typus: *S. quinquefaria* J. F. Gmel.

3450 ***Desmanthus*** Willd., Sp. Pl. 4: 888, 1044. Apr 1806.
Typus: *D. virgatus* (L.) Willd. (*Mimosa virgata* L.).

(≡) *Acuan* Medik., Theodora: 62. 1786.

3452 ***Dichrostachys*** (DC.) Wight & Arn., Prodr. Fl. Ind. Orient.: 271. Oct (prim.) 1834 (*Desmanthus* sect. *Dichrostachys* DC., Prodr. 2: 445. Nov (med.) 1825).
Typus: *Desmanthus cinereus* (L.) Willd. (*Mimosa cinerea* L., *Dichrostachys cinerea* (L.) Wight & Arn.).

(=) *Cailliea* Guill. & Perr. in Guillemin & al., Fl. Seneg. Tent.: 239. 2 Jul 1832.
Typus: *C. dicrostachys* Guill. & Perr., nom. illeg. (*Mimosa nutans* Pers.).

3466 ***Plathymenia*** Benth. in J. Bot. (Hooker) 2: 134. Apr 1840.
Typus: *P. foliolosa* Benth.

(=) *Echyrospermum* Schott in Schreibers, Nachr. Österr. Naturf. Bras. 2, App.: 55. 1822.
Typus: non designatus.

3468 ***Entada*** Adans., Fam. Pl. 2: 318, 554. Jul-Aug 1763.
Typus: *E. rheedei* Spreng. (Syst. Veg. 2: 325. Jan-Mai 1825) (*Mimosa entada* L.).

(=) *Gigalobium* P. Browne, Civ. Nat. Hist. Jamaica: 362. 10 Mar 1756.
Typus (vide Panigrahi in Taxon 34: 714. 1985): *Entada gigas* (L.) Fawc. & Rendle (*Mimosa gigas* L.).

3490 ***Copaifera*** L., Sp. Pl., ed. 2: 557. Sep 1762.
Typus: *C. officinalis* (Jacq.) L. (*Copaiva officinalis* Jacq.).

(≡) *Copaiva* Jacq., Enum. Syst. Pl.: 4, 21. Sep-Nov 1760.

(=) *Copaiba* Mill., Gard. Dict. Abr., ed. 4: [371]. 28 Jan 1754.
Typus: non designatus.

3495 ***Crudia*** Schreb., Gen. Pl.: 282. Apr 1789.
Typus: *C. spicata* (Aubl.) Willd. (Sp. Pl. 2: 539. Mar 1799) (*Apalatoa spicata* Aubl.) *(typ. cons.)*.

(≡) *Apalatoa* Aubl., Hist. Pl. Guiane: 382. Jun-Dec 1775.
(=) *Touchiroa* Aubl., Hist. Pl. Guiane: 384. 1775.
Typus: *T. aromatica* Aubl.

3500 ***Peltogyne*** Vogel in Linnaea 11: 410. Apr-Jul 1837.
Typus: *P. discolor* Vogel

(=) *Orectospermum* Schott in Schreibers, Nachr. Österr. Naturf. Bras. 2, App.: 54. 1822.
Typus: non designatus.

3506 ***Schotia*** Jacq. in Collect. Bot. Spectantia (Vienna) 1: 93. Jan-Sep 1787.
Typus: *S. speciosa* Jacq., nom. illeg. (*Guajacum afrum* L., *S. afra* (L.) Thunb.).

(≡) *Theodora* Medik., Theodora: 16. 1786.

3509 ***Afzelia*** Sm. in Trans. Linn. Soc. London 4: 221. 24 Mai 1798.
Typus: *A. africana* Sm. ex Pers. (Syn. Pl. 1: 455. 1 Apr - 15 Jun 1805).

(H) *Afzelia* J. F. Gmel., Syst. Nat. 2: 927. Apr (sero) - Oct 1792 [*Scrophular.*].
≡ *Seymeria* Pursh 1814 *(nom. cons.)* (7602).

3516 ***Berlinia*** Sol. ex Hook. f. in Hooker, Niger Fl.: 326. Nov-Dec 1849.
Typus: *B. acuminata* Sol. ex Hook. f.

(=) *Westia* Vahl in Skr. Naturhist.-Selsk. 6: 117. 1810.
Typus: non designatus.

3517 ***Macrolobium*** Schreb., Gen. Pl.: 30. Apr 1789.
Typus: *M. vuapa* J. F. Gmel. (Syst. Nat. 2: 93. Sep (sero) - Nov 1791), nom. illeg. (*Vouapa bifolia* Aubl., *M. bifolium* (Aubl.) Pers.) *(typ. cons.)*.

(≡) *Vouapa* Aubl., Hist. Pl. Guiane: 25. Jun-Dec 1775.
(=) *Outea* Aubl., Hist. Pl. Guiane: 28. Jun-Dec 1775.
Typus: *O. guianensis* Aubl.

3518 ***Humboldtia*** Vahl, Symb. Bot. 3: 106. 1794.
Typus: *H. laurifolia* Vahl

(H) *Humboltia* Ruiz & Pav., Fl. Peruv. Prodr.: 121. Oct (prim.) 1794 [*Orchid.*].
Typus (vide Garay & Sweet in J. Arnold Arbor. 53: 522. 1972): *H. purpurea* Ruiz & Pav. (Syst. Veg. Fl. Peruv. Chil.: 235. Dec (sero) 1798).

3524 ***Brownea*** Jacq., Enum. Syst. Pl.: 6, 26. Aug-Sep 1760 *('Brownaea') (orth. cons.)*.
Typus: *B. coccinea* Jacq.

(=) *Hermesias* Loefl., Iter Hispan.: 278. Dec 1758.
Typus: non designatus.

3528 ***Piliostigma*** Hochst. in Flora 29: 598. 14 Oct 1846.
Typus: *P. reticulatum* (DC.) Hochst. (*Bauhinia reticulata* DC.) *(typ. cons.)*.

(=) *Elayuna* Raf., Sylva Tellur.: 145. Oct-Dec 1838.
Typus: *E. biloba* Raf., nom. illeg. (*Bauhinia tamarindacea* Delile).

3532 ***Apuleia*** Mart. in Flora 20(2, Beibl.): 123. 21 Nov 1837 *('Apuleja') (orth. cons.)*.
Typus: *A. praecox* Mart.

(H) *Apuleja* Gaertn., Fruct. Sem. Pl. 2: 439. Sep-Dec 1791 [*Comp.*].
≡ *Berkheya* Ehrh. 1788 *(nom. cons.)* (9438).

3536 ***Cassia*** L., Sp. Pl.: 376. 1 Mai 1753.
Typus: *C. fistula* L. *(typ. cons.)*.

3553 ***Pterolobium*** R. Br. ex Wight & Arn., Prodr. Fl. Ind. Orient.: 283. Oct (prim.) 1834.
Typus: *P. lacerans* (Roxb.) Wight & Arn. (*Caesalpinia lacerans* Roxb.).

(H) *Pterolobium* Andrz. ex C. A. Mey., Verz. Pfl. Casp. Meer.: 185. Nov-Dec 1831 [*Cruc.*].
Typus: *P. biebersteinii* Andrz. ex C. A. Mey., nom. illeg. (*Thlaspi latifolium* M. Bieb.).

(=) *Cantuffa* J. F. Gmel., Syst. Nat. 2: 677. Sep (sero) - Nov 1791.
Typus: *C. excelsa* J. F. Gmel.

3557 ***Hoffmannseggia*** Cav., Icon. 4: 63. 14 Mai 1798 *('Hoffmanseggia') (orth. cons.)*.
Typus: *H. falcaria* Cav., nom. illeg. (*Larrea glauca* Ortega, *H. glauca* (Ortega) Eifert) (etiam vide 3973).

3558 ***Zuccagnia*** Cav., Icon. 5: 2. Jun-Sep 1799.
Typus: *Z. punctata* Cav.

(H) *Zuccangnia* Thunb., Nov. Gen. Pl.: 127. 17 Dec 1798 [*Lil.*].
Typus: *Z. viridis* (L.) Thunb. (*Hyacinthus viridis* L.).

3561 ***Peltophorum*** (Vogel) Benth. in J. Bot. (Hooker) 2: 75. Mar 1840 (*Caesalpinia* sect. *Peltophorum* Vogel in Linnaea 11: 406. Apr-Jul 1837).
Typus: *Caesalpinia dubia* Spreng. (*P. dubium* (Spreng.) Taub.).

(=) *Baryxylum* Lour., Fl. Cochinch.: 257, 266. Sep 1790.
Typus: *B. rufum* Lour.

3574 ***Swartzia*** Schreb., Gen. Pl.: 518. Mai 1791.
Typus: *S. alata* Willd. (Sp. Pl. 2: 1220. Dec 1799).

(=) *Possira* Aubl., Hist. Pl. Guiane: 934. Jun-Dec 1775.
Typus: *P. arborescens* Aubl.

(=) *Tounatea* Aubl., Hist. Pl. Guiane: 549. Jun-Dec 1775.
Typus: *T. guianensis* Aubl.

3575 ***Aldina*** Endl., Gen. Pl.: 1322. Oct 1840.
Typus: *A. insignis* (Benth.) Endl. (in Walpers, Repert. Bot. Syst. 1: 843. 26-29 Jan 1843) (*Allania insignis* Benth.).

(H) *Aldina* Adans., Fam. Pl. 2: 328, 514. Jul-Aug 1763 [*Legum.*].
≡ *Brya* P. Browne 1756.

3582 ***Sweetia*** Spreng., Syst. Veg. 2: 171, 213. Jan-Mai 1825.
Typus: *S. fruticosa* Spreng.

3584 ***Myroxylon*** L. f., Suppl. Pl.: 34, 233. Apr 1782.
Typus: *M. peruiferum* L. f.

(H) *Myroxylon* J. R. Forst. & G. Forst., Char. Gen. Pl.: 63. 29 Nov 1775 [*Flacourt.*].
≡ *Xylosma* G. Forst. 1786 *(nom. cons.)* (5320).

(=) *Toluifera* L., Sp. Pl.: 384. 1 Mai 1753.
Typus: *T. balsamum* L.

3589 ***Camoënsia*** Welw. ex Benth. & Hook. f., Gen. Pl. 1: 456, 557. 19 Oct 1865.
Typus: *C. maxima* Welw. ex Benth. (in Trans. Linn. Soc. London 25: 301. 30 Nov 1865) *(typ. cons.)*.

(=) *Giganthemum* Welw. in Ann. Cons. Ultramarino, ser. 1, 1858: 585. Dec 1859.
Typus: *G. scandens* Welw.

3597 ***Ormosia*** Jacks. in Trans. Linn. Soc. London 10: 360. 7 Sep 1811.
Typus: *O. coccinea* (Aubl.) Jacks. (*Robinia coccinea* Aubl.) *(typ. cons.)*.

(=) *Toulichiba* Adans., Fam. Pl. 2: 326, 612. Jul-Aug 1763.
Typus: non designatus.

3608 ***Virgilia*** Poir. in Lamarck, Encycl. 8: 677. 22 Aug 1808.
Typus: *V. capensis* (L.) Poir. (*Sophora capensis* L.) *(typ. cons.)*.

(H) *Virgilia* L'Hér., Virgilia: ad t. [1]. Jan-Jun 1788 [*Comp.*].
Typus: *V. helioides* L'Hér.

3619 ***Pickeringia*** Nutt. in Torrey & Gray, Fl. N. Amer. 1: 388. Jun 1840.
Typus: *P. montana* Nutt.

(H) *Pickeringia* Nutt. in J. Acad. Nat. Sci. Philadelphia 7: 95. 28 Oct 1834 [*Myrsin.*].
Typus: *P. paniculata* (Nutt.) Nutt. (*Cyrilla paniculata* Nutt.).

3621 ***Podalyria*** Willd., Sp. Pl. 2: 492, 501. Mar 1799.
Typus: *P. retzii* (J. F. Gmel.) Rickett & Stafleu (*Sophora retzii* J. F. Gmel., *Sophora biflora* Retz. 1799, non L. 1759) *(typ. cons.)*.

3624 ***Oxylobium*** Andrews in Bot. Repos.: ad t. 492. Nov 1807.
Typus: *O. cordifolium* Andrews

(=) *Callistachys* Vent., Jard. Malmaison: ad t. 115. Nov 1805.
Typus: *C. lanceolata* Vent.

3629 ***Burtonia*** R. Br. in Aiton, Hort. Kew., ed. 2, 3: 12. Oct-Nov 1811.
Typus: *B. scabra* (Sm.) R. Br. (*Gompholobium scabrum* Sm.).

(H) *Burtonia* Salisb., Parad. Lond.: ad t. 73. 1 Jun 1807 [*Dillen.*].
Typus: *B. grossulariifolia* Salisb.

3647 ***Walpersia*** Harv. in Harvey & Sonder, Fl. Cap. 2: 26. 16-31 Oct 1862.
Typus: *W. burtonioides* Harv.

(H) *Walpersia* Reissek ex Endl., Gen. Pl.: 1100. Apr 1840 [*Rhamn.*].
≡ *Trichocephalus* Brongn. 1827.

3657 ***Lotononis*** (DC.) Eckl. & Zeyh., Enum. Pl. Afric. Austral.: 176. Jan 1836 (*Ononis* sect. *Lotononis* DC., Prodr. 2: 166. Nov (med.) 1825).
Typus: *Crotalaria vexillata* E. Mey. (*L. vexillata* (E. Mey.) Eckl. & Zeyh.) *(typ. cons.)* [= *L. prostata* (L.) Benth. (*Ononis prostrata* L.)].

(=) *Amphinomia* DC., Prodr. 2: 522. Nov (med.) 1825.
Typus: *A. decumbens* (Thunb.) DC. (*Connarus decumbens* Thunb.).

(=) *Leobordea* Delile in Laborde, Voy. Arabie Pétrée: 82, 86. 1830.
Typus: *L. lotoidea* Delile

3659 ***Rothia*** Pers., Syn. Pl. 2: 638, [659]. Sep 1807.
Typus: *R. trifoliata* (Roth) Pers. (*Dillwynia trifoliata* Roth) *(typ. cons.)*.

(H) *Rothia* Schreb., Gen. Pl.: 531. Mai 1791 [*Comp.*].
Typus: non designatus.

3661 ***Wiborgia*** Thunb., Nov. Gen. Pl.: 137. 3 Jun 1800.
Typus: *W. obcordata* (Berg.) Thunb. (*Crotalaria obcordata* Berg.) *(typ. cons.)*.

(H) *Viborgia* Moench, Methodus: 132. 4 Mai 1794 [*Legum.*].
Typus: non designatus.

3673 ***Argyrolobium*** Eckl. & Zeyh., Enum. Pl. Afric. Austral.: 184. Jan 1836.
Typus: *A. argenteum* (Jacq.) Eckl. & Zeyh. (*Crotalaria argentea* Jacq.) *(typ. cons.)*.

(=) *Lotophyllus* Link, Handbuch 2: 156. Jan-Aug 1831.
Typus: *L. argenteus* Link

3675a ***Retama*** Raf., Sylva Tellur.: 22. Oct-Dec 1838.
Typus: *R. monosperma* (L.) Boiss. (Voy. Bot. Espagne 2: 144. 10 Feb 1840) (*Spartium monospermum* L.).

(≡) *Lygos* Adans., Fam. Pl. 2: 321, 573. Jul-Aug 1763.

3676 ***Petteria*** C. Presl in Abh. Königl. Böhm. Ges. Wiss., ser. 5, 3: 569. Jul-Dec 1845.
Typus: *P. ramentacea* (Sieber) C. Presl (*Cytisus ramentaceus* Sieber).

(H) *Pettera* Rchb., Icon. Fl. Germ. Helv. 5: 33. Mar 1841-Aug 1842 [*Caryophyll.*].
Typus: *P. graminifolia* (Ard.) Rchb. (*Arenaria graminifolia* Ard.).

3682 ***Cytisus*** Desf., Fl. Atlant. 2: 139. Nov 1798.
Typus: *C. triflorus* L'Hér. 1791, non Lam. 1786 *(typ. cons.)* [= *C. villosus* Pourr.].

(H) *Cytisus* L., Sp. Pl.: 739. 1 Mai 1753 [*Legum.*].
Typus (vide Green in Sprague & al., Nom. Prop. Brit. Bot.: 175. 1929): *C. sessilifolius* L.

3682a ***Sarothamnus*** Wimm., Fl. Schles.: 278. Feb-Jul 1832.
Typus: *S. vulgaris* Wimm., nom. illeg. (*Spartium scoparium* L., *Sarothamnus scoparius* (L.) W. D. J. Koch).

(≡) *Cytisogenista* Duhamel, Traité Arbr. Arbust. 1: 203. 1755.

3688 ***Medicago*** L., Sp. Pl.: 778. 1 Mai 1753.
Typus: *M. sativa* L. *(typ. cons.)*.

3693 ***Hymenocarpos*** Savi, Fl. Pis. 2: 205. 1798.
Typus: *H. circinnatus* (L.) Savi (*Medicago circinnata* L.).

(≡) *Circinnus* Medik. in Vorles. Churpfälz. Phys.-Öcon. Ges. 2: 384. 1787.

3694 ***Securigera*** DC. in Lamarck & Candolle, Fl. Franç., ed. 3, 4: 609. 17 Sep 1805.
Typus: *S. coronilla* DC., nom. illeg. (*Coronilla securidaca* L., *S. securidaca* (L.) Degen & Dörfl.).

(≡) *Bonaveria* Scop., Intr. Hist. Nat.: 310. Jan-Apr 1777.

3699 ***Tetragonolobus*** Scop., Fl. Carn., ed. 2, 2: 87, 507. Jan-Aug 1772.
Typus: *T. scandalida* Scop., nom. illeg. (*Lotus siliquosus* L., *T. siliquosus* (L.) Roth).

(≡) *Scandalida* Adans., Fam. Pl. 2: 326, 602. Jul-Aug 1763.

3708 ***Eysenhardtia*** Kunth in Humboldt & al., Nov. Gen. Sp. 6, ed. 4°: 489 [& ed. f°: 382]. Sep 1824.
Typus: *E. amorphoides* Kunth

(=) *Viborquia* Ortega, Nov. Pl. Descr. Dec.: 66. 1798.
Typus: *V. polystachya* Ortega

3709 ***Dalea*** L., Opera Var.: 244. 1758.
Typus: *D. cliffortiana* Willd. (*Psoralea dalea* L.).

(H) *Dalea* Mill., Gard. Dict. Abr., ed. 4: [433]. 28 Jan 1754 [*Solan.*].
≡ *Browallia* L. 1753.

3710 ***Petalostemon*** Michx., Fl. Bor.-Amer. 2: 48. 19 Mar 1803 *('Petalostemum')* *(orth. cons.)*.
Typus: *P. candidus* (Willd.) Michx. (*Dalea candida* Willd.).

(=) *Kuhnistera* Lam., Encycl. 3: 370. 13 Feb 1792.
Typus: *K. caroliniensis* Lam.

3718 ***Tephrosia*** Pers., Syn. Pl. 2: 328. Sep 1807.
Typus: *T. villosa* (L.) Pers. (*Cracca villosa* L.) (etiam vide 3745).

(=) *Erebinthus* Mitch., Diss. Princ. Bot.: 32. 1769.
Typus (vide Wood in Rhodora 51: 292. 1948): *Tephrosia spicata* (Walter) Torr. & A. Gray (*Galega spicata* Walter).

(=) *Needhamia* Scop., Intr. Hist. Nat.: 310. Jan-Apr 1777.
Typus: *Vicia littoralis* Jacq.

(=) *Reineria* Moench, Suppl. Meth.: 44. 2 Mai 1802.
Typus: *R. reflexa* Moench

3720 ***Millettia*** Wight & Arn., Prodr. Fl. Ind. Orient: 263. Oct (prim.) 1834.
Typus: *M. rubiginosa* Wight & Arn.

(=) *Pongamia* Adans., Fam. Pl. 2: 322, 593. Jul-Aug 1763 *(nom. cons.)* (3836).

3722 ***Wisteria*** Nutt., Gen. N. Amer. Pl. 2: 115. 14 Jul 1818.
Typus: *W. speciosa* Nutt., nom. illeg. (*Glycine frutescens* L., *W. frutescens* (L.) Poir.).

(≡) *Phaseoloides* Duhamel, Traité Arbr. Arbust. 2: 115. 1755.

(=) *Diplonyx* Raf., Fl. Ludov.: 101. Oct-Dec (prim.) 1817.
Typus: *D. elegans* Raf.

3745 ***Cracca*** Benth. in Vidensk. Meddel. Dansk Naturhist. Foren. Kjøbenhavn 1853: 8. 1853.
Typus: *C. glandulifera* Benth.

(H) *Cracca* L., Sp. Pl.: 752. 1 Mai 1753 [*Legum.*].
≡ *Tephrosia* Pers. 1807 *(nom. cons.)* (3718).

3747 ***Sesbania*** Scop., Intr. Hist. Nat.: 308. Jan-Apr 1777.
Typus: *S. sesban* (L.) Merr. (in Philipp. J. Sci. 7 (Bot.): 235. 1912 (*Aeschynomene sesban* L.).

(≡) *Sesban* Adans., Fam. Pl. 2: 327, 604. Jul-Aug 1763.

(=) *Agati* Adans., Fam. Pl. 2: 326, 513. Jul-Aug 1763.
Typus: *Robinia grandiflora* L.

3753 ***Clianthus*** Sol. ex Lindl. in Edwards's Bot. Reg.: ad t. 1775. 1 Jul 1835.
Typus: *C. puniceus* (G. Don) Sol. ex Lindl. (*Donia punicea* G. Don).

(=) *Sarcodum* Lour., Fl. Cochinch.: 425, 461. Sep 1790.
Typus: *S. scandens* Lour.

3754 ***Sutherlandia*** R. Br. in Aiton, Hort. Kew., ed. 2, 4: 327. Dec 1812.
Typus: *S. frutescens* (L.) R. Br. (*Colutea frutescens* L.).

(H) *Sutherlandia* J. F. Gmel., Syst. Nat. 2: 998, 1027. Apr (sero) - Oct 1792 [*Stercul.*].
≡ *Heritiera* Aiton 1789.

3756 ***Lessertia*** DC., Astragalogia, ed. 4°: 5, 19, 47 [& ed. f°: 4, 15, 37]. 15 Nov 1802.
Typus: *L. perennans* (Jacq.) DC. (*Colutea perennans* Jacq.).

(≡) *Sulitra* Medik. in Vorles. Churpfälz. Phys.-Öcon. Ges. 2: 366. 1787 (typ. des.: Brummitt in Regnum Veg. 40: 24. 1965).

(=) *Coluteastrum* Fabr., Enum., ed. 2: 317. Sep-Dec 1763.
Typus: *C. herbaceum* (L.) Kuntze (Revis. Gen. Pl. 1: 171. 5 Nov 1891) (*Colutea herbacea* L.).

3767 ***Oxytropis*** DC., Astragalogia, ed. 4°: 66 [& ed. f°: 53]. 15 Nov 1802.
Typus: *O. montana* (L.) DC. (*Astragalus montanus* L.) *(typ. cons.)*.

3784 ***Nissolia*** Jacq., Enum. Syst. Pl.: 7, 27. Aug-Sep 1760.
Typus: *N. fruticosa* Jacq. *(typ. cons.)*.

(H) *Nissolia* Mill., Gard. Dict. Abr., ed. 4: [954]. 28 Jan 1754 [*Legum.*].
Typus: non designatus.

3789 ***Poiretia*** Vent. in Mém. Cl. Sci. Math. Inst. Natl. France 1807(1): 4. Jul 1807.
Typus: *P. scandens* Vent.

(H) *Poiretia* J. F. Gmel., Syst. Nat. 2: 213, 263. Sep (sero) - Nov 1791 [*Rub.*].
Typus: non designatus.

3792 ***Ormocarpum*** P. Beauv., Fl. Oware 1: 95. 23 Feb 1807.
Typus: *O. verrucosum* P. Beauv.

(=) *Diphaca* Lour., Fl. Cochinch.: 424, 453. Sep 1790.
Typus: *D. cochinchinensis* Lour.

3796 ***Smithia*** Aiton, Hort. Kew. 3: 496. 7 Aug - 1 Oct 1789.
Typus: *S. sensitiva* Aiton

(H) *Smithia* Scop., Intr. Hist. Nat.: 322. Jan-Apr 1777 [*Clus.*].
≡ *Quapoya* Aubl. 1775.

(≡) *Damapana* Adans., Fam. Pl. 2: 323, 548. Jul-Aug 1763.

3800 ***Adesmia*** DC. in Ann. Sci. Nat. (Paris) 4: 94. Jan 1825.
Typus: *A. muricata* (Jacq.) DC. (*Hedysarum muricatum* Jacq.) *(typ. cons.)*.

(≡) *Patagonium* Schrank in Denkschr. Königl. Akad. Wiss. München 1808: 93. 1809.

3807 ***Desmodium*** Desv. in J. Bot. Agric. 1: 122. Feb 1813.
Typus: *D. scorpiurus* (Sw.) Desv. (*Hedysarum scorpiurus* Sw.) *(typ. cons.)*.

(=) *Meibomia* Heist. ex Fabr., Enum.: 168. 1759.
Typus: *Hedysarum canadense* L.

(=) *Grona* Lour., Fl. Cochinch.: 424, 459. Sep 1790.
Typus: *G. repens* Lour.

(=) *Pleurolobus* J. St.-Hil. in Nouv. Bull. Sci. Soc. Philom. Paris 3: 192. Dec 1812.
Typus: non designatus.

3810 ***Alysicarpus*** Desv. in J. Bot. Agric. 1: 120. Feb 1813.
Typus: *A. bupleurifolius* (L.) DC. (Prodr. 2: 352. Nov (med.) 1825) (*Hedysarum bupleurifolium* L.) *(typ. cons.)*.

3821 ***Dalbergia*** L. f., Suppl. Pl.: 52, 316. Apr 1782.
Typus: *D. lanceolaria* L. f.

(=) *Amerimnon* P. Browne, Civ. Nat. Hist. Jamaica: 288. 10 Mar 1756.
Typus: *A. brownei* Jacq. (Enum. Syst. Pl.: 27. Aug-Sep 1760).
(=) *Ecastaphyllum* P. Browne, Civ. Nat. Hist. Jamaica: 299. 10 Mar 1756.
Typus: *E. brownei* Pers. (Syn. Pl. 2: 277. Sep 1807) (*Hedysarum ecastaphyllum* L.).
(=) *Acouba* Aubl., Hist. Pl. Guiane: 753. Jun-Dec 1775.
Typus: *A. violacea* Aubl.

3823 ***Machaerium*** Pers., Syn. Pl. 2: 276. Sep 1807.
Typus: *M. ferrugineum* Pers. (*Nissolia ferruginea* Willd.).

(=) *Nissolius* Medik. in Vorles. Churpfälz. Phys.-Öcon. Ges. 2: 389. 1787.
Typus: *N. arboreus* (Jacq.) Medik. (*Nissolia arborea* Jacq.).
(=) *Quinata* Medik. in Vorles. Churpfälz. Phys.-Öcon. Ges. 2: 389. 1787.
Typus: *Q. violacea* Medik. (*Nissolia quinata* Aubl.).

3828 ***Pterocarpus*** Jacq., Sel. Stirp. Amer. Hist.: 283. 5 Jan 1763.
Typus: *P. officinalis* Jacq. *(typ. cons.)*.

(H) *Pterocarpus* L., Herb. Amb.: 10. 11 Mai 1754 [*Legum.*].
Typus: non designatus.

3834 ***Lonchocarpus*** Kunth in Humboldt & al., Nov. Gen. Sp. 6, ed. f°: 300. Apr 1824.
Typus: *L. sericeus* (Poir.) DC. (Prodr. 2: 260. Nov (med.) 1825) (*Robinia sericea* Poir.) *(typ. cons.)*.

(=) *Clompanus* Aubl., Hist. Pl. Guiane: 773. Jun-Dec 1775.
Typus: *C. paniculata* Aubl.
(=) *Coublandia* Aubl., Hist. Pl. Guiane 937. Jun-Dec 1775 (*nom. rej.* sub 3837).
Typus: *C. frutescens* Aubl.
(=) *Muellera* L. f., Suppl. Pl.: 53, 329. Apr 1782 *(nom. cons.)* (3837).

3836 ***Pongamia*** Adans., Fam. Pl. 2: 322, 593. Jul-Aug 1763 *('Pongam') (orth. cons.)*.
Typus: *P. pinnata* (L.) Pierre (Fl. Forest. Cochinch.: ad t. 385. 15 Apr 1899) (*Cytisus pinnatus* L.) *(typ. cons.)* (etiam vide 3720).

3837 ***Muellera*** L. f., Suppl. Pl.: 53, 329. Apr 1782.
Typus: *M. moniliformis* L. f. (etiam vide 3834).

(=) *Coublandia* Aubl., Hist. Pl. Guiane: 937. Jun-Dec 1775 (*nom. rej.* sub 3834).
Typus: *C. frutescens* Aubl.

3838 ***Derris*** Lour., Fl. Cochinch.: 423, 432. Sep 1790.
Typus: *D. trifoliata* Lour. *(typ. cons.)*.

(=) *Salken* Adans., Fam. Pl. 2: 322, 600. Jul-Aug 1763.
Typus: non designatus.
(=) *Solori* Adans., Fam. Pl. 2: 327, 606. Jul-Aug 1763.
Typus: non designatus.
(=) *Deguelia* Aubl., Hist. Pl. Guiane: 750. Jun-Dec 1775.
Typus: *D. scandens* Aubl.

3839 ***Piscidia*** L., Syst. Nat., ed. 10: 1151, 1155, 1376. 7 Jun 1759.
Typus: *P. erythrina* L., nom. illeg. (*Erythrina piscipula* L., *P. piscipula* (L.) Sarg.).

(≡) *Ichthyomethia* P. Browne, Civ. Nat. Hist. Jamaica: 296. 10 Mar 1756.

3841 ***Andira*** Juss., Gen. Pl.: 363. 4 Aug 1789.
Typus: *A. racemosa* Lam. ex J. St.-Hil. (in Cuvier, Dict. Sci. Nat. 2: 137. 12 Oct 1804).

(=) *Vouacapoua* Aubl., Hist. Pl. Guiane 2 [Misc.]: 9. Jun-Dec 1775.
Typus: *V. americana* Aubl.

3845 ***Dipteryx*** Schreb., Gen. Pl.: 485. Mai 1791.
Typus: *D. odorata* (Aubl.) Willd. (Sp. Pl. 3: 910. 1-10 Nov 1802) (*Coumarouna odorata* Aubl.) *(typ. cons.)*.

(≡) *Coumarouna* Aubl., Hist. Pl. Guiane: 740. Jun-Dec 1775.
(=) *Taralea* Aubl., Hist. Pl. Guiane: 745. Jun-Dec 1775.
Typus: *T. oppositifolia* Aubl.

3848 ***Inocarpus*** J. R. Forst. & G. Forst., Char. Gen. Pl.: 33. 29 Nov 1775.
Typus: *I. edulis* J. R. Forst. & G. Forst.

(=) *Aniotum* Parkinson, J. Voy. South Seas: 39. Jul 1773.
Typus: *A. fagiferum* Parkinson

3853 ***Lens*** Mill., Gard. Dict. Abr., ed. 4: [765]. 28 Jan 1754.
Typus: *L. culinaris* Medik. (in Vorles. Churpfälz. Phys.-Öcon. Ges. 2: 361. 1787) (*Ervum lens* L.).

3858 ***Centrosema*** (DC.) Benth., Comm. Legum. Gen.: 53. Jun 1837 (*Clitoria* sect. *Centrosema* DC., Prodr. 2: 234. Nov (med.) 1825).
Typus: *Clitoria brasiliana* L. (*Centrosema brasilianum* (L.) Benth.) *(typ. cons.)*.

(=) *Steganotropis* Lehm., Sem. Hort. Bot. Hamburg. 1826: 18. 1826.
Typus: *S. conjugata* Lehm.

3860 ***Amphicarpaea*** Elliott ex Nutt., Gen. N. Amer. Pl. 2: 113. 14 Jul 1818 *('Amphicarpa') (orth. cons.)*.
Typus: *A. monoica* Elliott ex Nutt., nom. illeg. (*Glycine monoica* L., nom. illeg., *Glycine bracteata* L., *A. bracteata* (L.) Fernald).

(=) *Falcata* J. F. Gmel., Syst. Nat. 2: 1131. Apr (sero) - Oct 1791.
Typus: *F. caroliniana* J. F. Gmel.

3863 ***Shuteria*** Wight & Arn., Prodr. Fl. Ind. Orient.: 207. Oct (prim.) 1834.
Typus: *S. vestita* Wight & Arn. *(typ. cons.)*.

(H) *Shutereia* Choisy, Convolv. Orient.: 103. Aug 1834 [*Convolvul.*].
Typus: *S. bicolor* Choisy (*Convolvulus bicolor* Vahl 1794, non Lam. 1788).

3864 ***Glycine*** Willd., Sp. Pl. 3: 854, 1053. 1-10 Nov 1802.
Typus: *G. clandestina* J. C. Wendl. (Bot. Beob.: 54. 1798).

(H) *Glycine* L., Sp. Pl.: 753. 1 Mai 1753 [*Legum.*].
Typus (vide Green in Sprague & al., Nom. Prop. Brit. Bot.: 176. 1929): *G. javanica* L.

(=) *Soja* Moench, Methodus: 153, index. 4 Mai 1794.
Typus: *S. hispida* Moench

3868 ***Kennedia*** Vent., Jard. Malmaison: ad t. 104. Jul 1805.
Typus: *K. rubicunda* (Schneev.) Vent. (*Glycine rubicunda* Schneev.).

3871 ***Rhodopis*** Urb., Symb. Antill. 2: 304. 20 Oct 1900.
Typus: *R. planisiliqua* (L.) Urb. (*Erythrina planisiliqua* L.).

(H) *Rhodopsis* Lilja, Fl. Sv. Odl. Vext., Suppl. 1: 42. 1840 [*Portulac.*].
≡ *Tegneria* Lilja 1839.

3874 ***Apios*** Fabr., Enum.: 176. 1759.
Typus: *A. americana* Medik. (in Vorles. Churpfälz. Phys.-Ökon. Ges. 2: 354. 1787) (*Glycine apios* L.).

3876 ***Butea*** Roxb. ex Willd., Sp. Pl. 3: 857, 917. 1-10 Nov 1802.
Typus: *B. frondosa* Roxb. ex Willd., nom. illeg. (*Erythrina monosperma* Lam., *B. monosperma* (Lam.) Taub.) *(typ. cons.)*.

(≡) *Plaso* Adans., Fam. Pl. 2: 325, 592. Jul-Aug 1763 (typ. des.: Panigrahi & Mishra in Taxon 33: 119. 1984).

3877 ***Mucuna*** Adans., Fam. Pl. 2: 325, 579. Jul-Aug 1763.
Typus: *M. urens* (L.) DC. (Prodr. 2: 405. Nov. (med.) 1825) (*Dolichos urens* L.) *(typ. cons.)*.

(≡) *Zoophthalmum* P. Browne, Civ. Nat. Hist. Jamaica: 295. 10 Mar 1756.

(=) *Stizolobium* P. Browne, Civ. Nat. Hist. Jamaica: 290. 10 Mar 1756.
Typus (vide Piper in U.S.D.A. Bur. Pl. Industr. Bull. 179: 9. 1910): *S. pruriens* (L.) Medik. (*Dolichos pruriens* L.).

3891 ***Canavalia*** Adans., Fam. Pl. 2: 325, 531. Jul-Aug 1763 *('Canavali') (orth. cons.)*.
Typus: *C. ensiformis* (L.) DC. (Prodr. 2: 404. Nov. (med.) 1825) (*Dolichos ensiformis* L.).

3892 ***Cajanus*** Adans., Fam. Pl. 2: 326, 529. Jul-Aug 1763 *('Cajan') (orth. cons.)*.
Typus: *C. cajan* (L.) Huth (in Helios 11: 133. 1893) (*Cytisus cajan* L.).

3897 ***Rhynchosia*** Lour., Fl. Cochinch.: 425, 460. Sep 1790.
Typus: *R. volubilis* Lour.

(=) *Dolicholus* Medik. in Vorles. Churpfälz. Phys.-Öcon. Ges. 2: 354. 1787.
Typus: *D. flavus* Medik., nom. illeg. (*Dolichos minimus* L.).

(=) *Cylista* Aiton, Hort. Kew. 3: 36, 512. 7 Aug - 1 Oct 1789.
Typus: *C. villosa* Aiton

3898 ***Eriosema*** (DC.) D. Don, Consp. Regni Veg.: 150. Dec 1828 - Mar 1829 (*Rhynchosia* sect. *Eriosema* DC., Prodr. 2: 388. Nov (med.) 1825).
Typus: *Rhynchosia rufa* (Kunth) DC. (*Glycine rufa* Kunth, *E. rufum* (Kunth) G. Don).

(=) *Euriosma* Desv. in Ann. Sci. Nat. (Paris) 9: 421. Dec 1826.
Typus (vide Pfeiffer, Nomencl. Bot. 1: 1306. 1874): *E. sessiliflora* (Poir.) Desv. (*Cytisus sessiliflorus* Poir.).

3899 ***Flemingia*** Roxb. ex W. T. Aiton, Hort. Kew., ed. 2, 4: 349. Dec 1812.
Typus: *F. strobilifera* (L.) W. T. Aiton (*Hedysarum strobiliferum* L.).

(H) *Flemingia* Roxb. ex Rottler in Ges. Naturf. Freunde Berlin Neue Schriften 4: 202. 1803 [*Acanth.*].
Typus: *F. grandiflora* Roxb. ex Rottler

(≡) *Luorea* Neck. ex J. St.-Hil. in Nouv. Bull. Sci. Soc. Philom. Paris 3: 193. Dec 1812.

3905 ***Vigna*** Savi in Nuov. Giorn. Lett. 8: 113. 1824.
Typus: *V. villosa* Savi, nom. illeg. (*Dolichos luteolus* Jacq., *V. luteola* (Jacq.) Benth.).

(=) *Voandzeia* Thouars, Gen. Nov. Madagasc.: 23. 17 Nov 1806.
Typus: *V. subterranea* (L. f.) DC. (Prodr. 2: 474. Nov (med.) 1825) (*Glycine subterranea* L. f.).

3908 ***Pachyrhizus*** Rich. ex DC., Prodr. 2: 402. Nov (med.) 1825.
Typus: *P. angulatus* Rich. ex DC., nom. illeg. (*Dolichos erosus* L., *P. erosus* (L.) Urban) *(typ. cons.)*.

(≡) *Cacara* Thouars in Cuvier, Dict. Sci. Nat. 6: 35. 1806.

3910 ***Dolichos*** L., Sp. Pl.: 725. 1 Mai 1753.
Typus: *D. trilobus* L. *(typ. cons.)*.

3910a ***Macrotyloma*** (Wight & Arn.) Verdc. in Kew Bull. 24: 322. 1 Apr 1970 (*Dolichos* sect. *Macrotyloma* Wight & Arn., Prodr. Fl. Ind. Orient.: 248. Oct 1834).
Typus: *Dolichos uniflorus* Lam. (*M. uniflorum* (Lam.) Verdc.).

(=) *Kerstingiella* Harms in Ber. Deutsch. Bot. Ges. 26a: 230. 23 Apr 1908.
Typus: *K. geocarpa* Harms

3914 ***Psophocarpus*** Neck. ex DC., Prodr. 2: 403. Nov (med.) 1825.
Typus: *P. tetragonolobus* (L.) DC. (*Dolichos tetragonolobus* L.).

(≡) *Botor* Adans., Fam. Pl. 2: 326, 527. Jul-Aug 1763.

GERANIACEAE

3931 ***Wendtia*** Meyen, Reise 1: 307. 23-31 Mai 1834.
Typus: *W. gracilis* Meyen

(H) *Wendia* Hoffm., Gen. Pl. Umbell.: 136. 1814 [*Umbell.*].
Typus: *W. chorodanum* Hoffm.

3932 ***Balbisia*** Cav. in Anales Ci. Nat. 7: 61. Feb 1804.
Typus: *B. verticillata* Cav.

(H) *Balbisia* Willd., Sp. Pl. 3: 1486, 2214. Apr-Dec 1803 [*Comp.*].
Typus: *B. elongata* Willd.

LINACEAE

3947 ***Durandea*** Planch. in London J. Bot. 6: 594. 1847.
Typus: *D. serrata* Planch. (in London J. Bot. 7: 528. 1848).

(H) *Durandea* Delarbre, Fl. Auvergne, ed. 2, 365. Aug 1800 [*Cruc.*].
Typus: *D. unilocularis* Delarbre, nom. illeg. (*Raphanus raphanistrum* L.).

HUMIRIACEAE

3953 ***Humiria*** Aubl., Hist. Pl. Guiane: 564. Jun-Dec 1775 *('Houmiri') (orth. cons.).*
Typus: *H. balsamifera* Aubl.

ZYGOPHYLLACEAE

3967 ***Augea*** Thunb., Prodr. Pl. Cap. 1: [viii], 80. 1794.
Typus: *A. capensis* Thunb.

(H) *Augia* Lour., Fl. Cochinch.: 327, 337. Sep 1790 [*Anacard.*].
Typus: *A. sinensis* Lour.

3973 ***Larrea*** Cav. in Anales Hist. Nat. 2: 119. Jun 1800.
Typus: *L. nitida* Cav. *(typ. cons.).*

(H) *Larrea* Ortega, Nov. Pl. Descr. Dec.: 15. 1797 [*Legum.*].
≡ *Hoffmannseggia* Cav. 1798 *(nom. cons.)* (3557).

3980	***Balanites*** Delile, Descr. Egypte, Hist. Nat. 2: 221. 1813 (sero) - 1814 (prim.). Typus: *B. aegyptiacus* (L.) Delile (*Ximenia aegyptiaca* L.).	(≡)	*Agialid* Adans., Fam. Pl. 2: 508, 514. Jul-Aug 1763.

RUTACEAE

3991	***Fagara*** L., Syst. Nat., ed. 10: 885, 897, 1362. 7 Jun 1759. Typus: *F. pterota* L.	(H)	*Fagara* Duhamel, Traité Arbr. Arbust. 1: 229. 1755 [*Rut.*]. Typus: non designatus.
		(≡)	*Pterota* P. Browne, Civ. Nat. Hist. Jamaica: 146. 10 Mar 1756.
3998	***Pentaceras*** Hook. f. in Bentham & Hooker, Gen. Pl. 1: 298. 7 Aug 1862. Typus: *P. australe* (F. Muell.) Benth. (Fl. Austral. 1: 365. 30 Mai 1863) (*Cookia australis* F. Muell.).	(H)	*Pentaceros* G. Mey., Prim. Fl. Esseq.: 136. Nov 1818 [*Byttner.*]. Typus: *P. aculeatum* G. Mey.
4011	***Boenninghausenia*** Rchb. ex Meisn., Pl. Vasc. Gen. 1: 60; 2: 44. 21-27 Mai 1837. Typus: *B. albiflora* (Hook.) Meisn. (*Ruta albiflora* Hook.).	(H)	*Boenninghausia* Spreng., Syst. Veg. 3: 153, 245. Jan-Mar 1826 [*Legum.*]. Typus: *B. vincentina* (Ker-Gawl.) Spreng. (*Glycine vincentina* Ker-Gawl.).
4012a	***Haplophyllum*** A. Juss. in Mém. Mus. Hist. Nat. 12: 464. 1825 *('Aplophyllum') (orth. cons.)*. Typus: *H. tuberculatum* (Forssk.) A. Juss. (*Ruta tuberculata* Forssk.).	(H)	*Aplophyllum* Cass. in Cuvier, Dict. Sci. Nat. 33: 463. Dec. 1824 [*Comp.*]. Typus: non designatus.
4020	***Myrtopsis*** Engl. in Engler & Prantl, Nat. Pflanzenfam. 3(4): 137. Mar 1896. Typus: *M. novae-caledoniae* Engl.	(H)	*Myrtopsis* O. Hoffm. in Linnaea 43: 133. Jan 1881 [*Lecythid.*]. Typus: *M. malangensis* O. Hoffm.
4031	***Correa*** Andrews in Bot. Repos.: ad t. 18. 1 Apr 1798. Typus: *C. alba* Andrews	(H)	*Correia* Vand., Fl. Lusit. Bras. Spec.: 28. 1788 [*Ochn.*]. Typus: non designatus.
4035	***Calodendrum*** Thunb., Nov. Gen. Pl.: 41. 10 Jul 1782. Typus: *C. capense* Thunb.	(=)	*Pallassia* Houtt., Nat. Hist. 2(4): 382. 4 Aug 1775. Typus: *P. capensis* Christm. (Vollst. Pflanzensyst. 3: 318. 1778).
4036	***Barosma*** Willd., Enum. Pl.: 257. Apr 1809. Typus: *B. serratifolia* (Curt.) Willd. (*Diosma serratifolia* Curt.).	(=)	*Parapetalifera* J. C. Wendl., Coll. Pl. 1: 49. 1806. Typus: *P. odorata* J. C. Wendl.

4037 ***Agathosma*** Willd., Enum. Pl.: 259. Apr 1809.
Typus: *A. villosa* (Willd.) Willd. (*Diosma villosa* Willd.).

(≡) *Bucco* J. C. Wendl., Coll. Pl. 1: 13. 1805.
(=) *Hartogia* L., Syst. Nat., ed. 10: 939, 1365. 7 Jun 1759.
Typus: *H. capensis* L.

4038 ***Adenandra*** Willd., Enum. Pl.: 256. Apr 1809.
Typus: *A. uniflora* (L.) Willd. (*Diosma uniflora* L.).

(=) *Haenkea* F. W. Schmidt, Neue Selt. Pfl.: 19. 1793 (ante 17 Jun).
Typus: non designatus.
(=) *Glandulifolia* J. C. Wendl., Coll. Pl. 1: 35. 1805.
Typus: *G. umbellata* J. C. Wendl.

4060 ***Naudinia*** Planch. & Linden in Ann. Sci. Nat., Bot., ser. 3, 19: 79. Feb 1853.
Typus: *N. amabilis* Planch. & Linden

(H) *Naudinia* A. Rich. in Sagra, Hist. Phys. Cuba, Bot. Pl. Vasc.: 561. 1846 [*Melastomat.*].
Typus (vide Mansfeld in Kew Bull. 1935: 438. 1935): *N. argyrophylla* A. Rich.

4063 ***Dictyoloma*** A. Juss. in Mém. Mus. Hist. Nat. 12: 499. 1825.
Typus: *D. vandellianum* A. Juss.

4065 ***Chloroxylon*** DC., Prodr. 1: 625. Jan (med.) 1824.
Typus: *C. swietenia* DC. (*Swietenia chloroxylon* Roxb.).

(H) *Chloroxylum* P. Browne, Civ. Nat. Hist. Jamaica: 187. 10 Mar 1756 [*Rhamn.*].
Typus: *Ziziphus chloroxylon* (L.) Oliv. (*Laurus chloroxylon* L.).

4066 ***Spathelia*** L., Sp. Pl., ed. 2: 386. Sep 1762.
Typus: *S. simplex* L.

4073 ***Araliopsis*** Engl. in Engler & Prantl, Nat. Pflanzenfam. 3(4): 175. Mar 1896.
Typus: *A. soyauxii* Engl.

4074 ***Sargentia*** S. Watson in Proc. Amer. Acad. Arts 25: 144. 25 Sep 1890.
Typus: *S. greggii* S. Watson

(H) *Sargentia* H. Wendl. & Drude ex Salomon, Palmen: 160. Sep-Oct 1887 [*Palm.*].
≡ *Pseudophoenix* H. Wendl. ex Sarg. 1886.

4077 ***Toddalia*** Juss., Gen. Pl.: 371. 4 Aug 1789.
Typus: *T. asiatica* (L.) Lam. (Tabl. Encycl. 2: 116. 6 Nov 1797) (*Paullinia asiatica* L.) (*typ. cons.:* [specimen] Herb. Hermann 3: 45, No. 143 (BM)).

4079 ***Acronychia*** J. R. Forst. & G. Forst., Char. Gen. Pl.: 27. 29 Nov 1775.
Typus: *A. laevis* J. R. Forst. & G. Forst.

(=) *Jambolifera* L., Sp. Pl.: 349. 1 Mai 1753.
Typus: *J. pedunculata* L.

(=) *Cunto* Adans., Fam. Pl. 2: 446, 547. Jul-Aug 1763.
Typus: non designatus.

4083 ***Skimmia*** Thunb., Nov. Gen. Pl.: 57. 18 Jun 1783.
Typus: *S. japonica* Thunb.

4085 ***Teclea*** Delile in Ann. Sci. Nat., Bot., ser. 2, 20: 90. Aug 1843.
Typus: *T. nobilis* Delile

(=) *Aspidostigma* Hochst in Schimper, Iter Abyss., Sect. 2: No. 1293. 1842-1843 [in sched.].
Typus: *A. acuminatum* Hochst.

4087 ***Glycosmis*** Corrêa in Ann. Mus. Natl. Hist. Nat. 6: 384. 1805.
Typus: *G. arborea* (Roxb.) DC. (Prodr. 1: 538. Jan (med.) 1824) (*Limonia arborea* Roxb.).

(=) *Panel* Adans., Fam. Pl. 2: 447, 587. Jul-Aug 1763.
Typus: *Limonia winterlia* Steud.

4089 ***Micromelum*** Blume, Bijdr.: 137. 20 Aug 1825.
Typus: *M. pubescens* Blume

(=) *Aulacia* Lour., Fl. Cochinch.: 258, 273. Sep 1790.
Typus: *A. falcata* Lour.

4090 ***Murraya*** J. König ex L., Mant. Pl.: 554, 563. Oct 1771 *('Murraea') (orth. cons.)*.
Typus: *M. exotica* L.

(=) *Bergera* J. König ex L., Mant. Pl.: 555, 563. Oct 1771.
Typus: *B. koenigii* L.

4096 ***Atalantia*** Corrêa in Ann. Mus. Natl. Hist. Nat. 6: 383, 385, 386. 1805.
Typus: *A. monophylla* (L.) DC. (Prodr. 1: 535. Jan (med.) 1824) (*Limonia monophylla* L.).

(=) *Malnaregam* Adans., Fam. Pl. 2: 345, 574. Jul-Aug 1763.
Typus: *M. malabarica* Raf. (Sylva Tellur.: 143. Oct-Dec 1838).

4099 ***Aegle*** Corrêa in Trans. Linn. Soc. London 5: 222. 1800.
Typus: *A. marmelos* (L.) Corrêa (*Crateva marmelos* L.).

(≡) *Belou* Adans., Fam. Pl. 2: 408, 525. Jul-Aug 1763.

SIMAROUBACEAE

4109 ***Samadera*** Gaertn., Fruct. Sem. Pl. 2: 352. Apr-Mai 1791.
Typus: *S. indica* Gaertn.

(=) *Locandi* Adans., Fam. Pl. 2: 449, 571. Jul-Aug 1763.
Typus: *Niota pentapetala* Poir. (in Lamarck, Encycl. 4: 490. 1 Nov 1798).

4111 ***Simarouba*** Aubl., Hist. Pl. Guiane: 859. Jun-Dec 1775.
Typus: *S. amara* Aubl.

(H) *Simaruba* Boehm. in Ludwig, Def. Gen. Pl., ed. 3: 513. 1760 [*Burser.*].
≡ *Bursera* Jacq. ex L. 1762 *(nom. cons.)* (4150).

4117 ***Harrisonia*** R. Br. ex A. Juss. in Mém. Mus. Hist. Nat. 12: 517. 1825.
Typus: *H. brownii* A. Juss.

(H) *Harissona* Adans. ex Léman in Cuvier, Dict. Sci. Nat. 20: 290. 29 Jun 1821 [*Musci*].
Typus: non designatus.

4118 ***Castela*** Turpin in Ann. Mus. Natl. Hist. Nat. 7: 78. 1806.
Typus: *C. depressa* Turpin *(typ. cons.)*.

(H) *Castelia* Cav. in Anales Ci. Nat. 3: 134. 1801 [*Verben.*].
Typus: *C. cuneato-ovata* Cav.

4120 ***Brucea*** J. F. Mill. [Icon. Anim. Pl.]: t. 25. 1779-1780.
Typus: *B. antidysenterica* J. F. Mill.

4124 ***Ailanthus*** Desf. in Mém. Acad. Sci. (Paris) 1786: 265. 1788.
Typus: *A. glandulosa* Desf.

(=) *Pongelion* Adans., Fam. Pl. 2: 319, 593. Jul-Aug 1763.
Typus: *Ailanthus triphysa* (Dennst.) Alston (*Adenanthera triphysa* Dennst.).

4131 ***Picramnia*** Sw., Prodr.: 2, 27. 20 Jun - 29 Jul 1788.
Typus: *P. antidesma* Sw.

(=) *Pseudo-brasilium* Adans., Fam. Pl. 2: 341, 595. Jul-Aug 1763.
Typus: *Brasiliastrum americanum* Lam.

(=) *Tariri* Aubl., Hist. Pl. Guiane: 37. Jun-Dec 1775.
Typus: *T. guianensis* Aubl.

4134 ***Picrodendron*** Griseb., Fl. Brit. W. I.: 176. Jun 1860.
Typus: [specimen] Jamaica, *Macfadyen s.n.* (K) *(typ. cons.)* [= *P. baccatum* (L.) Krug (*Juglans baccata* L.)].

(H) *Picrodendron* Planch. in London J. Bot. 5: 579. 1846 [*Sapind.*].
Typus: *P. arboreum* (Mill.) Planch. (*Toxicodendron arboreum* Mill.).

BURSERACEAE

4137 ***Protium*** Burm. f., Fl. Indica: 88. 1 Mar - 6 Apr 1768.
Typus: *P. javanicum* Burm. f. (*Amyris protium* L.).

4150 ***Bursera*** Jacq. ex L., Sp. Pl., ed. 2: 471. Sep 1762.
Typus: *B. gummifera* L., nom. illeg. (*Pistacia simaruba* L., *B. simaruba* (L.) Sarg.) (etiam vide 4111).

(=) *Elaphrium* Jacq., Enum. Syst. Pl.: 3, 19. Aug-Sep 1760.
Typus (vide Rose in N. Amer. Fl. 25: 241. 1911): *E. tomentosum* Jacq.

4151 ***Commiphora*** Jacq., Pl. Hort. Schoenbr. 2: 66. 1797.
Typus: *C. madagascarensis* Jacq.

(=) *Balsamea* Gled. in Schriften Berlin. Ges. Naturf. Freunde 3: 127. 1782.
Typus: *B. meccanensis* Gled.

MELIACEAE

4172 ***Naregamia*** Wight & Arn., Prodr. Fl. Ind. Orient.: 116. Oct (prim.) 1834.
Typus: *N. alata* Wight & Arn.

(≡) *Nelanaregam* Adans., Fam. Pl. 2: 343, 581. Jul-Aug 1763.

4189 ***Aglaia*** Lour., Fl. Cochinch.: 98, 173. Sep 1790.
Typus: *A. odorata* Lour.

(H) *Aglaia* F. Allam. in Nova Acta Phys.-Med. Acad. Caes. Leop.-Carol. Nat. Cur. 4: 93. 1770 [*Cyper.*].
Typus: non designatus.

(=) *Nialel* Adans., Fam. Pl. 2: 446, 582. Jul-Aug 1763.
Typus (vide Nicolson & Suresh in Taxon 35: 388. 1986): *Nyalel racemosa* Dennst. ex Kostel. (Allg. Med.-Pharm. Fl.: 2005. Jan-Sep 1836).

4190 ***Guarea*** F. Allam. ex L., Mant. Pl.: 150, 228. Oct 1771.
Typus: *G. trichilioides* L., nom. illeg. (*Melia guara* Jacq., *Guarea guara* (Jacq.) P. Wilson).

(=) *Elutheria* P. Browne, Civ. Nat. Hist. Jamaica: 369. 10 Mar 1756.
Typus: *E. microphylla* (Hook.) M. Roem. (Fam. Nat. Syn. Monogr. 1: 122. 14 Sep - 15 Oct 1846) (*Guarea microphylla* Hook.).

4195 ***Trichilia*** P. Browne, Civ. Nat. Hist. Jamaica: 278. 10 Mar 1756.
Typus: *T. hirta* L. (Syst. Nat., ed. 10: 1020. 7 Jun 1759) *(typ. cons.)*.

MALPIGHIACEAE

4208 ***Hiptage*** Gaertn., Fruct. Sem. Pl. 2: 169. Sep (sero) - Nov 1790.
Typus: *H. madablota* Gaertn., nom. illeg. (*Banisteria tetraptera* Sonn.) [= *H. benghalensis* (L.) Kurz (*Banisteria benghalensis* L.)] (etiam vide 8428).

4211 ***Triopterys*** L., Sp. Pl.: 428. 1 Mai 1753 *('Triopteris') (orth. cons.)*.
Typus: *T. jamaicensis* L.

4212 ***Tetrapterys*** Cav., Diss. 9: 433. Jan-Feb 1790 *('Tetrapteris') (orth. cons.)*.
Typus: *T. inaequalis* Cav.

4222 ***Ryssopterys*** Blume ex A. Juss. in Delessert, Icon. Sel. Pl. 3: 21. Feb 1838.
Typus: *R. timoriensis* (DC.) A. Juss. (*Banisteria timoriensis* DC.)

4226 ***Heteropterys*** Kunth in Humboldt & al., Nov. Gen. Sp. 5, ed. 4°: 163 [& ed. f°: 126] 25 Feb 1822 *('Heteropteris') (orth. cons.).*
Typus: *H. purpurea* (L.) Kunth (*Banisteria purpurea* L.) *(typ. cons.).*

(=) *Banisteria* L., Sp. Pl.: 427. 1 Mai 1753.
Typus (vide Sprague in Gard. Chron., ser. 3, 75: 104. 1924): *B. brachiata* L.

4234 ***Ptilochaeta*** Turcz. in Bull. Soc. Imp. Naturalistes Moscou 16: 52. 1843 (prim.).
Typus: *P. bahiensis* Turcz.

(H) *Ptilochaeta* Nees in Martius, Fl. Bras. 2(1): 147. 1 Apr 1842 [*Cyper.*].
Typus: *P. diodon* Nees

4244 ***Thryallis*** Mart., Nov. Gen. Sp. Pl. 3: 77. Jan-Jun 1829.
Typus: *T. longifolia* Mart.

(H) *Thryallis* L., Sp. Pl., ed. 2, 1: 554. Sep 1762 [*Malpigh.*].
Typus: *T. brasiliensis* L.

4247 ***Lophanthera*** A. Juss., Malpigh. Syn.: 53. Mai 1840.
Typus: *L. kunthiana* A. Juss., nom. illeg. (*Galphimia longifolia* Kunth, *L. longifolia* (Kunth) Griseb.).

(H) *Lophanthera* Raf., New Fl. 2: 58. Jul-Dec 1837 [*Scrophular.*].
Typus: *Gerardia delphiniifolia* L.

TRIGONIACEAE

4264 ***Trigoniastrum*** Miq., Fl. Ned. Ind., Eerste Bijv.: 394. Dec 1861.
Typus: *T. hypoleucum* Miq.

VOCHYSIACEAE

4266 ***Vochysia*** Aubl., Hist. Pl. Guiane: 18. Jun-Dec 1775 *('Vochy') (orth. cons.).*
Typus: *V. guianensis* Aubl.

POLYGALACEAE

4275 ***Securidaca*** L., Syst. Nat., ed. 10: 1151, 1155. 7 Jun 1759.
Typus: *S. volubilis* L. 1759, non L. 1753 [= *S. diversifolia* (L.) Blake (*Polygala diversifolia* L.)].

(H) *Securidaca* L., Sp. Pl.: 707. 1 Mai 1753 [*Legum.*].
Typus: *S. volubilis* L.

4277 ***Salomonia*** Lour., Fl. Cochinch.: 1, 14. Sep 1790.
Typus: *S. cantoniensis* Lour.

(H) *Salomonia* Heist. ex Fabr., Enum.: 20. 1759 [*Lil.*].
≡ *Polygonatum* Mill. 1754.

4278 ***Muraltia*** DC., Prodr. 1: 335. Jan (med.) 1824.
Typus: *M. heisteria* (L.) DC. (*Polygala heisteria* L.) *(typ. cons.)* (etiam vide 2147).
(H) *Muralta* Adans., Fam. Pl. 2: 460, 580. Jul-Aug 1763 [*Ranuncul.*].
Typus: *Clematis cirrhosa* L.

4281 ***Xanthophyllum*** Roxb., Pl. Coromandel 3: 81. 18 Feb 1820.
Typus: *X. flavescens* Roxb. *(typ. cons.)*.
(=) *Pelaë* Adans., Fam. Pl. 2: 448, 589. Jul-Aug 1763.
Typus: non designatus.
(=) *Eystathes* Lour., Fl. Cochinch.: 223, 234. Sep 1790.
Typus: *E. sylvestris* Lour.

EUPHORBIACEAE

4297 ***Securinega*** Comm. ex Juss., Gen. Pl.: 388. 4 Aug 1789.
Typus: *S. durissima* J. F. Gmel. (Syst. Nat. 2: 1008. Apr (sero) - Oct 1792) *(typ. cons.)*.

4299a ***Androstachys*** Prain in Bull. Misc. Inform. Kew 1908: 438. Dec 1908.
Typus: *A. johnsonii* Prain
(H) *Androstachys* Grand'Eury in Mém. Divers Savants Acad. Roy. Sci. Inst. Roy. France, Sci. Math. 24(1): 190. 1877 [Foss.].
Typus: *A. frondosa* Grand'Eury

4302 ***Glochidion*** J. R. Forst. & G. Forst., Char. Gen. Pl.: 57. 29 Nov 1775.
Typus: *G. ramiflorum* J. R. Forst. & G. Forst.
(=) *Agyneia* L., Mant. Pl.: 161, 296, 576. Oct 1771.
Typus (vide Regnum Veg. 46: 307. 1966): *A. pubera* L.

4303 ***Breynia*** J. R. Forst. & G. Forst., Char. Gen. Pl.: 73. 29 Nov 1775.
Typus: *B. disticha* J. R. Forst. & G. Forst.
(H) *Breynia* L., Sp. Pl.: 503. 1 Mai 1753 [*Cappar.*].
≡ *Linnaeobreynia* Hutch. 1967.

4331 ***Buraeavia*** Baill. in Adansonia 11: 83. 15 Nov 1873.
Typus: *B. carunculata* (Baill.) Baill. (*Baloghia carunculata* Baill.).
(H) *Bureava* Baill. in Adansonia 1: 71. Sep 1860-Aug 1861 [*Combret.*].
Typus: *B. crotonoides* Baill.

4349 ***Julocroton*** Mart. in Flora 20 (2, Beibl.): 119. 21 Nov 1837.
Typus: *J. phagedaenicus* Mart.
(=) *Cieca* Adans., Fam. Pl. 2: 356, 612. Jul-Aug 1763.
Typus: *Croton argenteus* L.

4355 ***Chrozophora*** A. Juss., Euphorb. Gen.: 27. 21 Feb 1824 *('Crozophora') (orth. cons.)*.
Typus: *C. tinctoria* (L.) A. Juss. (*Croton tinctorius* L.).
(≡) *Tournesol* Adans., Fam. Pl. 2: 356, 612. Jul-Aug 1763.

4397 ***Adelia*** L., Syst. Nat., ed. 10: 1285, 1298. 7 Jun 1759.
Typus: *A. ricinella* L. *(typ. cons.).*

(H) *Adelia* P. Browne, Civ. Nat. Hist. Jamaica: 361. 10 Mar 1756 [*Ol.*].
Typus: non designatus.

(=) *Bernardia* Mill., Gard. Dict. Abr., ed. 4: [185]. 28 Jan 1754.
Typus (vide Buchheim in Willdenowia 3: 217. 1962): *B. carpinifolia* Griseb. (Fl. Brit. W.I.: 45. Dec 1859) (*Adelia bernardia* L.).

4415 ***Acidoton*** Sw., Prodr.: 6, 83. 20 Jun - 29 Jul 1788.
Typus: *A. urens* Sw.

(H) *Acidoton* P. Browne, Civ. Nat. Hist. Jamaica: 355. 10 Mar 1756 [*Euphorb.*].
Typus: *Adelia acidoton* L.

4421 ***Pterococcus*** Hassk. in Flora 25 (2, Beibl.): 41. 7 Aug 1842.
Typus: *P. glaberrimus* Hassk., nom. illeg. (*Plukenetia corniculata* Sm., *Pterococcus corniculatus* (Sm.) Pax & K. Hoffm.).

(H) *Pterococcus* Pall., Reise Russ. Reich. 2: 738. 1773 [*Polygon.*].
Typus: *P. aphyllus* Pall.

4435 ***Micrandra*** Benth. in Hooker's J. Bot. Kew Gard. Misc. 6: 371. Dec 1854.
Typus: *M. siphonioides* Benth. *(typ. cons.).*

(H) *Micrandra* R. Br. in Bennett, Pl. Jav. Rar.: 237. 4 Jun 1844 [*Euphorb.*].
Typus: *M. ternata* R. Br.

4449 ***Trigonostemon*** Blume, Bijdr.: 600. 24 Jan 1826 *('Trigostemon') (orth. cons.).*
Typus: *T. serratus* Blume

(=) *Enchidium* Jack in Malayan Misc. 2(7): 89. 1822.
Typus: *E. verticillatum* Jack

4452 ***Sagotia*** Baill. in Adansonia 1: 53. 1 Dec 1860.
Typus: *S. racemosa* Baill.

(H) *Sagotia* Duchass. & Walp. in Linnaea 23: 737. Jan 1851 [*Legum.*].
Typus: *S. triflora* (L.) Duchass. & Walp. (*Hedysarum triflorum* L.).

4454 ***Codiaeum*** A. Juss., Euphorb. Gen.: 33. 21 Feb 1824.
Typus: *C. variegatum* (L.) A. Juss. (*Croton variegatus* L.) *(typ. cons.).*

(≡) *Phyllaurea* Lour., Fl. Cochinch.: 540, 575. Sep 1790.

4455 ***Galearia*** Zoll. & Moritzi in Moritzi, Syst. Verz.: 19. Mai-Jun 1846.
Typus: *G. pedicellata* Zoll. & Moritzi

(H) *Galearia* C. Presl, Symb. Bot. 1: 49. Sep-Dec 1831 [*Legum.*].
Typus (vide Hossain in Notes Roy. Bot. Gard. Edinburgh 23: 446. 1961): *G. fragifera* (L.) C. Presl (*Trifolium fragiferum* L.).

4459 ***Blachia*** Baill., Etude Euphorb.: 385. 1858.
Typus: *B. umbellata* (Willd.) Baill. (*Croton umbellatus* Willd.).

(=) *Bruxanellia* Dennst. ex Kostel., Allg. Med.-Pharm. Fl.: 2002. Jan-Sep 1836.
Typus: *B. indica* Dennst. ex Kostel.

4467 ***Chaetocarpus*** Thwaites in Hooker's J. Bot. Kew Gard. Misc. 6: 300. Oct 1854.
Typus: *C. castanicarpus* (Roxb.) Thwaites (Enum. Pl. Zeyl.: 275. 1861) (*Adelia castanicarpa* Roxb.).

(H) *Chaetocarpus* Schreb., Gen. Pl.: 75. Apr 1789 [*Sapot.*].
≡ *Pouteria* Aubl. 1775.

4470 ***Endospermum*** Benth., Fl. Hongk.: 304. Feb 1861.
Typus: *E. chinense* Benth.

(H) *Endespermum* Blume, Catalogus: 24. Feb-Sep 1823 [*Legum.*].
Typus: *E. scandens* Blume

4472 ***Omphalea*** L., Syst. Nat., ed. 10: 1254, 1264, 1378. 7 Jun 1759.
Typus: *O. triandra* L. *(typ. cons.)*.

(≡) *Omphalandria* P. Browne, Civ. Nat. Hist. Jamaica: 334. 10 Mar 1756.

4483 ***Sapium*** Jacq., Enum. Syst. Pl.: 9, 31. Aug-Sep 1760.
Typus: *S. aucuparium* Jacq., nom. illeg. (*Hippomane glandulosa* L., *S. glandulosum* (L.) Morong).

(H) *Sapium* P. Browne, Civ. Nat. Hist. Jamaica: 338. 10 Mar 1756 [*Euphorb.*].
Typus (vide Kruijt & Zijlstra in Taxon 38: 321. 1989): *Excoecaria glandulosa* Sw.

4498a ***Tithymalus*** Gaertn., Fruct. Sem. Pl. 2: 115. Sep (sero) - Nov 1790.
Typus: *T. peplus* (L.) Gaertn. (*Euphorbia peplus* L.).

(H) *Tithymalus* Mill., Gard. Dict. Abr., ed. 4: [1391]. 28 Jan 1754 [*Euphorb.*].
≡ *Pedilanthus* Poit. 1812 *(nom. cons.)* (4501).

4501 ***Pedilanthus*** Poit. in Ann. Mus. Natl. Hist. Nat. 19: 388. 1812.
Typus: *P. tithymaloides* (L.) Poit. (*Euphorbia tithymaloides* L.) (etiam vide 4498a).

(≡) *Tithymaloides* Ortega, Tab. Bot.: 9. 1773.

4516 ***Botryophora*** Hook. f., Fl. Brit. India 5: 476. Dec 1888.
Typus: *B. kingii* Hook. f.

(H) *Botryophora* Bompard in Hedwigia 6: 129. Sep 1867 [*Chloroph.*].
Typus: *B. dichotoma* Bompard

LIMNANTHACEAE

4542 ***Limnanthes*** R. Br. in London Edinburgh Philos. Mag. & J. Sci. 3: 70. Jul 1833.
Typus: *L. douglasii* R. Br.

(H) *Limnanthes* Stokes, Bot. Mat. Med. 1: 300. 1812 [*Gentian.*].
≡ *Limnanthemum* S. G. Gmel. 1770.

ANACARDIACEAE

4563 ***Lannea*** A. Rich. in Guillemin & al., Fl. Seneg. Tent.: 153. Sep 1831.
Typus: *L. velutina* A. Rich. *(typ. cons.)*.

(=) *Calesiam* Adans., Fam. Pl. 2: 446, 530. Jul-Aug 1763.
Typus: *Calesiam malabarica* Raf. (Sylva Tellur.: 12. Oct-Dec 1838).

4578 ***Campnosperma*** Thwaites in Hooker's J. Bot. Kew Gard. Misc. 6: 65. Mar 1854.
Typus: *C. zeylanicum* Thwaites

(=) *Coelopyrum* Jack in Malayan Misc. 2(7): 65. 1822.
Typus: *C. coriaceum* Jack

4600 ***Nothopegia*** Blume, Mus. Bot. 1: 203. Oct 1850.
Typus: *N. colebrookiana* (Wight) Blume (*Pegia colebrookiana* Wight).

(=) *Glycycarpus* Dalzell in J. Roy. Asiat. Soc. Bombay 3(1): 69. 1849.
Typus: *G. edulis* Dalzell

4604 ***Holigarna*** Buch.-Ham. ex Roxb., Pl. Coromandel 3: 79. 18 Feb 1820.
Typus: *H. longifolia* Buch.-Ham. ex Roxb.

(=) *Katou-tsjeroë* Adans., Fam. Pl. 2: 84, 534. Jul-Aug 1763.
Typus: non designatus.

AQUIFOLIACEAE

4615 ***Nemopanthus*** Raf. in Amer. Monthly Mag. & Crit. Rev. 4: 357. Mar 1819.
Typus: *N. fascicularis* Raf., nom. illeg. (*Ilex canadensis* Michx., *N. canadensis* (Michx.) DC.).

(=) *Ilicioides* Dum. Cours., Bot. Cult. 4: 27. 1-4 Jul 1802.
Typus: non designatus.

CELASTRACEAE

4618 ***Euonymus*** L., Sp. Pl.: 197. 1 Mai 1753 *('Evonymus') (orth. cons.)*.
Typus: *E. europaeus* L.

4621 ***Microtropis*** Wall. ex Meisn., Pl. Vasc. Gen. 1: 68; 2: 49. 27 Aug - 3 Sep 1837.
Typus: *M. discolor* (Wall.) Meisn. (*Cassine discolor* Wall.).

(H) *Microtropis* E. Mey., Comm. Pl. Afr. Austr.: 65. 14 Feb - 5 Jun 1836 [*Legum.*].
≡ *Euchlora* Eckl. & Zeyh. Jan-Feb 1836.

4623 ***Denhamia*** Meisn., Pl. Vasc. Gen. 1: 18; 2: 16. 26 Mar - 1 Apr 1837.
Typus: *D. obscura* (A. Rich.) Meisn. ex Walp. (Repert. Bot. Syst. 1: 203. 18-20 Sep 1842) (*Leucocarpum obscurum* A. Rich.).

(H) *Denhamia* Schott in Schott & Endlicher, Melet. Bot.: 19. 1832 [*Ar.*].
≡ *Culcasia scandens* P. Beauv. 1803 *(nom. cons.)* (690).

(≡) *Leucocarpum* A. Rich. in Urville, Voy. Astrolabe 2: 46. 1834.

4627 ***Gymnosporia*** (Wight & Arn.) Benth. & Hook. f., Gen. Pl. 1: 359, 365. 7 Aug 1862 (*Celastrus* sect. *Gymnosporia* Wight & Arn., Prodr. Fl. Ind. Orient.: 159. 10 Oct 1834).
Typus: *Celastrus montanus* Roth ex Roem. & Schult. (*G. montana* (Roth ex Roem. & Schult.) Benth.) *(typ. cons.)*.

(=) *Catha* Forssk. ex Scop., Intr. Hist. Nat.: 228. Jan-Apr 1777.
Typus (vide Friis in Taxon 33: 662. 1984): *C. edulis* (Vahl) Endl. (Ench. Bot.: 575. 15-21 Aug 1841) (*Celastrus edulis* Vahl).

(=) *Scytophyllum* Eckl. & Zeyh, Enum. Pl. Afric. Austral.: 124. Dec 1834 - Mar 1835.
Typus: non designatus.

(=) *Encentrus* C. Presl in Abh. Königl. Böhm. Ges. Wiss., ser 5, 3: 463. Jul-Dec 1845.
Typus: *E. linearis* (L. f.) C. Presl (*Celastrus linearis* L. f.).

(=) *Polyacanthus* C. Presl in Abh. Königl. Böhm. Ges. Wiss., ser. 5, 3: 463. Jul-Dec 1845.
Typus: *P. stenophyllus* (Eckl. & Zeyh.) C. Presl (*Celastrus stenophyllus* Eckl. & Zeyh.).

4637 ***Plenckia*** Reissek in Martius, Fl. Bras. 11(1): 29. 15 Feb 1861.
Typus: *P. populnea* Reissek

(H) *Plenckia* Raf. in Specchio Sci. 1: 194. 1 Jun 1814 [*Aiz.*].
Typus: *P. setiflora* (Forssk.) Rafin. (*Glinus setiflorus* Forssk.).

HIPPOCRATEACEAE

4662 ***Salacia*** L., Mant. Pl.: 159. Oct 1771.
Typus: *S. chinensis* L.

(=) *Courondi* Adans., Fam. Pl. 2: 446, 545. Jul-Aug 1763.
Typus (vide Nicolson & Suresh in Taxon 35: 181. 1986): *Christmannia corondi* Dennst. ex Kostel.

4662a ***Tontelea*** Miers in Trans. Linn. Soc. London 28: 384. 1872.
Typus: *T. attenuata* Miers *(typ. cons.)*.

(H) *Tontelea* Aubl., Hist. Pl. Guiane 1: 31. Jun-Dec 1775 [*Celastr.*].
Typus: *T. scandens* Aubl.

STAPHYLEACEAE

4666 ***Turpinia*** Vent. in Mém. Cl. Sci. Math. Inst. Natl. France 1807(1): 3. Jul 1807.
Typus: *T. paniculata* Vent.

(H) *Turpinia* Bonpl. in Humboldt & Bonpland, Pl. Aequinoct. 1: 113. Apr 1807 [*Comp.*].
Typus: *T. laurifolia* Bonpl.

(=) *Triceros* Lour., Fl. Cochinch.: 100, 184. Sep 1790.
Typus: *T. cochinchinensis* Lour.

4667 ***Euscaphis*** Siebold & Zucc., Fl. Jap. 1: 122. 1840.
Typus: *E. staphyleoides* Siebold & Zucc., nom. illeg. (*Sambucus japonica* Thunb., *E. japonica* (Thunb.) Kanitz).
(≡) *Hebokia* Raf., Alsogr. Amer.: 47. 1838.

ICACINACEAE

4693 ***Mappia*** Jacq., Pl. Hort. Schoenbr. 1: 22. 1797.
Typus: *M. racemosa* Jacq.
(H) *Mappia* Heist. ex Fabr., Enum.: 58. 1759 [*Lab.*].
≡ *Cunila* Mill. 1754.

4709 ***Pyrenacantha*** Wight in Bot. Misc. 2: 107. Nov-Dec 1830.
Typus: *P. volubilis* Wight

4712 ***Phytocrene*** Wall., Pl. Asiat. Rar. 3: 11. 10 Dec 1831.
Typus: *P. gigantea* Wall. [= *P. macrophylla* (Blume) Blume (*Gynocephalum macrophyllum* Blume)].
(=) *Gynocephalum* Blume, Bijdr. 483. 1825.
Typus: *G. macrophyllum* Blume

4713 ***Miquelia*** Meisn., Pl. Vasc. Gen. 1: 152; 2: 109. 16-22 Sep 1938.
Typus: *M. kleinii* Meisn.
(H) *Miquelia* Blume in Bull. Sci. Phys. Nat. Néerl. 1: 94. 30 Jun 1838 [*Gesner.*].
Typus: *M. caerulea* Blume

4715 ***Stachyanthus*** Engl. in Engler & Prantl, Nat. Pflanzenfam., Nachtr. 2-4, 1: 227. Oct 1897.
Typus: *S. zenkeri* Engl.
(H) *Stachyanthus* DC., Prodr. 5: 84. 1-10 Oct 1836 [*Comp.*].
≡ *Argyrovernonia* MacLeish 1984.

HIPPOCASTANACEAE

4722 ***Billia*** Peyr. in Bot. Zeitung (Berlin) 16: 153. 28 Mai 1858.
Typus: *B. hippocastanum* Peyr.
(H) *Billya* Cass. in Cuvier, Dict. Sci. Nat. 34: 38. Apr 1825 [*Comp.*].
Typus: *B. bergii* Cass.

SAPINDACEAE

4730 ***Bridgesia*** Bertero ex Cambess. in Nouv. Ann. Mus. Hist. Nat. 3: 234. 1834.
Typus: *B. incisifolia* Bertero ex Cambess.
(H) *Bridgesia* Hook. in Bot. Misc. 2: 222. 1831 (ante 11 Jun) [*Comp.*].
Typus: *B. echinopsoides* Hook.

4733 ***Thouinia*** Poit. in Ann. Mus. Natl. Hist. Nat. 3: 70. 1804.
Typus: *T. simplicifolia* Poit. *(typ. cons.)*.
(H) *Thouinia* Thunb. ex L. f., Suppl. Pl.: 9, 89. Apr 1782 [*Ol.*].
Typus: *T. nutans* L. f.

4747 ***Zollingeria*** Kurz in J. Asiat. Soc. Bengal, Pt. 2, Nat. Hist. 41: 303. 1872.
Typus: *Z. macrocarpa* Kurz

(H) *Zollingeria* Sch. Bip. in Flora 37: 274. 14 Mai 1854 [*Comp.*].
Typus: *Z. scandens* Sch. Bip.

4753 ***Pancovia*** Willd., Sp. Pl. 2: 280, 285. Mar 1799.
Typus: *P. bijuga* Willd.

(H) *Pancovia* Heist. ex Fabr., Enum.: 64. 1759 [*Ros.*].
≡ *Comarum* L. 1753.

4767 ***Schleichera*** Willd., Sp. Pl. 4: 892, 1096. Apr 1806.
Typus: *S. trijuga* Willd.

(=) *Cussambium* Lam., Encycl. 2: 230. 16 Oct 1786.
Typus: *C. spinosum* Buch.-Ham. (in Mem. Wern. Nat. Hist. Soc. 5: 357. 1826).

4820 ***Mischocarpus*** Blume, Bijdr.: 238. 20 Sep - 7 Dec 1825.
Typus: *M. sundaicus* Blume

(=) *Pedicellia* Lour., Fl. Cochinch.: 641, 655. Sep 1790.
Typus: *P. oppositifolia* Lour.

RHAMNACEAE

4862 ***Condalia*** Cav. in Anales Hist. Nat. 1: 39. Oct 1799.
Typus: *C. microphylla* Cav.

(H) *Condalia* Ruiz & Pav., Fl. Peruv. Prodr.: 11. Oct (prim.) 1794 [*Rub.*].
Typus: *C. repens* Ruiz & Pav. (Fl. Peruv. 1: 54. 1798 (med.)).

4868 ***Berchemia*** Neck. ex DC., Prodr. 2: 22. Nov (med.) 1825.
Typus: *B. volubilis* (L. f.) DC. (*Rhamnus volubilis* L. f.) [= *B. scandens* (Hill) K. Koch (*Rhamnus scandens* Hill)].

(≡) *Oenoplea* Michx. ex R. Hedw., Gen. Pl.: 151. Jul 1806.

4874 ***Scutia*** (Comm. ex DC.) Brongn. in Ann. Sci. Nat. (Paris) 10: 362. Apr 1827 (*Ceanothus* sect. *Scutia* Comm. ex DC., Prodr. 2: 29. Nov (med.) 1825).
Typus: *Ceanothus circumcissus* (L. f.) Gaertn. (*Rhamnus circumscissus* L. f., *S. circumcissa* (L. f.) W. Theob.).

(=) *Adolia* Lam., Encycl. 1: 44. 2 Dec 1783.
Typus: non designatus.

4882 ***Colubrina*** Rich. ex Brongn. in Ann. Sci. Nat. (Paris) 10: 368. Apr 1827.
Typus: *C. ferruginosa* Brongn. (*Rhamnus colubrinus* Jacq.) *(typ. cons.)*.

4899 ***Colletia*** Comm. ex Juss., Gen. Pl.: 380. 4 Aug 1789.
Typus: *C. spinosa* Lam. (Tabl. Encycl. 2: 91. 6 Nov 1797).

(H) *Colletia* Scop., Intr. Hist. Nat.: 207. Jan-Apr 1777 [*Ulm.*].
Typus: *Rhamnus iguanaeus* Jacq.

4905 ***Helinus*** E. Mey. ex Endl., Gen. Pl.: 1102. Apr 1840.
Typus: *H. mystacinus* (Aiton) E. Mey. ex Steud. (Nomencl. Bot., ed. 2, 1: 742. Nov 1840) (*Rhamnus mystacinus* Aiton).
(≡) *Mystacinus* Raf., Sylva Tellur.: 30. Oct-Dec 1838.

VITACEAE

4910 ***Ampelocissus*** Planch. in Vigne Amér. Vitic. Eur. 8: 371. Dec 1884.
Typus: *A. latifolia* (Roxb.) Planch. (*Vitis latifolia* Roxb.) *(typ. cons.)*.
(=) *Botria* Lour., Fl. Cochinch.: 96, 153. Sep 1790.
Typus: *B. africana* Lour.

4915 ***Parthenocissus*** Planch. in Candolle & Candolle, Monogr. Phan. 5: 447. Jul 1887.
Typus: *P. quinquefolia* (L.) Planch. (*Hedera quinquefolia* L.) *(typ. cons.)*.

4918a ***Cayratia*** Juss. in Cuvier, Dict. Sci. Nat. 10: 103. 23 Mai 1818.
Typus: *Columella pedata* Lour. [= *Cayratia pedata* (Lam.) Gagnep. (*Cissus pedata* Lam.)] (etiam vide 7897).
(=) *Lagenula* Lour., Fl. Cochinch.: 65, 88. Sep 1790.
Typus: *L. pedata* Lour.

4919 ***Leea*** D. Royen ex L., Syst. Nat., ed. 12, 2: 608, 627 [& Mant. Pl.: 17, 124]. 15-31 Oct 1767.
Typus: *L. aequata* L. *(typ. cons.)*.
(=) *Nalagu* Adans., Fam. Pl. 2: 445, 581. Jul-Aug 1763.
Typus: *Leea asiatica* (L.) Ridsdale (*Phytolacca asiatica* L.).

ELAEOCARPACEAE

4927 ***Aristotelia*** L'Hér., Stirp. Nov.: 31. Dec 1785 (sero) - Jan 1786.
Typus: *A. macqui* L'Hér.
(H) *Aristotela* Adans., Fam. Pl. 2: 125, 520. Jul-Aug 1763 [*Comp.*].
≡ *Othonna* L. 1753.

TILIACEAE

4938 ***Berrya*** Roxb., Pl. Coromandel 3: 60. 18 Feb 1820 *('Berria') (orth. cons.)*.
Typus: *B. ammonilla* Roxb.
(=) *Espera* Willd. in Ges. Naturf. Freunde Berlin Neue Schriften 3: 450. 1801 (post 21 Apr).
Typus: *E. cordifolia* Willd.

4943 ***Brownlowia*** Roxb., Pl. Coromandel 3: 61. 18 Feb 1820.
Typus: *B. elata* Roxb.
(=) *Glabraria* L., Mant. Pl. 156, 276. Oct 1771.
Typus: *G. tersa* L.

4948 **Ancistrocarpus** Oliv. in J. Linn. Soc., Bot. 9: 173. 12 Oct 1865.
Typus: *A. brevispinosus* Oliv. *(typ. cons.)*.

(H) *Ancistrocarpus* Kunth in Humboldt & al., Nov. Gen. Sp. 2: ed. 4°: 186 [& ed. f°: 149]. 8 Dec 1817 [*Phytolacc.*].
Typus: *A. maypurensis* Kunth

4957 **Sparrmannia** L. f., Suppl. Pl.: 41 *('Sparmannia')*, 265, [468]. Apr 1782 *(orth. cons.)*.
Typus: *S. africana* L. f.

(H) *Sparmannia* Buc'hoz, Pl. Nouv. Découv.: 3. 1779 [*Scrophular.*].
≡ *Rehmannia* Fisch. & C. A. Mey. 1835 *(nom. cons.)* (7592).

4959 **Luehea** Willd. in Ges. Naturf. Freunde Berlin Neue Schriften 3: 410. 1801 (post 21 Apr).
Typus: *L. speciosa* Willd.

(H) *Luehea* F. W. Schmidt, Neue Selt. Pfl.: 23. 1793 (ante 17 Jun) [*Verben.*].
Typus: *L. ericoides* F. W. Schmidt

4960 **Mollia** Mart., Nov. Gen. Sp. Pl. 1: 96. Jan-Mar 1826.
Typus: *M. speciosa* Mart.

(H) *Mollia* J. F. Gmel., Syst. Nat. 2: 303, 420. Sep (sero) - Nov 1791 [*Myrt.*].
Typus: *M. imbricata* (Gaertn.) J. F. Gmel. (*Jungia imbricata* Gaertn.).

MALVACEAE

4995 **Malvastrum** A. Gray in Mem. Amer. Acad. Arts, ser. 2, 4: 21. 10 Feb 1849.
Typus: *M. wrightii* A. Gray [= *M. aurantiacum* (Scheele) Walp., *Malva aurantiaca* Scheele].

(=) *Malveopsis* C. Presl. in Abh. Königl. Böhm. Ges. Wiss., ser. 5, 3: 449. Jul-Dec 1845.
Typus: *M. anomala* (Link & Otto) C. Presl (*Malva anomala* Link & Otto).

5004 **Cristaria** Cav., Icon. 5: 10. Jun-Sep 1799.
Typus: *C. glaucophylla* Cav.

(H) *Cristaria* Sonn., Voy. Indes Orient., ed. 4°, 2: 247 [& ed. 8°, 3: 284]. 1782 [*Combret.*].
Typus: *C. coccinea* Sonn.

5007 **Pavonia** Cav., Diss. 2, App.: [1]. Jan-Apr 1786.
Typus: *P. paniculata* Cav.

(=) *Lass* Adans., Fam. Pl. 2: 400, 568. Jul-Aug 1763.
Typus: *Hibiscus spinifex* L.

(=) *Malache* B. Vogel in Trew, Pl. Select.: 50. 1772.
Typus: *M. scabra* B. Vogel

5008a **Peltaea** (C. Presl) Standl. in Contr. U.S. Natl. Herb. 18: 113. 11 Feb 1916 (*Malachra* sect. *Peltaea* C. Presl, Reliq. Haenk. 2: 125. Jan-Jul 1835).
Typus: *Malachra ovata* C. Presl (*P. ovata* (C. Presl) Standl.).

(=) *Peltostegia* Turcz. in Bull. Soc. Imp. Naturalistes Moscou 31(1): 223. 27 Mai 1858.
Typus: *P. parviflora* Turcz.

5013 **Hibiscus** L., Sp. Pl.: 693. 1 Mai 1753.
Typus: *H. syriacus* L. *(typ. cons.)*.

5015 ***Kosteletzkya*** C. Presl, Reliq. Haenk. 2: 130. Jun-Jul 1835.
Typus: *K. hastata* C. Presl

(=) *Thorntonia* Rchb., Consp. Regni Veg.: 202. Dec 1828 - Mar 1829.
Typus: *Hibiscus pentaspermus* Bertero ex DC.

5018 ***Thespesia*** Sol. ex Corrêa in Ann. Mus. Natl. Hist. Nat. 9: 290. 1807.
Typus: *T. populnea* (L.) Sol. ex Corrêa (*Hibiscus populneus* L.).

(≡) *Bupariti* Duhamel, Semis Plantat. Arbr., Add.: 5. 1760.

TRIPLOCHITONACEAE

5022a ***Triplochiton*** K. Schum. in Bot. Jahrb. Syst. 28: 330. 22 Mai 1900.
Typus: *T. scleroxylon* K. Schum.

(H) *Triplochiton* Alef. in Oesterr. Bot. Z. 13: 13. Jan 1863 [*Bombac./Malv.*].
Typus: non designatus.

BOMBACACEAE

5024 ***Bombax*** L., Sp. Pl.: 511. 1 Mai 1753.
Typus: *B. ceiba* L. (*typ. cons.:* [icon] Rheede, Hort. Malab. 3: t. 52. 1682).

5035 ***Bernoullia*** Oliv. in Hooker's Icon. Pl.: ad t. 1169-1170. Dec 1873.
Typus: *B. flammea* Oliv.

(H) *Bernullia* Neck. ex Raf., Autik. Bot.: 173. 1840 [*Ros.*].
Typus: non designatus.

5036 ***Cumingia*** Vidal, Phan. Cuming. Philipp.: 211. Nov 1885.
Typus: *C. philippinensis* Vidal

(H) *Cummingia* D. Don in Sweet, Brit. Fl. Gard. 3: ad t. 257. Apr 1828 [*Haemodor.*].
Typus: *C. campanulata* (Lindl.) D. Don (*Conanthera campanulata* Lindl.).

5040 ***Neesia*** Blume in Nova Acta Phys.-Med. Acad. Caes. Leop.-Carol. Nat. Cur. 17: 83. 1835.
Typus: *N. altissima* (Blume) Blume (*Esenbeckia altissima* Blume).

(H) *Neesia* Spreng., Anleit. Kenntn. Gew., ed. 2, 2: 547. 31 Mar 1818 [*Comp.*].
Typus: non designatus.

5042a ***Bombacopsis*** Pittier in Contr. U.S. Natl. Herb. 18: 162. 3 Mar 1916.
Typus: *B. sessilis* (Benth.) Pittier (*Pachira sessilis* Benth.).

(=) *Pochota* Ram. Goyena, Fl. Nicarag. 1: 198. 1909.
Typus: *P. vulgaris* Ram. Goyena

STERCULIACEAE

5053 ***Dombeya*** Cav., Diss. 2, App.: [1]. Jan-Apr 1786.
Typus: *D. palmata* Cav.

(H) *Dombeya* L'Hér., Stirp. Nov.: 33. Dec (sero) 1785 - Jan 1786 [*Bignon.*].
≡ *Tourrettia* Foug. 1787 *(nom. cons.)* (7766).
(=) *Assonia* Cav., Diss. 2. App.: [2]. Jan-Apr 1786.
Typus: *A. populnea* Cav.

5060 ***Rulingia*** R. Br. in Bot. Mag.: ad t. 2191. 1820.
Typus: *R. pannosa* R. Br.

(H) *Ruelingia* Ehrh. in Neues Mag. Aerzte 6: 297. 12 Mai - 7 Sep 1784 [*Portulac.*].
≡ *Anacampseros* L. 1758 *(nom. cons.)* (2412).

5062 ***Byttneria*** Loefl., Iter Hispan.: 313. Dec 1758.
Typus: *B. scabra* L. (Syst. Nat., ed. 10: 939. 7 Jun 1759).

(H) *Butneria* Duhamel, Traité Arbr. Arbust. 1: 113. 1755 [*Calycanth.*].
Typus: non designatus.

5075 ***Seringia*** J. Gay in Mém. Mus. Hist. Nat. 7: 442. 1821.
Typus: *S. platyphylla* J. Gay, nom. illeg. (*Lasiopetalum arborescens* Aiton, *S. arborescens* (Aiton) Druce).

(H) *Seringia* Spreng., Anleit. Kenntn. Gew., ed. 2, 2: 694. 31 Mar 1818 [*Celastr.*].
≡ *Ptelidium* Thouars 1804.

5080 ***Pterospermum*** Schreb., Gen. Pl.: 461. Mai 1791.
Typus: *P. suberifolium* (L.) Willd. (Sp. Pl. 3: 728. 1800) (*Pentapetes suberifolia* L.).

5091 ***Cola*** Schott & Endl., Melet. Bot.: 33. 1832.
Typus: *C. acuminata* (P. Beauv.) Schott & Endl. (*Sterculia acuminata* P. Beauv.) *(typ. cons.)*.

(=) *Bichea* Stokes, Bot. Mater. Med. 2: 564. 1812.
Typus: *B. solitaria* Stokes

DILLENIACEAE

5109 ***Saurauia*** Willd. in Ges. Naturf. Freunde Berlin Neue Schriften 3: 407. 1801 (post 21 Apr) *('Saurauja')* *(orth. cons.)*.
Typus: *S. excelsa* Willd.

OCHNACEAE

5113 ***Ouratea*** Aubl., Hist. Pl. Guiane: 397. Jun-Dec 1775.
Typus: *O. guianensis* Aubl.

THEACEAE

5144 ***Bonnetia*** Mart., Nov. Gen. Sp. Pl. 1: 114. Jan-Mar 1826.
Typus: *B. anceps* Mart. *(typ. cons.).*

(H) *Bonnetia* Schreb., Gen. Pl.: 363. Apr 1789 [*The.*].
≡ *Mahurea* Aubl. 1775.

(=) *Kieseria* Nees in Wied-Neuwied, Reise Bras. 2: 338. Jan-Jun 1821.
Typus: *K. stricta* Nees

5148 ***Gordonia*** J. Ellis in Philos. Trans. 60: 520. 1771.
Typus: *G. lasianthus* (L.) J. Ellis (*Hypericum lasianthus* L.) (etiam vide 8412).

5149 ***Laplacea*** Kunth in Humboldt & al., Nov. Gen. Sp. 5, ed. 4°: 207 [& ed. f.°: 161]. 25 Feb 1822.
Typus: *L. speciosa* Kunth

5153 ***Ternstroemia*** Mutis ex L. f., Suppl. Pl.: 39, 264. Apr 1782.
Typus: *T. meridionalis* Mutis ex L. f.

(=) *Mokof* Adans., Fam. Pl. 2: 501, 578. Jul-Aug 1763.
Typus: non designatus.

(=) *Taonabo* Aubl., Hist. Pl. Guiane: 569. Jun-Dec 1775.
Typus: *T. dentata* Aubl.

5155 ***Anneslea*** Wall., Pl. Asiat. Rar. 1: 5. Sep 1829.
Typus: *A. fragrans* Wall.

(H) *Anneslia* Salisb., Parad. Lond.: ad t. 64. 1 Mar 1807 [*Legum.*].
Typus: *A. falcifolia* Salisb., nom. illeg. (*Gleditsia inermis* L.).

5157a ***Cleyera*** Thunb., Nov. Gen. Pl.: 68. 18 Jun 1783.
Typus: *C. japonica* Thunb.

(H) *Cleyera* Adans., Fam. Pl. 2: 224, 540. Jul-Aug 1763 [*Logan.*].
≡ *Polypremum* L. 1753.

5157b ***Freziera*** Willd., Sp. Pl. 2(2): 1122, 1179. Dec 1799.
Typus: *F. undulata* (Sw.) Willd. (*Eroteum undulatum* Sw.) *(typ. cons.).*

(≡) *Eroteum* Sw., Prodr.: 5, 85. 20 Jun - 29 Jul 1788.

(=) *Lettsomia* Ruiz & Pav., Fl. Peruv. Prodr.: 77. Oct (prim.) 1794.
Typus: non designatus.

GUTTIFERAE (CLUSIACEAE)

5171 ***Vismia*** Vand., Fl. Lusit. Bras. Spec.: 51. 1788.
Typus: *V. cayennensis* (Jacq.) Pers. (Syn. Pl. 2: 86. Nov 1806) (*Hypericum cayennense* Jacq.) *(typ. cons.)*.

(=) *Caopia* Adans., Fam. Pl. 2: 448. Jul-Aug 1763.
Typus: non designatus.

5195 ***Balboa*** Planchon & Triana in Ann. Sci. Nat., Bot., ser. 4, 13: 315. Mai 1860.
Typus: *B. membranaceum* Planchon & Triana

(H) *Balboa* Liebm. ex Didr. in Vidensk. Meddel. Dansk Naturhist. Foren. Kjøbenhavn 1853: 106. 1853 [*Legum.*].
Typus: *B. diversifolia* Liebm. ex Didr.

5205 ***Platonia*** Mart., Nov. Gen. Sp. Pl. 3: 168. Sep 1832.
Typus: *P. insignis* Mart., nom. illeg. (*Moronobea esculenta* Arruda, *P. esculenta* (Arruda) Rickett & Stafleu).

(H) *Platonia* Raf., Caratt. Nuov. Gen.: 73. 1810 [*Cist.*].
≡ *Helianthemum* Mill. 1754.

DIPTEROCARPACEAE

5214 ***Doona*** Thwaites in Hooker's J. Bot. Kew Gard. Misc. 3: t. 12. 1851 (sero).
Typus: *D. zeylanica* Thwaites

(=) *Caryolobis* Gaertn., Fruct. Sem. Pl. 1: 215. Dec 1788.
Typus: *C. indica* Gaertn.

5215 ***Hopea*** Roxb., Pl. Coromandel 3: 7. Jul 1811.
Typus: *H. odorata* Roxb.

(H) *Hopea* Garden ex L., Syst. Nat. Ed. 12, 2: 509 [& Mant. Pl.: 14, 105]. 15-31 Oct 1767 [*Symploc.*].
Typus: *H. tinctoria* L.

5221 ***Pierrea*** F. Heim in Bull. Mens. Soc. Linn. Paris: 958. 1891.
Typus: *P. pachycarpa* F. Heim

(H) *Pierrea* Hance in J. Bot. 15: 339. Nov 1877 [*Flacourt.*].
Typus: *P. dictyoneura* Hance

COCHLOSPERMACEAE

5250 ***Cochlospermum*** Kunth in Humboldt & al., Nov. Gen. Sp. 5, ed. 4°: 297 [& ed. f°: 231]. Jun 1822.
Typus: *Bombax gossypium* L., nom. illeg. (*Bombax religiosum* L., *C. religiosum* (L.) Alston).

CANELLACEAE

5254 ***Canella*** P. Browne, Civ. Nat. Hist. Jamaica: 275. 10 Mar 1756.
Typus: *C. winterana* (L.) Gaertn. (Fruct. Sem. Pl. 1: 373. Dec 1788) (*Laurus winterana* L.).

5256 ***Warburgia*** Engl., Pflanzenw. Ost-Afrikas C: 276. Jul 1895.
Typus: *W. stuhlmannii* Engl.

(=) *Chibaca* Bertol. in Mem. Reale Accad. Sci. Ist. Bologna 4: 545. 1853.
Typus: *C. salutaris* Bertol.

VIOLACEAE

5259 ***Amphirrhox*** Spreng., Syst. Veg. 4(2): 51, 99. Jan-Jun 1827.
Typus: *A. longifolia* (A. St.-Hil.) Spreng. (*Spathularia longifolia* A. St.-Hil.).

5262 ***Rinorea*** Aubl., Hist. Pl. Guiane: 235. Jun-Dec 1775.
Typus: *R. guianensis* Aubl.

(=) *Conohoria* Aubl., Hist. Pl. Guiane 239. Jun-Dec 1775.
Typus: *C. flavescens* Aubl.

5271 ***Hybanthus*** Jacq., Enum. Syst. Pl.: 2, 17. Aug-Sep 1760.
Typus: *H. havanensis* Jacq.

FLACOURTIACEAE

5278 ***Erythrospermum*** Thouars, Hist. Vég. Isles Austral. Afriq.: 65. Jan 1808.
Typus: *E. pyrifolium* Poir. (in Lamarck, Encycl., Suppl. 2: 584. 3 Jul 1812) *(typ. cons.)*.

(=) *Pectinea* Gaertn., Fruct. Sem. Pl. 2: 136. Sep (sero) - Nov 1790.
Typus: *P. zeylanica* Gaertn.

5304 ***Scolopia*** Schreb., Gen. Pl.: 335. Apr 1789.
Typus: *S. pusilla* (Gaertn.) Willd. (Sp. Pl. 2: 981. Dec 1799) (*Limonia pusilla* Gaertn.).

(=) *Aembilla* Adans., Fam. Pl. 2: 448, 513. Jul-Aug 1763.
Typus: non designatus.

5311 ***Byrsanthus*** Guill. in Delessert, Icon. Sel. Pl. 3: 30. Feb 1838.
Typus: *B. brownii* Guill.

(H) *Byrsanthes* C. Presl, Prodr. Monogr. Lobel.: 41. Jul-Aug 1836 [*Campanul.*].
Typus (vide Rickett & Stafleu in Taxon 8: 314. 1959): *B. humboldtiana* C. Presl, nom. illeg. (*Lobelia nivea* Willd.).

5320 ***Xylosma*** G. Forst., Fl. Ins. Austr.: 72. Oct-Nov 1786.
Typus: *X. orbiculata* (J. R. Forst. & G. Forst.) G. Forst. (*Myroxylon orbiculatum* J. R. Forst. & G. Forst.) (etiam vide 3584).

5331 ***Idesia*** Maxim. in Bull. Acad. Imp. Sci. Saint-Pétersbourg, ser. 3, 10: 485. 8 Sep 1866.
Typus: *I. polycarpa* Maxim.

(H) *Idesia* Scop., Intr. Hist. Nat.: 199. Jan-Apr 1777 [*Verben.*].
≡ *Ropourea* Aubl. 1775.

5334 ***Lunania*** Hook. in London J. Bot. 3: 317. 1844.
Typus: *L. racemosa* Hook.

(H) *Lunanea* DC., Prodr. 2: 92. Nov (med.) 1825 [*Stercul.*].
≡ *Bichea* Stokes 1812 (*nom. rej.* sub 5091).

5337 ***Samyda*** Jacq., Enum. Syst. Pl.: 4, 21. Aug-Sep 1760.
Typus: *S. dodecandra* Jacq. *(typ. cons.)*.

(H) *Samyda* L., Sp. Pl.: 443. 1 Mai 1753 [*Mel.*].
Typus: *S. guidonia* L.

5338 ***Laetia*** Loefl. ex L., Syst. Nat., ed. 10: 1068, 1074, 1373. 7 Jun 1759.
Typus: *L. americana* L.

5341 ***Ryania*** Vahl, Eclog. Amer. 1: 51. 1797.
Typus: *R. speciosa* Vahl

(=) *Patrisa* Rich. in Actes Soc. Hist. Nat. Paris 1: 110. 1792.
Typus: *P. pyrifera* Rich.

5353 ***Tetralix*** Griseb., Cat. Pl. Cub.: 8. Mai-Aug 1866.
Typus: *T. brachypetalus* Griseb.

(H) *Tetralix* Zinn, Cat. Pl. Hort. Gott.: 202. 20 Apr - 21 Mai 1757 [*Eric.*].
Typus: *Erica herbacea* L. *(nom. rej.)*.

LOASACEAE

5384 ***Eucnide*** Zucc., Delect. Sem. Hort. Monac. 1844: [4]. 28 Dec 1844.
Typus: *E. bartonioides* Zucc.

(=) *Microsperma* Hook. in Icon. Pl.: ad t. 234. Jan-Feb 1839.
Typus: *M. lobatum* Hook.

5392 ***Blumenbachia*** Schrad. in Gött. Gel. Anz. 1825: 1705. 24 Oct 1825.
Typus: *B. insignis* Schrad.

(H) *Blumenbachia* Koeler, Descr. Gram.: 28. 1802 [*Gram.*].
Typus: *B. halepensis* (L.) Koeler (*Holcus halepensis* L.).

ANCISTROCLADACEAE

5400 ***Ancistrocladus*** Wall., Numer. List: No. 1052. 1829.
Typus: *A. hamatus* (Vahl) Gilg (in Engler & Prantl, Nat. Pflanzenfam. 3(6): 276. 19 Feb 1895) (*Wormia hamata* Vahl).

(=) *Bembix* Lour., Fl. Cochinch.: 259, 282. Sep 1790.
Typus: *B. tectoria* Lour.

CACTACEAE

5401a ***Arthrocereus*** A. Berger, Kakteen: 337. Jul-Aug 1929.
Typus: [icon] *'Cereus damazoi'* in Monatsschr. Kakteenk. 28: [63]. 1918 *(typ. cons.)* [= *A. glaziovii* (K. Schum.) N. P. Taylor & Zappi (*Cereus glaziovii* K. Schum.)].

5409 ***Melocactus*** Link & Otto in Verh. Vereins Beförd. Gartenbaues Königl. Preuss. Staaten 3: 417. 1827.
Typus: *M. communis* Link & Otto (*Cactus melocactus* L.) *(typ. cons.)*.

(H) *Melocactus* Boehm. in Ludwig, Def. Gen. Pl., ed. 3: 79. 1760 [*Cact.*].
≡ *Cactus* L. 1753 (*nom. rej.* sub 5411).

5411 ***Mammillaria*** Haw., Syn. Pl. Succ.: 177. 1812.
Typus: *M. simplex* Haw., nom. illeg. (*Cactus mammillaris* L., *M. mammillaris* (L.) H. Karst.) *(typ. cons.)*.

(H) *Mammillaria* Stackh. in Mém. Soc. Imp. Naturalistes Moscou 2: 55, 74. 1809 [*Rhodoph.*].
Typus: non designatus.

(≡) *Cactus* L., Sp. Pl.: 466. 1 Mai 1753.

5411a ***Coryphantha*** (Engelm.) Lem., Cactées: 32. Aug 1868 (*Mammillaria* subg. *Coryphanta* Engelm. in Emory, Rep. U.S. Mex. Bound. 2: 10. 1858).
Typus: *Mammillaria sulcata* Engelm. (*C. sulcata* (Engelm.) Britton & Rose).

(=) *Aulacothele* Monv., Cat. Pl. Exot.: 21. 1846.
Typus: *Mammillaria aulacothele* Lem.

5416 ***Rhipsalis*** Gaertn., Fruct. Sem. Pl. 1: 137. Dec 1788.
Typus: *R. cassutha* Gaertn.

(=) *Hariota* Adans., Fam. Pl. 2: 243, 520. Jul-Aug 1763.
Typus: non designatus.

OLINIACEAE

5428 ***Olinia*** Thunb. in Arch. Bot. (Leipzig) 2(1): 4. Mai-Jul 1800.
Typus: *O. cymosa* (L. f.) Thunb. (*Sideroxylon cymosum* L. f.).

(=) *Plectronia* L., Syst. Nat., ed. 12, 2: 138, 183 [& Mant. Pl.: 6, 52]. 15-31 Oct 1767.
Typus: *P. ventosa* L.

THYMELAEACEAE

5430 ***Aquilaria*** Lam., Encycl. 1: 49. 2 Dec 1783.
Typus: *A. malaccensis* Lam.

(=) *Agallochum* Lam., Encycl. 1: 48. Dec 1783.
Typus: non designatus.

5436 ***Struthiola*** L., Syst. Nat., ed. 12, 2: 108, 127 [& Mant. Pl.: 4, 41]. 15-31 Oct 1767.
Typus: *S. virgata* L. *(typ. cons.)*.

(=) *Belvala* Adans., Fam. Pl. 2: 285, 525. Jul-Aug 1763.
Typus: *Passerina dodecandra* L.

5446 ***Wikstroemia*** Endl., Prodr. Fl. Norfolk.: 47. 1833 (post 12 Mai) *('Wickstroemia') (orth. cons.)*.
Typus: *W. australis* Endl.

(H) *Wikstroemia* Schrad. in Gött. Gel. Anz. 1821: 710. 5 Mai 1821 [*The.*].
Typus: *W. fruticosa* Schrad.

(=) *Capura* L., Mant. Pl.: 149, 225. Oct 1771.
Typus: *C. purpurata* L.

5453 ***Thymelaea*** Mill., Gard. Dict. Abr., ed. 4: [1381]. 28 Jan 1754.
Typus: *T. sanamunda* All. (Fl. Pedem. 1: 132. Apr-Jul 1785) (*Daphne thymelaea* L.) *(typ. cons.)*.

5457 ***Ovidia*** Meisn. in Candolle, Prodr. 14(2): 524. Nov (sero) 1857.
Typus: *O. pillopillo* (Gay) Meisn. (*Daphne pillopillo* Gay) *(typ. cons.)*.

(H) *Ovidia* Raf., Fl. Tellur. 3: 68. Nov-Dec 1837 [*Commelin.*].
Typus: *O. gracilis* (Ruiz & Pav.) Raf. (*Commelina gracilis* Ruiz & Pav.).

5467 ***Pimelea*** Banks ex Gaertn., Fruct. Sem. Pl. 1: 186. Dec 1788.
Typus: *P. laevigata* Gaertn., nom. illeg. (*Banksia prostrata* J. R. Forst. & G. Forst., *P. prostrata* (J. R. Forst. & G. Forst.) Willd.).

5467a ***Synandrodaphne*** Gilg in Bot. Jahrb. Syst. 53: 362. 19 Oct 1915.
Typus: *S. paradoxa* Gilg

(H) *Synandrodaphne* Meisn. in Candolle, Prodr. 15(1): 176. Mai (prim.) 1864 [*Laur.*].
≡ *Rhodostemonodaphne* Rohwer & Kubitzki 1985.

ELAEAGNACEAE

5471 ***Shepherdia*** Nutt., Gen. N. Amer. Pl. 2: 240. 14 Jul 1818.
Typus: *S. canadensis* (L.) Nutt. (*Hippophaë canadensis* L.) *(typ. cons.)*.

LYTHRACEAE

5486 ***Nesaea*** Comm. ex Kunth in Humboldt & al., Nov. Gen. Sp. 6, ed. f°: 151. 6 Aug 1823.
Typus: *N. triflora* (L. f.) Kunth (*Lythrum triflorum* L. f.).

(H) *Nesaea* J. V. Lamour. in Nouv. Bull. Sci. Soc. Philom. Paris 3: 185. Dec 1812 [*Chloroph.*].
Typus: non designatus.

SONNERATIACEAE

5497 ***Sonneratia*** L. f., Suppl. Pl.: 38, 252. Apr 1782.
Typus: *S. acida* L. f.

(=) *Blatti* Adans., Fam. Pl. 2: 88, 526. Jul-Aug 1763.
Typus: non designatus.

LECYTHIDACEAE

5505 ***Careya*** Roxb., Pl. Coromandel 3: 13. Jul 1811.
Typus: *C. herbacea* Roxb. *(typ. cons.)*.

5506 ***Barringtonia*** J. R. Forst. & G. Forst., Char. Gen. Pl.: 38. 29 Nov 1775.
Typus: *B. speciosa* J. R. Forst. & G. Forst.

(=) *Huttum* Adans., Fam. Pl. 2: 88, 616. Jul-Aug 1763.
Typus: non designatus.

5510 ***Gustavia*** L., Pl. Surin.: 12, 17, 18. 23 Jun 1775.
Typus: *G. augusta* L.

(=) *Japarandiba* Adans., Fam. Pl. 2: 448, 564. Jul-Aug 1763.
Typus (vide Miers in Trans. Linn. Soc. London 30: 183. 1874): *Gustavia marcgraviana* Miers

RHIZOPHORACEAE

5525 ***Carallia*** Roxb., Pl. Coromandel 3: 8. Jul 1811.
Typus: *C. lucida* Roxb.

(=) *Karekandel* Wolf, Gen. Pl.: 73. 1776.
Typus (vide Ross in Acta Bot. Neerl. 15: 158. 1966): *Karkandela malabarica* Raf.

(=) *Barraldeia* Thouars, Gen. Nov. Madagasc.: 24. 17 Nov 1806.
Typus: non designatus.

5528 ***Weihea*** Spreng., Syst. Veg. 2: 559, 594. Jan-Mai 1825.
Typus: *W. madagascarensis* Spreng.

(≡) *Richaeia* Thouars, Gen. Nov. Madagasc.: 25. 17 Nov 1806.

COMBRETACEAE

5538 ***Combretum*** Loefl., Iter Hispan.: 308. Dec 1758.
Typus: *C. fruticosum* (Loefl.) Stuntz (U.S.D.A. Bur. Pl. Industr. Invent. Seeds 31: 86. 1914) (*Gaura fruticosa* Loefl.).

(=) *Grislea* L., Sp. Pl.: 348. 1 Mai 1753.
Typus: *G. secunda* L.

5543 ***Bucida*** L., Syst. Nat., ed. 10: 1012, 1025, 1368. 7 Jun 1759.
Typus: *B. buceras* L.

(≡) *Buceras* P. Browne, Civ. Nat. Hist. Jamaica: 221. 10 Mar 1756.

5544 ***Terminalia*** L., Syst. Nat., ed. 12, 2: 665, 674 ('638') [& Mant. Pl.: 21, 128]. 15-31 Oct 1767.
Typus: *T. catappa* L.

(≡) *Adamaram* Adans., Fam. Pl. 2: (23), 445, 513. Jul-Aug 1763 (typ. des.: Exell in Index Nom. Gen.: No. 47/06700. 1958).

5545 ***Buchenavia*** Eichler in Flora 49: 164. 17 Apr 1866.
Typus: *B. capitata* (Vahl) Eichler (*Bucida capitata* Vahl).

(=) *Pamea* Aubl., Hist. Pl. Guiane: 946. Jun-Dec 1775.
Typus: *P. guianensis* Aubl.

MYRTACEAE

5557 ***Myrteola*** O. Berg in Linnaea 27: 348. Jan 1856.
Typus: *M. microphylla* (Humb. & Bonpl.) Berg (*Myrtus microphylla* Humb. & Bonpl.) *(typ. cons.)*.

(≡) *Amyrsia* Raf., Sylva Tellur.: 106. Oct-Dec 1838.

(=) *Cluacena* Raf., Sylva Tellur.: 104. Oct-Dec 1838.
Typus (vide McVaugh in Taxon 5: 139. 1956): *C. vaccinioides* (Kunth) Raf. (*Myrtus vaccinioides* Kunth).

5575 ***Calyptranthes*** Sw., Prodr.: 5, 79. 20 Jun - 29 Jul 1788.
Typus: *C. chytraculia* (L.) Sw. (*Myrtus chytraculia* L.) *(typ. cons.)*.

(≡) *Chytraculia* P. Browne, Civ. Nat. Hist. Jamaica: 239. 10 Mar 1756.

5582 ***Jambosa*** Adans., Fam. Pl. 2: 88, 564. Jul-Aug 1763 *('Jambos') (orth. cons.)*.
Typus: *J. vulgaris* DC., nom. illeg. (*Eugenia jambos* L., *J. jambos* (L.) Millsp.) *(typ. cons.)* (etiam vide 5583).

5583 ***Syzygium*** R. Br. ex Gaertn., Fruct. Sem. Pl. 1: 166. Dec 1788.
Typus: *S. caryophyllaeum* Gaertn. *(typ. cons.)*.

(H) *Suzygium* P. Browne, Civ. Nat. Hist. Jamaica: 240. 10 Mar 1756. [*Myrt.*].
Typus: *Myrtus zuzygium* L.

(=) *Caryophyllus* L., Sp. Pl.: 515. 1 Mai 1753.
Typus: *C. aromaticus* L.

(=) *Jambosa* Adans., Fam. Pl. 2: 88, 564. Jul-Aug 1763 *(nom. cons.)* (5582).

5585 ***Piliocalyx*** Brongn. & Gris in Bull. Soc. Bot. France 12: 185. 1865 (post Apr).
Typus: *P. robustus* Brongn. & Gris.

(H) *Pileocalyx* Gasp. in Ann. Sci. Nat., Bot., ser. 3, 9: 220. Apr 1848 [*Cucurbit.*].
≡ *Mellonia* Gasp. 1847.

5588 ***Metrosideros*** Banks ex Gaertn., Fruct. Sem. Pl. 1: 170. Dec 1788.
Typus: *M. spectabilis* Gaertn. *(typ. cons.)*.

(=) *Nani* Adans., Fam. Pl. 2: 88, 581. Jul-Aug 1763.
Typus: *Metrosideros vera* Lindl.

5594 ***Xanthostemon*** F. Muell. in Hooker's J. Bot. Kew Gard. Misc. 9: 17. Jan 1857.
Typus: *X. paradoxus* F. Muell.

(=) *Nani* Adans., Fam. Pl. 2: 88, 581. Jul-Aug 1763.
Typus: *Metrosideros vera* Lindl.

5599 ***Leptospermum*** J. R. Forst. & G. Forst., Char. Gen. Pl.: 36. 29 Nov 1775.
Typus: *L. scoparium* J. R. Forst. & G. Forst. *(typ. cons.)*.

5600 ***Agonis*** (DC.) Sweet, Hort. Brit., ed. 2: 209. Oct-Dec 1830 (*Leptospermum* sect. *Agonis* DC., Prodr. 3: 226. Mar (med.) 1828).
Typus: *Leptospermum flexuosum* (Willd.) Spreng. (*Metrosideros flexuosa* Willd., *A. flexuosa* (Willd.) Sweet) *(typ. cons.)*.

5601 ***Kunzea*** Rchb., Consp. Regni Veg.: 175. Dec 1828 - Mar 1829.
Typus: *K. capitata* (Sm.) Rchb. ex Heynh. (Nomencl. Bot. Hort.: 337. 27 Sep - 3 Oct 1840) (*Metrosideros capitata* Sm.).

(=) *Tillospermum* Salisb. in Monthly Rev. 75: 74. 1814.
Typus: *Leptospermum ambiguum* Sm.

5603 ***Melaleuca*** L., Syst. Nat., ed. 12, 2: 507, 509 . 15-31 Oct 1767.
Typus: *M. leucadendra* (L.) L. (*Myrtus leucadendra* L.) *(typ. cons.)*.

(≡) *Kajuputi* Adans., Fam. Pl. 2: 84, 530. Jul-Aug 1763.

5621 **Thryptomene** Endl. in Ann. Wiener Mus. Naturgesch. 2: 192. 1839.
Typus: *T. australis* Endl.

(=) *Gomphotis* Raf., Sylva Tellur.: 103. Oct-Dec 1838.
Typus: *G. saxicola* (A. Cunn. ex Hook.) Raf. (*Baeckea saxicola* A. Cunn. ex Hook.).

5625 **Verticordia** DC., Prodr. 3: 208. Mar (med.) 1828.
Typus: *V. fontanesii* DC., nom. illeg. (*Chamelaucium plumosum* Desf., *V. plumosa* (Desf.) Druce) *(typ. cons.)*.

MELASTOMATACEAE

5632 **Pterolepis** (DC.) Miq., Comm. Phytogr.: 72. 16-21 Mar 1840 (*Osbeckia* sect. *Pterolepis* DC., Prodr. 3: 140. Mar (med.) 1828).
Typus: *Osbeckia parnassiifolia* DC. (*P. parnassiifolia* (DC.) Triana) *(typ. cons.)*.

(H) *Pterolepis* Schrad. in Gött. Gel. Anz. 1821: 2071. 29 Dec 1821 [*Cyper.*].
Typus: *P. scirpoides* Schrad.

5648 **Microlepis** (DC.) Miq., Comm. Phytogr.: 71. 16-21 Mar 1840 (*Osbeckia* sect. *Microlepis* DC., Prodr. 3: 139. Mar (med.) 1828).
Typus: *Osbeckia oleifolia* DC. (*M. oleifolia* (DC.) Triana) *(typ. cons.)*.

5659 **Dissotis** Benth. in Hooker, Niger Fl.: 346. Nov-Dec 1849.
Typus: *D. grandiflora* (Sm.) Benth. (*Osbeckia grandiflora* Sm.).

(≡) *Hedusa* Raf., Sylva Tellur.: 101. Oct-Dec 1838.
(=) *Kadalia* Raf., Sylva Tellur.: 101. Oct-Dec 1838.
Typus: non designatus.

5665 **Monochaetum** (DC.) Naudin in Ann. Sci. Nat., Bot., ser. 3, 4: 48. Jul 1845 (*Arthrostemma* sect. *Monochaetum* DC., Prodr. 3: 135, 138. Mar (med.) 1828).
Typus: *Arthrostemma calcaratum* DC. (*M. calcaratum* (DC.) Triana) *(typ. cons.)*.

(=) *Ephynes* Raf., Sylva Tellur.: 101. Oct-Dec 1838.
Typus: *E. bonplandii* (Humb. & Bonpl.) Raf. (*Rhexia bonplandii* Humb. & Bonpl.).

5669 **Cambessedesia** DC., Prodr. 3: 110. Mar (med.) 1828.
Typus: *C. hilariana* (Kunth) DC. (*Rhexia hilariana* Kunth) *(typ. cons.)*.

(H) *Cambessedea* Kunth in Ann. Sci. Nat. (Paris) 2: 336. 1824 [*Anacard.*].
Typus: *Mangifera axillaris* Desr.

5676 ***Rhynchanthera*** DC., Prodr. 3: 106. Mar (med.) 1828.
Typus: *R. grandiflora* (Aubl.) DC. (*Melastoma grandiflorum* Aubl.) *(typ. cons.)*.

(H) *Rynchanthera* Blume, Tab. Pl. Jav. Orchid.: ad t. 78. 1-7 Dec 1825 [*Orchid.*].
Typus: *R. paniculata* Blume

5692 ***Meriania*** Sw., Fl. Ind. Occid.: 823. Jan-Jun 1798.
Typus: *M. leucantha* (Sw.) Sw. (*Rhexia leucantha* Sw.) *(typ. cons.)*.

(H) *Meriana* Trew, Pl. Select.: 11. 1754. [*Irid.*].
≡ *Watsonia* Mill. 1759 *(nom. cons.)* (1315).

5708 ***Bertolonia*** Raddi, Quar. Piant. Nuov. Bras.: 5. 1820.
Typus: *B. nymphaeifolia* Raddi

(H) *Bertolonia* Spin, Jard. St. Sébastien, ed. 1909: 24. 1809 [*Myopor.*].
Typus: *B. glandulosa* Spin

5729 ***Sonerila*** Roxb., Fl. Ind. 1: 180. Jan-Jun 1820.
Typus: *S. maculata* Roxb.

5759 ***Miconia*** Ruiz & Pav., Fl. Peruv. Prodr.: 60. Oct (prim.) 1794.
Typus: *M. triplinervis* Ruiz & Pav. (Syst. Veg. Fl. Peruv. Chil.: 104. Dec (sero) 1798) *(typ. cons.)*.

(=) *Leonicenia* Scop., Intr. Hist. Nat.: 212. Jan-Apr 1777.
Typus: *Fothergilla mirabilis* Aubl.

5768 ***Bellucia*** Neck. ex Raf., Sylva Tellur.: 92. Oct-Dec 1838.
Typus: *B. nervosa* Raf., nom. illeg. (*Blakea quinquenervia* Aubl., *Bellucia quinquenervia* (Aubl.) H. Karst.) [= *Bellucia grossularioides* (L.) Triana (*Melastoma grossularioides* L.)].

(H) *Belluccia* Adans., Fam. Pl. 2: 344, 525. Jul-Aug 1763 [*Rut.*].
≡ *Ptelea* L. 1753.

(≡) *Apatitia* Desv. ex Ham., Prodr. Pl. Ind. Occid.: 42. 1825.

5777a ***Astronidium*** A. Gray, U.S. Expl. Exped., Phan.: 581. Jun 1854.
Typus: *A. parviflorum* A. Gray

(=) *Lomanodia* Raf., Sylva Tellur.: 97. Oct-Dec 1838.
Typus (vide Veldkamp in Taxon 32: 134. 1983): *L. glabra* (G. Forst.) Raf. (*Melastoma glabrum* G. Forst.).

ONAGRACEAE

5826 ***Diplandra*** Hook. & Arn., Bot. Beechey Voy.: 291. Dec 1838.
Typus: *D. lopezioides* Hook. & Arn.

(H) *Diplandra* Bertero in Mercurio Chileno 13: 612. 15 Apr 1829 [*Hydrocharit.*].
Typus: *D. potamogeton* Bertero

ARALIACEAE

5852 ***Schefflera*** J. R. Forst. & G. Forst., Char. Gen. Pl.: 23. 29 Nov 1775.
Typus: *S. digitata* J. R. Forst. & G. Forst.

(=) *Sciodaphyllum* P. Browne, Civ. Nat. Hist. Jamaica: 190. 10 Mar 1756.
Typus: *S. brownii* Spreng.

UMBELLIFERAE (APIACEAE)

5938 ***Anthriscus*** Pers., Syn. Pl. 1: 320. 1 Apr - 15 Jun 1805.
Typus: *A. vulgaris* Pers. 1805, non Bernh. 1800 (*Scandix anthriscus* L., *A. caucalis* M. Bieb.) *(typ. cons.)*.

(H) *Anthriscus* Bernh., Syst. Verz.: 113. 1800 [*Umbell.*].
Typus: *A. vulgaris* Bernh. (*Tordylium anthriscus* L.).

(=) *Cerefolium* Fabr., Enum.: 36. 1759.
Typus: *Scandix cerefolium* L.

5941 ***Osmorhiza*** Raf. in Amer. Monthly Mag. & Crit. Rev. 4: 192. Jan 1819.
Typus: *O. claytonii* (Michx.) C. B. Clarke (in Hooker, Fl. Brit. India 2: 680. Mai 1879) (*Myrrhis claytonii* Michx.).

(≡) *Uraspermum* Nutt., Gen. N. Amer. Pl. 1: 192. 14 Jul 1818.

5956 ***Bifora*** Hoffm., Gen. Pl. Umbell., ed. 2: xxxiv, 191. 1816 (post 15 Mai).
Typus: *B. dicocca* Hoffm., nom. illeg. (*Coriandrum testiculatum* L., *B. testiculata* (L.) Spreng. ex Roem. & Schult.).

5964 ***Scaligeria*** DC., Coll. Mém. 5: 70. 12 Sep 1829.
Typus: *S. microcarpa* DC.

(H) *Scaligera* Adans., Fam. Pl. 2: 323, 601. Jul-Aug 1763 [*Legum.*].
≡ *Aspalathus* L. 1753.

5977 ***Tauschia*** Schltdl. in Linnaea 9: 607. 1835 (post Feb).
Typus: *T. nudicaulis* Schltdl.

(H) *Tauschia* Preissler in Flora 11: 44. 21 Jan 1828 [*Eric.*].
Typus: *T. hederifolia* Preissler

5990 ***Lichtensteinia*** Cham. & Schltdl. in Linnaea 1: 394. Aug-Oct 1826.
Typus: *L. lacera* Cham. & Schltdl. *(typ. cons.)*.

(H) *Lichtensteinia* Willd. in Ges. Naturf. Freunde Berlin Mag. Neuesten Entdeck. Gesammten Naturk. 2: 19. 1808 [*Lil.*].
Typus: non designatus.

5992 ***Heteromorpha*** Cham. & Schltdl. in Linnaea 1: 385. Aug-Oct 1826.
Typus: *H. arborescens* Cham. & Schltdl. (*Bupleurum arborescens* Thunb. 1794, non Jacq. 1788).

(H) *Heteromorpha* Cass. in Bull. Sci. Soc. Philom. Paris 1817: 12. Jan 1817 [*Comp.*].
≡ *Heterolepis* Cass. 1820 *(nom. cons.)* (9057).

5998 ***Trinia*** Hoffm., Gen. Pl. Umbell.: xxix, 92. 1814.
Typus: *T. glaberrima* Hoffm., nom. illeg. (*Seseli pumilum* L., *T. pumila* (L.) Rchb.) *(typ. cons.)*.

6004a ***Cyclospermum*** Lag., Amen. Nat. Españ. 1(2): 101. 1821.
Typus: *C. leptophyllum* (Pers.) Sprague ex Britton & P. Wilson (Bot. Porto Rico 6: 52. 14 Jan 1925) (*Pimpinella leptophylla* Pers.) *(typ. cons.)*.

6014 ***Trachyspermum*** Link, Enum. Hort. Berol. Alt. 1: 267. Mar-Jun 1821.
Typus: *T. copticum* (L.) Link (*Ammi copticum* L.).

(≡) *Ammios* Moench, Methodus: 99. 4 Mai 1794.

6015 ***Cryptotaenia*** DC., Coll. Mém. 5: 42. 12 Sep 1829.
Typus: *C. canadensis* (L.) DC. (*Sison canadense* L.).

(≡) *Deringa* Adans., Fam. Pl. 2: 498, 549. Jul-Aug 1763.

6018 ***Falcaria*** Fabr., Enum.: 34. 1759.
Typus: *F. vulgaris* Bernh. (Syst. Verz.: 176. 1800) (*Sium falcaria* L.).

6045 ***Polemannia*** Eckl. & Zeyh., Enum. Pl. Afric. Austral.: 347. Apr-Jun 1837.
Typus: *P. grossulariifolia* Eckl. & Zeyh.

(H) *Polemannia* K. Bergius ex Schltdl. in Linnaea 1: 250. Apr 1826 [*Lil.*].
Typus: *P. hyacinthiflora* K. Bergius ex Schltdl.

6052a ***Libanotis*** Haller ex Zinn, Cat. Pl. Hort. Gott.: 226. 20 Apr - 21 Mai 1757.
Typus: *L. montana* Crantz (Stirp. Austr. Fasc. 3: 117. 1767) (*Athamanta libanotis* L.).

(H) *Libanotis* Hill, Brit. Herb.: 420. Nov 1756 [*Umbell.*].
Typus (vide Rauschert in Taxon 31: 755. 1982): *Selinum cervaria* L.

6058 ***Schulzia*** Spreng. in Neue Schriften Naturf. Ges. Halle 2(1): 30. 1813.
Typus: *S. crinita* (Pall.) Spreng. (*Sison crinitum* Pall.).

(H) *Shultzia* Raf. in Med. Repos., ser. 2, 5: 356. Feb-Apr 1808 [*Gentian.*].
Typus: *S. obolarioides* Raf.

6064 ***Kundmannia*** Scop., Intr. Hist. Nat.: 116. Jan-Apr 1777.
Typus: *K. sicula* (L.) DC. (Prodr. 4: 143. Sep (sero) 1830) (*Sium siculum* L.).

(≡) *Arduina* Adans., Fam. Pl. 2: 499, 520. Jul-Aug 1763.

6070 **Selinum** L., Sp. Pl., ed. 2: 350. Sep 1762.
Typus: *S. carvifolia* (L.) L. (*Seseli carvifolia* L.) (*typ. cons.*).

(H) *Selinum* L., Sp. Pl.: 244. 1 Mai 1753 [*Umbell.*].
Typus (vide Hitchcock in Sprague & al., Nom. Prop. Brit. Bot.: 139. 1929): *S. sylvestre* L..

6083 **Levisticum** Hill, Brit. Herb.: 423. Nov 1756.
Typus: *L. officinale* W. D. J. Koch (*Ligusticum levisticum* L.).

(H) *Levisticum* Hill, Brit. Herb.: 410. Nov 1756 [*Umbell.*].
≡ *Ligusticum* L. 1753.

6099 **Bonannia** Guss., Fl. Sicul. Syn. 1: 355. Feb 1843.
Typus: *B. resinifera* Guss., nom. illeg. (*Ferula nudicaulis* Spreng., *B. nudicaulis* (Spreng.) Rickett & Stafleu).

(H) *Bonannia* Raf. in Specchio Sci. 1: 115. 1 Apr 1814 [*Sapind.*].
Typus: *B. nitida* Raf.

CORNACEAE

6154 **Alangium** Lam., Encycl. 1: 174. 2 Dec 1783.
Typus: *A. decapetalum* Lam. (*typ. cons.*).

(≡) *Angolam* Adans., Fam. Pl. 2: 85, 518. Jul-Aug 1763.
(=) *Kara-angolam* Adans., Fam. Pl. 2: 84, 532. Jul-Aug 1763.
Typus: *Alangium hexapetalum* Lam.

6156 **Curtisia** Aiton, Hort. Kew. 1: 162. 7 Aug - 1 Oct 1789.
Typus: *C. faginea* Aiton, nom. illeg. (*Sideroxylon dentatum* Burm. f., *C. dentata* (Burm. f.) C. A. Sm.).

(H) *Curtisia* Schreb., Gen. Pl.: 199. Apr 1789 [*Rut.*].
Typus: *C. schreberi* J. F. Gmel. (Syst. Nat. 2: 498. Sep (sero) - Nov 1791).

6157 **Helwingia** Willd., Sp. Pl. 4: 634, 716. Apr 1806.
Typus: *H. rusciflora* Willd., nom. illeg. (*Osyris japonica* Thunb., *H. japonica* (Thunb.) F. G. Dietr.).

(H) *Helvingia* Adans., Fam. Pl. 2: 345, 553. Jul-Aug 1763 [*Flacourt.*].
≡ *Thamnia* P. Browne 1756 (*nom. rej.* sub 3284).

ERICACEAE

6189 **Loiseleuria** Desv. in J. Bot. Agric. 1: 35. Jan 1813.
Typus: *L. procumbens* (L.) Desv. (*Azalea procumbens* L.).

(≡) *Azalea* L., Sp. Pl.: 150. 1 Mai 1753.

6191 ***Rhodothamnus*** Rchb. in Mössler, Handb. Gewächsk., ed. 2, 1: 667, 688. Jul-Dec 1827.
Typus: *R. chamaecistus* (L.) Rchb. (*Rhododendron chamaecistus* L.).

6195 ***Daboecia*** D. Don in Edinburgh New Philos. J. 17: 160. Jul 1834.
Typus: *D. polifolia* D. Don, nom. illeg. (*Vaccinium cantabricum* Huds., *D. cantabrica* (Huds.) K. Koch).

6200 ***Lyonia*** Nutt., Gen. N. Amer. Pl. 1: 266. 14 Jul 1818.
Typus: *L. ferruginea* (Walter) Nutt. (*Andromeda ferruginea* Walter).

(H) *Lyonia* Raf. in Med. Repos., ser. 2, 5: 353. Feb-Apr 1808 [*Polygon.*].
≡ *Polygonella* Michx. 1803.

6200a ***Chamaedaphne*** Moench, Methodus: 457. 4 Mai 1794.
Typus: *C. calyculata* (L.) Moench (*Andromeda calyculata* L.).

(H) *Chamaedaphne* Mitch., Diss. Princ. Bot.: 44. 1769 [*Rub.*].
≡ *Mitchella* L. 1753.

6208 ***Pernettya*** Gaudich. in Ann. Sci. Nat. (Paris) 5: 102. 1825 (*'Pernettia'*) (*orth. cons.*).
Typus: *P. empetrifolia* (Lam.) Gaudich. (*Andromeda empetrifolia* Lam.).

(H) *Pernetya* Scop., Intr. Hist. Nat.: 150. Jan-Apr 1777 [*Campanul.*].
≡ *Canarina* L. 1771 (*nom. cons.*) (8656).

6212 ***Arctostaphylos*** Adans., Fam. Pl. 2: 165, 520. Jul-Aug 1763.
Typus: *A. uva-ursi* (L.) Spreng. (*Arbutus uva-ursi* L.).

(≡) *Uva-ursi* Duhamel, Traité Arbr. Arbust. 2: 371. 1755.

6215 ***Gaylussacia*** Kunth in Humboldt & al., Nov. Gen. Sp. 3, ed. 4°: 275 [& ed. f°: 215]. 9 Jul 1819.
Typus: *G. buxifolia* Kunth

6232 ***Cavendishia*** Lindl. in Edwards's Bot. Reg.: ad t. 1791. 1 Sep 1835.
Typus: *C. nobilis* Lindl.

(H) *Cavendishia* Gray, Nat. Arr. Brit. Pl. 1: 678, 689. 1 Nov 1821 [*Hepat.*].
≡ *Antoiria* Raddi 1818.

(=) *Chupalon* Adans., Fam. Pl. 2: 164, 538. Jul-Aug 1763.
Typus: non designatus.

EPACRIDACEAE

6251 ***Lebetanthus*** Endl., Gen. Pl.: 1411. Feb-Mar 1841 *('Lebethanthus') (orth. cons.)*.
Typus: *L. americanus* (Hook.) Endl. ex Hook. f. (Fl. Antarct.: 327. 1846 (ante 1 Jun)) (*Prionitis americana* Hook.) *(typ. cons.)*.

(≡) *Allodape* Endl., Gen. Pl.: 749. Mar 1839.

6254 ***Richea*** R. Br., Prodr.: 555. 27 Mar 1810.
Typus: *R. dracophylla* R. Br.

(H) *Richea* Labill., Voy. Rech. Pérouse 1: 186. 22 Feb - 4 Mar 1800 [*Comp.*].
Typus: *R. glauca* Labill.

(=) *Cystanthe* R. Br., Prodr.: 555. 27 Mar 1810.
Typus: *C. sprengelioides* R. Br.

6260 ***Epacris*** Cav., Icon. 4: 25. Sep-Dec 1797.
Typus: *E. longiflora* Cav. *(typ. cons.)*.

(H) *Epacris* J. R. Forst. & G. Forst., Char. Gen. Pl.: 10. 29 Nov 1775 [*Epacrid.*].
Typus (vide Pichi Sermolli in Taxon 3: 119. 1954): *E. longifolia* J. R. Forst. & G. Forst.

6262a ***Leucopogon*** R. Br., Prodr.: 541. 27 Mar 1810.
Typus: *L. lanceolatus* R. Br., nom. illeg. (*Styphelia parviflora* Andrews, *L. parviflorus* (Andrews) Lindl.).

(=) *Perojoa* Cav., Icon. 4: 29. Sep-Dec 1797.
Typus: *P. microphylla* Cav.

DIAPENSIACEAE

6275 ***Shortia*** Torr. & A. Gray in Amer. J. Sci. Arts 42: 48. 1842.
Typus: *S. galacifolia* Torr. & A. Gray

(H) *Shortia* Raf., Autik. Bot.: 16. 1840 [*Cruc.*].
Typus: *S. dentata* Raf. (*Arabis dentata* Torr. & A. Gray 1838, non Clairv. 1811).

6277 ***Galax*** Sims in Bot. Mag.: ad t. 754. Jun 1804.
Typus: *G. urceolata* (Poir.) Brummitt (*Pyrola urceolata* Poir.).

(H) *Galax* L., Sp. Pl.: 200. 1 Mai 1753 [*Hydrophyll.*].
Typus: *G. aphylla* L.

THEOPHRASTACEAE

6282 ***Jacquinia*** L., Fl. Jamaica: 27. 22 Dec 1759 *('Jaquinia') (orth. cons.)*.
Typus: *J. ruscifolia* Jacq. (Enum. Syst. Pl.: 15. Aug-Sep 1760).

MYRSINACEAE

6285 ***Ardisia*** Sw., Prodr.: 3, 48. 20 Jun - 29 Jul 1788.
Typus: *A. tinifolia* Sw. *(typ. cons.).*

(=) *Katoutheka* Adans., Fam. Pl. 2: 159, 534. Jul-Aug 1763.
Typus (vide Ridsdale in Manilal, Bot. Hist. Hort. Malab.: 136. 1980): *Psychotria dalzellii* Hook. f.

(=) *Vedela* Adans., Fam. Pl. 2: 502, 617. Jul-Aug 1763.
Typus: non designatus.

(=) *Icacorea* Aubl., Hist. Pl. Guiane 2 [Misc.]: 1. Jun-Dec 1775.
Typus: *I. guianensis* Aubl.

(=) *Bladhia* Thunb., Nova Gen. Pl.: 6. 24 Nov 1781.
Typus: *B. japonica* Thunb.

6288 ***Heberdenia*** Banks ex A. DC. in Ann. Sci. Nat., Bot., ser. 2, 16: 79. Aug 1841.
Typus: *H. excelsa* Banks ex A. DC. (Prodr. 8: 106. Mar (med.) 1844), nom. illeg. (*Anguillaria bahamensis* Gaertn., *H. bahamensis* (Gaertn.) Sprague) (etiam vide 974).

6291 ***Labisia*** Lindl. in Edwards's Bot. Reg. 31: ad t. 48. Sep 1845.
Typus: *L. pothoina* Lindl.

(=) *Angiopetalum* Reinw. in Syll. Pl. Nov. 2: 7. 1825.
Typus: *A. punctatum* Reinw.

6301 ***Cybianthus*** Mart., Nov. Gen. Sp. Pl. 3: 87. Jan-Mar 1831.
Typus: *C. penduliflorus* Mart.

(=) *Peckia* Vell., Fl. Flumin.: 51. 7 Sep - 28 Nov 1829.
Typus: non designatus.

6304 ***Wallenia*** Sw., Prodr.: 2, 31. 20 Jun - 29 Jul 1788.
Typus: *W. laurifolia* Sw.

6310 ***Embelia*** Burm. f., Fl. Indica: 62. 1 Mar - 6 Apr 1768.
Typus: *E. ribes* Burm. f.

(≡) *Ghesaembilla* Adans., Fam. Pl. 2: 449, 561. Jul-Aug 1763.

(=) *Pattara* Adans., Fam. Pl. 2: 447, 588. Jul-Aug 1763.
Typus: *Ardisia tserim-cottam* Roem. & Schult.

PRIMULACEAE

6318 ***Douglasia*** Lindl. in Quart. J. Sci. Lit. Arts 1827: 385. Oct-Dec 1827.
Typus: *D. nivalis* Lindl.

(H) *Douglassia* Mill., Gard. Dict. Abr., ed. 4: [452]. 28 Jan 1754 [*Verben.*].
≡ *Volkameria* L. 1753.

(=) *Vitaliana* Sesl. in Donati, Essai Hist. Nat. Mer Adriat.: 69. Jan-Mar 1758.
Typus: *V. primuliflora* Bertol. (Fl. Ital. 2: 368. Mai 1835 - Mar 1836) (*Primula vitaliana* L.).

PLUMBAGINACEAE

6348 ***Acantholimon*** Boiss., Diagn. Pl. Orient. 7: 69. Jul-Oct 1846.
Typus: *A. glumaceus* (Jaub. & Spach) Boiss. (*Statice glumacea* Jaub. & Spach).

(≡) *Armeriastrum* (Jaub. & Spach) Lindl., Veg. Kingd.: 641. Jan-Mai 1846 (*Statice* subg. *Armeriastrum* Jaub. & Spach in Ann. Sci. Nat., Bot., ser. 2, 20: 248. Oct 1843).

6350 ***Armeria*** Willd., Enum. Pl.: 333. Apr 1809.
Typus: *A. vulgaris* Willd. (*Statice armeria* L.).

(≡) *Statice* L., Sp. Pl.: 274. 1 Mai 1753 (typ. des.: Hitchcock in Sprague & al., Nom. Prop. Brit. Bot.: 143. 1929).

6351 ***Limonium*** Mill., Gard. Dict. Abr., ed. 4: [1328]. 28 Jan 1754.
Typus: *L. vulgare* Mill. (*Statice limonium* L.) *(typ. cons.)*.

SAPOTACEAE

6365 ***Labatia*** Sw., Prodr.: 2, 32. 20 Jun - 29 Jul 1788.
Typus: *L. sessiliflora* Sw.

(H) *Labatia* Scop., Intr. Hist. Nat.: 197. Jan-Apr 1777 [*Aquifol.*].
≡ *Macoucoua* Aubl. 1775.

6368a ***Planchonella*** Pierre, Not. Bot. Sapot.: 34. 30 Dec 1890.
Typus: *P. obovata* (R. Br.) Pierre (*Sersalisia obovata* R. Br.) *(typ. cons.)*.

(=) *Hormogyne* A. DC., Prodr. 8: 176. Mar (med.) 1844.
Typus: *H. cotinifolia* A. DC.

6370 ***Argania*** Roem. & Schult., Syst. Veg. 4: xlvi, 502. Mar-Jun 1819.
Typus: *A. sideroxylon* Roem. & Schult., nom. illeg. (*Sideroxylon spinosum* L., *A. spinosa* (L.) Skeels).

6373 ***Dipholis*** A. DC., Prodr. 8: 188. Mar (med.) 1844.
Typus: *D. salicifolia* (L.) A. DC. (*Achras salicifolia* L.).
(=) *Spondogona* Raf., Sylva Tellur.: 35. Oct-Dec 1838.
Typus: *S. nitida* Raf., nom. illeg. (*Bumelia pentagona* Sw.).

6374 ***Bumelia*** Sw., Prodr.: 3, 49. 20 Jun - 29 Jul 1788.
Typus: *B. retusa* Sw. *(typ. cons.).*
(=) *Robertia* Scop., Intr. Hist. Nat.: 154. Jan-Apr 1777.
Typus: *Sideroxylon decandrum* L.

6382 ***Niemeyera*** F. Muell., Fragm. 7: 114. Dec 1870.
Typus: *N. prunifera* (F. Muell.) F. Muell. (*Chrysophyllum pruniferum* F. Muell.).
(H) *Niemeyera* F. Muell., Fragm. 6: 96. Dec 1867 [*Orchid.*].
Typus: *N. stylidioides* F. Muell.

6384 ***Cryptogyne*** Hook. f. in Bentham & Hooker, Gen. Pl. 2: 652, 656. Mai 1876.
Typus: *C. gerrardiana* Hook f.
(H) *Cryptogyne* Cass. in Cuvier, Dict. Sci. Nat. 50: 491, 493, 498. Nov 1827 [*Comp.*].
Typus: *C. absinthioides* Cass.

6386a ***Manilkara*** Adans., Fam. Pl. 2: 166, 574. Jul-Aug 1763.
Typus: *M. kauki* (L.) Dubard (in Ann. Inst. Bot.-Géol. Colon. Marseille, ser. 3, 3: 9. 1915) (*Mimusops kauki* L.) *(typ. cons.).*
(=) *Achras* L., Sp. Pl.: 1190. 1 Mai 1753.
Typus: *A. zapota* L.

EBENACEAE

6408 ***Brachynema*** Benth. in Trans. Linn. Soc. London 22: 126. 21 Nov 1857.
Typus: *B. ramiflorum* Benth.
(H) *Brachynema* Griff., Not. Pl. Asiat. 4: 176. 1854 [*Verben.*].
Typus: *B. ferrugineum* Griff.

STYRACACEAE

6410 ***Halesia*** J. Ellis ex L., Syst. Nat., ed. 10: 1041, 1044, 1369. 7 Jun 1759.
Typus: *H. carolina* L.
(H) *Halesia* P. Browne, Civ. Nat. Hist. Jamaica: 205. 10 Mar 1756 [*Rub.*].
Typus: *Guettarda argentea* Lam.

OLEACEAE

6421 ***Forsythia*** Vahl, Enum. Pl. 1: 39. Jul-Dec 1804.
Typus: *F. suspensa* (Thunb.) Vahl (*Ligustrum suspensum* Thunb.).
(H) *Forsythia* Walter, Fl. Carol.: 153. Apr-Jun 1788 [*Saxifrag.*].
Typus: *F. scandens* Walter

6422 ***Schrebera*** Roxb., Pl. Coromandel 2: 1. Mai 1799.
Typus: *S. swietenioides* Roxb.
(H) *Schrebera* L., Sp. Pl., ed. 2, 2: 1662. Jul-Aug 1763 [*Convolvul.*].
Typus: *S. schinoides* L., nom. illeg. (*Schinus myricoides* L.).

6427 ***Forestiera*** Poir. in Lamarck, Encycl., Suppl. 1: 132. 3 Sep 1810.
Typus: *F. cassinoides* (Willd.) Poir. (in Lamarck, Encycl., Suppl. 2: 665. 3 Jul 1812) (*Borya cassinoides* Willd.).

(≡) *Adelia* P. Browne, Civ. Nat. Hist. Jamaica: 361. 10 Mar 1756.

6428 ***Linociera*** Sw. ex Schreb., Gen. Pl.: 784. Mai 1791.
Typus: *L. ligustrina* (Sw.) Sw. (*Thouinia ligustrina* Sw.).

(=) *Mayepea* Aubl., Hist. Pl. Guiane: 81. Jun-Dec 1775.
Typus: *M. guianensis* Aubl.

(=) *Ceranthus* Schreb., Gen. Pl.: 14. Apr 1789.
Typus: *C. schreberi* J. F. Gmel. (Syst. Nat. 2: 26. Sep (sero) - Nov 1791).

LOGANIACEAE

6450 ***Logania*** R. Br., Prodr.: 454. 27 Mar 1810.
Typus: *L. floribunda* R. Br., nom. illeg. (*Euosma albiflora* Andrews, *L. albiflora* (Andrews) Druce) *(typ. cons.)*.

(H) *Loghania* Scop., Intr. Hist. Nat.: 236. Jan-Apr 1777 [*Marcgrav.*].
≡ *Souroubea* Aubl. 1775.

(≡) *Euosma* Andrews in Bot. Repos.: ad t. 520. Mai 1808.

6468 ***Peltanthera*** Benth. in Bentham & Hooker, Gen. Pl. 2: 788, 797. Mai 1876.
Typus: *P. floribunda* Benth. & Hook. f.

(H) *Peltanthera* Roth, Nov. Pl. Sp.: 132. Apr 1821 [*Apocyn.*].
Typus: *P. solanacea* Roth

GENTIANACEAE

6483 ***Belmontia*** E. Mey., Comment. Pl. Afr. Austr.: 183. 1-8 Jan 1838.
Typus: *B. cordata* E. Mey., nom. illeg. (*Sebaea cordata* Roem. & Schult., nom. illeg., *Gentiana exacoides* L., *B. exacoides* (L.) Druce) *(typ. cons.)*.

(≡) *Parrasia* Raf., Fl. Tellur. 3: 78. Nov-Dec 1837.

6484 ***Enicostema*** Blume, Bijdr.: 848. Jul-Dec 1826.
Typus: *E. littorale* Blume

6501 ***Bartonia*** Muhl. ex Willd. in Ges. Naturf. Freunde Berlin Neue Schriften 3: 444. 1801 (post 21 Apr).
Typus: *B. tenella* Willd. [= *B. virginica* (L.) Britton & al. (*Sagina virginica* L.)].

6504 ***Orphium*** E. Mey., Comment. Pl. Afr. Austr.: 181. 1-8 Jan 1838.
Typus: *O. frutescens* (L.) E. Mey. (*Chironia frutescens* L.).

6509a ***Gentianella*** Moench, Methodus: 482. 4 Mai 1794.
Typus: *G. tetrandra* Moench, nom. illeg. (*Gentiana campestris* L., *Gentianella campestris* (L.) Börner).
(=) *Amarella* Gilib., Fl. Lit. Inch. 1: 36. 1782.
Typus (vide Rauschert in Taxon 25: 192. 1976): *Gentiana amarella* L.

6513 ***Halenia*** Borkh. in Arch. Bot. (Leipzig) 1(1): 25. 1796.
Typus: *H. sibirica* Borkh., nom. illeg. (*Swertia corniculata* L., *H. corniculata* (L.) Cornaz).

6526 ***Schultesia*** Mart., Nov. Gen. Sp. Pl. 2: 103. Jan-Jul 1827.
Typus: *S. crenuliflora* Mart. *(typ. cons.)*.
(H) *Schultesia* Spreng., Pl. Min. Cogn. Pug. 2: 17. 1815 [*Gram.*].
≡ *Eustachys* Desv. 1810.

6544 ***Villarsia*** Vent., Choix Pl.: ad t. 9. 1803.
Typus: *V. ovata* (L. f.) Vent. (*Menyanthes ovata* L. f.) *(typ. cons.)*.
(H) *Villarsia* J. F. Gmel., Syst. Nat. 2: 306, 447. Sep (sero) - Nov 1791 [*Gentian.*].
Typus: *V. aquatica* J. F. Gmel.

APOCYNACEAE

6559 ***Carissa*** L., Syst. Nat., ed. 12, 2: 135, 189 [& Mant. Pl.: 7, 52]. 15-31 Oct 1767.
Typus: *C. carandas* L.
(≡) *Carandas* Adans., Fam. Pl. 2: 171, 532. Jul-Aug 1763.

6562 ***Landolphia*** P. Beauv., Fl. Oware 1: 54. Mai 1806.
Typus: *L. owariensis* P. Beauv.
(=) *Pacouria* Aubl., Hist. Pl. Guiane: 268. Jun-Dec 1775.
Typus: *P. guianensis* Aubl.

6564 ***Willughbeia*** Roxb., Pl. Coromandel 3: 77. 18 Feb 1820.
Typus: *W. edulis* Roxb.
(H) *Willughbeja* Scop. ex Schreb., Gen. Pl.: 162. Apr 1789 [*Apocyn.*].
≡ *Ambelania* Aubl. 1775.

6566 ***Urnularia*** Stapf in Hooker's Icon. Pl.: ad t. 2711. Sep 1901.
Typus: *U. beccariana* (Kuntze) Stapf (*Ancylocladus beccarianus* Kuntze).
(H) *Urnularia* P. Karst. in Not. Sällsk. Fauna Fl. Fenn. Förh. 8: 209. 1866 [*Fungi*].
Typus: *U. boreella* P. Karst.

6583 ***Alstonia*** R. Br., Asclepiadeae: 64. 3 Apr 1810.
Typus: *A. scholaris* (L.) R. Br. (*Echites scholaris* L.) *(typ. cons.)*.
(H) *Alstonia* Scop., Intr. Hist. Nat.: 198. Jan-Apr 1777 [*Apocyn.*].
≡ *Pacouria* Aubl. 1775 (*nom. rej.* sub 6562).

6588 ***Aspidosperma*** Mart. & Zucc. in Flora 7 (1, Beil.): 135. Mai-Jun 1824.
Typus: *A. tomentosum* Mart. & Zucc. *(typ. cons.)*.

(=) *Coutinia* Vell., Quinogr. Port.: 166. 1799.
Typus: *C. illustris* Vell.

(=) *Macaglia* Rich. ex Vahl in Skr. Naturhist.-Selsk. 6: 107. 1810.
Typus (vide Woodson in Ann. Missouri Bot. Gard. 38: 136. 1951): *M. alba* Vahl

6616 ***Alyxia*** Banks ex R. Br., Prodr.: 469. 27 Mar 1810.
Typus: *A. spicata* R. Br. *(typ. cons.)*.

(≡) *Gynopogon* J. R. Forst. & G. Forst., Char. Gen. Pl.: 18. 29 Nov 1775.

6626 ***Kopsia*** Blume, Catalogus: 12. Feb-Sep 1823.
Typus: *K. arborea* Blume

(H) *Kopsia* Dumort., Comment. Bot.: 16. Nov (sero) - Dec (prim.) 1822 [*Orobanch.*].
Typus (vide Bullock in Taxon 5: 197. 1956): *K. ramosa* (L.) Dumort. (*Orobanche ramosa* L.).

6632 ***Thevetia*** L., Opera Var.: 212. 1758.
Typus: *T. ahouai* (L.) A. DC. (Prodr. 8: 345. Mar (med.) 1844) (*Cerbera ahouai* L.) *(typ. cons.)*.

(≡) *Ahouai* Mill., Gard. Dict. Abr., ed. 4: [42]. 28 Jan 1754.

6639 ***Urceola*** Roxb. in Asiat. Res. 5: 169. 1799.
Typus: *U. elastica* Roxb.

(H) *Urceola* Vand., Fl. Lusit. Bras. Spec.: 8. 1788 [*Spermatoph.*].
Typus: non designatus.

6670 ***Spirolobium*** Baill. in Bull. Mens. Soc. Linn. Paris: 773. 1889.
Typus: *S. cambodianum* Baill.

6677 ***Chonemorpha*** G. Don, Gen. Hist. 4: 69, 76. 1837.
Typus: *C. macrophylla* G. Don, nom. illeg. (*Echites fragrans* Moon, *C. fragrans* (Moon) Alston).

(≡) *Belutta-kaka* Adans., Fam. Pl. 2: 172, 525. Jul-Aug 1763.

6683 ***Ichnocarpus*** R. Br., Asclepiadeae: 50. 3 Apr 1810.
Typus: *I. frutescens* (L.) W. T. Aiton (Hort. Kew., ed. 2, 2: 69. Feb-Mai 1811) (*Apocynum frutescens* L.).

6691 ***Parsonsia*** R. Br., Asclepiadeae: 53. 3 Apr 1810.
Typus: *P. capsularis* (G. Forst.) R. Br. ex Endl. (in Ann. Wiener Mus. Naturgesch. 1: 175. 1836) (*Periploca capsularis* G. Forst.) *(typ. cons.)*.

(H) *Parsonsia* P. Browne, Civ. Nat. Hist. Jamaica: 199. 10 Mar 1756 [*Lythr.*].
Typus: *P. herbacea* J. St.-Hil. (Expos. Fam. Nat. 2: 178. Feb-Apr 1805) (*Lythrum parsonsia* L.).

6702 ***Prestonia*** R. Br., Asclepiadeae: 58. 3 Apr 1810.
Typus: *P. tomentosa* R. Br.

(H) *Prestonia* Scop., Intr. Hist. Nat.: 281. Jan-Apr 1777 [*Malv.*].
≡ *Lass* Adans. 1763.

ASCLEPIADACEAE

6726 ***Camptocarpus*** Decne. in Candolle, Prodr. 8: 493. Mar (med.) 1844.
Typus: *C. mauritianus* (Lam.) Decne. (*Cynanchum mauritianum* Lam.) *(typ. cons.)*.

(H) *Camptocarpus* K. Koch in Linnaea 17: 304. Jan 1844 [*Boragin.*].
≡ *Oskampia* Moench 1794.

6772 ***Schubertia*** Mart., Nov. Gen. Sp. Pl. 1: 55. 1 Oct 1824.
Typus: *S. multiflora* Mart. *(typ. cons.)*.

(H) *Schubertia* Mirb. in Nouv. Bull. Sci. Soc. Philom. Paris 3: 123. Aug 1812 [*Pin.*].
≡ *Taxodium* Rich. 1810.

6857 ***Oxypetalum*** R. Br., Asclepiadeae: 30. 3 Apr 1810.
Typus: *O. banksii* Schult. (in Roemer & Schultes, Syst. Veg. 6: 91. Aug-Dec 1820).

(=) *Gothofreda* Vent., Choix Pl.: ad t. 60. 1808.
Typus: *G. cordifolia* Vent.

6870 ***Brachystelma*** R. Br. in Bot. Mag.: ad t. 2343. 2 Sep 1822.
Typus: *B. tuberosum* (Meerb.) R. Br. (*Stapelia tuberosa* Meerb.).

(=) *Microstemma* R. Br., Prodr.: 459. 27 Mar 1810.
Typus: *M. tuberosum* R. Br.

6877a ***Rhytidocaulon*** P. R. O. Bally in Candollea 18: 335. Mar 1962.
Typus: *R. subscandens* P. R. O. Bally

(H) *Rhytidocaulon* Nyl. ex Elenkin in Izv. Imp. Bot. Sada Petra Velikago 16: 263. 1916 [*Fungi*].
≡ *Chlorea* Nyl. 1855 *(nom. rej.* sub *Letharia)*.

6889 ***Pectinaria*** Haw., Suppl. Pl. Succ.: 14. Mai 1819.
Typus: *P. articulata* (Aiton) Haw. (*Stapelia articulata* Aiton).

(H) *Pectinaria* Bernh., Syst. Verz.: 113, 221. 1800 [*Umbell.*].
≡ *Scandix* L. 1753 (typ. des.: Hitchcock in Sprague & al., Nom. Prop. Brit. Bot.: 141. 1929).

6911 ***Marsdenia*** R. Br., Prodr.: 460. 27 Mar 1810.
Typus: *M. tinctoria* R. Br.

(=) *Stephanotis* Thouars, Gen. Nov. Madagasc.: 11. 17 Nov 1806.
Typus (vide Forster in Taxon 39: 364. 1990): *S. thouarsii* Brongn.

6914 ***Dregea*** E. Mey., Comment. Pl. Afr. Austr.: 199. 1-8 Jan 1838.
Typus: *D. floribunda* E. Mey.

(H) *Dregea* Eckl. & Zeyh., Enum. Pl. Afric. Austral.: 350. Apr 1837 [*Umbell.*].
Typus: non designatus.

CONVOLVULACEAE

6979 ***Bonamia*** Thouars, Hist. Vég. Iles France: 33. 1804 (ante 22 Sep).
Typus: *B. alternifolia* J. St.-Hil. (Expos. Fam. Nat. 2: 349. Feb-Aug 1805).

6994 ***Calystegia*** R. Br., Prodr.: 483. 27 Mar 1810.
Typus: *C. sepium* (L.) R. Br. (*Convolvulus sepium* L.) *(typ. cons.)*.

(≡) *Volvulus* Medik., Philos. Bot. 2: 42. Mai 1791.

6997 ***Merremia*** Dennst. ex Endl., Gen. Pl.: 1403. Feb-Mar 1841.
Typus: *M. hederacea* (Burm. f.) Hallier f. (in Bot. Jahrb. Syst. 18: 118. 22 Dec 1893) (*Evolvulus hederaceus* Burm. f.).

(=) *Operculina* Silva Manso, Enum. Subst. Braz.: 16. 1836.
Typus: *O. turpethum* (L.) Silva Manso (*Convolvulus turpethum* L.).

(=) *Camonea* Raf., Fl. Tellur. 4: 81. 1838 (med.).
Typus: *C. bifida* (Vahl) Raf. (*Convolvulus bifidus* Vahl).

7003 ***Ipomoea*** L., Sp. Pl.: 159. 1 Mai 1753.
Typus: *I. pes-tigridis* L. *(typ. cons.)*.

7003a ***Pharbitis*** Choisy in Mém. Soc. Phys. Genève 6: 438. 1833.
Typus: *P. hispida* Choisy, nom. illeg. (*Convolvulus purpureus* L., *P. purpurea* (L.) Voigt) *(typ. cons.)*.

(≡) *Convolvuloides* Moench, Methodus: 451. 4 Mai 1794.

(=) *Diatremis* Raf. in Ann. Gén. Sci. Phys. 8: 271. 1821.
Typus: *Convolvulus nil* L.

(=) *Diatrema* Raf., Herb. Raf.: 80. 1833.
Typus: *D. trichocarpa* Raf., nom. illeg. (*Convolvulus carolinus* L.).

HYDROPHYLLACEAE

7022 ***Nemophila*** Nutt. in J. Acad. Nat. Sci. Philadelphia 2: 179. 1822 (med.).
Typus: *N. phacelioides* Nutt.

(=) *Viticella* Mitch., Diss. Princ. Bot.: 42. 1769.
Typus: non designatus.

7023 ***Ellisia*** L., Sp. Pl., ed. 2: 1662. Jul-Aug 1763.
Typus: *E. nyctelea* (L.) L. (*Ipomoea nyctelea* L.).

(H) *Ellisia* P. Browne, Civ. Nat. Hist. Jamaica: 262. 10 Mar 1756 [*Verben.*].
Typus: *E. acuta* L. (Syst. Nat., ed. 10, 2: 1121. 7 Jun 1759).

7029 ***Hesperochiron*** S. Watson, Botany [Fortieth Parallel]: 281. Sep-Dec 1871.
Typus: *H. californicus* (Benth.) S. Watson (*Ourisia californica* Benth.).

(=) *Capnorea* Raf., Fl. Tellur. 3: 74. Nov-Dec 1837.
Typus: *C. nana* (Lindl.) Raf. (*Nicotiana nana* Lindl.).

7033 ***Nama*** L., Syst. Nat., ed. 10: 908, 950. 7 Jun 1759.
Typus: *N. jamaicensis* L. *(typ. cons.)*.
(H) *Nama* L., Sp. Pl.: 226. 1 Mai 1753 [*Hydrophyll.*].
Typus: *N. zeylanica* L.

7035 ***Wigandia*** Kunth in Humboldt & al., Nov. Gen. Sp. 3, ed. 4°: 126 [& ed. f°: 98]. 8 Feb 1819.
Typus: *W. caracasana* Kunth. *(typ. cons.)*.

7037 ***Hydrolea*** L., Sp. Pl., ed. 2: 328. Sep 1762.
Typus: *H. spinosa* L.

BORAGINACEAE

7042 ***Bourreria*** P. Browne, Civ. Nat. Hist. Jamaica: 168. 10 Mar 1756.
Typus: *B. baccata* Raf. (Sylva Tellur.: 42. Oct-Dec 1838) (*Cordia bourreria* L.).
(H) *Beureria* Ehret, Pl. Papil. Rar.: ad t. 13. 1755 [*Calycanth.*].
Typus: non designatus.

7056 ***Trichodesma*** R. Br., Prodr.: 496. 27 Mar 1810.
Typus: *T. zeylanicum* (Burm. f.) R. Br. (*Borago zeylanica* Burm. f.) *(typ. cons.)*.
(=) *Borraginoides* Boehm. in Ludwig, Def. Gen. Pl., ed. 3: 18. 1760.
Typus: *Borago indica* L.

7082 ***Amsinckia*** Lehm., Sem. Hort. Bot. Hamburg. 1831: 3, 7. 1831.
Typus: *A. lycopsoides* Lehm.

7097 ***Alkanna*** Tausch in Flora 7: 234. 21 Apr 1824.
Typus: *A. tinctoria* Tausch *(typ. cons.)*.
(H) *Alkanna* Adans., Fam. Pl. 2: 444, 514. Jul-Aug 1763 [*Lythr.*].
≡ *Lawsonia* L. 1753.

7102 ***Mertensia*** Roth, Catal. Bot. 1: 34. Jan-Feb 1797.
Typus: *M. pulmonarioides* Roth
(=) *Pneumaria* Hill, Veg. Syst. 7: 40. 1764.
Typus: non designatus.

7124 ***Rochelia*** Rchb. in Flora 7: 243. 28 Apr 1824.
Typus: *R. saccharata* Rchb., nom. illeg. (*Lithospermum dispermum* L. f., *R. disperma* (L. f.) Wettst.).
(H) *Rochelia* Roem. & Schult., Syst. Veg. 4: xi, 108. Jan-Jun 1819 [*Boragin.*].
≡ *Lappula* Gilib. 1792.

7124a ***Vaupelia*** Brand in Repert. Spec. Nov. Regni Veg. 13: 82. 30 Jan 1914.
Typus: *V. barbata* (Vaupel) Brand (*Trichodesma barbatum* Vaupel).
(H) *Vaupellia* Griseb., Fl. Brit. W. I.: 460. Mai 1862 [*Gesner.*].
Typus: *V. calycina* (Sw.) Griseb. (*Gesneria calycina* Sw.).

VERBENACEAE

7139	***Urbania*** Phil., Verz. Antofagasta Pfl.: 60. Sep-Oct 1891. Typus: *U. pappigera* Phil. *(typ. cons.)*.	(H)	*Urbania* Vatke in Oesterr. Bot. Z. 25: 10. Jan 1875 [*Scrophular.*]. Typus: *U. lyperiifolia* Vatke
7148	***Bouchea*** Cham. in Linnaea 7: 252. 1832. Typus: *B. pseudogervao* (A. St.-Hil.) Cham. (*Verbena pseudogervao* A. St.-Hil.) *(typ. cons.)*.		
7148a	***Chascanum*** E. Mey., Comment. Pl. Afr. Austr.: 275. 1-8 Jan 1838. Typus: *C. cernuum* (L.) E. Mey. (*Buchnera cernua* L.) *(typ. cons.)*.	(=)	*Plexipus* Raf., Fl. Tellur. 2: 104. Jan-Mar 1837. Typus: *P. cuneifolius* (L. f.) Raf. (*Buchnera cuneifolia* L. f.).
7151	***Stachytarpheta*** Vahl, Enum. Pl. 1: 205. Jul-Dec 1804. Typus: *S. jamaicensis* (L.) Vahl (*Verbena jamaicensis* L.) *(typ. cons.)*.	(≡)	*Valerianoides* Medik., Philos. Bot. 1: 177. Apr 1789.
		(=)	*Vermicularia* Moench, Suppl. Meth.: 150. 2 Mai 1802. Typus: non designatus.
7156	***Amasonia*** L. f., Suppl. Pl.: 48, 294. Apr 1782. Typus: *A. erecta* L. f.	(=)	*Taligalea* Aubl., Hist. Pl. Guiane: 625. Jun-Dec 1775. Typus: *T. campestris* Aubl.
7157	***Casselia*** Nees & Mart. in Nova Acta Phys.-Med. Acad. Caes. Leop.-Carol. Nat. Cur. 11: 73. 1823. Typus: *C. serrata* Nees & Mart. *(typ. cons.)*.	(H)	*Casselia* Dumort., Comment. Bot.: 21. Nov (sero) - Dec (prim.) 1822 [*Boragin.*]. ≡ *Mertensia* Roth 1797 *(nom. cons.)* (7102).
7181	***Tectona*** L. f., Suppl. Pl.: 20, 151. Apr 1782. Typus: *T. grandis* L. f.	(≡)	*Theka* Adans., Fam. Pl. 2: 445, 610. Jul-Aug 1763.
7182a	***Xerocarpa*** H. J. Lam, Verben. Malay. Archip.: 98. 7 Apr 1919. Typus: *X. avicenniifoliola* H. J. Lam	(H)	*Xerocarpa* (G. Don) Spach, Hist. Nat. Vég. 9: 583. 15 Aug 1840 (*Scaevola* sect. *Xerocarpa* G. Don, Gen. Hist. 3: 728. 8-15 Nov. 1834) [*Gooden.*]. Typus: non designatus.
7185	***Premna*** L., Mant. Pl.: 154, 252. Oct 1771. Typus: *P. serratifolia* L. *(typ. cons.)*.	(=)	*Appella* Adans., Fam. Pl. 2: 445, 519. Jul-Aug 1763. Typus: non designatus.

LABIATAE (LAMIACEAE)

7227 ***Stenogyne*** Benth. in Edwards's Bot. Reg.: ad t. 1292. 1 Jan 1830.
Typus: *S. rugosa* Benth. *(typ. cons.)*.

7249 ***Glechoma*** L., Sp. Pl.: 578. 1 Mai 1753 *('Glecoma') (orth. cons.)*.
Typus: *G. hederacea* L.

7250 ***Dracocephalum*** L., Sp. Pl.: 594. 1 Mai 1753.
Typus: *D. moldavica* L. *(typ. cons.)*.

7299 ***Sphacele*** Benth. in Edwards's Bot. Reg.: ad t. 1289. 1 Dec 1829.
Typus: *S. lindleyi* Benth., nom. illeg. (*Stachys salviae* Lindl., *Sphacele salviae* (Lindl.) Briq.) *(typ. cons.)*.

(=) *Alguelaguen* Adans., Fam. Pl. 2: 505, 515. Jul-Aug 1763.
Typus: non designatus.

(=) *Phytoxis* Molina, Sag. Stor. Nat. Chili, ed. 2: 145. 1810.
Typus: *P. sideritifolia* Molina

7305 ***Micromeria*** Benth. in Edwards's Bot. Reg.: ad t. 1282. Mar-Dec 1829.
Typus: *M. juliana* (L.) Benth. ex Rchb. (Fl. Germ. Excurs.: 311. Jul 1831 - Jul 1832 (*Satureja juliana* L.).

(=) *Xenopoma* Willd. in Ges. Naturf. Freunde Berlin Mag. Neuesten Entdeck. Gesammten Naturk. 5: 399. 1811.
Typus: *X. obovatum* Willd.

(=) *Zygis* Desv. ex Ham., Prodr. Pl. Ind. Occid.: 40. 1825.
Typus: *Z. aromatica* Ham.

7306 ***Saccocalyx*** Coss. & Durieu in Ann. Sci. Nat., Bot., ser. 3, 20: 80. Aug 1853.
Typus: *S. satureioides* Coss. & Durieu

7312 ***Amaracus*** Gled., Syst. Pl. Stamin. Situ: 189. 1764 (ante 13 Sep).
Typus: *A. dictamnus* (L.) Benth. (Labiat. Gen. Sp.: 323. Mai 1834) (*Origanum dictamnus* L.) *(typ. cons.)*.

(H) *Amaracus* Hill, Brit. Herb.: 381. 13 Oct 1756 [*Lab.*].
≡ *Majorana* Mill. 1754 *(nom. cons.)* (7314).

(=) *Hofmannia* Heist. ex Fabr., Enum.: 61. 1759.
Typus: *Origanum sipyleum* L.

7314 ***Majorana*** Mill., Gard. Dict. Abr., ed. 4: [829]. 28 Jan 1754.
Typus: *M. hortensis* Moench (Methodus: 406. 4 Mai 1794) (*Origanum majorana* L.) *(typ. cons.)*.

7317 ***Pycnanthemum*** Michx., Fl. Bor.-Amer. 2: 7. 19 Mar 1803.
Typus: *P. incanum* (L.) Michx. (*Clinopodium incanum* L.) *(typ. cons.)*.

(=) *Furera* Adans., Fam. Pl. 2: 193, 560. Jul-Aug 1763.
Typus: *Satureja virginiana* L.

7321 ***Bystropogon*** L'Hér., Sert. Angl.: 19. Jan (prim.) 1789.
Typus: *B. plumosus* L'Hér. *(typ. cons.)*.

7327 ***Cunila*** L., Syst. Nat., ed. 10: 1359. 7 Jun 1759.
Typus: *C. mariana* L., nom. illeg. (*Satureja origanoides* L., *C. origanoides* (L.) Britton).

(H) *Cunila* L. ex Mill., Gard. Dict. Abr., ed. 4: [414]. 28 Jan 1754 [*Lab.*].
Typus (vide Reveal & Strachan in Taxon 29: 333. 1980): *Sideritis romana* L.

7342 ***Hyptis*** Jacq. in Collect. Bot. Spectentia (Vienna) 1: 101, 103. Jan-Sep 1787.
Typus: *H. capitata* Jacq. *(typ. cons.)*.

(=) *Mesosphaerum* P. Browne, Civ. Nat. Hist. Jamaica: 257. 10 Mar 1756.
Typus: *M. suaveolens* (L.) Kuntze (Revis. Gen. Pl. 2: 525. 5 Nov 1891) (*Ballota suaveolens* L.).

(=) *Condea* Adans., Fam. Pl. 2: 504, 542. Jul-Aug 1763.
Typus (vide Kuntze, Revis. Gen. Pl. 2: 525. 1891): *Satureja americana* Poir.

7346 ***Alvesia*** Welw. in Trans. Linn. Soc. London 27: 55. 24 Dec 1869.
Typus: *A. rosmarinifolia* Welw.

(H) *Alvesia* Welw. in Ann. Cons. Ultramarino, ser. 1: 587. Dec 1859 [*Legum.*].
Typus: *A. bauhinioides* Welw.

7350 ***Plectranthus*** L'Hér., Stirp. Nov.: 84. Mar-Apr 1788.
Typus: *P. fruticosus* L'Hér. *(typ. cons.)*.

SOLANACEAE

7377 ***Nicandra*** Adans., Fam. Pl. 2: 219, 582. Jul-Aug 1763.
Typus: *N. physalodes* (L.) Gaertn. (Fruct. Sem. Pl. 2: 237. Apr-Mai 1791) (*Atropa physalodes* L.).

(≡) *Physalodes* Boehm. in Ludwig, Def. Gen. Pl., ed. 3: 41. 1760.

7380 ***Dunalia*** Kunth in Humboldt & al., Nov. Gen. Sp. 3, ed. 4°: 55 [& ed. f°: 43]. Sep (sero) 1818.
Typus: *D. solanacea* Kunth

(H) *Dunalia* Spreng., Pl. Min. Cogn. Pug. 2: 25. 1815 [*Rub.*].
≡ *Lucya* DC. 1830 *(nom. cons.)* (8140).

7382 ***Iochroma*** Benth. in Edwards's Bot. Reg. 31: ad t. 20. 1 Apr 1845.
Typus: *I. tubulosum* Benth., nom. illeg. (*Habrothamnus cyaneus* Lindl., *I. cyaneum* (Lindl.) M. L. Green) *(typ. cons.)*.

(=) *Diplukion* Raf., Sylva Tellur.: 53. Oct-Dec 1838.
Typus (vide D'Arcy, Solanaceae Newslett. 2(4): 21. 1986): *D. cornifolium* (Kunth) Raf. (*Lycium cornifolium* Kunth).

(=) *Trozelia* Raf., Sylva Tellur.: 54. Oct-Dec 1838.
Typus: *T. umbellata* (Ruiz & Pav.) Raf. (*Lycium umbellatum* Ruiz & Pav.).

(=) *Valteta* Raf., Sylva Tellur.: 53. Oct-Dec 1838.
Typus (vide D'Arcy in Solanaceae Newslett. 2(4): 21. 1986): *V. gesnerioides* (Kunth) Raf. (*Lycium gesnerioides* Kunth).

7388 ***Hebecladus*** Miers in London J. Bot. 4: 321. 1845.
Typus: *H. umbellatus* (Ruiz & Pav.) Miers (*Atropa umbellata* Ruiz & Pav.) *(typ. cons.)*.

(≡) *Kokabus* Raf., Sylva Tellur.: 55. Oct-Dec 1838.

(=) *Ulticona* Raf., Sylva Tellur.: 55. Oct-Dec 1838.
Typus (vide D'Arcy in Solanaceae Newslett. 2(4): 20. 1986): *U. biflora* (Ruiz & Pav.) Raf. (*Atropa biflora* Ruiz & Pav.).

(=) *Kukolis* Raf., Sylva Tellur.: 55. Oct-Dec 1838.
Typus: *K. bicolor* Raf.

7392 ***Triguera*** Cav., Diss. 2, App.: [1]. Jan-Apr 1786.
Typus: *T. ambrosiaca* Cav. *(typ. cons.)*.

(H) *Triguera* Cav., Diss. 1: 41. 15 Apr 1785 [*Bombac.*].
Typus: *T. acerifolia* Cav.

7393 ***Scopolia*** Jacq., Observ. Bot. 1: 32. 1764 *('Scopola') (orth. cons.)*.
Typus: *S. carniolica* Jacq.

(H) *Scopolia* Adans., Fam. Pl. 2: 419, 603. Jul-Aug 1763 [*Cruc.*].
≡ *Ricotia* L., Jul-Aug 1763 *(nom. cons.)* (2968).

7398 ***Athenaea*** Sendtn. in Martius, Fl. Bras. 10: 133. 1 Jul 1846.
Typus: *A. picta* (Mart.) Sendtn. (*Witheringia picta* Mart.) *(typ. cons.)*.

(H) *Athenaea* Adans., Fam. Pl. 2: 121, 522. Jul-Aug 1763 [*Comp.*].
≡ *Struchium* P. Browne 1756.

(=) *Deprea* Raf., Sylva Tellur.: 57. Oct-Dec 1838.
Typus (vide D'Arcy in Ann. Missouri Bot. Gard. 60: 624. 1974): *D. orinocensis* (Kunth) Raf. (*Physalis orinocensis* Kunth).

7400 ***Withania*** Pauquy, Belladone: 14. Apr 1825.
Typus: *W. frutescens* (L.) Pauquy (*Atropa frutescens* L.) *(typ. cons.)*.

7407a ***Lycianthes*** (Dunal) Hassl. in Annuaire Conserv. Jard. Bot. Genève 20: 180. 1 Oct 1917 (*Solanum* subsect. *Lycianthes* Dunal in Candolle, Prodr. 13(1): 29. 10 Mai 1852).
Typus: *Solanum lycioides* L. (*L. lycioides* (L.) Hassl.).

(≡) *Otilix* Raf., Med. Fl. 2: 87. 1830.
(=) *Parascopolia* Baill., Hist. Pl. 9: 338. Feb-Mar 1888.
Typus: *P. acapulcensis* Baill.

7414 ***Solandra*** Sw. in Kongl. Vetensk. Acad. Nya Handl. 8: 300. 1787.
Typus: *S. grandiflora* Sw.

(H) *Solandra* L., Syst. Nat. Ed. 10, 2: 1269. 7 Jun 1759 [*Umbell.*].
Typus: *S. capensis* L.

7421 ***Goetzea*** Wydler in Linnaea 5: 423. Jul 1830.
Typus: *G. elegans* Wydler

(H) *Goetzea* Rchb., Consp. Regni Veg.: 150. Dec 1828 - Mar 1829 [*Legum.*].
≡ *Rothia* Pers. 1807 *(nom. cons.)* (3659).

7436 ***Petunia*** Juss. in Ann. Mus. Natl. Hist. Nat. 2: 215. 1803.
Typus: *P. nyctaginiflora* Juss. *(typ. cons.)*.

7450 ***Brunfelsia*** L., Sp. Pl.: 191. 1 Mai 1753 (*'Brunsfelsia'*) *(orth. cons.)*.
Typus: *B. americana* L.

SCROPHULARIACEAE

7467 ***Aptosimum*** Burch. ex Benth. in Edwards's Bot. Reg.: ad t. 1882. 1 Aug 1836.
Typus: *A. depressum* Burch. ex Benth., nom. illeg. (*Ohlendorffia procumbens* Lehm., *A. procumbens* (Lehm.) Steud.).

(≡) *Ohlendorffia* Lehm., Sem. Hort. Bot. Hamburg. 1835: 7. 1835.

7472 ***Hemimeris*** L. f., Suppl. Pl.: 45, 280. Apr 1782.
Typus: *H. montana* L. f. *(typ. cons.)*.

(H) *Hemimeris* L., Pl. Rar. Afr.: 8. 20 Dec 1760 [*Scrophular.*].
Typus: *H. bonae-spei* L.

7474 ***Calceolaria*** L. in Kongl. Vetensk. Acad. Handl. 31: 286. Oct-Dec 1770.
Typus: *C. pinnata* L.

(H) *Calceolaria* Loefl., Iter Hispan.: 183, 185. Dec 1758 [*Viol.*].
Typus: non designatus.

7485 ***Anarrhinum*** Desf., Fl. Atlant. 2: 51. Oct 1798.
Typus: *A. pedatum* Desf. *(typ. cons.)*.

(=) *Simbuleta* Forssk., Fl. Aegypt.-Arab.: 115. 1 Oct 1775.
Typus: *S. forskaohlii* J. F. Gmel. (Syst. Nat. 2: 242. Sep (sero) - Nov 1791).

7510 ***Tetranema*** Benth. in Edwards's Bot. Reg. 29: ad t. 52. 1 Oct 1843.
Typus: *T. mexicanum* Benth.

7517	***Manulea*** L., Syst. Nat., ed. 12, 2: 385, 419 [& Mant. Pl.: 12, 88]. 15-31 Oct 1767. Typus: *M. cheiranthus* (L.) L. (*Lobelia cheiranthus* L.).	(≡)	*Nemia* P. J. Bergius, Descr. Pl. Cap.: 160, 162. Sep 1767.
7518	***Chaenostoma*** Benth. in Companion Bot. Mag. 1: 374. 1 Jul 1836. Typus: *C. aethiopicum* (L.) Benth. (*Buchnera aethiopica* L.) *(typ. cons.)*.	(=)	*Palmstruckia* Retz., Obs. Bot. Pugill.: 15. 14 Nov 1810. Typus: *Manulea foetida* (Andrews) Pers. (*Buchnera foetida* Andrews).
7523	***Zaluzianskya*** F. W. Schmidt, Neue Selt. Pfl.: 11. 1793 (ante 17 Jun). Typus: *Z. villosa* F. W. Schmidt	(H)	*Zaluzianskia* Neck. in Hist. & Commentat. Acad. Elect. Sci. Theod.-Palat. 3: 303. 1775 [*Pteridoph.*]. ≡ *Marsilea* L. 1753 *(nom. cons.)*.
7532	***Limnophila*** R. Br., Prodr.: 442. 27 Mar 1810. Typus: *L. gratioloides* R. Br., nom. illeg. (*Hottonia indica* L., *L. indica* (L.) Druce).	(≡)	*Hydropityon* C. F. Gaertn., Suppl. Carp.: 19. 24-26 Jun 1805.
		(=)	*Ambuli* Adans., Fam. Pl. 2: 208, 516. Jul-Aug 1763. Typus: *A. aromatica* Lam. (Encycl. 1: 128. 2 Dec 1783).
		(=)	*Diceros* Lour., Fl. Cochinch.: 358, 381, post 722. Sep 1790. Typus: *D. cochinchinesis* Lour.
7534	***Stemodia*** L., Syst. Nat., ed. 10: 1091, 1118, 1374. 7 Jun 1759. Typus: *S. maritima* L.	(≡)	*Stemodiacra* P. Browne, Civ. Nat. Hist. Jamaica: 261. 10 Mar 1756.
7546	***Bacopa*** Aubl., Hist. Pl. Guiane: 128. Jun-Dec 1775. Typus: *B. aquatica* Aubl.	(=)	*Moniera* P. Browne, Civ. Nat. Hist. Jamaica: 269. 10 Mar 1756. Typus: non designatus.
		(=)	*Brami* Adans., Fam. Pl. 2: 208, 527. Jul-Aug 1763. Typus: *B. indica* Lam. (Encycl. 1: 456. 1 Aug 1785).
7549	***Micranthemum*** Michx., Fl. Bor.-Amer. 1: 10. 19 Mar 1803. Typus: *M. orbiculatum* Michx., nom. illeg. (*Globifera umbrosa* J. F. Gmel., *M. umbrosum* (J. F. Gmel.) Blake).	(≡)	*Globifera* J. F. Gmel., Syst. Nat. 2: 32. Sep (sero) - Nov 1791.
7556	***Glossostigma*** Wight & Arn. in Nova Acta Phys.-Med. Acad. Caes. Leop.-Carol. Nat. Cur. 18: 355. 1836. Typus: *G. spathulatum* Arn., nom. illeg. (*Limosella diandra* L., *F. diandrum* (L.) Kuntze).	(≡)	*Peltimela* Raf. in Atlantic J. 1: 199. 1833 (aut.).

7559 ***Artanema*** D. Don in Sweet, Brit. Fl. Gard. 7: ad t. 234. 1 Apr 1834.
Typus: *A. fimbriatum* (Hook. ex Graham) D. Don (*Torenia fimbriata* Hook. ex Graham).

(=) *Bahel* Adans., Fam. Pl. 2: 210, 523. Jul-Aug 1763.
Typus: *Columnea longifolia* L.

7579a ***Parahebe*** W. R. B. Oliv. in Rec. Domin. Mus. 1: 229. 1944.
Typus: *P. catarractae* (G. Forst.) W. R. B. Oliv. (*Veronica catarractae* G. Forst.).

(=) *Derwentia* Raf., Fl. Tellur. 4: 55. 1838 (med.).
Typus (vide Garnock-Jones & al. in Taxon 39: 536. 1990): *D. suaveolens* Raf., nom. illeg. (*Veronica derwentiana* Andrews).

7592 ***Rehmannia*** Libosch. ex Fisch. & C. A. Mey., Index Sem. Hort. Petrop. 1: 36. Jan 1835.
Typus: *R. sinensis* (Buc'hoz) Libosch. ex Fisch. & C. A. Mey. (*Sparmannia sinensis* Buc'hoz) (etiam vide 4957).

7602 ***Seymeria*** Pursh, Fl. Amer. Sept. 2: 736. Dec (sero) 1813-Jan 1814.
Typus: *S. tenuifolia* Pursh, nom. illeg. (*Afzelia cassioides* J. F. Gmel., *S. cassioides* (J. F. Gmel.) Blake) *(typ. cons.)* (etiam vide 3509).

7604a ***Agalinis*** Raf., New Fl. 2: 61. Jul-Dec 1837.
Typus: *A. palustris* Raf., nom. illeg. (*Gerardia purpurea* L., *A. purpurea* (L.) Pennell) *(typ. cons.)*.

(=) *Virgularia* Ruiz & Pav., Fl. Peruv. Prodr.: 92. Oct (prim.) 1794.
Typus (vide D'Arcy in Taxon 28: 419-420. 1979): *V. lanceolata* Ruiz & Pav. (Syst. Veg. Fl. Peruv. Chil.: 161. Dec (sero) 1798).

(=) *Chytra* C. F. Gaertn., Suppl. Carp.: 184. 1807.
Typus: *C. anomala* C. F. Gaertn.

(=) *Tomanthera* Raf., New Fl. 2: 65. Jul-Dec 1837.
Typus (vide D'Arcy in Taxon 28: 419-420. 1979): *T. lanceolata* Raf.

7632 ***Cordylanthus*** Nutt. ex Benth. in Candolle, Prodr. 10: 597. 8 Apr 1846.
Typus: *C. filifolius* Nutt. ex Benth., nom. illeg. (*Adenostegia rigida* Benth., *C. rigidus* (Benth.) Jeps.).

(≡) *Adenostegia* Benth. in Lindley, Intr. Nat. Syst. Bot., ed. 2: 445. Jul 1836.

7645 ***Bartsia*** L., Sp. Pl.: 602. 1 Mai 1753.
Typus: *B. alpina* L. *(typ. cons.).*

7649 ***Rhynchocorys*** Griseb., Spicil. Fl. Rumel. 2: 12. Jul 1844.
Typus: *R. elephas* (L.) Griseb. (*Rhinanthus elephas* L.).

(≡) *Elephas* Mill., Gard. Dict. Abr., ed. 4: [461]. 28 Jan 1754.

7650 ***Lamourouxia*** Kunth in Humboldt & al., Nov. Gen. Sp. 2, ed. 4°: 335 [& ed. f°: 269]. 8 Jun 1818.
Typus: *L. multifida* Kunth

(H) *Lamourouxia* C. Agardh, Syn. Alg. Scand.: xiv. Mai-Dec 1817 [*Rhodoph.*].
≡ *Claudea* J. V. Lamour. 1813.

BIGNONIACEAE

7665 ***Anemopaegma*** Mart. ex Meisn., Pl. Vasc. Gen. 1: 300; 2: 208. 25-31 Oct 1840 *('Anemopaegmia') (orth. cons.).*
Typus: *A. mirandum* (Cham.) DC. (*Bignonia miranda* Cham.).

(=) *Cupulissa* Raf., Fl. Tellur. 2: 57. Jan-Mar 1837.
Typus: *C. grandifolia* (Jacq.) Raf. (*Bignonia grandifolia* Jacq.).

(=) *Platolaria* Raf., Sylva Tellur.: 78. Oct-Dec 1838.
Typus: *P. flavescens* Raf., nom. illeg. (*Bignonia orbiculata* Jacq.).

7668 ***Cuspidaria*** DC. in Biblioth. Universelle Genève, ser. 2, 17: 125. Sep 1838.
Typus: *C. pterocarpa* (Cham.) DC. (Prodr. 9: 178. 1 Jan 1845) (*Bignonia pterocarpa* Cham.).

(H) *Cuspidaria* (DC.) Besser, Enum. Pl.: 104. 1822 (post 25 Mai) (*Erysimum* sect. *Cuspidaria* DC., Syst. Nat. 2: 493. Mai 1821) [*Cruc.*].
≡ *Acachmena* H. P. Fuchs 1960.

7673 ***Haplolophium*** Cham. in Linnaea 7: 556. 1832 *('Aplolophium') (orth. cons.).*
Typus: *H. bracteatum* Cham.

7679 ***Phaedranthus*** Miers in Proc. Hort. Soc. London, ser. 2, 3: 182. 1863.
Typus: *P. lindleyanus* Miers

(=) *Sererea* Raf., Sylva Tellur.: 107. Oct-Dec 1838.
Typus: *S. heterophylla* Raf., nom. illeg. (*Bignonia heterophylla* Willd., nom. illeg., *Bignonia kerere* Aubl.).

7697 ***Lundia*** DC. in Biblioth. Universelle Genève, ser. 2, 17: 127. Sep 1838.
Typus: *L. glabra* DC. (Prodr. 9: 180. 1 Jan 1845) *(typ. cons.).*

(H) *Lundia* Schumach., Beskr. Guin. Pl.: 231. 1827 [*Flacourt.*].
Typus: *L. monacantha* Schumach.

7705 ***Bignonia*** L., Sp. Pl.: 622. 1 Mai 1753.
Typus: *B. capreolata* L. *(typ. cons.)*.

7714 ***Campsis*** Lour., Fl. Cochinch.: 358, 377. Sep 1790.
Typus: *C. adrepens* Lour.

(=) *Notjo* Adans., Fam. Pl. 2: 226, 582. Jul-Aug 1763.
Typus: non designatus.

7741 ***Dolichandrone*** (Fenzl) Seem. in Ann. Mag. Nat. Hist., ser. 3, 10: 31. 1862 (*Dolichandra* [unranked] *Dolichandrone* Fenzl in Denkschr. Königl.-Baier. Bot. Ges. Regensburg 3: 265. 1841).
Typus: *Dolichandrone spathacea* (L. f.) K. Schum. (*Bignonia spathacea* L. f.).

(≡) *Pongelia* Raf., Sylva Tellur.: 78. Oct-Dec 1838.

7757 ***Enallagma*** (Miers) Baill., Hist. Pl. 10: 54. Nov-Dec 1888 (*Crescentia* sect. *Enallagma* Miers in Trans. Linn. Soc. London 26: 174. 1868).
Typus: *Crescentia cucurbitina* L. (*E. cucurbitinum* (L.) Baill. ex K. Schum.).

(≡) *Dendrosicus* Raf., Sylva Tellur.: 80. Oct-Dec 1838.

7760 ***Colea*** Bojer ex Meisn., Pl. Vasc. Gen. 1: 301; 2: 210. 25-31 Oct 1840.
Typus: *C. colei* (Bojer ex Hook.) M. L. Green (in Sprague & al., Nom. Prop. Brit. Bot.: 107. Aug 1929) (*Bignonia colei* Bojer ex Hook.) *(typ. cons.)*.

(≡) *Odisca* Raf., Sylva Tellur.: 80. Oct-Dec 1838.

(=) *Uloma* Raf., Fl. Tellur. 2: 62. Jan-Mar 1837.
Typus: *U. telfairiae* (Bojer) Raf. (*Bignonia telfairiae* Bojer).

7766 ***Tourrettia*** Foug. in Mém. Acad. Sci. (Paris) 1784: 205. 1787 *('Tourretia') (orth. cons.)*.
Typus: *T. lappacea* (L'Hér.) Willd. (Sp. Pl. 3: 263. 1800) (*Dombeya lappacea* L'Hér.) (etiam vide 5053).

OROBANCHACEAE

7792 ***Epifagus*** Nutt., Gen. N. Amer. Pl. 2: 60. 14 Jul 1818.
Typus: *E. americanus* Nutt., nom. illeg. (*Orobanche virginiana* L., *E. virginianus* (L.) Barton).

GESNERIACEAE

7800 ***Ramonda*** Rich. in Persoon, Syn. Pl. 1: 216. 1 Apr - 15 Jun 1805.
Typus: *R. pyrenaica* Pers., nom. illeg. (*Verbascum myconi* L., *R. myconi* (L.) Rchb.).

(H) *Ramondia* Mirb. in Bull. Sci. Soc. Philom. Paris 2: 179. 20 Jan - 21 Feb 1801 [*Pteridoph.*].
Typus: *R. flexuosa* (L.) Mirb. (*Ophioglossum flexuosum* L.).

7808 ***Oreocharis*** Benth. in Bentham & Hooker, Gen. Pl. 2: 995, 1021. Mai 1876.
Typus: *O. benthamii* C. B. Clarke (in Candolle & Candolle, Monogr. Phan. 5: 63. Jul 1883) (*Didymocarpus oreocharis* Hance) *(typ. cons.)*.

(H) *Oreocharis* (Decne.) Lindl., Veg. Kingd.: 656. Jan-Mai 1846 (*Lithospermum* subg. *Oreocharis* Decne. in Jacquemont, Voy. Inde 4, Bot.: 122. 1843) [*Boragin.*].
Typus: non designatus.

7809 ***Didissandra*** C. B. Clarke in Candolle & Candolle, Monogr. Phan. 5: 65. Jul 1883.
Typus: *D. elongata* (Jack) C. B. Clarke (*Didymocarpus elongata* Jack) *(typ. cons.)*.

(=) *Ellobum* Blume, Bijdr.: 746. 1826.
Typus: *E. montanum* Blume

7810 ***Didymocarpus*** Wall. in Edinburgh Philos. J. 1: 378. 1819.
Typus: *D. primulifolius* D. Don (Prodr. Fl. Nepal.: 123. 26 Jan - 1 Feb 1825) *(typ. cons.)*.

(=) *Henckelia* Spreng., Anleit. Kenntn. Gew., ed. 2, 2: 402. 20 Apr 1817.
Typus: *H. incana* (Vahl) Spreng. (*Roettlera incana* Vahl).

7824 ***Aeschynanthus*** Jack in Trans. Linn. Soc. London 14: 42. 28 Mai - 12 Jun 1823.
Typus: *A. volubilis* Jack *(typ. cons.)*.

(=) *Trichosporum* D. Don in Edinburgh Philos. J. 7: 84. 1822.
Typus: non designatus.

7833 ***Rhynchoglossum*** Blume, Bijdr.: 741. Jul-Dec 1826 *('Rhinchoglossum') (orth. cons.)*.
Typus: *R. obliquum* Blume

7835 ***Acanthonema*** Hook. f. in Bot. Mag.: ad t. 5339. 1 Oct 1862.
Typus: *A. strigosum* Hook. f.

(H) *Acanthonema* J. Agardh in Öfvers. Förh. Kongl. Svenska Vetensk.-Akad. 3: 104. 1846 [*Rhodoph.*].
≡ *Camontagnea* Pujals 1981.

7853 ***Mitraria*** Cav. in Anales Ci. Nat. 3: 230. Mar 1801.
Typus: *M. coccinea* Cav.

(H) *Mitraria* J. F. Gmel., Syst. Nat. 2: 771, 799. Sep (sero) - Nov 1791 [*Barrington.*].
≡ *Commersona* Sonn. 1776, non *Commersonia* J. R. Forst. & G. Forst. 1775.

7854 ***Sarmienta*** Ruiz & Pav., Fl. Peruv. Prodr.: 4. Oct (prim.) 1794.
Typus: *S. repens* Ruiz & Pav. (Fl. Peruv. 1: 8. 1798 (med.)), nom. illeg. (*Urceolaria scandens* J. D. Brandis, *S. scandens* (J. D. Brandis) Pers.).

(≡) *Urceolaria* Molina ex J. D. Brandis in Molina, Naturgesch. Chili: 133. 1786.

7857a ***Nautilocalyx*** Linden ex Hanst. in Linnaea 26: 181, 206-207. Apr 1854.
Typus: *N. hastatus* Linden ex Hanst., nom. illeg. (*Centrosolenia bractescens* Hook.) [= *N. bracteatus* (Planch.) Sprague (*Centrosolenia bracteata* Planch.)].

(=) *Centrosolenia* Benth. in London J. Bot. 5: 362. 1846.
Typus: *C. hirsuta* Benth.

7860 ***Alloplectus*** Mart., Nov. Gen. Sp. Pl. 3: 53. Jan-Jun 1829.
Typus: *A. hispidus* (Kunth) Mart. (*Besleria hispida* Kunth) *(typ. cons.)*.

(=) *Crantzia* Scop., Intr. Hist. Nat.: 173. Jan-Apr 1777.
Typus: *Besleria cristata* L.

(=) *Vireya* Raf. in Specchio Sci. 1: 194. 1 Jun 1814.
Typus: *V. sanguinolenta* Raf.

7864 ***Nematanthus*** Schrad. in Gött. Gel. Anz. 1821: 718. 5 Mai 1821.
Typus: *N. corticicola* Schrad.

(=) *Orobanchia* Vand., Fl. Lusit. Bras. Spec.: 41. 1788.
Typus (vide Chautems in Taxon 36: 656. 1987): *O. radicans* Poir. (in Lamarck, Encycl. Suppl. 4: 202. 29 Jun 1816).

7866 ***Codonanthe*** (Mart.) Hanst. in Linnaea 26: 209. Apr 1854 (*Hypocyrta* sect. *Codonanthe* Mart., Nov. Gen. Sp. Pl. 3: 50. Jan-Jun 1829).
Typus: *Hypocyrta gracilis* Mart. (*C. gracilis* (Mart.) Hanst.) *(typ. cons.)*.

(H) *Codonanthus* G. Don, Gen. Hist. 4: 164, 166. 1837 [*Logan.*].
Typus: *C. africana* G. Don

7874 ***Achimenes*** Pers., Syn. Pl. 2: 164. Nov 1806.
Typus: *A. coccinea* (Scop.) Pers. (*Buchnera coccinea* Scop.) *(typ. cons.)*.

(H) *Achimenes* P. Browne, Civ. Nat. Hist. Jamaica: 270. 10 Mar 1756 [*Gesner.*].
Typus: non designatus.

7878 ***Seemannia*** Regel in Gartenflora 4: 183. 1855.
Typus: *S. ternifolia* Regel

7887a ***Rechsteineria*** Regel in Flora 31: 247. 21 Apr 1848.
Typus: *R. allagophylla* (Mart.) Regel (*Gesneria allagophylla* Mart.).

(≡) *Alagophyla* Raf., Fl. Tellur. 2: 33. Jan-Mar 1837.
(=) *Megapleilis* Raf., Fl. Tellur. 2: 57. Jan-Mar 1837.
Typus: *M. tuberosa* Raf., nom. illeg. (*Gesneria bulbosa* Ker Gawl.).
(=) *Styrosinia* Raf., Fl. Tellur. 2: 95. Jan-Mar 1837.
Typus: *S. coccinea* Raf., nom. illeg. (*Gesneria aggregata* Ker Gawl.).
(=) *Tulisma* Raf., Fl. Tellur. 2: 98. Jan-Mar 1837.
Typus: *T. verticillata* (Hook.) Raf. (*Gesneria verticillata* Hook.).

7892 ***Rhytidophyllum*** Mart., Nov. Gen. Sp. Pl. 3: 38. Jan-Jun 1829 *('Rytidophyllum') (orth. cons.)*.
Typus: *R. tomentosum* (L.) Mart. (*Gesneria tomentosa* L.).

COLUMELLIACEAE

7897 ***Columellia*** Ruiz & Pav., Fl. Peruv. Prodr.: 3. Oct (prim.) 1794.
Typus: *C. oblonga* Ruiz & Pav. (Fl. Peruv. 1: 28. 1798 (med.)) *(typ. cons.)*.

(H) *Columella* Lour., Fl. Cochinch.: 64, 85. Sep 1790 [*Vit.*].
≡ *Cayratia* Juss. 1818 *(nom. cons.)* (4918a).

LENTIBULARIACEAE

7900 ***Polypompholyx*** Lehm. in Bot. Zeitung (Berlin) 2: 109. 9 Feb 1844.
Typus: *P. tenella* Lehm. (Nov. Stirp. Pug. 8: 48. Apr-Mai (prim.) 1844) *(typ. cons.)*.

(=) *Cosmiza* Raf., Fl. Tellur. 4: 110. 1838 (med.).
Typus: *C. coccinea* Raf., nom. illeg. (*Utricularia multifida* R. Br.).

ACANTHACEAE

7908 ***Elytraria*** Michx., Fl. Bor.-Amer. 1: 8. 19 Mar 1803.
Typus: *E. virgata* Michx., nom. illeg. (*Tubiflora caroliniensis* J. F. Gmel., *E. caroliniensis* (J. F. Gmel.) Pers.).

(≡) *Tubiflora* J. F. Gmel., Syst. Nat. 2: 27. Sep (sero) - Nov 1791.

7914 ***Thunbergia*** Retz. in Physiogr. Sälsk. Handl. 1(3): 163. 1780.
Typus: *T. capensis* Retz.

(H) *Thunbergia* Montin in Kongl. Vetensk. Acad. Handl. 34: 288. 1773 [*Rub.*].
≡ *Pleimeris* Raf. 1838.

7932	***Phaulopsis*** Willd., Sp. Pl. 3: 4, 342. 1800 *('Phaylopsis') (orth. cons.).* Typus: *P. parviflora* Willd., nom. illeg. (*Micranthus oppositifolius* J. C. Wendl., *P. oppositifolia* (J. C. Wendl.) Lindau) (etiam vide 1313).		
7972	***Crabbea*** Harv. in London J. Bot. 1: 27. 1 Jan 1842. Typus: *C. hirsuta* Harv.	(H)	*Crabbea* Harv., Gen. S. Afr. Pl.: 276. Aug-Dec 1838 [*Acanth.*]. Typus: *C. pungens* Harv.
7990	***Stenandrium*** Nees in Lindley, Intr. Nat. Syst. Bot., ed. 2: 444. Jul 1836. Typus: *S. mandioccanum* Nees	(=)	*Gerardia* L., Sp. Pl.: 610. 1 Mai 1753. Typus: *G. tuberosa* L.
8014	***Carlowrightia*** A. Gray in Proc. Amer. Acad. Arts 13: 364. 5 Apr 1878. Typus: *C. linearifolia* (Torr.) A. Gray (*Schaueria linearifolia* Torr.).	(=)	*Cardiacanthus* Nees & Schauer in Candolle, Prodr. 11: 331. 25 Nov 1847. Typus: *C. neesianus* Schauer ex Nees
8028	***Tetramerium*** Nees in Bentham, Bot. Voy. Sulphur: 147. 8 Mai 1846. Typus: *T. polystachyum* Nees	(H)	*Tetramerium* C. F. Gaertn., Suppl. Carp.: 90. Mai 1806 [*Rub.*]. Typus: *T. odoratissimum* C. F. Gaertn., nom. illeg. (*Ixora americana* L.).
		(=)	*Henrya* Nees in Bentham, Bot. Voy. Sulphur: t. 49. 14 Apr 1845. Typus: *H. insularis* Nees
8031	***Dicliptera*** Juss. in Ann. Mus. Natl. Hist. Nat. 9: 267. Jul 1807. Typus: *D. chinensis* (L.) Juss. (*Justicia chinensis* L.) *(typ. cons.).*	(≡)	*Diapedium* K. D. Koenig in Ann. Bot. (König & Sims) 2: 189. 1 Jun 1805.
8037	***Odontonema*** Nees in Linnaea 16: 300. Jul-Aug (prim.) 1842. Typus: [specimen cult. sine anno, sine coll.] (GZU) *(typ. cons.)* [= *O. rubrum* (Vahl) Kuntze (*Justicia rubra* Vahl)].	(H)	*Odontonema* Nees ex Endl., Gen. Pl., Suppl. 2: 63. Mar-Jun 1842 [*Acanth.*]. Typus (vide Baum & Reveal in Taxon 29: 336. 1980): *Justicia lucida* Andrews
8039	***Mackaya*** Harv., Thes. Cap. 1: 8. 1859. Typus: *M. bella* Harv.	(H)	*Mackaia* Gray, Nat. Arr. Brit. Pl. 1: 320, 391. 1 Nov. 1821 [*Phaeoph.*]. Typus: non designatus.
8042	***Schaueria*** Nees, Del. Sem. Hort. Vratisl. 1838: [3]. 1838. Typus: *S. calycotricha* (Link & Otto) Nees (*Justicia calycotricha* Link & Otto).	(H)	*Schauera* Nees in Lindley, Intr. Nat. Syst. Bot., ed. 2: 202. Jul 1836 [*Laur.*]. ≡ *Endlicheria* Nees 1833 *(nom. cons.)* (2811a).

8069 ***Fittonia*** Coem. in Fl. Serres Jard. Eur. 15: 185. 1865.
Typus: *F. verschaffeltii* (Lem.) Van Houtte (*Gymnostachyum verschaffeltii* Lem.).

(=) *Adelaster* Lindl. ex Veitch in Gard. Chron. 1861: 499. 1 Jun 1861.
Typus: *A. albivenis* Lindl. ex Veitch

8079 ***Isoglossa*** Oerst. in Vidensk. Meddel. Dansk Naturhist. Foren. Kjøbenhavn 1854: 155. 1854.
Typus: *I. origanoides* (Nees) S. Moore (in Trans. Linn. Soc. London, Bot. 4: 34. 1894) (*Rhytiglossa origanoides* Nees).

(≡) *Rhytiglossa* Nees in Lindley, Intr. Nat. Syst. Bot., ed. 2: 444. Jul 1836.

8096 ***Anisotes*** Nees in Candolle, Prodr. 11: 424. 25 Nov 1847.
Typus: *A. trisulcus* (Forssk.) Nees (*Dianthera trisulca* Forssk.).

(H) *Anisotes* Lindl. ex Meisn., Pl. Vasc. Gen. 1: 117, 2: 84. 8-14 Apr 1838 [*Lythr.*].
Typus: *A. hilariana* Meisn., nom. illeg. (*Lythrum anomalum* A. St.-Hil.).

(≡) *Calasias* Raf., Fl. Tellur. 4: 64. 1838 (med.).

8097 ***Jacobinia*** Nees ex Moricand, Pl. Nouv. Amér.: 156. Jan-Jun 1847.
Typus: *J. lepida* Nees ex Moricand

8100 ***Trichocalyx*** Balf. f. in Proc. Roy. Soc. Edinburgh 12: 87. 1883.
Typus: *T. obovatus* Balf. f. *(typ. cons.).*

RUBIACEAE

8126 ***Bikkia*** Reinw. in Syll. Pl. Nov. 2: 8. 1825.
Typus: *B. grandiflora* Reinw., nom. illeg. (*Portlandia tetrandra* L. f., *B. tetrandra* (L. f.) A. Gray).

8130 ***Lerchea*** L., Mant. Pl.: 155, 256. Oct 1771.
Typus: *L. longicauda* L.

(H) *Lerchia* Haller ex Zinn, Cat. Pl. Hort. Gott.: 30. 20 Apr - 21 Mai 1757 [*Chenopod.*].
Typus: non designatus.

8136 ***Kohautia*** Cham. & Schltdl. in Linnaea 4: 156. Apr 1829.
Typus: *K. senegalensis* Cham. & Schltdl.

(=) *Duvaucellia* Bowdich in Bowdich & Bowdich, Exc. Madeira: 259. 1825.
Typus: *D. tenuis* Bowdich

8140 ***Lucya*** DC., Prodr. 4: 434. Sep (sero) 1830.
Typus: *L. tuberosa* DC., nom. illeg. (*Peplis tetrandra* L., *L. tetrandra* (L.) K. Schum.) (etiam vide 7380).

8158 ***Cruckshanksia*** Hook. & Arn. in Bot. Misc. 3: 361. 1 Aug 1833.
Typus: *C. hymenodon* Hook. & Arn.

(H) *Cruckshanksia* Hook. in Bot. Misc. 2: 211. 1831 (ante 11 Jun) [*Geran.*].
Typus: *C. cistiflora* Hook.

8162 ***Payera*** Baill. in Bull. Mens. Soc. Linn. Paris: 178. 1878.
Typus: *P. conspicua* Baill.

(H) *Payeria* Baill. in Adansonia 1: 50. 1 Oct 1860 [*Mel.*].
Typus: *P. excelsa* Baill.

8181 ***Wendlandia*** Bartl. ex DC., Prodr. 4: 411. Sep (sero) 1830.
Typus: *W. paniculata* (Roxb.) DC. (*Rondeletia paniculata* Roxb.) *(typ. cons.)*.

(H) *Wendlandia* Willd., Sp. Pl. 2: 6, 275. Mar 1799 [*Menisperm.*].
≡ *Androphylax* J. C. Wendl. 1798 (*nom. rej.* sub 2570).

8183 ***Augusta*** Pohl, Pl. Bras. Icon. Descr. 2: 1. 1828 (sero) - Feb 1829.
Typus: *A. lanceolata* Pohl *(typ. cons.)* [= *A. longifolia* (Spreng.) Rehder (*Ucriana longifolia* Spreng.)].

(H) *Augusta* Leandro in Denkschr. Königl. Akad. Wiss. München 7: 235. Jul-Dec 1821 [*Comp.*].
Typus: non designatus.

8197 ***Hymenodictyon*** Wall. in Roxburgh, Fl. Ind. 2: 148. Mar-Jun 1824.
Typus: *H. excelsum* (Roxb.) DC. (Prodr. 4: 358. Sep (sero) 1824) (*Cinchona excelsa* Roxb.).

(=) *Benteca* Adans., Fam. Pl. 2: 166, 525. Jul-Aug 1763.
Typus: *B. rheedei* Roem. & Schult. (Syst. Veg. 4: 706. Mar-Jun 1819).

8204 ***Manettia*** Mutis ex L., Mant. Pl.: 553, 558. Oct 1771.
Typus: *M. reclinata* L.

(H) *Manettia* Boehm. in Ludwig, Def. Gen. Pl., ed. 3: 99. 1760 [*Scrophular.*].
≡ *Selago* L. 1753.

(=) *Lygistum* P. Browne, Civ. Nat. Hist. Jamaica: 142. 10 Mar 1756.
Typus: *Petesia lygistum* L.

8209 ***Cosmibuena*** Ruiz & Pav., Fl. Peruv. 3: 2. Aug 1802.
Typus: *C. obtusifolia* Ruiz & Pav., nom. illeg. (*Cinchona grandiflora* Ruiz & Pav., *Cosmibuena grandiflora* (Ruiz & Pav.) Rusby) *(typ. cons.)*.

(H) *Cosmibuena* Ruiz & Pav., Fl. Peruv. Prodr.: 10. Oct (prim.) 1794 [*Ros.*].
Typus (vide Pichi Sermolli in Taxon 3: 121. 1954): *Hirtella cosmibuena* Lam.

8215 ***Schizocalyx*** Wedd. in Ann. Sci. Nat., Bot., ser. 4, 1: 73. Feb 1854.
Typus: *S. bracteosus* Wedd.

(H) *Schizocalyx* Scheele in Flora 26: 568. 14 Sep 1843 [*Lab.*].
Typus: non designatus.

8227 ***Mitragyna*** Korth., Observ. Naucl. Indic.: 19. 1839.
Typus: *M. parvifolia* (Roxb.) Korth. (*Nauclea parvifolia* Roxb.).

(H) *Mitragyne* R. Br., Prodr.: 452. 27 Mar 1810 [*Logan./Gentian.*].
≡ *Mitrasacme* Labill. 1804.
(=) *Mamboga* Blanco, Fl. Filip.: 140. 1837.
Typus: *M. capitata* Blanco

8228 ***Uncaria*** Schreb., Gen. Pl.: 125. Apr 1789.
Typus: *U. guianensis* (Aubl.) J. F. Gmel. (Syst. Nat. 2: 370. Sep (sero) - Nov 1791) (*Ourouparia guianensis* Aubl.).

(≡) *Ourouparia* Aubl., Hist. Pl. Guiane: 177. Jun-Dec 1775.

8237 ***Acranthera*** Arn. ex Meisn., Pl. Vasc. Gen. 1: 162; 2: 115. 16-22 Sep 1838.
Typus: *A. ceylanica* Arn. ex Meisn.

(=) *Psilobium* Jack in Malayan Misc. 2(7): 84. 1822.
Typus: *P. nutans* Jack

8241 ***Schradera*** Vahl, Eclog. Amer. 1: 35. 1797.
Typus: *S. capitata* Vahl, nom. illeg. (*Fuchsia involucrata* Sw., *S. involucrata* (Sw.) K. Schum.).

(H) *Schraderia* Heist. ex Medik., Philos. Bot. 2: 40. Mai 1791 [*Lab.*].
≡ *Arischrada* Pobed. 1972.

8244 ***Coptophyllum*** Korth. in Ned. Kruidk. Arch. 2(2): 161. 1851.
Typus: *C. bracteatum* Korth.

(H) *Coptophyllum* Gardner in London J. Bot. 1: 133. 1 Mar 1842 [*Pteridoph.*].
Typus: *C. buniifolium* Gardner

8250 ***Coccocypselum*** P. Browne, Civ. Nat. Hist. Jamaica: 144. 10 Mar 1756 (*'Coccocipsilum') (orth. cons.)*.
Typus: *C. repens* Sw. (Prodr.: 31. 20 Jun - 29 Jul 1788) *(typ. cons.)*.

(=) *Sicelium* P. Browne, Civ. Nat. Hist. Jamaica: 144. 10 Mar 1756.
Typus: non designatus.

8265 ***Pentagonia*** Benth., Bot. Voy. Sulphur: t. 39. 25 Oct 1844.
Typus: *P. macrophylla* Benth.

(H) *Pentagonia* Heist. ex Fabr., Enum., ed. 2: 336. Sep-Dec 1763 [*Solan.*].
≡ *Nicandra* Adanson, Jul-Aug 1763 *(nom. cons.)* (7377).

8285 ***Gardenia*** J. Ellis in Philos. Trans. 51: 935. 1761.
Typus: *G. jasminoides* J. Ellis

(H) *Gardenia* Colden in Essays Observ. Phys. Lit. Soc. Edinburgh 2: 2. 1756 [*Gutt.*].
Typus: non designatus.

8296 ***Villaria*** Rolfe in J. Linn. Soc., Bot. 21: 311. 12 Dec 1884.
Typus: *V. philippinensis* Rolfe

(H) *Vilaria* Guett. in Mém. Minéral. Dauphiné 1: clxx. 1779 [*Comp.*].
Typus: *V. subacaulis* Guett.

8312 ***Zuccarinia*** Blume, Bijdr.: 1006. Oct 1826 - Mar 1827.
Typus: *Z. macrophylla* Blume

(H) *Zuccarinia* Maerkl. in Ann. Wetterauischen Ges. Gesammte Naturk. 2: 252. 28 Apr 1811 [*Spermatoph.*].
Typus: *Z. verbenacea* Maerkl.

8316 ***Duroia*** L. f., Suppl. Pl.: 30, 209. Apr 1782.
Typus: *D. eriopila* L. f.

(=) *Pubeta* L., Pl. Surin.: 16. 23 Jun 1775.
Typus: non designatus.

8353 ***Mesoptera*** Hook. f. in Bentham & Hooker, Gen. Pl. 2: 25, 130. 7-9 Apr 1873.
Typus: *M. maingayi* Hook. f.

(H) *Mesoptera* Raf., Herb. Raf.: 73. 1833 [*Orchid.*].
≡ *Liparis* Rich. 1818 *(nom. cons.)* (1556).

8357 ***Cuviera*** DC. in Ann. Mus. Natl. Hist. Nat. 9: 222. 30 Apr 1807.
Typus: *C. acutiflora* DC.

(H) *Cuviera* Koeler, Descr. Gram.: 328 ('382'). 1802 [*Gram.*].
≡ *Hordelymus* (Jessen) Jessen 1885 (*Hordeum* subg. *Hordelymus* Jessen 1863).

8365 ***Timonius*** DC., Prodr. 4: 461. Sep (sero) 1830.
Typus: *T. rumphii* DC., nom. illeg. (*Erithalis timon* Spreng., *T. timon* (Spreng.) Merr.) *(typ. cons.)*.

(=) *Porocarpus* Gaertn., Fruct. Sem. Pl. 2: 473. Sep-Dec 1791.
Typus: *P. helminthotheca* Gaertn.
(=) *Polyphragmon* Desf. in Mém. Mus. Hist. Nat. 6: 5. 1820.
Typus: *P. sericeum* Desf.
(=) *Helospora* Jack in Trans. Linn. Soc. London 14: 127. 28 Mai - 12 Jun 1823.
Typus: *H. flavescens* Jack
(=) *Burneya* Cham. & Schltdl. in Linnaea 4: 188. Apr 1829.
Typus: *B. forsteri* Cham. & & Schltdl., nom. illeg. (*Erithalis polygama* G. Forst.).

8366 ***Chomelia*** Jacq., Enum. Syst. Pl.: 1, 12. Aug-Sep 1760.
Typus: *C. spinosa* Jacq.

(H) *Chomelia* L., Opera Var.: 210. 1758 [*Rub.*].
Typus (vide Dandy in Taxon 18: 470. 1969): *Rondeletia asiatica* L.

8388 ***Psilanthus*** Hook. f. in Bentham & Hooker, Gen. Pl. 2: 23, 115. 7-9 Apr 1873.
Typus: *P. mannii* Hook. f. (in Hooker's Icon. Pl. 12: 28. Apr 1873).

(H) *Psilanthus* (DC.) Juss. ex M. Roem., Fam. Nat. Syn. Monogr. 2: 132, 198. Dec 1846 (*Tacsonia* sect. *Psilanthus* DC., Prodr. 3: 335. Mar (med.) 1828) [*Passiflor.*].
≡ *Synactila* Raf. 1838.

8397 ***Trichostachys*** Hook. f. in Bentham & Hooker, Gen. Pl. 2: 24, 128. 7-9 Apr 1873.
Typus: *T. longifolia* Hiern (in Oliver, Fl. Trop. Afr. 3: 227. Oct 1877).

(H) *Trichostachys* Welw., Syn. Madeir. Drog. Med.: 19. 1862 [*Prot.*].
Typus: *T. speciosa* Welw.

8399 ***Psychotria*** L., Syst. Nat., ed. 10: 906, 929, 1364. 7 Jun 1759.
Typus: *P. asiatica* L.

(=) *Psychotrophum* P. Browne, Civ. Nat. Hist. Jamaica: 160. 10 Mar 1756.
Typus: non designatus.

(=) *Myrstiphyllum* P. Browne, Civ. Nat. Hist. Jamaica: 152. 10 Mar 1756.
Typus: *Psychotria myrstiphyllum* Sw.

8410 ***Geophila*** D. Don, Prodr. Fl. Nepal.: 136. 26 Jan - 1 Feb 1825.
Typus: *G. reniformis* D. Don, nom. illeg. (*Psychotria herbacea* Jacq., *G. herbacea* (Jacq.) K. Schum.).

(H) *Geophila* Bergeret, Fl. Basses-Pyrénées 2: 184. 1803 [*Lil.*].
Typus: *G. pyrenaica* Bergeret

8411 ***Cephaëlis*** Sw., Prodr.: 3, 45. 20 Jun - 29 Jul 1788.
Typus: *C. muscosa* (Jacq.) Sw. (*Morinda muscosa* Jacq.) *(typ. cons.)*.

(=) *Evea* Aubl., Hist. Pl. Guiane: 100. Jun-Dec 1775.
Typus: *E. guianensis* Aubl.

(=) *Carapichea* Aubl., Hist. Pl. Guiane: 167. Jun-Dec 1775.
Typus: *C. guianensis* Aubl.

(=) *Tapogomea* Aubl., Hist. Pl. Guiane: 157. Jun-Dec 1775.
Typus: *T. violacea* Aubl.

8412 ***Lasianthus*** Jack in Trans. Linn. Soc. London 14: 125. 28 Mai - 12 Jun 1823.
Typus: *L. cyanocarpus* Jack *(typ. cons.)*.

(H) *Lasianthus* Adans., Fam. Pl. 2: 398, 568. Jul-Aug 1763 [*The.*].
≡ *Gordonia* J. Ellis 1771 *(nom. cons.)* (5148).

(=) *Dasus* Lour., Fl. Cochinch.: 96, 141. Sep 1790.
Typus: *D. verticillata* Lour.

8428 ***Gaertnera*** Lam., Tabl. Encycl. 1: 379. 30 Jul 1792.
Typus: *G. vaginata* Lam. (Tabl. Encycl. 2: 273. 31 Oct 1819).

(H) *Gaertnera* Schreb., Gen. Pl.: 290. Apr 1789 [*Malpigh.*].
≡ *Hiptage* Gaertn. 1790 *(nom. cons.)* (4208).

(H) *Gaertneria* Medik., Philos. Bot. 1: 45. Apr 1789 [*Comp.*].
≡ *Franseria* Cav. 1793 *(nom. cons.)* (9147).

8430 ***Paederia*** L., Syst. Nat., ed. 12, 2: 135, 189 [& Mant. Pl.: 7, 52]. 15-31 Oct 1767.
Typus: *P. foetida* L.

(≡) *Daun-contu* Adans., Fam. Pl. 2: 146, 549. Jul-Aug 1763.

(=) *Hondbessen* Adans., Fam. Pl. 2: 158, 584. Jul-Aug 1763.
Typus: *Paederia valli-kara* Juss.

8445 ***Nertera*** Banks ex Gaertn., Fruct. Sem. Pl. 1: 124. Dec 1788.
Typus: *N. depressa* Gaertn.

(=) *Gomozia* Mutis ex L. f., Suppl. Pl.: 17, 129. Apr 1782.
Typus: *G. granadensis* L. f.

8473 ***Borreria*** G. Mey., Prim. Fl. Esseq.: 79. Nov 1818.
Typus: *B. suaveolens* G. Mey. *(typ. cons.).*

(H) *Borrera* Ach., Lichenogr. Universalis: 93, 496. Apr-Mai 1810 [*Fungi*].
Typus: non designatus.

(=) *Tardavel* Adans., Fam. Pl. 2: 145, 609. Jul-Aug 1763.
Typus: non designatus.

8473a ***Robynsia*** Hutch. in Hutchinson & Dalziel, Fl. W. Trop. Afr. 2: 68, 108. Mar 1931.
Typus: *R. glabrata* Hutch.

(H) *Robynsia* Drap. in Lemaire, Hort. Universel 2: 127. Aug-Oct 1840 [*Amaryllid.*].
Typus: *R. geminiflora* Drap.

8485 ***Asperula*** L., Sp. Pl.: 103. 1 Mai 1753.
Typus: *A. arvensis* L. *(typ. cons.).*

VALERIANACEAE

8530 ***Fedia*** Gaertn., Fruct. Sem. Pl. 2: 36. Sep (sero) - Nov 1790.
Typus: *F. cornucopiae* (L.) Gaertn. (*Valeriana cornucopiae* L.) *(typ. cons.).*

(H) *Fedia* Adans., Fam. Pl. 2: 152, 557. Jul-Aug 1763 [*Valerian.*].
Typus: *Valeriana ruthenica* Willd.

8535 ***Patrinia*** Juss. in Ann. Mus. Natl. Hist. Nat. 10: 311. Oct 1807.
Typus: *P. sibirica* (L.) Juss. (*Valeriana sibirica* L.) *(typ. cons.).*

DIPSACACEAE

8541 ***Cephalaria*** Schrad. in Roemer & Schultes, Syst. Veg. 3: 1, 43. Apr-Jul 1818.
Typus: *C. alpina* (L.) Roem. & Schult. (*Scabiosa alpina* L.) *(typ. cons.).*

(=) *Lepicephalus* Lag., Gen. Sp. Pl.: 7. Jun-Dec 1816.
Typus: non designatus.

CUCURBITACEAE

8596 ***Ecballium*** A. Rich. in Bory, Dict. Class. Hist. Nat. 6: 19. 9 Oct 1824.
Typus: *E. elaterium* (L.) A. Rich. (*Momordica elaterium* L.).

(≡) *Elaterium* Mill., Gard. Dict. Abr., ed. 4: [459]. 28 Jan 1754.

8598 ***Citrullus*** Schrad. ex Eckl. & Zeyh., Enum. Pl. Afric. Austral.: 279. Jan 1836.
Typus: *C. vulgaris* Schrad. ex Eckl. & Zeyh. (*Cucurbita citrullus* L.) *(typ. cons.).*

(≡) *Anguria* Mill., Gard. Dict. Abr., ed. 4: [93]. 28 Jan 1754.

(=) *Colocynthis* Mill., Gard. Dict. Abr., ed. 4: [357]. 28 Jan 1754.
Typus: non designatus.

8627 ***Cayaponia*** Silva Manso, Enum. Subst. Braz.: 31. 1836.
Typus: *C. diffusa* Silva Manso *(typ. cons.)*.

8629 ***Echinocystis*** Torr. & A. Gray, Fl. N. Amer. 1: 542. Jun 1840.
Typus: *E. lobata* (Michx.) Torr. & A. Gray (*Sicyos lobata* Michx.).

(=) *Micrampelis* Raf. in Med. Repos., ser. 2, 5: 350. Feb-Apr 1808.
Typus: *M. echinata* Raf.

8636 ***Sechium*** P. Browne, Civ. Nat. Hist. Jamaica: 355. 10 Mar 1756.
Typus: *S. edule* (Jacq.) Sw. (Fl. Ind. Occid.: 1150. Oct 1800) (*Sicyos edulis* Jacq.) *(typ. cons.)*.

CAMPANULACEAE

8651 ***Michauxia*** L'Hér., Michauxia: ad t. 1. Mar-Apr 1788.
Typus: *M. campanuloides* L'Hér. *(typ. cons.)*.

8656 ***Canarina*** L., Mant. Pl.: 148, 225, 588. Oct 1771.
Typus: *C. campanula* L., nom. illeg. (*Campanula canariensis* L., *Canarina canariensis* (L.) Vatke).

(≡) *Mindium* Adans., Fam. Pl. 2: 134, 578. Jul-Aug 1763 (per typ. des.).

8660 ***Cyananthus*** Wall. ex Benth. in Royle, Ill. Bot. Himal. Mts.: 309. Mai 1836.
Typus: *C. lobatus* Wall. ex Benth.

(H) *Cyananthus* Raf., Anal. Nat.: 192. Apr-Jul 1815 [*Comp.*].
≡ *Cyanus* P. Mill. 1754.

8663 ***Prismatocarpus*** L'Hér., Sert. Angl.: 1. Jan (prim.) 1789.
Typus: *P. paniculatus* L'Hér. *(typ. cons.)*.

8668 ***Wahlenbergia*** Schrad. ex Roth, Nov. Pl. Sp.: 399. Apr 1821.
Typus: *W. elongata* (Willd.) Schrad. ex Roth (*Campanula elongata* Willd.) [= *W. capensis* (L.) A. DC. (*Campanula capensis* L.)].

(=) *Cervicina* Delile, Descr. Egypte, Hist. Nat. 2: 150. 1813 (sero) - 1814 (prim.).
Typus: *C. campanuloides* Delile

8680 ***Sphenoclea*** Gaertn., Fruct. Sem. Pl. 1: 113. Dec 1788.
Typus: *S. zeylanica* Gaertn.

8706 ***Downingia*** Torr. in Rep. Explor. Railroad Pacif. Ocean 4(1,4): 116. Aug-Sep 1857.
Typus: *D. elegans* (Douglas ex Lindl.) Torr. (*Clintonia elegans* Douglas ex Lindl.).

(≡) *Bolelia* Raf. in Atlantic J. 1: 120. 1832 (aut.).

GOODENIACEAE

8716 ***Scaevola*** L., Mant. Pl.: 145. Oct 1771.
Typus: *S. lobelia* L. (Syst. Veg., ed. 13: 178. Apr-Jun 1774), nom. illeg. (*Lobelia plumieri* L., *S. plumieri* (L.) Vahl).

STYLIDIACEAE

8724 ***Stylidium*** Sw. ex Willd., Sp. Pl. 4: 7, 146. 1805.
Typus: *S. graminifolium* Sw. *(typ. cons.)*.

(H) *Stylidium* Lour., Fl. Cochinch.: 219, 220. Sep 1790 [*Alang.*].
Typus: *S. chinense* Lour.

CALYCERACEAE

8726 ***Calycera*** Cav., Icon. 4: 34. Sep-Dec 1797 *('Calicera') (orth. cons.)*.
Typus: *C. herbacea* Cav.

COMPOSITAE (ASTERACEAE)

8751 ***Vernonia*** Schreb., Gen. Pl.: 541. Mai 1791.
Typus: *V. noveboracensis* (L.) Willd. (Sp. Pl. 3: 1632. Apr-Dec 1803) (*Serratula noveboracensis* L.) *(typ. cons.)*.

8761 ***Piptolepis*** Sch. Bip. in Jahresber. Pollichia 20-21: 380. Jul-Dec 1863.
Typus: *P. ericoides* Sch. Bip.

(H) *Piptolepis* Benth., Pl. Hartw.: 29. Feb 1840 [*Ol.*].
Typus: *P. phillyreoides* Benth.

8772 ***Soaresia*** Sch. Bip. in Jahresber. Pollichia 20-21: 376. Jul-Dec 1863.
Typus: *S. velutina* Sch. Bip.

(H) *Soaresia* Allemão in Trab. Soc. Vellosiana Rio de Janeiro 1851: 72. 1851 [*Mor.*].
Typus: *S. nitida* Allemão

8775a ***Pseudelephantopus*** Rohr in Skr. Naturhist.-Selsk. 2(1). 214. 1792 *('Pseudo-elephantopus') (orth. cons.)*.
Typus: *P. spicatus* (Aubl.) C. F. Baker (in Trans. Acad. Sci. St. Louis 12: 45, 54, 56. 20 Mai 1902) (*Elephantopus spicatus* Aubl.).

8808 ***Brachyandra*** Phil., Fl. Atacam.: 34. 1860.
Typus: *B. macrogyne* Phil.

(H) *Brachyandra* Naudin in Ann. Sci. Nat., Bot., ser. 3, 2: 143. Sep 1844 [*Melastomat.*].
Typus: *B. perpusilla* Naudin

8818 ***Mikania*** Willd., Sp. Pl. 3: 1481, 1742. Apr-Dec 1803.
Typus: *M. scandens* (L.) Willd. (*Eupatorium scandens* L.) *(typ. cons.)*.

8823 ***Brickellia*** Elliott, Sketch Bot. S.-Carolina 2: 290. 1823.
Typus: *B. cordifolia* Elliott

(H) *Brickellia* Raf. in Med. Repos., ser. 2, 5: 353. Feb-Apr 1808 [*Polemon.*].
≡ *Ipomopsis* Michx. 1803.

(=) *Kuhnia* L., Sp. Pl., ed. 2: 1662. Jul-Aug 1763.
Typus: *K. eupatorioides* L.

(=) *Coleosanthus* Cass. in Bull. Sci. Soc. Philom. Paris 1817: 67. Apr 1817.
Typus: *C. cavanillesii* Cass.

8826 ***Liatris*** Gaertn. ex Schreb., Gen. Pl.: 542. Mai 1791.
Typus: *L. squarrosa* (L.) Michx. (Fl. Bor.-Amer. 2: 92. 19 Mar 1803) (*Serratula squarrosa* L.) *(typ. cons.)*.

(≡) *Lacinaria* Hill, Veg. Syst. 4: 49. 1762.

8840 ***Bradburia*** Torr. & A. Gray, Fl. N. Amer. 2: 250. Apr 1842.
Typus: *B. hirtella* Torr. & A. Gray

(H) *Bradburya* Raf., Fl. Ludov.: 104. Oct-Dec (prim.) 1817 [*Legum.*].
Typus (vide Cowan in Regnum Veg. 100: 232. 1979): *B. scandens* Raf.

8843 ***Chiliophyllum*** Phil. in Linnaea 33: 132. Aug 1864.
Typus: *C. densifolium* Phil.

(H) *Chiliophyllum* DC., Prodr. 5: 554. 1-10 Oct 1836 [*Comp.*].
≡ *Hybridella* Cass. 1817.

8844 ***Chrysopsis*** (Nutt.) Elliott, Sketch Bot. S. Carolina 2: 333. 1823 (*Inula* sect. *Chrysopsis* Nutt., Gen. N. Amer. Pl. 2: 150. 14 Jul 1818).
Typus: *Inula mariana* L. (*C. mariana* (L.) Elliott) *(typ. cons.)*.

(≡) *Diplogon* Raf. in Amer. Monthly Mag. & Crit. Rev. 4: 195. Jan 1819.

8852 ***Haplopappus*** Cass. in Cuvier, Dict. Sci. Nat. 56: 168. Sep 1828 *('Aplopappus') (orth. cons.).*
Typus: *H. glutinosus* Cass.

8855 ***Bigelowia*** DC., Prodr. 5: 329. 1-10 Oct 1836.
Typus: *B. nudata* (Michx.) DC. (*Chrysocoma nudata* Michx.) *(typ. cons.).*

(H) *Bigelowia* Raf. in Amer. Monthly Mag. & Crit. Rev. 1: 442. Oct 1817 [*Caryophyll.*].
Typus: *B. montana* Raf.

(≡) *Pterophora* L., Pl. Rar. Afr.: 17. 20 Dec 1760.

8862 ***Pteronia*** L., Sp. Pl., ed. 2: 1176. Jul-Aug 1763.
Typus: *P. camphorata* (L.) L. (*Pterophora camphorata* L.).

(≡) *Pterophora* L., Pl. Rar. Afr.: 17. 20 Dec 1760.

(=) *Pterophorus* Boehm. in Ludwig, Def. Gen. Pl., ed. 3: 165. 1760.
Typus: non designatus.

8887 ***Amellus*** L., Syst. Nat., ed. 10: 1189, 1225, 1377. 7 Jun 1759.
Typus: *A. lychnitis* L. *(typ. cons.).*

(H) *Amellus* P. Browne, Civ. Nat. Hist. Jamaica: 317. 10 Mar 1756 [*Comp.*].
Typus: *Santolina amellus* L.

8898 ***Callistephus*** Cass. in Cuvier, Dict. Sci. Nat. 37: 491. Dec 1825.
Typus: *C. chinensis* (L.) Nees (Gen. Sp. Aster.: 222. Jul-Dec 1832) (*Aster chinensis* L.).

(≡) *Callistemma* Cass. in Bull. Sci. Soc. Philom. Paris 1817: 32. Feb 1817.

8909 ***Celmisia*** Cass. in Cuvier, Dict. Sci. Nat. 37: 259. Dec 1825.
Typus: *C. longifolia* Cass. *(typ. cons.).*

(H) *Celmisia* Cass. in Bull. Sci. Soc. Philom. Paris 1817: 32. Feb 1817 [*Comp.*].
Typus: *C. rotundifolia* Cass. (in Cuvier, Dict. Sci. Nat. 7: 356. 24 Mai 1817).

8916 ***Olearia*** Moench, Suppl. Meth.: 254. 2 Mai 1802.
Typus: *O. dentata* Moench, nom. illeg. (*Aster tomentosus* J. C. Wendl., *O. tomentosa* (J. C. Wendl.) DC.).

(=) *Shawia* J. R. Forst. & G. Forst., Char. Gen. Pl.: 48. 29 Nov 1775.
Typus: *S. paniculata* J. R. Forst. & G. Forst.

8918 ***Sommerfeltia*** Less., Syn. Gen. Compos.: 189. Jul-Aug 1832.
Typus: *S. spinulosa* (Spreng.) Less. (*Conyza spinulosa* Spreng.).

(H) *Sommerfeltia* Flörke ex Sommerf. in Kongel. Norske Videnskabersselsk. Skr. 19de Aarhundr. 2(2): 60. 1827 [*Fungi*].
Typus: *S. arctica* Flörke

8919 ***Felicia*** Cass. in Bull. Sci. Soc. Philom. Paris 1818: 165. Nov 1818.
Typus: *F. tenella* (L.) Nees (Gen. Sp. Aster.: 208. Jul-Dec 1832) (*Aster tenellus* L.).

(=) *Detris* Adans., Fam. Pl. 2: 131, 549. Jul-Aug 1763.
Typus: non designatus.

8926 ***Conyza*** Less., Syn. Gen. Compos.: 203. Jul-Aug 1832.
Typus: *C. chilensis* Spreng. (Novi Provent.: 14. Dec 1818) *(typ. cons.).*

(H) *Conyza* L., Sp. Pl.: 861. 1 Mai 1753 [*Comp.*].
Typus (vide Green in Sprague & al., Nom. Prop. Brit. Bot.: 181. 1929): *C. squarrosa* L.

(=) *Eschenbachia* Moench, Methodus: 573. 4 Mai 1794.
Typus: *E. globosa* Moench, nom. illeg. (*Erigeron aegyptiacus* L.).

(=) *Dimorphanthes* Cass. in Bull. Sci. Soc. Philom. Paris 1818: 30. Feb 1818.
Typus: non designatus.

(=) *Laënnecia* Cass. in Cuvier, Dict. Sci. Nat. 25: 91. 1822.
Typus: *L. gnaphalioides* (Kunth) Cass. (*Conyza gnaphalioides* Kunth).

8933 ***Baccharis*** L., Sp. Pl.: 860. 1 Mai 1753.
Typus: *B. halimifolia* L. *(typ. cons.).*

8939 ***Blumea*** DC. in Arch. Bot. (Paris) 2: 514. 23 Dec 1833.
Typus: *B. balsamifera* (L.) DC. (Prodr. 5: 447. 1-10 Oct 1836) (*Conyza balsamifera* L.) *(typ. cons.).*

(H) *Blumia* Nees in Flora 8: 152. 14 Mar 1825 [*Magnol.*].
Typus: *B. candollei* (Blume) Nees (*Talauma candollei* Blume).

(=) *Placus* Lour., Fl. Cochinch.: 475, 496. Sep 1790.
Typus (vide Merrill in Trans. Amer. Philos. Soc., ser. 2, 24: 387. 1935): *P. tomentosus* Lour.

8969 ***Filago*** L., Sp. Pl.: 927, 1199, [add. post indicem]. 1 Mai 1753.
Typus: *F. pyramidata* L. *(typ. cons.).*

8978 ***Antennaria*** Gaertn., Fruct. Sem. Pl. 2: 410. Sep-Dec 1791.
Typus: *A. dioica* (L.) Gaertn. (*Gnaphalium dioicum* L.).

(H) *Antennaria* Link in Neues J. Bot. 3(1,2): 16. Apr 1809 : Fr., Syst. Mycol. 1: xlvii. 1 Jan 1821 [*Fungi*].
Typus: *A. ericophila* Link : Fr.

8994 ***Cassinia*** R. Br., Observ. Compos.: 126. 1817 (ante Sep).
Typus: *C. aculeata* (Labill.) R. Br. (*Calea aculeata* Labill.) *(typ. cons.).*

(H) *Cassinia* R. Br. in Aiton, Hort. Kew., ed. 2, 5: 184. Nov 1813 [*Comp.*].
Typus: *C. aurea* R. Br.

9006 ***Helichrysum*** Mill., Gard. Dict. Abr., ed. 4: [462]. 28 Jan 1754 *('Elichrysum') (orth. cons.).*
Typus: *H. orientale* (L.) Gaertn. (Fruct. Sem. Pl. 2: 404. Sep-Dec 1791) (*Gnaphalium orientale* L.) *(typ. cons.).*

9009 ***Podotheca*** Cass. in Cuvier, Dict. Sci. Nat. 23: 561. Nov 1822.
Typus: *P. angustifolia* (Labill.) Less. (*Podosperma angustifolium* Labill.).
(≡) *Podosperma* Labill., Nov. Holl. Pl. 2: 35. Apr 1806.

9028 ***Angianthus*** J. C. Wendl., Coll. Pl. 2: 31. 1808.
Typus: *A. tomentosus* J. C. Wendl.
(=) *Siloxerus* Labill, Nov. Holl. Pl. 2: 57. Jun 1806.
Typus: *S. humifusus* Labill.

9039 ***Disparago*** Gaertn., Fruct. Sem. Pl. 2: 463. Sep-Dec 1791.
Typus: *D. ericoides* (P. J. Bergius) Gaertn. (*Stoebe ericoides* P. J. Bergius).

9050 ***Relhania*** L'Hér., Sert. Angl.: 22. Jan (prim.) 1789.
Typus: *R. paleacea* (L.) L'Hér. (*Leysera paleacea* L.).
(=) *Osmites* L., Sp. Pl., ed. 2: 1285. Jul-Aug 1763.
Typus (vide Bremer in Taxon 28: 412. 1979): *O. bellidiastrum* L., nom illeg. (*Anthemis bellidiastrum* L., nom. illeg., *Anthemis fruticosa* L.).

9054 ***Podolepis*** Labill., Nov. Holl. Pl. 2: 56. Jun 1806.
Typus: *P. rugata* Labill.

9057 ***Heterolepis*** Cass. in Bull. Sci. Soc. Philom. Paris 1820: 26. Feb 1820.
Typus: *H. decipiens* Cass., nom. illeg. (*Arnica inuloides* Vahl) [= *H. aliena* (L. f.) Druce (*Oedera aliena* L. f.)] (etiam vide 5992).

9059 ***Printzia*** Cass. in Cuvier, Dict. Sci. Nat. 37: 488. Dec 1825.
Typus: *P. cernua* (P. J. Bergius) Druce (Bot. Soc. Exch. Club Brit. Isles 1916: 642. 1917) (*Inula cernua* P. J. Bergius).

9065 ***Iphiona*** Cass. in Bull. Sci. Soc. Philom. Paris 1817: 153. Oct 1817.
Typus: *I. dubia* Cass., nom. illeg. (*Conyza pungens* Lam.) *(typ. cons.)* [= *I. mucronata* (Forssk.) Asch. & Schweinf. (*Chrysocoma mucronata* Forssk.)].

9091 ***Pallenis*** Cass. in Bull. Sci. Soc. Philom. Paris 1818: 166. Nov 1818.
Typus: *P. spinosa* (L.) Cass. (in Cuvier, Dict. Sci. Nat. 37: 276. Dec 1825) (*Buphthalmum spinosum* L.).

9101 ***Lagascea*** Cav. in Anales Ci. Nat. 6: 331. Jun 1803 *('Lagasca') (orth. cons.)*.
Typus: *L. mollis* Cav.

(=) *Nocca* Cav., Icon. 3: 12. Apr 1795.
Typus: *N. rigida* Cav.

9147 ***Franseria*** Cav., Icon. 2: 78. Dec 1793 -Jan 1794.
Typus: *F. ambrosioides* Cav., nom. illeg. (*Ambrosia arborescens* Mill., non *Franseria arborescens* Brandegee 1903) (etiam vide 8428) [nom. legit. sub *Franseria* deest].

9150 ***Podanthus*** Lag., Gen. Sp. Pl.: 24. Jun-Dec 1816.
Typus: *P. ovatifolius* Lag.

(H) *Podanthes* Haw., Syn. Pl. Succ.: 32. 1812 [*Asclepiad.*].
Typus (vide Mansfeld in Bull. Misc. Inform. Kew 1935: 451. 31 Aug 1935): *P. pulchra* Haw.

9155 ***Zinnia*** L., Syst. Nat., ed. 10: 1189, 1221, 1377. 7 Jun 1759.
Typus: *Z. peruviana* (L.) L. (*Chrysogonum peruvianum* L.).

(≡) *Crassina* Scepin, Acid. Veg.: 42. 19 Mai 1758.

(=) *Lepia* Hill, Exot. Bot.: 29. Feb-Sep 1759.
Typus: non designatus.

9157 ***Heliopsis*** Pers., Syn. Pl. 2: 473. Sep 1807.
Typus: *H. helianthoides* (L.) Sweet (Hort. Brit. 2: 487. Sep-Oct 1826) (*Buphthalmum helianthoides* L.) *(typ. cons.)*.

9166 ***Eclipta*** L., Mant. Pl.: 157, 286. Oct 1771.
Typus: *E. erecta* L., nom. illeg. (*Verbesina alba* L., *E. alba* (L.) Hassk.) *(typ. cons.)*.

(≡) *Eupatoriophalacron* Mill., Gard. Dict. Abr., ed. 4: [479]. 28 Jan 1754.

9168 ***Selloa*** Kunth in Humboldt & al., Nov. Gen. Sp. 4, ed. f°: 208. 26 Oct 1818.
Typus: *S. plantaginea* Kunth

(H) *Selloa* Spreng., Novi Provent: 36. Dec 1818 [*Comp.*].
Typus: *S. glutinosa* Spreng.

9192 ***Wedelia*** Jacq., Enum. Syst. Pl.: 8, 28. Aug-Sep 1760.
Typus: *W. fruticosa* Jacq.

(H) *Wedelia* Loefl., Iter Hispan.: 180. Dec 1758 [*Nyctagin.*].
≡ *Allionia* L. 1759 *(nom. cons.)* (2348).

9208	***Salmea*** DC., Cat. Pl. Horti Monsp.: 140. Feb-Mar 1813. Typus: *S. scandens* (L.) DC. (*Bidens scandens* L.) *(typ. cons.)*.	(H)	*Salmia* Cav., Icon. 3: 24. Apr 1795 [*Lil.*]. Typus (vide Rickett & Stafleu in Taxon 9: 156. 1960): *S. spicata* Cav.
9215	***Actinomeris*** Nutt., Gen. N. Amer. Pl. 2: 181. 14 Jul 1818. Typus: *A. squarrosa* Nutt., nom. illeg. (*Coreopsis alternifolia* L., *A. alternifolia* (L.) DC.) *(typ. cons.)*.	(≡)	*Ridan* Adans., Fam. Pl. 2: 130, 598. Jul-Aug 1763.
9218	***Verbesina*** L., Sp. Pl.: 901. 1 Mai 1753. Typus: *V. alata* L. *(typ. cons.)*.		
9222	***Guizotia*** Cass. in Cuvier, Dict. Sci. Nat. 59: 237, 247, 248. Jun 1829. Typus: *G. abyssinica* (L. f.) Cass. (*Polymnia abyssinica* L. f.).		
9224	***Synedrella*** Gaertn., Fruct. Sem. Pl. 2: 456. Sep-Dec 1791. Typus: *S. nodiflora* (L.) Gaertn. (*Verbesina nodiflora* L.).	(≡)	*Ucacou* Adans., Fam. Pl. 2: 131, 615. Jul-Aug 1763.
9241	***Balduina*** Nutt., Gen. N. Amer. Pl. 2: 175. 14 Jul 1818. Typus: *B. uniflora* Nutt. *(typ. cons.)*.	(=)	*Mnesiteon* Raf., Fl. Ludov.: 67. Oct - Dec (prim.) 1817. Typus: non designatus.
9247	***Marshallia*** Schreb., Gen. Pl.: 810. Mai 1791. Typus: *M. obovata* (Walter) Beadle & F. E. Boynton (in Biltmore Bot. Stud. 1: 5. 8 Apr 1901) (*Athanasia obovata* Walter) *(typ. cons.)*.		
9258	***Layia*** Hook. & Arn. ex DC., Prodr. 7: 294. Apr (sero) 1838. Typus: *L. gaillardioides* (Hook. & Arn.) DC. (*Tridax gaillardioides* Hook. & Arn.).	(H) (=)	*Layia* Hook. & Arn., Bot. Beechey Voy.: 182. Oct 1833 [*Legum.*]. Typus: *L. emarginata* Hook. & Arn. *Blepharipappus* Hook., Fl. Bor.-Amer. 1: 316. 1833 (sero). Typus: *B. scaber* Hook.
9285	***Villanova*** Lag., Gen. Sp. Pl.: 31. Jun-Dec 1816. Typus: *V. alternifolia* Lag. *(typ. cons.)*.	(H) (=)	*Villanova* Ortega, Nov. Pl. Descr. Dec.: 47. 1797 [*Comp.*]. Typus: *V. bipinnatifida* Ortega *Unxia* L. f., Suppl. Pl.: 56, 368. Apr 1782. Typus: *U. camphorata* L. f.

9289 ***Thymopsis*** Benth. in Bentham & Hooker, Gen. Pl. 2: 201, 407. 7-9 Apr 1873.
Typus: *T. wrightii* Benth., nom. illeg. (*Tetranthus thymoides* Griseb., *Thymopsis thymoides* (Griseb.) Urb.).

(H) *Thymopsis* Jaub. & Spach, Ill. Pl. Orient. 1: 72. Oct 1842 [*Gutt.*].
Typus: *T. aspera* Jaub. & Spach

9291 ***Schkuhria*** Roth, Catal. Bot. 1: 116. Jan-Feb 1797.
Typus: *S. abrotanoides* Roth

(H) *Sckuhria* Moench, Methodus: 566. 4 Mai 1794 [*Comp.*].
Typus: *S. dichotoma* Moench, nom. illeg. (*Siegesbeckia flosculosa* L'Hér.).

9322 ***Oedera*** L., Mant. Pl.: 159, 291. Oct 1771.
Typus: *O. prolifera* L., nom. illeg. (*Buphthalmum capense* L., *O. capensis* (L.) Druce).

(H) *Oedera* Crantz, Duab. Drac. Arbor.: 30. 1768 [*Lil.*].
Typus: *O. dragonalis* Crantz

9341a ***Prolongoa*** Boiss., Voy. Bot. Espagne: 320. 23 Sep 1840.
Typus: *P. hispanica* G. López & C. E. Jarvis (in Anales Jard. Bot. Madrid 40: 343. Mai 1984) *(typ. cons.)*.

9365 ***Peyrousea*** DC., Prodr. 6: 76. Jan (prim.) 1838.
Typus: *P. oxylepis* DC., nom. illeg. (*Cotula umbellata* L. f., *P. umbellata* (L. f.) Fourc.) *(typ. cons.)*.

(H) *Peyrousia* Poir. in Cuvier, Dict. Sci. Nat. 39: 363. Apr 1826 [*Irid.*].
≡ *Lapeirousia* Pourr. 1788.

9382 ***Robinsonia*** DC. in Arch. Bot. (Paris) 2: 333. 21 Oct 1833.
Typus: *R. gayana* Decne. (in Ann. Sci. Nat., Bot., ser. 2, 1: 28. Jan 1834) *(typ. cons.)*.

(H) *Robinsonia* Scop., Intr. Hist. Nat.: 218. Jan-Apr 1777 [*Quiin.*].
≡ *Touroulia* Aubl. 1775.

9405 ***Gynura*** Cass. in Cuvier, Dict. Sci. Nat. 34: 391. Apr 1825.
Typus: *G. auriculata* Cass. (Opusc. Phytol. 3: 100. 19 Apr 1834) *(typ. cons.)*.

(=) *Crassocephalum* Moench, Methodus: 516. 4 Mai 1794.
Typus: *C. cernuum* Moench, nom. illeg. (*Senecio rubens* B. Juss. ex Jacq., *C. rubens* (B. Juss. ex Jacq.) S. Moore).

9412 ***Ligularia*** Cass. in Bull. Sci. Soc. Philom. Paris 1816: 198. Dec 1816.
Typus: *L. sibirica* (L.) Cass. (in Cuvier, Dict. Sci. Nat. 26: 402. Mai 1823) (*Othonna sibirica* L.).

(H) *Ligularia* Duval, Pl. Succ. Horto Alencon.: 11. 1809 [*Saxifrag.*].
≡ *Sekika* Medik. 1791.

(=) *Senecillis* Gaertn., Fruct. Sem. Pl. 2: 453. Sep-Dec 1791.
Typus: non designatus.

9425 ***Dimorphotheca*** Moench, Methodus: 585. 4 Mai 1794.
Typus: *D. pluvialis* (L.) Moench (*Calendula pluvialis* L.) *(typ. cons.)*.

9428 ***Tripteris*** Less. in Linnaea 6: 95. 1831 (post Mar).
Typus: *T. arborescens* (Jacq.) Less. (*Calendula arborescens* Jacq.) *(typ. cons.)*.

9431 ***Ursinia*** Gaertn., Fruct. Sem. Pl. 2: 462. Sep-Dec 1791.
Typus: *U. paradoxa* (L.) Gaertn. (*Arctotis paradoxa* L.).

9434 ***Gazania*** Gaertn., Fruct. Sem. Pl. 2: 451. Sep-Dec 1791.
Typus: *G. rigens* (L.) Gaertn. (*Othonna rigens* L.).

(=) *Meridiana* Hill, Veg. Syst. 2: 121**. Oct 1761.
Typus: *M. tesselata* Hill (Hort. Kew.: 26. 1768).

9438 ***Berkheya*** Ehrh. in Neues Mag. Aerzte 6: 303. 12 Mai - 7 Sep 1784.
Typus: *B. fruticosa* (L.) Ehrh. (*Atractylis fruticosa* L.).

(≡) *Crocodilodes* Adans., Fam. Pl. 2: 127, 545. Jul-Aug 1763.

9439 ***Didelta*** L'Hér., Stirp. Nov.: 55. Mar 1786.
Typus: *D. tetragoniifolia* L'Hér.

(=) *Breteuillia* Buc'hoz, Grand Jard.: t. 62. 1785.
Typus: *B. trianensis* Buc'hoz

9446 ***Siebera*** J. Gay in Mém. Soc. Hist. Nat. Paris 3: 344. 1827.
Typus: *S. pungens* (Lam.) DC. (Prodr. 6: 531. Jan (prim.) 1838) (*Xeranthemum pungens* Lam.).

(H) *Sieberia* Spreng., Anleit. Kenntn. Gew., ed. 2, 2: 282. 20 Apr 1817 [*Orchid.*].
Typus: non designatus.

9457 ***Saussurea*** DC. in Ann. Mus. Natl. Hist. Nat. 16: 156, 198. Jul-Dec 1810.
Typus: *S. alpina* (L.) DC. (*Serratula alpina* L.) *(typ. cons.)*.

(H) *Saussuria* Moench, Methodus: 388. 4 Mai 1794 [*Lab.*].
Typus: *S. pinnatifida* Moench, nom. illeg. (*Nepeta multifida* L.).

9464 ***Silybum*** Adans., Fam. Pl. 2: 116, 605. Jul-Aug 1763.
Typus: *S. marianum* (L.) Gaertn. (Fruct. Sem. Pl. 2: 378. Sep-Dec 1791) (*Carduus marianus* L.) *(typ. cons.)*.

(≡) *Mariana* Hill, Veg. Syst. 4: 19. 1762.

9466 ***Galactites*** Moench, Methodus: 558. 4 Mai 1794.
Typus: *G. tomentosus* Moench (*Centaurea galactites* L.) *(typ. cons.)*.

9476 ***Amberboa*** (Pers.) Less., Syn. Gen. Compos.: 8. Jul-Aug 1832 (*Centaurea* subg. *Amerboa* Pers., Syn. Pl. 1: 481. 1 Apr - 15 Jun 1805).
Typus: *Centaurea moschata* L. (*A. moschata* (L.) DC.) *(typ. cons.)*.

(=) *Amberboi* Adans., Fam. Pl. 2: 117, 516. Jul-Aug 1763.
Typus: *Centaurea lippii* L.

(=) *Chryseis* Cass. in Bull. Sci. Soc. Philom. Paris 1817: 33. Feb 1817.
Typus: *C. odorata* Cass. (in Cuvier, Dict. Sci. Nat. 9: 154. 16 Dec 1817) nom. illeg. (*Centaurea amberboi* Mill.).

(=) *Lacellia* Viv., Fl. Libyc. Sp.: 58. 1824.
Typus: *L. libyca* Viv.

9479 ***Cnicus*** L., Sp. Pl.: 826. 1 Mai 1753.
Typus: *C. benedictus* L. *(typ. cons.)*.

9483 ***Moquinia*** DC., Prodr. 7: 22. Apr (sero) 1838.
Typus: *M. racemosa* (Spreng.) DC. (*Conyza racemosa* Spreng.) *(typ. cons.)*.

(H) *Moquinia* A. Spreng., Tent. Suppl.: 9. 20 Sep 1828 [*Loranth.*].
≡ *Moquiniella* Balle 1954.

9490 ***Stifftia*** J. C. Mikan, Del. Fl. Faun. Bras.: ad t. 1. 1820 (sero).
Typus: *S. chrysantha* J. C. Mikan

9511 ***Schlechtendalia*** Less. in Linnaea 5: 242. Apr 1830.
Typus: *S. luzulifolia* Less.

(H) *Schlechtendalia* Willd., Sp. Pl. 3: 1486, 2125. Apr-Dec 1803 [*Comp.*].
Typus: *S. glandulosa* (Cav.) Willd. (*Willdenowa glandulosa* Cav.).

9528 ***Gerbera*** L., Opera Var.: 247. 1758.
Typus: *G. linnaei* Cass. (in Cuvier, Dict. Sci. Nat. 18: 460. 6 Apr 1821) (*Arnica gerbera* L.) *(typ. cons.)*.

9529 ***Chaptalia*** Vent., Descr. Pl. Nouv.: ad t. 61. 22 Mar 1802.
Typus: *C. tomentosa* Vent.

9545 ***Moscharia*** Ruiz & Pav., Fl. Peruv. Prodr.: 103. Oct (prim.) 1794.
Typus: *M. pinnatifida* Ruiz & Pav. (Syst. Veg. Fl. Peruv. Chil.: 186. Dec (sero) 1798).

(H) *Moscharia* Forssk., Fl. Aegypt.-Arab.: 158. 1 Oct 1775 [*Lab.*].
Typus: *M. asperifolia* Forssk.

9560 ***Krigia*** Schreb., Gen. Pl.: 532. Mai 1791.
Typus: *K. virginica* (L.) Willd. (Sp. Pl. 3: 1618. Apr-Dec 1803) (*Tragopogon virginicus* L.) *(typ. cons.)*.

9566 ***Rhagadiolus*** Juss., Gen. Pl.: 168. 4 Aug 1789.
Typus: *R. edulis* Gaertn. (Fruct. Sem. Pl. 2: 354. Sep-Dec 1791) (*Lapsana rhagadiolus* L.).

(H) *Rhagadiolus* Zinn, Cat. Pl. Hort. Gott.: 436. 20 Apr - 21 Mai 1757 (typ. des.: Meikle in Taxon 29: 159. 1980) [*Comp.*].
≡ *Hedypnois* Mill. 1754.

9574 ***Leontodon*** L., Sp. Pl.: 798. 1 Mai 1753.
Typus: *L. hispidus* L. *(typ. cons.)*.

9576 ***Stephanomeria*** Nutt. in Trans. Amer. Philos. Soc., ser. 2, 7: 427. 2 Apr 1841.
Typus: *S. minor* (Hook.) Nutt. (*Lygodesmia minor* Hook.) *(typ. cons.)*.

(=) *Ptiloria* Raf. in Atlantic J. 1: 145. 1832 (sero).
Typus: non designatus.

9578 ***Rafinesquia*** Nutt. in Trans. Amer. Philos. Soc., ser. 2, 7: 429. 2 Apr 1841.
Typus: *R. californica* Nutt.

(H) *Rafinesquia* Raf., Fl. Tellur. 2: 96. Jan-Mar 1837 [*Legum.*].
≡ *Hosackia* Benth. ex Lindl. 1829.

9581 ***Podospermum*** DC. in Lamarck & Candolle, Fl. Franç., ed. 3, 4: 61. 17 Sep 1805.
Typus: *P. laciniatum* (L.) DC. (*Scorzonera laciniata* L.) *(typ. cons.)*.

(≡) *Arachnospermum* F. W. Schmidt in Samml. Phys.-Oekon. Aufsätze 1: 274. 1795.

9592 ***Taraxacum*** F. H. Wigg., Prim. Fl. Holsat.: 56. 29 Mar 1780.
Typus: *T. officinale* F. H. Wigg. (*Leontodon taraxacum* L.) *(typ. cons.)*.

(H) *Taraxacum* Zinn, Cat. Pl. Hort. Gott.: 425. 20 Apr - 21 Mai 1757 [*Comp.*].
≡ *Leontodon* L. 1753 *(nom. cons.)* (9574).

9604 ***Pyrrhopappus*** DC., Prodr. 7: 144. Apr (sero) 1838.
Typus: *P. carolinianus* (Walter) DC. (*Leontodon carolinianus* Walter) *(typ. cons.)*.

9604a ***Thorelia*** Gagnep. in Notul. Syst. (Paris) 4: 18. 28 Nov 1920.
Typus: *T. montana* Gagnep.

(H) *Thorelia* Hance in J. Bot. 15: 268. Sep 1877 [*Myrt.*].
Typus: *T. deglupta* Hance

[HETEROPYXIDACEAE]

9712 ***Heteropyxis*** Harv., Thes. Cap. 2: 18. 1863.
Typus: *H. natalensis* Harv.

(H) *Heteropyxis* Griff., Not. Pl. Asiat. 4: 524. 1854 [*Bombac.*].
Typus: *Boschia griffithii* Mast.

F. PLANTAE FOSSILES (EXCL. BACILLARIOPH.)

Asterophyllites Brongn., Prodr. Hist. Vég. Foss.: 159. Dec 1828 [*Calamit.*].
Typus: *A. equisetiformis* (Sternb.) Brongn. (*Bornia equisetiformis* Sternb.).

(H) *Asterophyllites* Brongn. in Mém. Mus. Hist. Nat. 8: 210. Mai 1822 [Foss.].
Typus: *A. radiatus* Brongn.

(≡) *Bornia* Sternb., Vers. Fl. Vorwelt 4: xxviii. 1825.

(=) *Bechera* Sternb., Vers. Fl. Vorwelt 4: xxx. 1825.
Typus: *B. ceratophylloides* Sternb.

(=) *Brukmannia* Sternb., Vers. Fl. Vorwelt 4: xxix. 1825.
Typus: *B. tenuifolia* (Sternb.) Sternb. (*Schlotheimia tenuifolia* Sternb.).

Baiera Braun in Beitr. Petrefacten-Kunde 6: 20. 1843 [*Ginkg.*].
Typus: *B. muensteriana* (C. Presl) Heer (in Mém. Acad. Imp. Sci. Saint Pétersbourg, ser. 7, 22: 52. 1876) (*Spaerococcites muensterianus* C. Presl) *(typ. cons.)*.

(H) *Bajera* Sternb., Vers. Fl. Vorwelt 4: xxviii. 1825 [Foss.].
Typus: B. scanica Sternb.

Calamites Brongn., Prodr. Hist. Vég. Foss.: 121. Dec 1828 [*Calamit.*].
Typus: *C. suckowii* Brongn. *(typ. cons.)*.

(H) *Calamitis* Sternb., Vers. Fl. Vorwelt 1(1): 22. 31 Dec 1820 [Foss.].
Typus: *C. pseudobambusia* Sternb.

Cardiocarpus Brongn., Rech. Graines Foss. Silic.: 20. 1 Dec 1880 [*Cordaitales*].
Typus: *C. drupaceus* Brongn.

(H) *Cardiocarpus* Reinw. in Syll. Pl. Nov. 2: 14. 1825 [*Simaroub.*].
Typus: *C. amarus* Reinw.

Cordaianthus Grand'Eury in Mém. Acad. Roy. Sci. Inst. France 24: 227. 1877 [*Cordaitales*].
Typus: *C. gemmifer* Grand'Eury

(=) *Botryoconus* Göpp. in Palaeontographica 12: 152. 1864.
Typus: *B. goldenbergii* Göpp.

Cordaites Unger, Gen. Sp. Pl. Foss.: 277. 17-20 Apr 1850 [*Cordaitales*].
Typus: *C. borassifolius* (Sternb.) Unger (*Flabellaria borassifolia* Sternb.).

(≡) *Neozamia* Pomel in Bull. Soc. Géol. France, ser. 2, 3: 655. 1846 (post 15 Jun).

Dolerotheca T. Halle in Kongl. Svenska Vetenskapsakad. Handl., ser. 3, 12(6): 42. Jul-Dec 1933 [*Medullos.*].
Typus: *D. fertilis* (Renault) T. Halle (*Dolerophyllum fertile* Renault).

(=) *Discostachys* Grand'Eury, Géol. Paléontol. Bassin Houillier Gard: t. 8, f. 2. 1890.
Typus: *D. cebennensis* Grand'Eury

Glossopteris Brongn., Prodr. Hist. Vég. Foss.: 54. Dec 1828 [*Glossopterid.*].
Typus: *G. browniana* Brongn. (Hist. Vég. Foss. 1: 222. 6 Jun 1831).

(H) *Glossopteris* Raf., Anal. Nat.: 205. Apr-Jul 1815 [*Pteridoph.*].
≡ *Phyllitis* Hill 1757.

Lycopodites Lindl. & Hutton, Foss. Fl. Gr. Brit. 1: 171. Jan-Apr 1833 [*Lycopod.*].
Typus: *L. falcatus* Lindl. & Hutton *(typ. cons.).*

(H) *Lycopodites* Brongn. in Mém. Mus. Hist. Nat. 8: 231. Mai 1822 [Foss.].
Typus: *L. taxiformis* Brongn.

Megalopteris (J. W. Dawson) E. B. Andrews in Rep. Geol. Surv. Ohio 2(2): 415. 1875 (*Neuropteris* subg. *Megalopteris* J. W. Dawson, Fossil Pl. Canada: 51. 1871) [*Megalopterid.*].
Typus: *M. dawsonii* (Hartt) E. B. Andrews (*Neuropteris dawsonii* Hartt).

(=) *Cannophyllites* Brongn., Prodr. Hist. Vég. Foss.: 130. Dec 1828.
Typus: *C. virletii* Brongn.

Odontopteris (Brongn.) Sternb., Vers. Fl. Vorwelt 4: xxi. 1825 (*Filicites* sect. *Odontopteris* Brongn. in Mém. Mus. Hist. Nat. 8: 234. Mai 1822) [*Pteridoph.*].
Typus: *Filicites brardii* Brongn. (*O. brardii* (Brongn.) Sternb.).

(H) *Odontopteris* Bernh. in J. Bot. (Schrader) 1800(2): 7, 106. Oct-Dec 1801 [*Pteridoph.*].
≡ *Lygodium* Bernh. 1801.

Sphenophyllum Brongn., Prodr. Hist. Vég. Foss.: 68. Dec 1828 [*Sphenophyll.*].
Typus: *S. emarginatum* (Brongn.) Brongn. (*Sphenophyllites emarginatus* Brongn.).

(≡) *Sphenophyllites* Brongn. in Mém. Mus. Hist. Nat. 8: 209, 234. Mai 1822.

(=) *Rotularia* Sternb., Vers. Fl. Vorwelt 1(2): 33. Jan-Aug 1821.
Typus: *R. marsiliifolia* Sternb.

APPENDIX IIIB

NOMINA SPECIFICA CONSERVANDA ET REJICIENDA

In the following list the ***nomina conservanda*** have been inserted in the left column, in ***bold-face italics***. They are arranged in alphabetical sequence. Rejected synonyms *(nomina rejicienda)* are listed in the right column.

Names listed in this Appendix fall under the special provisions of Art. 14.4.

Neither a rejected name, nor any combination based on a rejected name, may be used for a taxon which includes the type of the corresponding conserved name (Art. 14.7; see also Art. 14 Note 2). Combinations based on a conserved name are therefore, in effect, similarly conserved. When such a later combination is in current use, it is cross-referenced to its conserved basionym.

typ. cons. typus conservandus, type to be conserved (Art. 14.9; see also Art. 14.3 and 10.4); as by Art. 14.8, listed types of conserved names may not be changed even if they are not explicitly designated as *typ. cons.*

(≡) nomenclatural synonym, based on the same nomenclatural type as the conserved name, only the earliest legitimate one, if any, being listed (Art. 14.4).

(=) taxonomic synonym, based on a type different from that of the conserved name, to be rejected only in favour of the conserved name (Art. 14.6 and 14.7).

Some names listed as conserved have no corresponding *nomina rejicienda* because they were conserved solely to maintain a particular type.

B. FUNGI

Aspergillus niger Tiegh. in Ann. Sci. Nat., Bot., ser. 5, 8: 240. Oct 1867.
Typus: No. 554.65 (CBS); isotypus: No. 50566 (IMI) *(typ. cons.).*

(=) *Ustilago phoenicis* Corda, Icon. Fung. 4: 9. 1840 (*Aspergillus phoenicis* (Corda) Thom).
Typus: non designatus.

(=) *Ustilago ficuum* Reichardt in Verh. K.K. Zool.-Bot. Ges. Wien 17, Abh.: 335. 1867 (*Aspergillus ficuum* (Reichardt) Thom & Church).
Typus: No. 91881 (IMI).

Penicillium chrysogenum Thom in U.S.D.A. Bur. Anim. Industr. Bull. 118: 58. 1910.
Typus: No. 24314 (IMI) *(typ. cons.).*

(=) *Penicillium griseoroseum* Dierckx in Ann. Soc. Sci. Bruxelles 25: 86. 1901.
Neotypus (vide Pitt, Genus Penicillium: 249. 1980): No. 92220 (IMI).

(=) *Penicillium citreoroseum* Dierckx in Ann. Soc. Sci. Bruxelles 25: 86. 1901.
Neotypus (vide Pitt, Genus Penicillium: 250. 1980): No. 889 (NRRL).

(=) *Penicillium brunneorubrum* Dierckx in Ann. Soc. Sci. Bruxelles 25: 88. 1901.
Neotypus (vide Pitt, Genus Penicillium: 250. 1980): No. 92198 (IMI).

C. BRYOPHYTA

Bryum hyperboreum Dicks., Fasc. Pl. Crypt. Brit. 4: [29]. 4 Oct 1801 (*Arctoa hyperborea* (Dicks.) Bruch & Schimp.) [*Musci*].
Typus: "Trockene Felsen unterhalb Kongsvold Dovrefjeld", 23 Jul 1843, *W. P. Schimper* (BM, herb. Schimper) *(typ. cons.).*

Grimmia schisti Gunnerus ex F. Weber & D. Mohr, Index Mus. Pl. Crypt.: 2. Aug-Dec 1803 (*Cnestrum schisti* (Gunnerus ex F. Weber & D. Mohr) I. Hagen) [*Musci*].
Typus: "*Weissia schisti*, e Lapponia", 1802, *Wahlenberg* per Weber & Mohr 1804 ex herb. Bridel (B) *(typ. cons.).*

Mnium fissum L., Sp. Pl.: 1114. 1 Mai 1753 (*Calypogeia fissa* (L.) Raddi) [*Hepat.*].
Typus: [icon] Micheli, Nov. Pl. Gen.: t. 5, f. 14. 1729 *(typ. cons.).*

E. SPERMATOPHYTA

Acalypha virginica L., Sp. Pl.: 1003. 1 Mai 1753 [*Euphorb.*].
Typus: *Clayton 201* (BM) *(typ. cons.).*

Allasia payos Lour., Fl. Cochinch.: 85. Sep 1790 [*Verben.*].
Typus: Tanzania, Tanga, Jan 1893, *Volkens 1* (BM) *(typ. cons.).*

Bactris gasipaes Kunth in Humb. & al., Nov. Gen. Sp. 1, ed 4°: 302 [& ed. f°: 242]. 1816 [*Palm.*].
Typus: Colombia, Ibagué, *Bonpland s.n.* (P).

(=) *Martinezia ciliata* Ruiz & Pav., Syst. Veg. Fl. Peruv. Chil.: 275. Dec 1798 (*Bactris ciliata* (Ruiz & Pav.) Mart.).
Typus: Peru, Pozuzo, *Pavón s.n.* (MA).

Bromus sterilis L., Sp. Pl.: 77. 1 Mai 1753 [*Gram.*].
Typus: England, Surrey, Tothill, near Hedley, 15 Jun 1932, *C. E. Hubbard 9045* in Gram. Brit. Exsicc. Herb. Kew. Distrib. No. 69 (E; iso- K) *(typ. cons.)*.

Cereus jamacaru DC., Prodr. 3: 467. Mar 1828 [*Cact.*].
Typus: Brazil, Bahia, Mun. Curaçá, north of Barro Vermelho towards Curaçá, caatinga, 395 m, 7 Jan 1991, *N. P. Taylor, D. C. Zappi & U. Eggli 1369* (CEPC; iso- K, ZSS, HRCB) *(typ. cons.)*.

Commelina benghalensis L., Sp. Pl.: 41. 1 Mai 1753 [*Commelin.*].
Typus: India, Herb. Linn. No. 65.16 (LINN) *(typ. cons.)*.

Erica carnea L., Sp. Pl.: 355. 1 Mai 1753 [*Eric.*].
Typus: [icon] "*Erica Coris folio* IX", Clusius, Rar. Pl. Hist.: 44. 1601.

(=) *Erica herbacea* L., Sp. Pl.: 352. 1 Mai 1753.
Typus (vide Brickell & McClintock in Taxon 36: 480. 1987): [icon] "*Erica Coris folio* VII", Clusius, Rar. Pl. Hist.: 44. 1601.

Erica vagans L., Mant. Pl.: 230. 15-31 Oct 1767 [*Eric.*].
Typus: U.K., Cornwall, Goonhilly, 12 Jul 1932, *W. B. Turrill s.n.* (K) *(typ. cons.)*.

Gomphrena ficoidea L., Sp. Pl.: 225. 1 Mai 1753 [*Amaranth.*].
Typus: Herb. Linn. No. 290.23 (LINN) *(typ. cons.)*.

Grewia mollis Juss. in Ann. Mus. Natl. Hist. Nat. 4: 91. 1804 [*Til.*].
Typus: Nigeria, Nupe, *Barter 1097* (K) *(typ. cons.)*.

Fraxinus angustifolia Vahl, Enum. Pl. 1: 52. Jul-Dec 1804 [*Ol.*].
Typus: Spain, *Schousboe* (C).

(=) *Fraxinus rotundifolia* Mill., Gard. Dict., ed. 8: *Fraxinus* No. 2. 16 Apr 1768.
Lectotypus (vide Green in Kew Bull. 40: 131. 1985): [icon] *"Fraxinus rotundiore folio"*, Bauhin, Hist. Pl. Univ. 1(2): 177. 1650.

Lycopersicon esculentum Mill., Gard. Dict., ed. 8: *Lycopersicon* No. 1. 16 Apr 1768 (*Solanum lycopersicum* L., Sp. Pl.: 185. 1 Mai 1753) [*Solan.*].
Typus: Herb. Linn. No. 248.16 (LINN).

(≡) *Lycopersicon lycopersicum* (L.) H. Karsten, Deut. Fl.: 966. Mai 1882 *('Lycopersicum lycopersicum')*.

Maerua crassifolia Forssk., Fl. Aegypt.-Arab.: 104. 1 Oct 1775 [*Cappar.*].
Typus: Yemen, between Watadah and Sirwah, *J. R. I. Wood 3153* (BM) *(typ. cons.)*.

Mimosa pigra L., Cent. Pl. I: 13. 19 Feb 1755 [*Legum.*].
Typus: Mozambique, Gaza District, between Chibuto and Canicado by R. Limpopo, *Barbosa & Lemos 7999* (K; iso- COI, LISC, LMJ) *(typ. cons.)*.

Ononis spinosa L., Sp. Pl.: 716. 1 Mai 1753 [*Boragin.*].
Typus: Herb. Burser XXI: 79 (UPS) *(typ. cons.)*.

Tetracera volubilis L., Sp. Pl.: 533. 1 Mai 1753 [*Dillen.*].
Typus: Mexico, Veracruz, Zacupan, Sulphur Spring, Dec 1906, *Purpus 2206* (F; iso- US No. 840326) *(typ. cons.)*.

Triticum aestivum L., Sp. Pl.: 85. 1 Mai 1753 [*Gram.*].
Typus: Herb. Clifford: 24, *Triticum* No. 3. (BM).

(=) *Triticum hybernum* L., Sp. Pl.: 86. 1 Mai 1753.
Lectotypus (vide Hanelt & al. in Taxon 32: 492. 1983): Herb. Clifford: 24, *Triticum* No. 2. (BM).

Uvularia perfoliata L., Sp. Pl.: 304. 1 Mai 1753 [*Lil.*].
Typus: Virginia, *Clayton 258* (BM) *(typ. cons.)*.

Vitex payos (Lour.) Merr., vide *Allasia payos*.

APPENDIX IV

NOMINA UTIQUE REJICIENDA

The names printed in ***bold-face italics***, and all combinations based on these names, are ruled as rejected under Art. 56, and none is to be used. The rejected names are arranged in alphabetical sequence, irrespective of rank, within each major group. Cross-references to the rejected basionyms are provided under some well-known combinations based on them.

The rejected names are neither illegitimate nor invalid under Art. 6. Later homonyms of a rejected name (Art. 53), and names illegitimate because of inclusion of the type of a rejected name (Art. 52), are not to be used unless they are conserved.

B. FUNGI

Arthonia lurida Ach., Lichenogr. Universalis: 143. Apr-Mai 1810.
Typus: Helvetia, *Schleicher* (UPS).

Helotium Tode, Fungi Mecklenb. Sel. 1: 22. 1790 : Fr., Syst. Mycol. 3, index: 94. 1832.
Typus (vide Donk in Beih. Nova Hedwigia 5: 123. 1962): *H. glabrum* Tode : Fr.

Lecanora subfusca (L.) Ach., vide *Lichen subfuscus*.

Lecidea synothea Ach. in Kongl. Vetensk. Acad. Nya Handl. 29: 236. 1808.
Lectotypus (vide Cannon & Hawksworth in Taxon 32: 479. 1983): Suecia, 1807, *Swartz* (BM).

Lichen jubatus L., Sp. Pl.: 1155. 1 Mai 1753.
Lectotypus (vide Hawksworth in Taxon 19: 238. 1970): Herb. Linn. No. 1273.281 (LINN).

Lichen subfuscus L., Sp. Pl.: 1142. 1 Mai 1753.
Lectotypus (vide Brodo & Vitikainen in Mycotaxon 21: 294. 1984): [icon] Dillenius, Hist. Musc.: t. 18, f. 16B. 1742.

Lycoperdon aurantium L., Sp. Pl.: 1053. 1 Mai 1753.
Lectotypus (vide Demoulin in Bull. Jard. Bot. Belg. 37: 297. 1967): [icon] Vaillant, Bot. Paris.: t. 16, f. 9-10. 1727.

Peziza [unranked] ***Phialea*** Pers., Mycol. Eur. 1: 276. 1 Jan-14 Apr 1822 (*Peziza* ser. *Phialea* (Pers. : Fr.) Fr., Syst. Mycol. 2: 116. 1822).
Typus: *Peziza phiala* Vahl : Fr.

Phacidium musae Lév. in Ann. Sci. Nat., Bot., ser. 3, 5: 303. 1846.
Typus: *M. Bonpland* (PC, Herb. Amér. Equat.).

Phialea (Pers. : Fr.) Gillet, vide *Peziza* [unranked] *Phialea.*

Scleroderma aurantium (L. : Pers.) Pers., vide *Lycoperdon aurantium.*

Stilbum cinnabarinum Mont. in Ann. Sci. Nat., Bot., ser. 2, 8: 360. 1837.
Typus: Cuba, *M. Ramon* ex Herb. Montagne (PC).

C. BRYOPHYTA

Calypogeia trichomanis (L.) Corda, vide *Mnium trichomanis.*

Mnium trichomanis L., Sp. Pl.: 1114. 1 Mai 1753 [*Hepat.*].
Lectotypus (vide Isoviita in Acta Bot. Fenn. 89: 15. 1970): [icon] Dillenius, Hist. Musc.: t. 31, f. 5. 1742.

E. SPERMATOPHYTA

Anthospermum ciliare L., Sp. Pl., ed. 2: 1512. Jul-Aug 1763 [*Rub.*].
Lectotypus (vide Brummitt in Taxon 36: 73-74. 1987): Herb. Linn. No. 1233.4 (LINN).

Bromus purgans L., Sp. Pl.: 76. 1 Mai 1753 [*Gram.*].
Lectotypus (vide Hitchcock in Contr. U.S. Natl. Herb. 12: 122. 1908): *P. Kalm*, Herb. Linn. No. 93.11 (LINN).

Buchnera euphrasioides Vahl, Symb. Bot. 3: 81. 1794 [*Scrophular.*].
Lectotypus (vide Hepper in Taxon 35: 390. 1986): Ghana, *König* ex Herb. Vahl (C).

Centaurium minus Moench, Methodus: 349. 4 Mai 1794 [*Gentian.*].
Lectotypus (vide Melderis in Bot. J. Linn. Soc. 65: 229. 1972): *"Gentiana Centaurium minus* C.B.", Herb. Clifford: 278 (BM).

Citta nigricans Lour., Fl. Cochinch.: 456. Sep 1790 [*Legum.*].
Lectotypus (vide Wilmott-Dear in Taxon 40: 517. 1991): Vietnam, *Loureiro s.n.* (BM).

Crataegus oxyacantha L., Sp. Pl.: 477. 1 Mai 1753 [*Ros.*].
Lectotypus (vide Dandy in Bot. Soc. Exch. Club Brit. Isles 12: 867. 1946): Herb. Linn. No. 643.12 (LINN).

Echium lycopsis L., Fl. Angl.: 12. 3 Apr 1754 [*Boragin.*].
Lectotypus (vide Stearn in Ray Soc. Publ. 148, Introd.: 65. 1973): [icon] *"Echii altera species",* Dodoens, Stirp. Hist. Pempt.: 620. 1583.

Epilobium junceum Spreng. in Biehler, Pl. Nov. Herb. Spreng.: 17. 30 Mai 1807 [*Onagr.*].
Lectotypus (vide Garnock-Jones in Taxon 32: 656. 1983): New Zealand, *Forster s.n.* (K).

Festuca elatior L., Sp. Pl.: 75. 1 Mai 1753 [*Gram.*].
Lectotypus (vide Linder in Bothalia 16: 59. 1986): Herb. Linn. No. 92.17 (LINN).

Fumaria bulbosa L., Sp. Pl.: 699. 1 Mai 1753 [*Papaver.*].
Typus: Herb. Linn. No. 881.5 (LINN) (*typ. cons.* sub 2858).

Heliconia humilis (Aubl.) Jacq., vide *Musa humilis.*

Hypericum quadrangulare L., Sp. Pl.: 785. 1 Mai 1753 [*Gutt.*].
Lectotypus (vide Robson in Taxon 39: 135. 1990): Herb. Clifford.: 380, *Hypericum* No. 5 (BM).

Hypoëstes verticillaris (L. f.) Sol. ex Roem. & Schult., vide *Justicia verticillaris.*

Justicia verticillaris L. f., Suppl. Pl.: 85. Apr 1782 [*Acanth.*].
Lectotypus (vide Brummitt & al. in Taxon 32: 658. 1983): Herb. Thunberg No. 427 (UPS).

Lupinus hirsutus L., Sp. Pl.: 721. 1 Mai 1753 [*Legum.*].
Lectotypus (vide Lee & Gladstone in Taxon 28: 618. 1979): "*Lupinus hirsutus* L. - 1015 ns. Roy. prodr. 367" ex Herb. van Royen (L No. 908.119-125).

Melianthus minor L., Sp. Pl.: 639. 1 Mai 1753 [*Melianth.*].
Lectotypus (vide Wijnands in Taxon 34: 314. 1985): [icon] Commelijn, Hort. Med. Amstel. Pl.: t. 4. 1706.

Mucuna nigricans (Lour.) Steud., vide *Citta nigricans.*

Musa humilis Aubl., Hist. Pl. Guiane: 931. Jun-Dec 1775 [*Mus.*].
Lectotypus (vide Andersson in Taxon 33: 524. 1984): French Guiana, *Aublet* (BM).

Pterocephalus papposus (L.) Coult., vide *Scabiosa papposa.*

Rotala decussata DC., Prodr. 3: 76. Mar 1828 [*Lythr.*].
Typus: Endeavour River, Queensland, Australia, *R. Brown* (G-DC).

Scabiosa papposa L., Sp. Pl.: 101. 1 Mai 1753 [*Dipsac.*].
Lectotypus (vide Meikle in Taxon 31: 542. 1982): "14. *Scabiosa*" ex Herb. van Royen (L No. 902.125-731).

Silene rubella L., Sp. Pl.: 419. 1 Mai 1753 [*Caryophyll.*].
Lectotypus (vide Oxelman & Lidén in Taxon 36: 477. 1987): Herb. Linn. No. 583.43 (LINN).

Solanum indicum L., Sp. Pl.: 187. 1 Mai 1753 [*Solan.*].
Lectotypus (vide Hepper in Bot. J. Linn. Soc. 76: 288. 1978): Herb. Hermann 3: 16, No. 94 (BM).

Solanum sodomeum L., Sp. Pl.: 187. 1 Mai 1753 [*Solan.*].
Lectotypus (vide Hepper in Bot. J. Linn. Soc. 76: 290. 1978): Herb. Hermann 3: 30, No. 95 (BM).

Solanum verbascifolium L., Sp. Pl.: 184. 1 Mai 1753 [*Solan.*].
Lectotypus (vide Roe in Taxon 17: 177. 1968): Herb. Linn. No. 248.1 (LINN).

Stipa columbiana Macoun, Cat. Canad. Pl. 4-5: 101. 1888 [*Gram.*].
Lectotypus (vide Hitchcock in Contr. U.S. Natl. Herb. 24: 253. 1925): British Columbia, Yale, 17 Mai 1875, *Macoun* (CAN No. 9899).

Striga euphrasioides (Vahl) Benth., vide *Buchnera euphrasioides.*

Trigonella hamosa L., Syst. Nat., ed. 10: 1180. 7 Jun 1759 [*Legum.*].
Lectotypus (vide Lassen in Taxon 36: 478. 1987): Herb. Linn. No. 932.5 (LINN).

APPENDIX V

OPERA UTIQUE OPPRESSA

Publications are listed alphabetically by authors. Numbers of the relevant entries in TL-2 (Stafleu & Cowan, *Taxonomic literature* 1-7; in Regnum Veg. 94, 98, 105, 110, 112, 115, 116. 1976-1988) are added parenthetically in **bold-face-type** when available. Names appearing in the listed publications in any of the ranks specified in square brackets at the end of each entry are not accepted as validly published under the present *Code* (Art. 32.8).

Agosti, J. 1770. *De re botanica tractatus.* Belluno. (TL-2 No. **66**.) [Genera.]

Buc'hoz, P. J. 1762-1770. *Traité historique des plantes qui croissent dans la Lorraine et les Trois Evêchés.* 10 vol. Nancy & Paris. (TL-2 No. **872**.) [Species and infraspecific taxa.]

Buc'hoz, P. J. 1764. *Tournefortius Lotharingiae, ou catalogue des plantes qui croissent dans la Lorraine et les Trois Evêchés.* Paris & Nancy. (TL-2 No. **873**.) [Species and infraspecific taxa.]

Buc'hoz, P. J. 1770. *Dictionnaire raisonné universel des plantes, arbres et arbustes de France.* Vol. 1, 2, and 3 (pp. 1-528) [but not vol. 3 (pp. 529-643), nor vol. 4]. Paris. (TL-2 No. **874**.) [Species and infraspecific taxa.]

Donati, V. 1753. *Auszug seiner Naturgeschichte des adriatischen Meers.* Halle. (TL-2 sub No. **1500**.) [All ranks.]

Ehrhart, J. B. 1753-1762. *Oeconomische Pflanzenhistorie.* 12 vol. Ulm & Memmingen. (TL-2 No. **1647**.) [Genera.]

Ehrhart, J. F. 1780-1785. *Phytophylacium ehrhartianum.* 10 decades. Hannover. (vide TL-2, 1: 731.) [Genera.]

Ehrhart, J. F. 1789. Index phytophylacii ehrhartiani. Pp. 145-150 *in:* Ehrhart, J. F.: *Beiträge zur Botanik,* 4. Hannover & Osnabrück. (TL-2 sub No. **1645**.) [Genera.]

Feuillée, L. 1756-1757, 1766. *Beschreibung zur Arzeney dienlicher Pflanzen, welche in den Reichen des mittägigen America in Peru und Chily vorzüglich im Gebrauch sind.* 2 vol. [and re-issue]. Nürnberg. (TL-2 sub No. **1767**.) [Genera.]

Gandoger, M. 1883-1891. *Flora Europae terrarumque adjacentium.* 27 vol. Paris, London & Berlin. (TL-2 No. **1942**.) [Species.]

Garsault, F. A. P. de, 1764. *Les figures de plantes et animaux d'usage en médecine, décrites dans la matière médicale de Mr. Geoffroy.* 5 vol. Paris. (TL-2 No. **1959**.) [Species and infraspecific taxa.]

Garsault, F. A. P. de, 1764-1767. *Description, vertus et usages de sept cent dix-neuf plantes, tant étrangères que de nos climats.* 5 vol. Paris. (TL-2 No. **1961**.) [Genera, species and infraspecific taxa.]

Garsault, F. A. P. de, 1765. *Explication abrégée de sept cent dix-neuf plantes, tant étrangères que de nos climats.* Paris. (TL-2 No. **1960**.) [Species and infraspecific taxa.]

Gilibert, J. E. 1782. *Flora lituanica inchoata.* 2 vol. Grodno. (TL-2 No. **2012**.) [Species and infraspecific taxa.]

Gilibert, J. E. 1782. *Exercitium botanicum, in schola vilnensi peractum, seu enumeratio methodica plantarum tam indigenarum quam exoticarum quas proprio marte determinaverunt alumni in campis vilniensibus aut in horto botanico universitatis.* Wilnius. (TL-2 No. **2013**.) [Species and infraspecific taxa.]

Gilibert, J. E. 1785-1787. *Caroli Linnaei botanicorum principis systema plantarum Europae.* 7 vol. Vienne. (TL-2 No. **2014**.) [Species and infraspecific taxa.]

Gilibert, J. E. 1792. *Exercitia phytologica, quibus omnes plantae Europae, quas vivas invenit in variis herbationibus, seu in Lithuania, Gallia, Alpibus, analysi nova proponuntur.* 2 vol. Lyon. (TL-2 No. **2015**.) [Species and infraspecific taxa.]

Gleditsch, J. G. 1753. Observation sur la pneumonanthe, nouveau genre de plante, dont le caractère diffère essentiellement de celui de la gentiane. *Hist. Acad. Roy. Sci. (Berlin)* 1751: 158-166. [All ranks.]

Guettard, J. E. 1755. Cinquième [Sixième] mémoire sur les glandes des plantes, et le quatrième [cinquième] sur l'usage que l'on peut faire de ces parties dans l'établissement des genres des plantes. *Hist. Acad. Roy. Sci. Mém. Math. Phys. (Paris, 4°)* 1749: 322-377, 392-443. (TL-2 No. **2208**.) [All ranks.]

Haller, A. von, 1753. *Enumeratio plantarum horti regii et agri gottingensis.* Göttingen. (TL-2 No. **2309**.) [All ranks.]

Heister, L. 1753. *Descriptio novi generis plantae rarissimae et speciosissimae africanae.* Braunschweig. (TL-2 No. **2592**.) [All ranks.]

Hill, J. 1753. [Entries on natural history.] *In:* Scott, G. L. (ed.), *A supplement of Mr. Chambers's Cyclopaedia: or a universal dictionary of the arts and sciences.* 2 vol. London. [All ranks.]

Hill, J. 1753-1754. [Entries on natural history.] *In:* Society of Gentlemen (ed.), *A new and complete dictionary of arts and sciences.* 4 vol. London. [All ranks.]

Hill, J. 1754, etc. *The useful family herbal.* [Including subsequent re-issues and editions]. London. (TL-2 No. **2768**.) [Genera, species and infraspecific taxa.]

Hill, J. 1756-1757 *The British herbal.* London. (TL-2 No. **2769**.) [Species and infraspecific taxa.]

Necker, N. J. de, 1790-1791, 1808. *Elementa botanica.* [All issues and editions]. Neuwied, Paris, Strasbourg, Mainz. (TL-2 No. **6670**.) [Genera.]

Rumphius, G. E. 1755. *Herbarii amboinensis auctuarium.* Amsterdam. (TL-2 No. **9785**.) [Genera; this does not affect species names published by Burman in the "Index", pp. 75-94, in that work.]

Secretan, L. 1833. *Mycographie suisse, ou description des champignons qui croissent en Suisse.* 3 vol. Genève. (TL-2 No. **11595**.) [Species and infraspecific taxa.]

Trew, C. J. [and others] [1747-]1753-1773. *Herbarium blackwellianum emendatum et auctum.* 6 vol. Nürnberg. (TL-2 No. **546**.) [Genera.]

INDEX TO APPENDIX IIIA

In this Index, the letter A refers to *Algae* (A1 to A11 to the various algal classes), B to the *Fungi,* C to the *Bryophyta* (C1: *Hepaticae,* C2: *Musci*), D to the *Pteridophyta*, and F to the fossil plants. Under each of these headings, the conserved generic names are listed alphabetically.

The Arabic numerals refer to section E, *Spermatophyta,* where the names are arranged according to the Dalla Torre & Harms system.

Names printed in roman type refer to conserved names (left columns). Names printed in *italics* refer to rejected names (right columns). Rejected homonyms (or names treated as such) are not listed. For sections A-D and F, the italicized names will be found under the heading of the corresponding conserved name (added in parentheses): "*Acetabulum* A2 (Acetabularia)" means that the rejected name *Acetabulum* will be found under A2, *Chlorophyceae*, following the conserved name ***Acetabularia***.

Abedinium A6 (Trichodesmium)
Aberemoa 2679
Abrotanifolia A8 (Cystoseira)
Abumon 1046
Acampe 1824
Acantholimon 6348
Acanthonema 7835
Acetabularia A3
Acetabulum A3 (Acetabularia)
Achimenes 7874
Achras 6386a
Achroanthes 1553
Achyrodes 374
Acidodontium C2
Acidoton 4415
Acinaria A8 (Sargassum)
Aconiopteris D (Elaphoglossum)
Acouba 3821
Acranthera 8237
Acrolejeunea C1
Acronychia 4079
Acrosporium B (Oidium)
Actinella A1
Actinodontium C2 (Lepidopilum)
Actinomeris 9215
Acuan 3450
Acyntha 1110
Adamaram 5544
Adelanthus C1
Adelaster 8069
Adelia 4397
Adelia 6427
Adenandra 4038
Adenostegia 7632
Adesmia 3800
Adlumia 2857
Adolia 4874
Aduseton 3013
Aechmea 861
Aedycia B (Mutinus)
Aegle 4099
Aembilla 5304
Aerva 2317
Aeschynanthus 7824
Afzelia 3509
Agalinis 7604a
Agallochum 5430
Agapanthus 1046
Agardhia A3 (Mougeotia)
Agaricus B
Agarum A8
Agathis 20
Agathosma 4037
Agati 3747
Agialid 3980
Aglaia 4189
Agonis 5600
Agrimonoides 3377
Agyneia 4302
Ahouai 6632
Ailanthus 4124
Alagophyla 7887a
Alangium 6154
Alaria A8
Albina 1328
Aldina 3575
Aleurodiscus B
Alga 57
Alguelaguen 7299
Alicastrum 1957
Alina 2611

International Code of Botanical Nomenclature (Tokyo Code) – 1994

Errata Corrige

Due to an unfortunate oversight a cancel page was printed instead of the correct p. 29 of the Tokyo Code. Please correct your copy accordingly:

Delete the two lines at the top of the page: "family ... 32.7)."

Renumber the three following paragraphs (now 19.6, 19.7, 19.8) to: 19.5, 19.6, 19.7.

With apologies for any inconvenience caused

The Editors

Alismorkis 1631
Alkanna 7097
Allionia 2348
Allodape 6251
Alloplectus 7860
Allosorus D (Cheilanthes)
Alocasia 752
Aloidella C2 (Aloina)
Aloina C2
Alpinia 1328
Alstonia 6583
Alternaria B
Alvesia 7346
Alysicarpus 3810
Alyxia 6616
Amanita B
Amanitopsis B
Amaracus 7312
Amarella 6509a
Amaryllis 1176
Amasonia 7156
Amberboa 9476
Amberboi 9476
Amblostima 1006
Amblyodon C2
Ambuli 7532
Amellus 8887
Amerimnon 3821
Amianthium 955
Ammios 6014
Amomum 1344
Amorphophallus 723
Ampelocissus 4910
Amphibia A9 (Bostrychia)
Amphicarpaea 3860
Amphidium C2
Amphilothus A6
Amphinomia 3657
Amphirrhox 5259
Amphisphaeria B
Amphitholus A6 (Amphilothus)
Amphithrix A5 (Homoeothrix)
Amsinckia 7082
Amyrsia 5557
Anabaena A5
Anacampseros 2412
Anacolia C2
Anadyomene A3
Anarrhinum 7485
Anastomaria B (Gyrodon)
Ancistrocarpus 4948
Ancistrocladus 5400
Andira 3841
Androgyne 1714
Androphylax 2570
Androstachys 4299a
Anema B
Anemia D
Anemopaegma 7665
Anepsa 957
Angianthus 9028
Angiopetalum 6291
Angiopoma B (Drechslera)
Angiopteris D
Angolam 6154
Anguillaria 974
Anguria 8598
Anictangium C2 (Anoectangium)
Aniotum 3848
Anisomeridium B
Anisonema A7
Anisotes 8096
Anneslea 5155
Annulina A3 (Cladophora)
Anodontium C2 (Drummondia)
Anoectangium C2
Anoectochilus 1500
Antennaria 8978
Anthophysa A4
Anthriscus 5938
Antiaris 1956
Anzia B
Apalatoa 3495
Apatitia 5768
Aphananthe 1904
Aphanochaete A3
Aphanothece A5
Aphoma 975
Apios 3874
Apona A9 (Lemanea)
Aponogeton 65
Apora A9 (Phymatolithon)
Aposphaeria B
Appella 7185
Aptosimum 7467
Apuleia 3532
Aquilaria 5430
Arabidopsis 2999
Arachnitis 1386
Arachnoidiscus A1
Arachnospermum 9581
Araiostegia D
Araliopsis 4073
Arceuthobium 2091
Arctostaphylos 6212
Ardisia 6285
Arduina 6064
Aremonia 3377
Arenga 575
Areschougia A9
Argania 6370
Argolasia 1236
Argyrolobium 3673
Aristotelia 4927
Armeria 6350
Armeriastrum 6348
Armoracia 2965a
Aronia 3338a
Arrhenopterum C2 (Aulacomnium)
Artanema 7559
Arthonia B
Arthrocereus 5401a
Artocarpus 1946
Artotrogus B (Pythium)
Aschersonia B
Aschistodon C2 (Ditrichum)
Ascidium B (Ocellularia)
Ascolepis 454
Ascophora B (Rhizopus)
Ascophyllum A8
Asperula 8485
Aspicilia B
Aspidosperma 6588
Aspidostigma 4085
Asplundia 678a
Assonia 5053
Astasia A7
Astelia 1111
Asterella C1
Asteristion B (Phaeotrema)
Asterophyllites F
Astrocaryum 668
Astronidium 5777a
Atalantia 4096
Atamosco 1181
Athenaea 7398
Atherurus 787
Atractylocarpus C2
Atrichum C2
Atropis 384
Atylus 2026
Audouinella A9
Augea 3967
Augusta 8183
Aulacia 4089
Aulacodiscus A1
Aulacomnium C2

Aulacothele 5411a
Auricula A1
Avoira 668
Aytonia C1 (Plagiochasma)
Azalea 6189

Babiana 1310
Baccharis 8933
Baccifer A8 (Cystoseira)
Bacopa 7546
Bahel 7559
Baiera F
Baillouviana A9 (Dasya)
Baitaria 2407
Balanites 3980
Balanoplis 1891a
Balbisia 3932
Balboa 5195
Balduina 9241
Balsamea 4151
Bambos 424
Bambusa 424
Bambusina A3
Banisteria 4226
Banksia 2068
Barbarea 2961
Barbula C2
Barclaya 2515
Barosma 4036
Barraldeia 5525
Barringtonia 5506
Bartonia 6501
Bartramia C2
Bartramidula C2
Bartsia 7645
Baryxylum 3561
Basilaea 1088
Basteria 2663
Bathelium B (Trypethelium)
Baumgartia 2570
Baxteria 1044
Bazzania C1
Bechera F (Asterophyllites)
Belamcanda 1285
Belis 31
Bellevalia 1093
Bellucia 5768
Belmontia 6483
Belou 4099
Belutta-kaka 6677
Belvala 5436
Bembix 5400
Benteca 8197
Benzoin 2821
Berchemia 4868
Bergenia 3182
Bergera 4090
Berkheya 9438
Berlinia 3516
Bernardia 4397
Bernieria 2804
Bernoullia 5035
Berrya 4938
Bertolonia 5708
Bessera 1055
Beverna 1310
Biarum 784
Biblarium A1 (Tetracyclus)
Bichatia A5 (Gloeocapsa)
Bichea 5091
Bifida A9 (Rhodophyllis)
Bifora 5956
Bigelowia 8855
Bignonia 7705
Bihai 1321
Bikkia 8126
Bikukulla 2856
Bilimbiospora B (Leptosphaeria)
Billia 4722
Bistella 3201
Bivonaea 2902
Blachia 4459
Bladhia 6285
Blandfordia 1021
Blasteniospora B (Xanthoria)
Blatti 5497
Blepharipappus 9258
Bletilla 1533
Blossevillea A8 (Cystophora)
Blumea 8939
Blumenbachia 5392
Blysmus 468a
Bobartia 1284
Boehmia B (Leptoglossum)
Boenninghausenia 4011
Boldu 2759
Bolelia 8706
Boletus B
Bombacopsis 5042a
Bombax 5024
Bonamia 6979
Bonannia 6099
Bonaveria 3694
Bonnetia 5144
Bornia F (Asterophyllites)
Borraginoides 7056
Borreria 8473
Borrichius A9 (Dudresnaya)
Boscia 3106
Bostrychia A9
Botor 3914
Botria 4910
Botrydiopsis A11
Botryocladia A9
Botryoconus F (Cordaianthus)
Botryonipha B (Stilbella)
Botryophora 4516
Bouchea 7148
Bougainvillea 2350
Bourreria 7042
Bouteloua 295
Bowiea 1011
Boykinia 3185
Brachtia 1751
Brachyandra 8808
Brachynema 6408
Brachysteleum C2 (Ptychomitrium)
Brachystelma 6870
Bradburia 8840
Brami 7546
Brassavola 1619
Braunea 2577
Brebissonia A1
Breteuillia 9439
Brexia 3225
Breynia 4303
Brickellia 8823
Bridgesia 4730
Brigantiaea B (Lopadium)
Brodiaea 1053
Brosimum 1957
Broussonetia 1923
Brownea 3524
Brownlowia 4943
Brucea 4120
Brukmannia F (Asterophyllites)
Brunfelsia 7450
Brunia 3292
Bruxanellia 4459
Bryantea 2797
Bryoxiphium C2
Bucco 4037
Bucephalon 1917
Buceras 5543
Buchenavia 5545
Buchloë 308
Bucida 5543
Buckleya 2109
Buda 2450

Buekia 1328
Buellia B
Bulbine 985
Bulbophyllum 1705
Bulbostylis 471a
Bullardia B (Melanogaster)
Bumelia 6374
Bupariti 5018
Buraeavia 4331
Burchardia 968
Burneya 8365
Bursa-pastoris 2986
Bursera 4150
Burtonia 3629
Butea 3876
Byrsanthus 5311
Byssus A3 (Trentepohlia)
Bystropogon 7321
Byttneria 5062

Cacara 3908
Cactus 5411
Cailliea 3452
Cajanus 3892
Calacinum 2208
Calamites F
Calandrinia 2407
Calanthe 1631
Calasias 8096
Calceolaria 7474
Caldesiella B (Tomentella)
Calesiam 4563
Caliphruria 1196
Calliandra 3444
Calliblepharis A9
Callicostella C2
Callista 1694
Callistachys 3624
Callistemma 8898
Callistephus 8898
Callixene 1146
Calodendrum 4035
Caloplaca B
Calopogon 1534
Calorophus 815
Calucechinus 1889
Calusparassus 1889
Calvatia B
Calycanthus 2663
Calycera 8726
Calypogeia C1
Calypso 1559
Calyptranthes 5575
Calystegia 6994
Camassia 1087
Cambessedesia 5669
Cammarum 2528
Camoënsia 3589
Camonea 6997
Camphora 2782
Campnosperma 4578
Campsis 7714
Camptocarpus 6726
Campulosus 286
Campylobasidium B (Septobasidium)
Campylus 2583
Cananga 2684
Canarina 8656
Canavalia 3891
Candida B
Canella 5254
Cannophyllites F (Megalopteris)
Cansjera 2124
Cantuffa 3553
Caopia 5171
Capnoides 2858
Capnorchis 2856
Capnorea 7029
Capriola 282
Capsella 2986
Capura 5446
Carallia 5525
Carandas 6559
Carapichea 8411
Cardaminum 2965
Cardiacanthus 8014
Cardiocarpus F
Careya 5505
Carissa 6559
Carlowrightia 8014
Carpolepidum C1 (Plagiochila)
Carpomitra A8
Carrichtera 2936
Carrodorus A4 (Hydrurus)
Carya 1882
Caryolobis 5214
Caryophyllus 5583
Cassebeera D (Doryopteris)
Casselia 7157
Cassia 3536
Cassinia 8994
Castanopsis 1891a
Castela 4118
Catenella A9
Catevala 1029
Catha 4627
Catharinea C2 (Atrichum)
Cavendishia 6232
Cayaponia 8627
Caylusea 3122
Cayratia 4918a
Cebatha 2570
Cecropia 1971
Cedrus 23
Celmisia 8909
Centotheca 357
Centritractus A11
Centrophorum 134c
Centrosema 3858
Centrosolenia 7857a
Cephaëlis 8411
Cephalaria 8541
Cephalotus 3176
Cephaloziella C1
Ceraia 1694
Ceramianthemum A9 (Gracilaria)
Ceramium A9
Cerania B (Thamnolia)
Ceranthus 6428
Cerataulina A1
Ceratophora B (Gloeophyllum)
Ceratostoma B (Melanospora)
Cerefolium 5938
Cervicina 8668
Cesius C1 (Gymnomitrion)
Ceterach D
Cetraria B
Ceuthospora B
Chaenomeles 3336a
Chaenostoma 7518
Chaetangium A9 (Suhria)
Chaetocarpus 4467
Chaetomorpha A3
Chaetoporellus B (Hyphodontia)
Chamaedaphne 6200a
Chamaedorea 594
Chamissoa 2297
Chaos A9 (Porphyridium)
Chaptalia 9529
Chascanum 7148a
Cheilanthes D
Chibaca 5256
Chiliophyllum 8843
Chilomastix A10
Chiloscyphus C1

Chimonanthus 2663a
Chionographis 951
Chlamydomonas A3
Chlamysporum 992
Chlorangium B (Aspicilia)
Chlorea B (Letharia)
Chlorociboria B
Chlorococcum A3
Chlorogalum 1007
Chloromonas A3
Chloronitum A3 (Chaetomorpha)
Chloronotus C2 (Crossidium)
Chloroxylon 4065
Choaspis A3 (Sirogonium)
Chomelia 8366
Chondria A9
Chondropsis B
Chondrospora B (Anzia)
Chonemorpha 6677
Chordaria A8
Chorispermum 3051
Chorispora 3051
Chrosperma 955
Chrozophora 4355
Chryseis 9476
Chrysopogon 134c
Chrysopsis 8844
Chrysothrix B
Chupalon 6232
Chylocladia A9
Chytra 7604a
Chytraculia 5575
Chytraphora A8 (Carpomitra)
Cieca 4349
Ciliaria A9 (Calliblepharis)
Cinnamomum 2782
Circinaria B (Aspicilia)
Circinnus 3693
Cirrhopetalum 1704
Cistella B
Citrullus 8598
Cladaria B (Ramaria)
Claderia 1569
Cladonia B
Cladophora A3
Cladophoropsis A3
Cladoporus B (Laetiporus)
Clarisia 1937
Clathrospermum 2691a
Clathrospora B (Pleospora)
Clavaria B
Clavatula A9 (Catenella)
Cleyera 5157a
Clianthus 3753
Clisosporium B (Coniothyrium)
Clompanus 3834
Cluacena 5557
Cluzella A4 (Hydrurus)
Cnicus 9479
Coccochloris A5 (Aphanothece)
Coccocypselum 8250
Coccoloba 2209
Cocculus 2570
Cochlospermum 5250
Codiaeum 4454
Codiolum A3 (Urospora)
Codonanthe 7866
Coelopyrum 4578
Coeomurus B (Uromyces)
Coilotapalus 1971
Cola 5091
Colea 7760
Coleanthus 228
Coleochaete A3
Coleosanthus 8823
Collea 1488
Collema B
Colletia 4899
Collinella A6 (Dogelodinium)
Collybia B
Colocasia 755
Colocynthis 8598
Colophermum A8 (Ectocarpus)
Colubrina 4882
Columellia 7897
Coluteastrum 3756
Combretum 5538
Commiphora 4151
Compsoa 967
Condalia 4862
Condea 7342
Conferva A3 (Cladophora)
Conia B (Lepraria)
Coniocarpon B (Arthonia)
Coniogramme D
Coniothyrium B
Conjugata A3 (Spirogyra)
Conocephalum C1
Conocybe B
Conohoria 5262
Convolvuloides 7003a
Conyza 8926
Copaiba 3490
Copaifera 3490
Copaiva 3490
Coprinarius B (Panaeolus)
Coptophyllum 8244
Cordaianthus F
Cordaites F
Cordula 1393a
Cordyceps B
Cordylanthus 7632
Cordyline 1108
Corinophoros B (Peccania)
Cornea A9 (Gelidium)
Coronopus 2884
Correa 4031
Cortaderia 329
Cortinarius B
Corydalis 2858
Corynephorus 269
Corynomorpha A9
Coryphantha 5411a
Coscinodiscus A1
Cosmibuena 8209
Cosmiza 7900
Coublandia 3834
Coublandia 3837
Coumarouna 3845
Courondi 4662
Coutinia 6588
Crabbea 7972
Cracca 3745
Crantzia 7860
Crassina 9155
Crassocephalum 9405
Craterellus B
Crepidopus B (Pleurotus)
Cristaria 5004
Crocodia B (Pseudocyphellaria)
Crocodilodes 9438
Crocynia B
Crossidium C2
Cruckshanksia 8158
Crudia 3495
Crumenula A7 (Lepocinclis)
Crypsis 221
Cryptanthus 846
Cryptocarya 2813
Cryptococcus B
Cryptogyne 6384
Cryptopleura A9
Cryptosphaeria B
Cryptotaenia 6015
Cryptothecia B
Ctenium 286

Ctenodus A9 (Phacelocarpus)
Cudrania 1942
Culcasia 690
Cumingia 5036
Cunila 7327
Cunninghamia 31
Cunonia 3275
Cunto 4079
Cupulissa 7665
Curcuma 1351
Curtisia 6156
Cuspidaria 7668
Cussambium 4767
Cuviera 8357
Cyananthus 8660
Cyanotis 904
Cyanotris 1087
Cyathophora C1 (Mannia)
Cyathula 2312
Cybele 2066
Cybianthus 6301
Cyclobalanopsis 1893
Cyclospermum 6004a
Cyclotella A1
Cylindrocarpon B
Cylista 3897
Cymatopleura A1
Cymbella A1
Cymodocea 60
Cynodon 282
Cynodontium C2
Cynontodium C2
(Distichium)
Cynophallus B (Mutinus)
Cyphella B (Aleurodiscus)
Cyrtanthus 1191
Cystanthe 6254
Cysticapnos 2858
Cystodium D
Cystophora A8
Cystopteris D
Cystoseira A8
Cytinus 2180
Cytisogenista 3682a
Cytisus 3682

Daboecia 6195
Dactilon 282
Dactylicapnos 2856
Dalbergia 3821
Daldinia B
Dalea 3709
Daltonia C2
Damapana 3796
Danaea D
Danthonia 280
Darlingtonia 3131
Dasus 8412
Dasya A9
Dasyphylla A9 (Chondria)
Daun-contu 8430
Debaryomyces B
Debaryozyma B
(Debaryomyces)
Decaisnea 2551
Deguelia 3838
Delesseria A9
Dendrella A1
(Didymosphenia)
Dendrobium 1694
Dendrorkis 1565
Dendrosicus 7757
Denhamia 4623
Deprea 7398
Deringa 6015
Derminus B (Pholiota)
Derris 3838
Derwentia 7579a
Descurainia 2997
Desmanthus 3450
Desmarestia A8
Desmodium 3807
Desmoncus 670
Desmotrichum A8
Detris 8919
Diamorpha 3172
Diapedium 8031
Diaphanophyllum C2
(Ditrichum)
Diarrhena 356
Diatoma A1
Diatomella A1
Diatrema 7003a
Diatremis 7003a
Dicentra 2856
Diceros 7532
Dichiton C1 (Cephaloziella)
Dichorisandra 909
Dichotomocladia A8
(Carpomitra)
Dichromena 492
Dichrostachys 3452
Dicliptera 8031
Diclytra 2856
Dictyogramme D
(Coniogramme)
Dictyoloma 4063
Dictyopteris A8
Dictyosiphon A8
Dictyota A8
Didelta 9439
Didissandra 7809
Didymocarpus 7810
Didymosphenia A1
Diectomis 134a
Dietes 1265a
Digitaria 166a
Dilasia 899a
Dimorphanthes 8926
Dimorphotheca 9425
Dinamoeba A6
(Dinamoebidium)
Dinamoebidium A6
Diomphala A1
(Didymosphenia)
Dioon 4
Dipetalia 3126
Diphaca 3792
Dipholis 6373
Diphryllum 1494
Diplandra 5826
Diplodium 1449
Diplogon 8844
Diplonyx 3722
Diplophyllum C1
Diploschistella B
(Gyalideopsis)
Diplostromium A8
(Desmotrichum)
Diplukion 7382
Dipteryx 3845
Diracodes 1337a
Disarrenum 206
Disceraea A3
(Haematococcus)
Discostachys F (Dolerotheca)
Disiphonia A1 (Diatomella)
Disparago 9039
Dissotis 5659
Distichium C2
Ditrichum C2
Dogelodinium A6
Dolerotheca F
Dolichandrone 7741
Dolicholus 3897
Dolichos 3910
Dombeya 5053
Donatia 3204
Dontostemon 3050
Doona 5214
Doryopteris D
Dothiora B

Douglasia 6318
Downingia 8706
Dracocephalum 7250
Drechslera B
Dregea 6914
Drepano-hypnum C2 (Drepanocladus)
Drepanocladus C2
Drepanophyllaria C2 (Hygroamblystegium)
Drimys 2658
Drummondia C2
Dryandra 2069
Drymoglossum D
Drynaria D
Dryopteris D
Duchekia 894
Dudresnaya A9
Dufourea B (Xanthoria)
Duguetia 2680
Dunalia 7380
Dupathya 830
Durandea 3947
Duroia 8316
Duvaucellia 8136

Ecastaphyllum 3821
Ecballium 8596
Echinaria 320
Echinochloa 166
Echinocystis 8629
Echyrospermum 3466
Eclipta 9166
Ectocarpus A8
Ehrharta 201
Eichhornia 921
Elachista A8
Elaphoglossum D
Elaphrium 4150
Elaterium 8596
Elatostema 1988
Elayuna 3528
Elephas 7649
Eleutherine 1292
Ellimia 3126
Ellisia 7023
Ellisius A9 (Heterosiphonia)
Ellobum 7809
Elutheria 4190
Elytraria 7908
Elytrospermum 468b
Embelia 6310
Emex 2194
Enallagma 7757
Enargea 1146
Encentrus 4627
Enchidium 4449
Encoelia B
Endlicheria 2811a
Endophis B (Leptorhaphis)
Endosigma A1 (Pleurosigma)
Endospermum 4470
Enicostema 6484
Enneastemon 2691a
Entada 3468
Enteromorpha A3
Epacris 6260
Ephedrosphaera B (Nectria)
Ephemerella C2
Ephemerum C2
Ephippium 1704
Ephynes 5665
Epibaterium 2570
Epicostorus 3316
Epidendrum 1614
Epidermophyton B
Epifagus 7792
Epineuron A9 (Vidalia)
Epipactis 1482
Epiphylla A9 (Phyllophora)
Eranthis 2528
Erebinthus 3718
Eria 1697
Ericaria A8 (Cystoseira)
Eriocladus B (Lachnocladium)
Eriosema 3898
Erophila 2989a
Eroteum 5157b
Erporkis 1516
Erythrospermum 5278
Erythrotrichia A9
Eschenbachia 8926
Espera 4938
Etlingera 1344
Eucharis 1196
Euclidium 3038
Eucnide 5384
Eucomis 1088
Eulophia 1648
Euonymus 4618
Euosma 6450
Eupatoriophalacron 9166
Euphyllodium A1 (Podocystis)
Eupodiscus A1
Euriosma 3898
Euscaphis 4667
Eusideroxylon 2793
Eustichium C2 (Bryoxiphium)
Euterpe 631
Eutypella B
Evea 8411
Exilaria A1 (Licmophora)
Exocarpos 2097
Eysenhardtia 3708
Eystathes 4281

Fagara 3991
Fagaster 1889
Fagopyrum 2202
Falcaria 6018
Falcata 3860
Fasciata A8 (Petalonia)
Fastigiaria A9 (Furcellaria)
Fedia 8530
Felicia 8919
Ferolia 1957
Ficinia 465
Filago 8969
Filaspora B (Rhabdospora)
Filix D (Dryopteris)
Fimbriaria A9 (Odonthalia)
Fimbristylis 471
Fittonia 8069
Flemingia 3899
Forestiera 6427
Forsythia 6421
Franseria 9147
Freesia 1316
Freziera 5157b
Frustulia A1
Funckia 1111
Funicularius A8 (Himanthalia)
Furcellaria A9
Furera 7317
Fuscaria A9 (Rhodomela)
Fusidium B (Cylindrocarpon)

Gabura B (Collema)
Gaertnera 8428
Galactites 9466
Galax 6277
Galearia 4455
Gamoscyphus C1 (Heteroscyphus)
Gansblum 2989a
Gardenia 8285
Gasparrinia B (Caloplaca)
Gassicurtia B (Buellia)

Gastrochilus 1822
Gastroclonium A9
Gausapia B (Septobasidium)
Gautieria B
Gaylussacia 6215
Gazania 9434
Gelidium A9
Gelona B (Pleurotus)
Genosiris 1289
Gentianella 6509a
Genyorchis 1704a
Geophila 8410
Gerardia 7990
Gerbera 9528
Ghesaembilla 6310
Gigalobium 3468
Giganthemum 3589
Glabraria 4943
Glandulifolia 4038
Glechoma 7249
Gleichenia D
Glenospora B
(Septobasidium)
Globifera 7549
Glochidion 4302
Gloeocapsa A5
Gloeococcus A3
Gloeophyllum B
Glossopteris F
Glossostigma 7556
Glossula 1403a
Glyceria 383
Glycine 3864
Glycosmis 4087
Glycycarpus 4600
Glyphocarpa C2 (Anacolia)
Glyphocarpa C2
(Bartramidula)
Goetzea 7421
Goldbachia 2923
Gomozia 8445
Gomphonema A1
Gomphotis 5621
Gongolaria A8 (Cystoseira)
Gongrosira A3
Goniotrichum A9
(Erythrotrichia)
Gordonia 5148
Gothofreda 6857
Gracilaria A9
Grammita A9 (Polysiphonia)
Grandinia B (Hyphodontia)
Granularius A8
(Dictyopteris)
Graphorkis 1648
Graphorkis 1648a
Grateloupia A9
Gratelupella A9
(Polysiphonia)
Grevillea 2045
Griffithsia A9
Grislea 5538
Grona 3807
Guaiabara 2209
Guarea 4190
Guatteria 2679
Guignardia B
Guizotia 9222
Gustavia 5510
Gyalideopsis B
Gymnocephalus C2
(Aulacomnium)
Gymnoderma B
Gymnogrammitis D
(Araiostegia)
Gymnomitrion C1
Gymnopus B (Collybia)
Gymnoscyphus C1
(Solenostoma)
Gymnosporia 4627
Gymnostomum C2
Gymnozyga A3 (Bambusina)
Gynandropsis 3087
Gynocephalum 4712
Gynopogon 6616
Gynura 9405
Gyrocephalus B (Gyromitra)
Gyrodinium A6
Gyrodon B
Gyromitra B
Gyrosigma A1
Gyrosigma A1 (Pleurosigma)
Gyroweisia C2

Haematococcus A3
Haenkea 4038
Halenia 6513
Halesia 6410
Halidrys A8
Halimeda A3
Halymenia A9
Hantzschia A1
Hapale 756
Hapaline 756
Haplohymenium C2
Haplolophium 7673
Haplomitrium C1
Haplopappus 8852
Haplophyllum 4012a
Hariota 5416
Harrisonia 4117
Hartogia 4037
Haworthia 1029
Hebecladus 7388
Heberdenia 6288
Hebokia 4667
Hedusa 5659
Hedwigia C2
Heisteria 2147
Heleophylax 468b
Helichrysum 9006
Helicodiceros 779
Heliconia 1321
Helinus 4905
Heliopsis 9157
Helleborine 1482
Helminthocladia A9
Helminthora A9
Helminthosporium B
Helodium C2
Heloniopsis 952
Helosis 2163
Helospora 8365
Helwingia 6157
Hemiaulus A1
Hemichlaena 465
Hemieva 3187
Hemimeris 7472
Hemiptychus A1
(Arachnoidiscus)
Henckelia 7810
Hendersonia B
(Stagonospora)
Henrya 8028
Hepetis 878
Herbacea A8 (Desmarestia)
Hermesias 3524
Hermupoa 3103
Herposteiron A3
(Aphanochaete)
Hesperochiron 7029
Hesperophycus A8
Hessea 1166
Hetaeria 1507
Heteranthera 924
Heteranthus 272
Heterolepis 9057
Heteromorpha 5992
Heteropterys 4226
Heteropyxis 9712
Heteroschisma A6
(Latifascia)

Heteroscyphus C1
Heterosiphonia A9
Hexagonia B
Hexalepis 891
Hexonix 952
Heydia 2103
Hibiscus 5013
Hicorius 1882
Hierochloë 206
Hierochontis 3038
Hildenbrandia A9
Himanthalia A8
Himantoglossum 1399
Hippeastrum 1208
Hippoperdon B (Calvatia)
Hippurina A8 (Desmarestia)
Hiptage 4208
Hirneola B
Hoffmannseggia 3557
Hofmannia 7312
Hohenbuehelia B (Pleurotus)
Hoiriri 861
Holcus 257
Holigarna 4604
Hollandella A6 (Gyrodinium)
Holodiscus 3332
Holomitrium C2
Holothrix 1408
Homaid 784
Homalocenchrus 194
Homoeocladia A1 (Nitzschia)
Homoeothrix A5
Hondbessen 8430
Hookeria C2
Hopea 5215
Hormiscia A3 (Urospora)
Hormogyne 6368a
Hormosira A8
Hosta 1018
Houttuynia 1857
Hugueninia 2997
Humboldtia 3518
Humida A3 (Prasiola)
Humida A3 (Schizogonium)
Humiria 3953
Huttum 5506
Hyalina A8 (Desmarestia)
Hyaloscypha B
Hybanthus 5271
Hydnum B
Hydrodictyon A3
Hydrolapatha A9 (Delesseria)
Hydrolea 7037
Hydrophora B (Mucor)
Hydropisphaera B (Nectria)
Hydropityon 7532
Hydrostemma 2515
Hydrurus A4
Hygroamblystegium C2
Hylogyne 2062
Hymenocarpos 3693
Hymenochaete B
Hymenodictyon 8197
Hypaelyptum 452
Hyperbaena 2611
Hyphodontia B
Hypholoma B
Hypnum C2
Hypocistis 2180
Hypoderma B
Hypodiscus 816
Hypolaena 815
Hypopeltis D (Polystichum)
Hypoxylon B
Hyptis 7342

Icacorea 6285
Ichnocarpus 6683
Ichthyomethia 3839
Idesia 5331
Ilicioides 4615
Ilmu 1261
Imhofia 1175
Inocarpus 3848
Inochorion A9 (Rhodophyllis)
Iochroma 7382
Ioxylon 1918
Iphigenia 975
Iphiona 9065
Ipo 1956
Ipomoea 7003
Iresine 2339
Iria 471
Iridaea A9
Iridorkis 1558
Isidium B (Pertusaria)
Isoglossa 8079
Isopogon 2026
Ithyphallus B (Mutinus)
Ixia 1302

Jacobinia 8097
Jacquinia 6282
Jambolifera 4079
Jambosa 5582
Jambosa 5583
Jamesia 3209
Japarandiba 5510
Jimensia 1533
Johnsonia 1037
Josephia 2069
Jubula C1
Julocroton 4349
Juncoides 937

Kadalia 5659
Kajuputi 5603
Kalawael 3424a
Kaliformis A9 (Chylocladia)
Kara-angolam 6154
Karekandel 5525
Karkinetron 2208
Karotomorpha A2
Karstenia B
Katou-tsjeroë 4604
Katoutheka 6285
Kennedia 3868
Keppenodinium A6
Kernera 2908
Kerstingiella 3910a
Kieseria 5144
Kneiffiella B (Hyphodontia)
Knightia 2064
Kniphofia 1024
Kohautia 8136
Kokabus 7388
Kokera 2297
Kolman B (Collema)
Kopsia 6626
Kosteletzkya 5015
Kozola 952
Krempelhuberia B (Pseudographis)
Krigia 9560
Kuhnia 8823
Kuhnistera 3710
Kukolis 7388
Kundmannia 6064
Kunzea 5601
Kyllinga 462

Labatia 6365
Labisia 6291
Lacellia 9476
Lachnanthes 1161
Lachnocladium B
Lacinaria 8826
Lactarius B
Laelia 1617
Laënnecia 8926

Laetia 5338
Laetinaevia B
Laetiporus B
Lagascea 9101
Lagenula 4918a
Lamarckia 374
Laminaria A8
Lamourouxia 7650
Lampranthus 2405a
Lanaria 1236
Landolphia 6562
Langermannia B (Calvatia)
Lannea 4563
Laothoë 1007
Laplacea 5149
Laportea 1980
Larrea 3973
Laschia B (Hirneola)
Lasianthus 8412
Lass 5007
Latifascia A6
Laurelia 2775
Laurencia A9
Laxmannia 1032
Layia 9258
Leaeba 2570
Lebetanthus 6251
Lecanactis B
Leda A3 (Zygogonium)
Leea 4919
Leersia 194
Leiotheca C2 (Drummondia)
Lejeunea C1
Lemanea A9
Lembidium C1
Lenormandia A9
Lens 3853
Leobordea 3657
Leonicenia 5759
Leontodon 9574
Leontopetaloides 1248
Leopoldia 1095a
Leopoldia 1208
Leperiza 1211
Lepia 9155
Lepicaulon 985a
Lepicephalus 8541
Lepidanthus 816
Lepidocarpus 2035
Lepidopilum C2
Lepidostemon 3022
Lepidozia C1
Lepiota B
Lepocinclis A7
Lepra B (Pertusaria)
Lepraria B
Leproncus B (Pertusaria)
Leptaxis 3196
Leptocarpus 808
Leptodon C2
Leptoglossum B
Leptohymenium C2 (Platygyrium)
Leptonematella A8
Leptophyllus A6 (Trichodesmium)
Leptorhaphis B
Leptorkis 1556
Leptospermum 5599
Leptosphaeria B
Leptostomum C2
Lerchea 8130
Lessertia 3756
Letharia B
Lethocolea C1
Lettsomia 5157b
Leucadendron 2037
Leucocarpum 4623
Leucoloma C2
Leucopogon 6262a
Leucospermum 2036
Levisticum 6083
Liatris 8826
Libanotis 6052a
Libertia 1283
Lichen B (Parmelia)
Lichina B
Lichtensteinia 5990
Licmophora A1
Ligularia 9412
Liliastrum 982
Limnanthes 4542
Limnophila 7532
Limodorum 1483
Limonium 6351
Lindera 2821
Lindleya 3328
Linkia 2023
Linociera 6428
Liparis 1556
Lipocarpha 452
Lippius C1 (Saccogyna)
Lissochilus 1648
Listera 1494
Lithophragma 3197
Lithothamnion A9
Litsea 2798
Lloydia 1077
Lobularia 3013
Locandi 4109
Logania 6450
Lohmannia A6 (Sphaeripara)
Loiseleuria 6189
Lomanodia 5777a
Lomatia 2063
Lonchocarpus 3834
Lonchostoma 3286
Lopadium B
Lophanthera 4247
Lophidium D (Schizaea)
Lophiodon C2 (Ditrichum)
Lophiostoma B
Lophodermium B
Lopholejeunea C1
Loranthus 2074
Lorea A8 (Himanthalia)
Lorentzia C2 (Pelekium)
Lotononis 3657
Lotophyllus 3673
Loudetia 278a
Lucernaria A3 (Zygnema)
Lucya 8140
Ludovia 682
Luehea 4959
Lunania 5334
Lundia 7697
Luorea 3899
Luzula 937
Luzuriaga 1146
Lychnis 2490
Lycianthes 7407a
Lycopodioides D (Selaginella)
Lycopodites F
Lyginia 800
Lygistum 8204
Lygodium D
Lygos 3675a
Lyngbya A5
Lyomices B (Hyphodontia)
Lyonia 6200
Lysanthe 2045
Lysigonium A1 (Melosira)

Macaglia 6588
Machaerium 3823
Mackaya 8039
Maclura 1918
Macrodon C2 (Leucoloma)
Macrolobium 3517
Macroplodia B (Sphaeropsis)
Macrostoma A10 (Chilomastix)

Macrotyloma 3910a
Mahonia 2566
Maianthemum 1119
Majorana 7314
Malache 5007
Malapoënna 2798
Malcolmia 3032
Malnaregam 4096
Malvastrum 4995
Malveopsis 4995
Mamboga 8227
Mammillaria 5411
Mancoa 2973
Manettia 8204
Manilkara 6386a
Manisuris 127
Mannia C1
Manulea 7517
Mappia 4693
Marasmius B
Marchesinia C1
Mariana 9464
Mariscus 459
Marsdenia 6911
Marshallia 9247
Marsilea D
Martensia A9
Martinellius C1 (Radula)
Martinezia 612
Martinezia 631
Mastigophora C1
Matteuccia D
Matthiola 3042
Maximiliana 660
Mayepea 6428
Medicago 3688
Meesia C2
Megalangium C2 (Acidodontium)
Megalopteris F
Megapleilis 7887a
Megathecium B (Melanospora)
Megotigea 779
Meibomia 3807
Meistera 1344
Melaleuca 5603
Melancranis 465
Melanogaster B
Melanoleuca B
Melanospora B
Melocactus 5409
Melosira A1
Membranifolia A9 (Phyllophora)
Meratia 2663a
Meriania 5692
Meridiana 9434
Merkia C1 (Pellia)
Merremia 6997
Mertensia 7102
Mesembryanthemum 2405
Mesoptera 8353
Mesosphaerum 7342
Metasequoia 32a
Metrosideros 5588
Metroxylon 565
Micarea B
Michauxia 8651
Miconia 5759
Micrampelis 8629
Micrandra 4435
Micranthemum 7549
Micranthus 1313
Microchaete A5
Microcystis A5
Microlepis 5648
Micromelum 4089
Micromeria 7305
Micromphale B (Marasmius)
Microsperma 5384
Microspora A3
Microstemma 6870
Microstylis 1553
Microthelia B (Anisomeridium)
Microtropis 4621
Mikania 8818
Millettia 3720
Milligania 1112
Miltonia 1778
Mindium 8656
Miquelia 4713
Mischocarpus 4820
Mison B (Phellinus)
Mitragyna 8227
Mitraria 7853
Mittenothamnium C2
Mnesiteon 9241
Mniobryum C2
Mnium C2
Moenchia 2432
Mokof 5153
Mollia 4960
Mollisia B
Mondo 1140
Moniera 7546
Monilia B
Moniliformia A8 (Hormosira)
Monochaetum 5665
Monodus A11
Monotris 1408
Monstera 700
Montrichardia 730
Moorea 329
Moquinia 9483
Moraea 1265
Morenia 594
Moscharia 9545
Mougeotia A3
Mucor B
Mucuna 3877
Muehlenbeckia 2208
Muellera 3834
Muellera 3837
Muelleriella C2
Muraltia 4278
Murdannia 899a
Murraya 4090
Musaefolia A8 (Alaria)
Mutinus B
Mycoblastus B
Mycobonia B
Mycoporum B
Mylia C1
Myridium B (Laetinaevia)
Myrinia C2
Myriophylla A9 (Botryocladia)
Myriostigma B (Cryptothecia)
Myristica 2750
Myroxylon 3584
Myrstiphyllum 8399
Myrteola 5557
Myrtopsis 4020
Mystacinus 4905
Myxolibertella B (Phomopsis)
Myxonema A3 (Stigeoclonium)

Nageia 13
Nalagu 4919
Nama 7033
Nani 5588
Nani 5594
Naravelia 2542
Nardia C1
Naregamia 4172
Naron 1265a
Narthecium 944
Nasturtium 2965

Naudinia 4060
Nautilocalyx 7857a
Nazia 143
Neckera C2
Nectandra 2790
Nectria B
Needhamia 3718
Neesia 5040
Nelanaregam 4172
Nemastoma A9
Nematanthus 7864
Nemia 7517
Nemopanthus 4615
Nemophila 7022
Neolitsea 2797
Neottia 1495
Neozamia F (Cordaites)
Nephroia 2570
Nereidea A9 (Plocamium)
Nerine 1175
Nertera 8445
Nervilia 1468
Nesaea 5486
Neslia 2988
Nestronia 2109
Neurocarpus A8
(Dictyopteris)
Neurocaulon A9
Nialel 4189
Nicandra 7377
Nicolaia 1337a
Nidularia B
Niemeyera 6382
Nissolia 3784
Nissolius 3823
Nitophyllum A9
Nitzschia A1
Nocca 9101
Nodularia A5
Nodularius A8
(Ascophyllum)
Nodulosphaeria B
Nodulosphaeria B
(Leptosphaeria)
Nomochloa 468a
Nothofagus 1889
Nothopegia 4600
Nothoscordum 1050
Notjo 7714
Nunnezharia 594
Nuphar 2514
Nyctophylax 1332
Nymphaea 2513
Nymphozanthus 2514

Oberonia 1558
Obsitila 985a
Ocellularia B
Odisca 7760
Odonthalia A9
Odontia B (Tomentella)
Odontonema 8037
Odontopteris F
Oedera 9322
Oenoplea 4868
Oeonia 1834
Ohlendorffia 7467
Oidium B
Olearia 8916
Oligomeris 3126
Olinia 5428
Omphalandria 4472
Omphalaria B (Anema)
Omphalea 4472
Oncidium 1779
Opa 3339
Opegrapha B
Operculina 6997
Ophiocytium A11
Ophiopogon 1140
Ophthalmidium B (Porina)
Oplismenus 169
Opospermum A8 (Elachista)
Orbignya 657
Orchiastrum 1490
Orectospermum 3500
Oreocharis 7808
Oreodoxa 612
Oreodoxa 631
Ormocarpum 3792
Ormosia 3597
Ornithopteris D (Anemia)
Orobanchia 7864
Orphium 6504
Orthopixis C2
(Aulacomnium)
Orthopixis C2
(Leptostomum)
Orthopogon 169
Orthothecium C2
Oscularia 2405a
Osmites 9050
Osmorhiza 5941
Osmundea A9 (Laurencia)
Ostrya 1885
Otilix 7407a
Ouratea 5113
Ouret 2317
Ourouparia 8228

Outea 3517
Ovidia 5457
Oxylobium 3624
Oxypetalum 6857
Oxytria 1006
Oxytropis 3767

Pachyrhizus 3908
Pacouria 6562
Padina A8
Paederia 8430
Paepalanthus 830
Palisota 894
Pallassia 4035
Pallavicinia C1
Pallenis 9091
Palma-filix 7
Palmstruckia 7518
Paludana 1344
Pamea 5545
Panaeolus B
Pancheria 3277
Pancovia 4753
Panel 4087
Panicastrella 320
Panisea 1714
Pantocsekia A1
Panus B
Paphiopedilum 1393a
Papillaria C2
Papyracea A9 (Cryptopleura)
Paradisea 982
Parahebe 7579a
Parapetalifera 4036
Paraphysorma B
(Staurothele)
Parasaccharomyces B
(Candida)
Parascopolia 7407a
Parendomyces B (Candida)
Parmelia B
Parmeliopsis B
Parmostictina B
(Pseudocyphellaria)
Parrasia 6483
Parsonsia 6691
Parthenocissus 4915
Patagonium 3800
Patella B (Scutellinia)
Patersonia 1289
Patrinia 8535
Patrisa 5341
Pattara 6310
Pavonia 5007

Payera 8162
Peccania B
Peckia 6301
Pectinaria 6889
Pectinea 5278
Pedicellaria 3087
Pedicellia 4820
Pedilanthus 4501
Pelaë 4281
Pelekium C2
Pelexia 1488
Pellaea D
Pellia C1
Pellionia 1987
Peltaea 5008a
Peltandra 747
Peltanthera 6468
Peltigera B
Peltimela 7556
Peltogyne 3500
Peltophorum 3561
Peltostegia 5008a
Pentaceras 3998
Pentagonia 8265
Pentapodiscus A1 (Aulacodiscus)
Peribotryon B (Chrysothrix)
Pericampylus 2568
Peridermium B
Peripherostoma B (Daldinia)
Peristylus 1403a
Pernettya 6208
Perojoa 6262a
Peronia A1
Persea 2783
Persoonia 2023
Pertusaria B
Perytis 1893
Petalonia A8
Petalostemon 3710
Petermannia 1258
Petteria 3676
Petunia 7436
Peumus 2759
Peyrousea 9365
Pezicula B
Phacelocarpus A9
Phacidium B
Phacus A7
Phaedranthus 7679
Phaeocollybia B
Phaeosticta B (Pseudocyphellaria)
Phaeotrema B
Pharbitis 7003a
Pharomitrium C2 (Pterygoneurum)
Phaseoloides 3722
Phaulopsis 7932
Phellinus B
Phibalis B (Encoelia)
Phillipsia B
Philodendron 739
Phlyctis B
Pholiota B
Pholiotella B (Conocybe)
Pholiotina B (Conocybe)
Phoma B
Phomopsis B
Phragmipedium 1393b
Phrynium 1368
Phucagrostis 60
Phyllachora B
Phyllactidium A3 (Coleochaete)
Phyllaurea 4454
Phyllocladus 15
Phyllodes 1368
Phyllona A9 (Porphyra)
Phyllophora A9
Phyllorkis 1705
Phyllostachys 417
Phyllosticta B
Phymatolithon A9
Physalodes 7377
Physconia B
Physedium C2 (Ephemerella)
Physocarpus 3316
Phytocrene 4712
Phytoxis 7299
Piaropus 921
Pickeringia 3619
Picramnia 4131
Picrodendron 4134
Pierrea 5221
Pigafetta 567
Pilea 1984
Piliocalyx 5585
Piliostigma 3528
Pimelea 5467
Pinellia 787
Pinnularia A1
Piptocalyx 2758
Piptochaetium 212
Piptolepis 8761
Piratinera 1957
Piscidia 3839
Pistolochia 2858
Pitcairnia 878
Pithecellobium 3441
Pittosporum 3252
Placodion B (Peltigera)
Placus 8939
Plagiochasma C1
Plagiochila C1
Planchonella 6368a
Plaso 3876
Platanthera 1410
Plathymenia 3466
Platolaria 7665
Platonia 5205
Platychloris A3 (Chloromonas)
Platygyrium C2
Platylepis 1516
Platylophus 3269
Platyphyllum B (Cetraria)
Platysphaera B (Lophiostoma)
Plaubelia C2 (Trichostomum)
Plectranthus 7350
Plectronia 5428
Plenckia 4637
Pleonosporium A9
Pleospora B
Pleurendotria 3197
Pleuridium C2
Pleurochaete C2 (Tortella)
Pleurolobus 3807
Pleuropus B (Pleurotus)
Pleurosigma A1
Pleurospora 730
Pleurotus B
Pleurozium C2
Plexipus 7148a
Plocamium A9
Plocaria A9 (Gracilaria)
Plumaria A9
Pneumaria 7102
Pochota 5042a
Podalyria 3621
Podanthe C1 (Lethocolea)
Podanthus 9150
Podocarpus 13
Podocratera B (Tholurna)
Podocystis A1
Podolepis 9054
Podopogon 212
Podosperma 9009
Podospermum 9581
Podospora B
Podotheca 9009

Pogomesia 910
Poiretia 3789
Polemannia 6045
Polia 2455
Pollichia 2467
Pollinia 134c
Polyacanthus 4627
Polyblastia B
Polycarpaea 2455
Polychroa 1987
Polycystis B (Urocystis)
Polyedrium A11 (Tetraedriella)
Polygonastrum 1118
Polygonum 2201
Polyneura A9
Polyphragmon 8365
Polypompholyx 7900
Polyschidea A8 (Saccorhiza)
Polysiphonia A9
Polystachya 1565
Polystichum D
Pongamia 3720
Pongamia 3836
Pongelia 7741
Pongelion 4124
Porina B
Porocarpus 8365
Porphyra A9
Porphyridium A9
Porphyrostromium A9 (Erythrotrichia)
Posidonia 57
Possira 3574
Prasiola A3
Premna 7185
Prestoea 612
Prestonia 6702
Printzia 9059
Prionitis A9
Prismatocarpus 8663
Prismatoma A9 (Corynomorpha)
Pritchardia 542
Prolongoa 9341a
Protea 2035
Protium 4137
Protococcus A3 (Chlamydomonas)
Psammospora B (Melanoleuca)
Pselium 2568
Pseudelephantopus 8775a
Pseudo-brasilium 4131
Pseudo-fumaria 2858
Pseudocyphellaria B
Pseudographis B
Pseudolarix 25
Pseudomonilia B (Candida)
Psilanthus 8388
Psilobium 8237
Psophocarpus 3914
Psora B
Psychotria 8399
Psychotrophum 8399
Pteretis D (Matteuccia)
Pteridium D
Pterigynandrum C2 (Platygyrium)
Pterocarpus 3828
Pterococcus 4421
Pterogonium C2 (Platygyrium)
Pterolepis 5632
Pterolobium 3553
Pteronia 8862
Pterophora 8855
Pterophorus 8862
Pterophyllus B (Pleurotus)
Pteropsis D (Drymoglossum)
Pterospermum 5080
Pterostylis 1449
Pterota 3991
Pterygoneurum C2
Ptilochaeta 4234
Ptiloria 9576
Ptilota A9
Ptychomitrium C2
Ptyxostoma 3286
Pubeta 8316
Puccinellia 384
Pucciniola B (Uromyces)
Pulina B (Lepraria)
Pulparia B (Pulvinula)
Pulveraria B (Chrysothrix)
Pulvinula B
Pupal 2314
Pupalia 2314
Pycnanthemum 7317
Pycnoporus B
Pygmaea B (Lichina)
Pyrenacantha 4709
Pyrenodesmia B (Caloplaca)
Pyrenopsis B
Pyrenotea B (Lecanactis)
Pyrenula B
Pyrrhopappus 9604
Pythion 723
Pythium B
Pyxidicula A1 (Rhopalodia)

Quercella B (Phaeocollybia)
Quinata 3823
Quinchamalium 2120

Racodium B
Raddetes B (Conocybe)
Radula C1
Rafinesquia 9578
Ramalina B
Ramaria B
Ramonda 7800
Ramularia B
Raphanis 2965a
Raphanozon B (Telamonia)
Rapistrum 2956
Ravensara 2813
Razoumofskya 2091
Reboulia C1
Rechsteineria 7887a
Rehmannia 7592
Reineckea 1129
Reineria 3718
Relhania 9050
Renealmia 1331
Restio 804
Resupinatus B (Pleurotus)
Retama 3675a
Reticula A3 (Hydrodictyon)
Reussia 923
Rhabdonema A1
Rhabdospora B
Rhagadiolus 9566
Rhaphiolepis 3339
Rhaphis 134c
Rhipidium B
Rhipsalis 5416
Rhizopus B
Rhizosolenia A1
Rhodobryum C2
Rhodomela A9
Rhodophyllis A9
Rhodopis 3871
Rhodothamnus 6191
Rhodymenia A9
Rhopalodia A1
Rhynchanthera 5676
Rhynchocorys 7649
Rhynchoglossum 7833
Rhynchosia 3897
Rhynchospora 492
Rhytidocaulon 6877a

Rhytidophyllum 7892
Rhytiglossa 8079
Riccardia C1
Riccia C1
Richaeia 5528
Richea 6254
Ricotia 2968
Ridan 9215
Riedelia 1332
Rinorea 5262
Rivularia A5
Robertia 6374
Robillarda B
Robinsonia 9382
Robynsia 8473a
Roccella B
Rochea 3171
Rochelia 7124
Romulea 1261
Rothia 3659
Rottboellia 127
Rotularia F (Sphenophyllum)
Rourea 3424
Rulingia 5060
Rutstroemia B
Ryania 5341
Rymandra 2064
Ryssopterys 4222

Saccardoa B (Pseudocyphellaria)
Saccharina A8 (Laminaria)
Saccidium 1408
Saccocalyx 7306
Saccogyna C1
Saccolabium 1822
Saccorhiza A8
Sagedia B (Aspicilia)
Sagotia 4452
Saguerus 575
Sagus 565
Salacia 4662
Salken 3838
Salmea 9208
Salomonia 4277
Samadera 4109
Samyda 5337
Sanseverinia 1110
Sansevieria 1110
Santaloides 3424a
Sapium 4483
Sarcanthus 1824
Sarcinanthus 678a
Sarcoderma A9 (Porphyridium)
Sarcodum 3753
Sargassum A8
Sargentia 4074
Sarmienta 7854
Sarothamnus 3682a
Satyrium 1430
Saurauia 5109
Saussurea 9457
Savastana 206
Scaevola 8716
Scaligeria 5964
Scalius C1 (Haplomitrium)
Scalptrum A1 (Gyrosigma)
Scalptrum A1 (Pleurosigma)
Scandalida 3699
Scapania C1
Schaereria B
Schaueria 8042
Schefflera 5852
Schelhammera 962
Schisandra 2656
Schistidium C2
Schizaea D
Schizocalyx 8215
Schizogonium A3
Schizomitrium C2 (Callicostella)
Schizonotus 3323
Schizothecium B (Podospora)
Schizymenia A9
Schkuhria 9291
Schlechtendalia 9511
Schleichera 4767
Schmidtia 312
Schoenodum 808
Schoenolirion 1006
Schoenoplectus 468b
Schotia 3506
Schouwia 2940
Schradera 8241
Schrankia 3448
Schrebera 6422
Schubertia 6772
Schultesia 6526
Schulzia 6058
Sciadiodaphne 2789
Sciodaphyllum 5852
Scirpus 468
Sclerodontium C2 (Leucoloma)
Scleropyrum 2103
Sclerotinia B
Scolochloa 381
Scolopia 5304
Scopolia 7393
Scoptria B (Eutypella)
Scopularia 1408
Scurrula 2074
Scutarius A9 (Nitophyllum)
Scutellinia B
Scutia 4874
Scytophyllum 4627
Scytosiphon A8
Sechium 8636
Securidaca 4275
Securigera 3694
Securinega 4297
Sedoidea A9 (Gastroclonium)
Seemannia 7878
Segestria B (Porina)
Selaginella D
Selaginoides D (Selaginella)
Selinum 6070
Selloa 9168
Senecillis 9412
Septobasidium B
Septoria B
Sequoia 32
Serapias 1397
Serda B (Gloeophyllum)
Sererea 7679
Seringia 5075
Serpentinaria A3 (Mougeotia)
Sertularia A3 (Halimeda)
Sesban 3747
Sesbania 3747
Sesia B (Gloeophyllum)
Setaria 171
Seymeria 7602
Shawia 8916
Shepherdia 5471
Shortia 6275
Shuteria 3863
Sicelium 8250
Siebera 9446
Sieglingia 280
Sigmatella A1 (Nitzschia)
Silene 2490
Siliquarius A8 (Halidrys)
Siloxerus 9028
Silybum 9464
Simarouba 4111
Simbuleta 7485
Simethis 987
Simocybe B

Siphonychia 2477
Siphula B
Siraitos 951
Sirogonium A3
Sitodium 1946
Skimmia 4083
Smilacina 1118
Smithia 3796
Soaresia 8772
Soja 3864
Solandra 7414
Solenostoma C1
Solori 3838
Somion B (Spongipellis)
Sommerfeltia 8918
Sonerila 5729
Sonneratia 5497
Sophia 2997
Soranthe 2028
Sorbaria 3323
Sordaria B
Sorghum 134b
Soria 3038
Sorocephalus 2028
Sparrmannia 4957
Spathelia 4066
Spergularia 2450
Spermatochnus A8
Sphacele 7299
Sphaerella A3 (Chlamydomonas)
Sphaeria B (Hypoxylon)
Sphaeripara A6
Sphaerophorus B
Sphaeropsis B
Sphaerothallia B (Aspicilia)
Sphaerotheca B
Sphaerozosma A3
Sphenoclea 8680
Sphenomeris D
Sphenophyllites F (Sphenophyllum)
Sphenophyllum F
Sphinctocystis A1 (Cymatopleura)
Spiranthes 1490
Spirhymenia A9 (Vidalia)
Spirodinium A6 (Gyrodinium)
Spirodiscus A11 (Ophiocytium)
Spirogyra A3
Spirolobium 6670
Spirostylis 2078
Splaknon A3 (Enteromorpha)
Spondogona 6373
Spongipellis B
Spongocladia A3 (Cladophoropsis)
Spongopsis A3 (Chaetomorpha)
Sporodictyon B (Polyblastia)
Stachyanthus 4715
Stachygynandrium D (Selaginella)
Stachytarpheta 7151
Stagonospora B
Statice 6350
Stauroptera A1 (Pinnularia)
Staurothele B
Steganotropis 3858
Stelis 1587
Stellandria 2656
Stellorkis 1468
Stemodia 7534
Stemodiacra 7534
Stenandrium 7990
Stenanthium 957
Stenocarpus 2066
Stenogyne 7227
Stenoloma D (Sphenomeris)
Stenophyllus 471a
Stephanomeria 9576
Stephanotis 6911
Stereocaulon B
Stereococcus A3 (Gongrosira)
Steriphoma 3103
Stictina B (Pseudocyphellaria)
Stifftia 9490
Stigeoclonium A3
Stilbella B
Stilophora A8
Stimegas 1393a
Stizolobium 3877
Streptylis 899a
Stromatosphaeria B (Daldinia)
Struthanthus 2078
Struthiola 5436
Struvea A3
Stylidium 8724
Styllaria A1 (Licmophora)
Stylurus 2045
Styrosinia 7887a
Suaeda 2261
Suhria A9
Suksdorfia 3187
Sulitra 3756
Sutherlandia 3754
Swartzia 3574
Sweetia 3582
Sychnogonia B (Thelopsis)
Symphyglossum 1834a
Symplocarpus 708
Symplocia B (Crocynia)
Synandrodaphne 5467a
Synedrella 9224
Syringidium A1 (Cerataulina)
Syringodea 1260
Syringospora B (Candida)
Syzygium 5583

Tacca 1248
Taetsia 1108
Taligalea 7156
Talinum 2406
Taonabo 5153
Tapeinochilos 1360
Tapesia B (Mollisia)
Tapinanthus 2074a
Tapinothrix A5 (Homoeothrix)
Tapogomea 8411
Taralea 3845
Taraxacum 9592
Tardavel 8473
Tariri 4131
Tauschia 5977
Taxilejeunea C1
Teclea 4085
Tectona 7181
Tekel 1283
Telamonia B
Telopea 2062
Tema 166
Tephrosia 3718
Terminalia 5544
Ternstroemia 5153
Tessella A1 (Rhabdonema)
Tetracyclus A1
Tetradonta A3 (Chloromonas)
Tetraedriella A11
Tetragonolobus 3699
Tetralix 5353
Tetramastix A2 (Karotomorpha)
Tetramerium 8028
Tetranema 7510

Tetrapodiscus A1 (Aulacodiscus)
Tetrapterys 4212
Thamnea 3284
Thamnium B (Roccella)
Thamnolia B
Theka 7181
Thelopsis B
Thelypteris D
Theodora 3506
Thespesia 5018
Thevetia 6632
Tholurna B
Thomsonia 723
Thorelia 9604a
Thorntonia 5015
Thouinia 4733
Thryallis 4244
Thryptomene 5621
Thunbergia 7914
Thymelaea 5453
Thymopsis 9289
Thysanotus 992
Tiliacora 2577
Tillospermum 5601
Timmia C2
Timonius 8365
Tinantia 910
Tinospora 2583
Tissa 2450
Tithymaloides 4501
Tithymalus 4498a
Tittmannia 3285
Tobira 3252
Toddalia 4077
Tolmiea 3196
Toluifera 3584
Tomanthera 7604a
Tomentella B
Tontelea 4662a
Torresia 206
Torreya 17
Tortella C2
Tortula C2
Touchiroa 3495
Toulichiba 3597
Tounatea 3574
Tournesol 4355
Tourrettia 7766
Tovaria 3081
Trachyandra 985a
Trachylejeunea C1
Trachyspermum 6014
Tragus 143
Tremella B
Trentepohlia A3
Treubia C1
Triceros 4666
Trichilia 4195
Trichocalyx 8100
Trichocolea C1
Trichodesma 7056
Trichodesmium A5
Tricholoma B
Trichophorum 466a
Trichosporum 7824
Trichostachys 8397
Trichostomum C2
Tricondylus 2063
Tricyrtis 967
Trigoniastrum 4264
Trigonostemon 4449
Triguera 7392
Trimenia 2758
Trinia 5998
Triopterys 4211
Triplochiton 5022a
Tripodiscus A1 (Aulacodiscus)
Tripteris 9428
Trochera 201
Trombetta B (Craterellus)
Trophis 1917
Trozelia 7382
Trypethelium B
Tryphia 1408
Tsjeru-caniram 2124
Tubercularia B
Tubiflora 7908
Tuburcinia B (Urocystis)
Tulbaghia 1047
Tulisma 7887a
Tumboa 48
Turpinia 4666

Ucacou 9224
Ugena D (Lygodium)
Uloma 7760
Ulticona 7388
Ulva A3
Umbellularia 2789
Uncaria 8228
Unxia 9285
Uraspermum 5941
Urbania 7139
Urceola 6639
Urceolaria 7854
Urceolina 1211
Urnularia 6566
Urocystis B
Uromyces B
Uropedium 1393b
Urospora A3
Ursinia 9431
Urticastrum 1980
Uva-ursi 6212

Vaginarius B (Amanitopsis)
Vaginata B (Amanitopsis)
Vagnera 1118
Vahlia 3201
Valerianoides 7151
Vallota 1178
Valsa B
Valteta 7382
Vanieria 1942
Variolaria B (Pertusaria)
Vaupelia 7124a
Vedela 6285
Veitchia 639
Venana 3225
Ventenata 272
Venturia B
Verbesina 9218
Vermicularia 7151
Vernonia 8751
Verrucaria B
Versipellis B (Xerocomus)
Vertebrata A9 (Polysiphonia)
Verticordia 5625
Vibo 2194
Viborquia 3708
Vidalia A9
Vigna 3905
Villanova 9285
Villaria 8296
Villarsia 6544
Vireya 7860
Virgilia 3608
Virgularia 7604a
Vismia 5171
Vitaliana 6318
Viticella 7022
Voandzeia 3905
Vochysia 4266
Volubilaria A9 (Vidalia)
Volutella B
Volvariella B
Volvarius B (Volvariella)
Volvulus 6994
Vossia 124

Vouacapoua 3841
Vouapa 3517
Vriesea 891

Wahlenbergia 8668
Wallenia 6304
Walpersia 3647
Warburgia 5256
Warmingia 1739
Washingtonia 543
Watsonia 1315
Wedelia 9192
Weihea 5528
Weingaertneria 269
Weinmannia 3276
Weisiodon C2 (Gyroweisia)
Welwitschia 48
Wendlandia 8181
Wendtia 3931
Westia 3516
Wiborgia 3661
Wigandia 7035
Wikstroemia 5446
Wilckia 3032
Willughbeia 6564
Windmannia 3276
Wisteria 3722
Withania 7400
Wolffia 796
Wurfbainia 1344

Xanthocarpia B (Caloplaca)
Xanthophyllum 4281
Xanthoria B
Xanthostemon 5594
Xenopoma 7305
Xerocarpa 7182a
Xerocomus B
Xylaria B
Xylometron B (Pycnoporus)
Xylophylla 2097
Xylopia 2717
Xylopicrum 2717
Xylosma 5320

Zaluzianskya 7523
Zamia 7
Zantedeschia 748
Zelkova 1901
Zephyranthes 1181
Zerumbet 1328
Zeugites 358
Zeuxine 1502
Zingiber 1324
Zinnia 9155
Zollingeria 4747
Zonaria A8
Zoophthalmum 3877
Zoysia 150
Zuccagnia 3558
Zuccarinia 8312
Zygia 3441
Zygis 7305
Zygnema A3
Zygoglossum 1704
Zygogonium A3

Index of Scientific Names

Abies balsamea (L.) Mill. 23.Ex.4
Acaena anserinifolia (J. R. Forst. & G. Forst.) Druce H.10.Ex.3
– ×*anserovina* Orchard H.10.Ex.3
– *ovina* A. Cunn. H.10.Ex.3
Acanthococcus Hook. f. & Harv. 53.Ex.12
Acanthococos Barb. Rodr. 53.Ex.12
Acanthoeca W. N. Ellis 53.Ex.7
Acanthoica Lohmann 53.Ex.7
Acer pseudoplatanus L., not *"pseudo-platanus"* 60.Ex.13
Aceras R. Br. 62.Ex.5
"*Acosmus* Desv." (Desfontaines, 1829) 34.Ex.4
Adenanthera bicolor Moon 7.Ex.4
Adiantum capillus-veneris 23.Ex.1
Adonis L. 62.Ex.1
Aecidium Pers. : Pers. 59.Ex.2
Aesculus L. 11.Ex.5
Aextoxicaceae 18.Ex.1
Aextoxicon, *Aextoxicou* 18.Ex.1
Agaricus 32.Ex.8
– "tribus" *Hypholoma* Fr. : Fr. 32.Ex.8
– "tribus" *Pholiota* Fr. : Fr. 33.Ex.13
– *atricapillus* Batsch 15.Ex.3
– *cervinus* Hoffm. : Fr., non Schaeff. 15.Ex.3
– *cinereus* Schaeff. 23.Ex.9
– *equestris* L. 15.Ex.4
– *ericetorum* Fr. 15.Ex.1
– *flavorineus* Pers. 15.Ex.4
– *"octogesimus nonus"* (Schaeffer, 1763) 23.Ex.9
– *umbelliferus* L. 15.Ex.1
Agathopyllum Juss. 55.Ex.1
– *neesianum* Blume 55.Ex.1
Agati Adans. 62.Ex.7
×*Agroelymus* A. Camus 11.Ex.28, H.8.Ex.1
×*Agrohordeum* A. Camus H.8.Ex.1, H.9.Ex.1
×*Agropogon* P. Fourn. H.3.Ex.1, H.6.Ex.1
– *littoralis* (Sm.) C. E. Hubb. H.3.Ex.1
Agropyron Gaertn. 11.Ex.28, H.8.Ex.1, H.9.Ex.1
Agrostis L. H.2.Ex.1, H.6.Ex.1
– *radiata* L. 52.Ex.11
– *stolonifera* L. H.2.Ex.1
Albizia, not *"Albizzia"* Rec.60H.Ex.1
Aletris punicea Labill. 52.Ex.10
Alexitoxicum 51.Ex.1
Algae 13.1(e)
Alpinia Roxb., *nom. cons.*, non L. 55.Ex.2
– *galanga* (L.) Willd. 55.Ex.2
– *languas* J. F. Gmel. 55.Ex.2
Alsophila kalbreyeri Baker 1892, non C. Chr. 1905 33.Ex.6
Alyssum flahaultianum Emb. 36.Ex.3
Alyxia ceylanica Wight, not *"zeylanica"* 60.Ex.1
Amaranthus L., not *"Amarantus"* 60.Ex.1
×*Amarcrinum* Coutts H.6.Ex.2
Amaryllidaceae 53.Ex.1, 53.Ex.12
Amaryllis L. H.6.Ex.2
Amblyanthera Müll. Arg., non Blume 53.Ex.2
Amorphophallus campanulatus Decne. 48.Ex.2
Amphiprora Ehrenb. 45.Ex.4
Amphitecna Miers 14.Ex.6
Anacamptis Rich. H.6.Ex.1
Anacyclus L. 10.Ex.1
– *valentinus* L. 10.Ex.1
Anagallis arvensis subsp. *caerulea* Hartm. 53.Ex.14
– – var. *caerulea* (L.) Gouan 53.Ex.14
– *caerulea* Schreber, non L. 53.Ex.14
Andreaea angustata Lindb. ex Limpr. 46.Ex.22
Andromeda polifolia L., not *"poliifolia"* 60.Ex.11
Andropogon L. 62.Ex.3
– *fasciculatus* L. 52.Ex.11
– *martini* Roxb. 32.Ex.6
– *sorghum* subsp. *halepensis* (L.) Hack. 53.Ex.13
– – var. *halepensis* (L.) Hack. 53.Ex.13
"Anema" (Nylander, 1879) 42.Ex.1
Anema nummulariellum Forsell, not "Nyl." 42.Ex.1
Anemone ×*elegans* Decne. 11.Ex.27
– *hupehensis* (Lemoine & E. Lemoine) Lemoine & E. Lemoine 11.Ex.27

– ×*hybrida* Paxton 11.Ex.27
– *vitifolia* Buch.-Ham. ex DC. 11.Ex.27
Annona, not *"Anona"* Rec.60H.Ex.1
"Anonymos" (Walter, 1788) 20.Ex.8
– *"aquatica"* (Walter, 1788) 43.Ex.3
Anthemis valentina L. 10.Ex.1
"Anthopogon" 20.4.Note.2
Anthyllis sect. *Aspalathoides* DC. 41.Ex.2, 49.Ex.2
– *barba-jovis* L., not *"Barba jovis"* 23.Ex.11
Antidesmatinae Pax Rec.19A.Ex.1
Antidesmatoideae (Pax) Hurus. Rec.19A.Ex.1
Antirrhinum spurium L. 11.Ex.7
Apiaceae 18.5
Apium L.
Apocynum androsaemifolium L., not *"fol. [foliis] androsaemi"* 23.Ex.13
×*Arabidobrassica* Gleba & Fr. Hoffm. H.9.Ex.3
Arabidopsis Heynh. H.9.Ex.3
– *thaliana* (L.) Heynh. H.9.Ex.3
Arabis beckwithii S. Watson 47.Ex.1
– "Sekt. *Brassicararbis*" (Schulz, 1936) 36.Ex.1
– "Sekt. *Brassicoturritis*" (Schulz, 1936) 36.Ex.1
– *shockleyi* Munz 47.Ex.1
Arachnis Blume H.8.Ex.2
Arctostaphylos uva-ursi (L.) Spreng. 60.Ex.14
Ardisia pentagona A. DC. 51.Ex.2
– *quinquegona* Blume 51.Ex.2
Areca L. 18.5
Arecaceae 18.5
Arenaria L. 11.Ex.10
– ser. *Anomalae* 21.Ex.1
– *stricta* Michx. 11.Ex.10
– *uliginosa* Schleich. ex Schltdl. 11.Ex.10
Aronia arbutifolia var. *nigra* (Willd.) F. Seym. 33.Ex.5
Artemisia nova A. Nelson 23.Ex.7
Arum campanulatum Roxb. 48.Ex.2
– *dracunculus* L. 11.Ex.11
Arytera sect. *Mischarytera* 6.Ex.2
Ascocentrum Schltr. ex J. J. Sm. H.6.Ex.6
Aspalathoides (DC.) K. Koch 41.Ex.2, 49.Ex.2
Aspicarpa Rich. 34.Ex.4
Aspidium berteroanum Rec.60C.1(c)
Asplenium dentatum L., not *"Trich. dentatum"* 23.Ex.14
Aster L. 11.Ex.25, 18.5, 19.Ex.4
– *novae-angliae* L. 60.Ex.14
Asteraceae Dumort. 18.5, 19.Ex.4
×*Asterago* Everett 11.Ex.25
Astereae Cass. 19.Ex.4
Asterinae Less. 19.Ex.4
Asteroideae Asch. 19.Ex.4
Asterostemma Decne. 53.Ex.5
Astragalus cariensis Boiss. 53.Ex.4
– *(Cycloglottis) contortuplicatus* Rec.21A.Ex.1
– *matthewsiae* Podlech & Kirchhoff, not *"matthewsii"* 60.Ex.19
– *matthewsii* S. Watson 60.Ex.19
– *rhizanthus* Boiss., non Royle 53.Ex.4
– *(Phaca) umbellatus* Rec.21A.Ex.1
Astrostemma Benth. 53.Ex.5
"Atherospermaceae" (Lindley, 1846) 18.Ex.6
Atherospermataceae R. Br., *"Atherospermeae"* 18.Ex.6
Athyrium austro-occidentale Ching 60.Ex.14
Atriplex L. 23.Ex.7, 62.Ex.1
– *"nova"* (Winterl, 1788) 23.Ex.7
Atropa bella-donna 23.Ex.1

Balardia Cambess. 53.Ex.11
Ballardia Montrouz. 53.Ex.11
Bartlingia Brongn., non Rchb. nec F. Muell. Rec.50C.Ex.1
Bartramia 20.Ex.1
Behen Moench 11.Ex.12
– *"behen"* 11.Ex.12
– *vulgaris* Moench 11.Ex.12
Belladonna Sweet H.6.Ex.2
Berberis L. 14.Ex.3
Blandfordia grandiflora R. Br. 52.Ex.10
Blephilia ciliata 33.Ex.2
Boletellus Murrill Rec.62A.Ex.1
Boletus L. : Fr. Rec.62A.Ex.1
– *piperatus* Bull. : Fr. Rec.50E.Ex.2
– *ungulatus* Schaeff. 23.Ex.9
– *"vicesimus sextus"* (Schaeffer, 1763) 23.Ex.9
Bouchea Cham. Rec.60B.Ex.1
Bougainvillea 60.Ex.5

Brachystelma R. Br. 46.Ex.5
Braddleya Vell. 53.Ex.6
Bradlea Adans. 53.Ex.6
Bradleja Banks ex Gaertn. 53.Ex.6
Brassavola R. Br. H.6.Ex.5
Brassica L. 18.5, H.9.Ex.3
– *campestris* L. H.9.Ex.3
– *napus* 53.Ex.9
– *nigra* (L.) W. D. J. Koch 23.Ex.4
Brassicaceae 18.5
Brazzeia Baill. 34.Ex.7
"tribus *Brevipedunculata*" (Huth, 1895) 33.Ex.12
Bromeliineae 17.Ex.2
Bromus inermis subsp. *pumpellianus* (Scribn.) Wagnon 34.Ex.12
– – var. *pumpellianus* (Scribn.) C. L. Hitchc. 34.Ex.12
– *pumpellianus* Scribn. 34.Ex.12
– *sterilis* L. 14.Ex.8
Brosimum Sw. 34.Ex.9

Cacalia napaeifolia DC., not "*napeaefolia*" 60.Ex.12
Cactaceae Juss. 22.Ex.2
Cactus L. 22.Ex.2, 32.Ex.7
– [unranked] *Melocactus* L. 22.Ex.2
– *ficus-indica* L. 32.Ex.7
– *mammillaris* L. 22.Ex.2
– *melocactus* L. 22.Ex.2
– *opuntia* L. 32.Ex.7
Cainito Adans. 52.Ex.1
Calandrinia 58.Ex.2
– *polyandra* Benth., not "(Hook.) Benth." 58.Ex.2
"*Calicium debile* Turn. and. Borr. Mss." (Smith, 1812, pro syn.) 46.Ex.9
Callicarpa L. 62.Ex.2
Callistemon Rec.60G.1(c)
Callixene Comm. ex Juss. 14.Ex.5
Calluna Salisb. H.9.Ex.2
– *vulgaris* (L.) Hull H.9.Ex.2
Calothyrsus Spach 11.Ex.5
Calyptridium Nutt. 46.Ex.4
– *monandrum* Nutt. 46.Ex.4
Cambogia gummi-gutta L., not "*G. gutta*" 23.Ex.14
Camellia L. 13.Ex.3
Campanopsis (R. Br.) Kuntze 11.Ex.1
Campanula sect. *Campanopsis* R. Br. 11.Ex.1
Candida populi Hagler & al. 8.Ex.1
Cardamine L. 11.Ex.14
Cardaminum Moench 14.Ex.4
Carex L. 35.Ex.2
– sect. *Eucarex* 21.Ex.1
– [unranked] *Scirpinae* Tuck. 35.Ex.2
– sect. *Scirpinae* (Tuck.) Kük. 35.Ex.2
– "*bebbii*" (Olney, 1871) Rec.50B.Ex.1
Carphalea Juss. 41.Ex.2
Caryophyllaceae Juss. 18.Ex.4, 19.Ex.5, 53.Ex.11
Caryophylloideae (Juss.) Rabeler & Bittrich. 19.Ex.5
Caryophyllus Mill. non L. 18.Ex.4, 19.Ex.5
Cassipourea Aubl. 14.Ex.2
Castanella Spruce ex Benth. & Hook. f. 10.Ex.2
– *granatensis* Triana & Planch. 10.Ex.2
Cathaya Chun & Kuang 53.Ex.12
Cathayeia Ohwi 53.Ex.12
Cattleya Lindl. H.6.Ex.5, H.6.Ex.6
Cedrus Duhamel 52.Ex.7
Cedrus Trew, *nom. cons.*, non Duhamel 62.Ex.1
Celsia sect. *Aulacospermae* Murb. 53.Ex.16
Cenomyce ecmocyna Ach. 48.Ex.3
Centaurea amara L. 47.Ex.3
– *jacea* L. 47.Ex.3
Centrospermae 17.Ex.1
Cephaëlis 60.6
Cephalotos Adans. 53.Ex.10
Cephalotus Labill. 53.Ex.10
Cercospora aleuritidis Miyake 59.Ex.5
Cereus jamacaru DC., not "*mandacaru*" 60.Ex.6
Cervicina Delile 11.Ex.1
Chamaecyparis Spach H.6.Ex.1
Chloris *52.Ex.11*
– radiata (L.) Sw. 52.Ex.11
Chlorosarcina Gerneck 7.Ex.6
– *elegans* 7.Ex.6
– *minor* 7.Ex.6
Chlorosphaera G. A. Klebs 7.Ex.6
Christella H. Lév. 53.Ex.12
Chrysophyllum L. 52.Ex.1
– *cainito* L. 52.Ex.2
– *sericeum* Salisb. 52.Ex.2
Cichoriaceae Juss. 19.Ex.4
Cichorieae D. Don 19.Ex.4
Cichoriinae Sch. Bip. 19.Ex.4

Cichorioideae W. D. J. Koch 19.Ex.4
Cichorium L. 19.Ex.4
Cineraria sect. *Eriopappus* Dumort. 49.Ex.3
Cistus aegyptiacus L. 49.Ex.4
Cladonia ecmocyna Leight. 48.Ex.3
Claudopus Gillet 11.Ex.15
Clianthus 11.Ex.19
– *dampieri* Lindl. 11.Ex.19
– *formosus* (D. Don) Ford & Vickery 11.Ex.19
– *oxleyi* Lindl. 11.Ex.19
– *speciosus* (D. Don) Asch. & Graebn., non (Endl.) Steud. 11.Ex.19
Climacioideae Grout, not *"Climacieae"* 19.Ex.6
Clusia L. 18.5
Clusiaceae 18.5
Clutia L., not *"Cluytia"* 60.Ex.9
"Clypeola minor" (Linnaeus, 1756) 45.Ex.1
Cnidium peucedanoides Kunth 33.Ex.2
Cochlioda Lindl. H.6.Ex.6
Codium geppiorum O. C. Schmidt, not *"geppii"* 60.Ex.20
Coeloglossum viride (L.) Hartm. H.11.Ex.2
Coffea 59.Ex.1
×*Cogniauxara* Garay & H. R. Sweet H.8.Ex.2
Coix lacryma-jobi L. 60.Ex.14
Collaea Rec.60B.1(a)
Collema cyanescens Rabenh. 58.Ex.3
– *tremelloides* var. *caesium* Ach. 58.Ex.3
– – var. *cyanescens* Ach. 58.Ex.3
Columella Lour. 53.Ex.10
Columellia Ruiz & Pav. 53.Ex.10
Colura (Dumort.) Dumort. 53.Ex.12
Coluria R. Br. 53.Ex.12
Combretum Loefl. Rec.50E.Ex.1
Comparettia Poepp. & Endl. H.6.Ex.6
Compositae Giseke 18.5, 19.Ex.4
Coniferopsida 3.Ex.1
Conophytum N. E. Br., not "Haw." 34.Ex.3
Convolvulus L. 20.Ex.1, 35.Ex.1
– [unranked] *"Occidentales"* (House, 1908) 35.Ex.1
– [unranked] *"Sepincoli"* (House, 1908) 35.Ex.1
– [unranked] *"Soldanellae"* (House, 1908) 35.Ex.1
Cornus "gharaf" (Forsskål, 1775) 23.Ex.8
– *sanguinea* 23.Ex.1
Correa Rec.60B.1(a)
Corticium microsclerotium G. F. Weber, non (Matz) G. F. Weber 59.Ex.6
Corydalis DC. 49.Ex.5
– *solida* (L.) Clairv., not "(Mill.) Clairv." 49.Ex.5
Coscinodiscaceae Kütz., *"Coscinodisceae"* 18.Ex.5
Costus subg. *Metacostus* 21.Ex.1
"×*Crindonna*" (Ragionieri, 1921) H.6.Ex.2
Crinum L. H.6.Ex.2
Cristella Pat. 53.Ex.12
Croton ciliatoglandulifer Ortega, not *"ciliato-glandulifer"* 60.Ex.13
Cruciferae 18.5
Cucubalus angustifolius Mill. 52.Ex.5
– *behen* L. 11.Ex.12, 52.Ex.5
– *latifolius* Mill. 52.Ex.5
×*Cupressocyparis* Dallim. H.6.Ex.1
Cupressus L. H.6.Ex.1
Curculigo Gaertn., not *"Cvrcvligo"* 60.Ex.7
Cuviera DC., *nom. cons.*, non Koeler 52.Ex.12
Cyanobacteria Pre.7
Cylindrocladiella infestans Boesw. 33.Ex.4
Cylindrocladium "infestans" (Peerally, 1991) 33.Ex.4
Cymbidium iansonii Rolfe, not *"i'ansonii"* 60.Ex.16
Cymbopogon martini (Roxb.) W. Watson 32.Ex.6
Cyperaceae Juss., "ordo *Cyperoideae*" 18.Ex.3
Cyperus heyneanus Rec.60C.1(c)
Cytisus Desf. 11.Ex.9
– *biflorus* L'Hér. 11.Ex.9
– *fontanesii* Spach 11.Ex.9

×*Dactyloglossum mixtum* (Asch. & Graebn.) Rauschert H.11.Ex.2
Dactylorhiza fuchsii (Druce) Soó H.11.Ex.2
Dadoxylon Endl. 3.Ex.1
Damapana Adans. 14.Ex.10
Delphinium L. 10.Ex.6, 33.Ex.12
– "tribus *Brevipedunculata*" (Huth, 1895) 33.Ex.12
– "tribus *Involuta*" (Huth, 1895) 33.Ex.12
– *consolida* L. 10.Ex.6
– *peregrinum* L. 10.Ex.6

Dendromecon Benth. 62.Ex.4
Dendrosicus Raf. 14.Ex.6
Dentaria L. 11.Ex.14
Desmidiaceae 13.1(e)
Desmodium griffithianum Rec.60C.1(d)
Desmostachya (Stapf) Stapf 53.Ex.9
Desmostachys Miers 53.Ex.9
Dianthus monspessulanus 23.Ex.1
Dichelodontium Hook. f. & Wilson ex Broth. 46.Ex.11
– *nitidulum* (Hook. f. & Wilson) Broth. 46.Ex.11
Didymopanax gleasonii Britton & Wilson Rec.46C.Ex.1
Digitalis grandiflora L. H.3.Ex.3
– *mertonensis* B. H. Buxton & C. D. Darl. H.3.Ex.3
– *purpurea* L. H.3.Ex.3
Dillenia Rec.60B.1(c)
Dionysia Fenzl 11.Ex.6
– sect. *Ariadne* Wendelbo 11.Ex.6
– sect. *Dionysiopsis* (Pax) Melch. 11.Ex.6
Diospyros L. 62.Ex.1
Dipterocarpus C. F. Gaertn. 62.Ex.2
Dodecatheon sect. *"Etubulosa"* (Kunth, 1905) 22.Ex.1
– *meadia* L. 22.Ex.1
Donia formosa D. Don 11.Ex.19
– *speciosa* D. Don 11.Ex.19
Dracunculus Mill. 11.Ex.11
– *vulgaris* Schott 11.Ex.11
Drimys J. R. Forst. & G. Forst. 18.Ex.4
Drypeteae (Pax) Hurus. Rec.19A.Ex.1
Drypetinae Pax Rec.19A.Ex.1
Durvillaea Bory 53.Ex.9, Rec.60B.Ex.1
Dussia Krug & Urb. ex Taub. 6.Ex.1
– *martinicensis* Krug & Urb. ex Taub. 6.Ex.1

Eccilia (Fr. : Fr.) P. Kumm. 11.Ex.15
"Echii altera species" (Dodonaeus, 1583) 7.Ex.5
Echium lycopsis L. 7.Ex.5
Eclipta erecta L. 11.Ex.18
– *prostrata* (L.) L. 11.Ex.18
Ectocarpus mucronatus D. A. Saunders 33.Ex.3
"Egeria" (Néraud, 1826) 32.Ex.1
Elcaja "roka" (Forsskål, 1775) 23.Ex.8
"Elodes" (Clusius, 1601) 10.Ex.3
Elodes Adans. 10.Ex.3
×*Elyhordeum* Mansf. ex Tsitsin & Petrova H.8.Ex.1
×*Elymopyrum* Cugnac 11.Ex.28
×*Elymotriticum* P. Fourn. H.8.Ex.1
Elymus L. 11.Ex.28, Rec.20A.Ex.1, H.3.Ex.2, H.8.Ex.1
– *europaeus* L. 52.Ex.12
– *farctus* (Viv.) Melderis H.5.Ex.1
– – subsp. *boreoatlanticus* (Simonet & Guin.) Melderis H.5.Ex.1
– ×*laxus* (Fr.) Melderis & D. C. McClint. H.5.Ex.1
– *repens* (L.) Gould H.5.Ex.1
Embelia sarasiniorum 23.Ex.1
Enallagma Baill. 14.Ex.6
Enantioblastae 17.Ex.1
Enargea Banks ex Gaertn. 14.Ex.5
Englerastrum Briq. Rec.60B.Ex.1
Englerella Pierre Rec.60B.Ex.1
Engleria O. Hoffm. Rec.60B.Ex.1
Entoloma (Fr. ex Rabenh.) P. Kumm. 11.Ex.15
Epiphyllum Haw. H.6.Ex.1
Equisetum palustre var. *americanum* 6.Ex.2
– – f. *fluitans* 6.Ex.2
Erica L. 19.Ex.3, Rec.19A.Ex.2, H.9.Ex.2
– *cinerea* L. H.9.Ex.2
Ericaceae Juss. 19.Ex.3, Rec.19A.Ex.2
×*Ericalluna* Krüssm. H.9.Ex.2
– *bealei* Krüssm. H.9.Ex.2
Ericeae D. Don 19.Ex.3
Ericoideae Endl. 19.Ex.3
Erigeron L. 62.Ex.1
Eryngium nothosect. *Alpestria* Burdet & Miège pro sect. H.9.Ex.1
– sect. *Alpina* H. Wolff H.9.Ex.1
– sect. *Campestria* H. Wolff H.9.Ex.1
– *amorginum* Rech. f. Rec.60D.Ex.1
Erysimum hieraciifolium var. *longisiliquum* Rouy & Foucard 53.Ex.18
Erythrina poeppigiana (Walp.) O. F. Cook 34.Ex.6
– "*micropteryx* Poepp." 34.Ex.6
Eschweilera DC. 53.Ex.5
Eschweileria Boerl. 53.Ex.5
Euanthe Schltr. H.8.Ex.2
– *sanderiana* (Rchb.) Schltr. H.8.Ex.2
Eucalyptus L'Hér. 62.Ex.1
Eulophus 33.Ex.2

– *peucedanoides* 33.Ex.2
Eunotia gibbosa Grunow 44.Ex.2
Eupenicillium brefeldianum (B. O. Dodge) Stolk & D. B. Scott 59.Ex.3
Euphorbia amygdaloides L. H.5.Ex.2
– subg. *Esula* Pers. 22.Ex.5
– subsect. *Tenellae* 21.Ex.1
– sect. *Tithymalus* 21.Ex.1
– *characias* L. H.5.Ex.2
– – subsp. *wulfenii* (W. D. J. Koch) Radcl.-Sm. H.5.Ex.2
– ×*cornubiensis* Radcl.-Sm. H.5.Ex.2
– *esula* L. 22.Ex.5
– *"jaroslavii"* (Poljakova, 1953) 34.Ex.10
– ×*martini* Rouy H.5.Ex.2
– – nothosubsp. *cornubiensis* (Radcl.-Sm.) Radcl.-Sm H.5.Ex.2
– *peplis* L. 53.Ex.9
– *peplus* L. 22.Ex.5, 53.Ex.9
– *wulfenii* W. D. J. Koch H.5.Ex.2
– *yaroslavii* Poljak. 34.Ex.10
Euphorbiaceae Rec.19A.Ex.1
Excoecaria 11.Ex.17

Faba Mill. 18.5
Fabaceae 18.5, 19.7
Faboideae 19.7
Fagaceae 3.Ex.2
Fagus sylvatica L., not *"silvatica"* 60.Ex.1
Farinosae 17.Ex.1
Festuca myuros L. 26.Ex.3
Ficus "exasperata" auct., non Vahl Rec.50D.Ex.1
– *gameleira* Standl. 53.Ex.11
– *gomelleria* Kunth 53.Ex.11
– *irumuënsis* De Wild. Rec.50D.Ex.1
– *neoëbudarum* Summerh., not *"neo-ebudarum"* 60.Ex.13
– *stortophylla* Warb. Rec.50D.Ex.1
Filago 20.Ex.1
Flacourtiaceae 53.Ex.12
Fucales 17.Ex.2
Fuirena Rottb. 20.Ex.9
– *umbellata* Rottb. 43.Ex.4
Fumaria bulbosa var. *solida* L. 49.Ex.5
– *densiflora* 40.Ex.3
– *gussonei* 23.Ex.1
– *officinalis* 40.Ex.3
– "×*salmonii*" (Druce, 1908) 40.Ex.3
– *solida* (L.) Mill. 49.Ex.5
Fungi 13.1(d), 53.Ex.12
Fusarium stilboides Wollenw. 59.Ex.1

Galium tricorne Stokes 52.Ex.6
– *tricornutum* Dandy 52.Ex.6
Gasteromycetes 13.1(d)
×*Gaulnettya* Marchant 11.Ex.26
Gaultheria L. 11.Ex.26
×*Gaulthettya* Camp 11.Ex.26
Geaster Fr. 61.Ex.1
Geastrum Pers. : Pers. 61.Ex.1
– *hygrometricum* Pers., not *"Geastrvm hygrometricvm"* 60.Ex.8
Gentiana lutea 6.Ex.2
– *tenella* var. *occidentalis* 6.Ex.2
Geranium robertianum 23.Ex.1
Gerardia L. Rec.60B.Ex.1
Gerardiina Engl. 53.Ex.9
Gerrardina Oliv. 53.Ex.9
Gibberella stilboides W. L. Gordon & C. Booth 59.Ex.1
Giffordia 33.Ex.3
– *mucronata* (D. A. Saunders) Kjeldsen & Phinney 33.Ex.3
Gigartina cordata var. *splendens* (Setch. & N. L. Gardner) D. H. Kim 7.Ex.3
Ginkgo 18.Ex.2
Ginkgoaceae 18.Ex.2
Gleditsia L., not *"Gleditschia"* 60.Ex.9
Globba trachycarpa Baker, not *"brachycarpa"* 60.Ex.2
Globularia cordifolia L. excl. var. (emend. Lam.) Rec.47A.Ex.1
Gloeosporium balsameae Davis 23.Ex.4
Gloriosa 20.Ex.1
Gluta renghas L., not *"benghas"* 45.Ex.3, 60.Ex.3
Gnaphalium "fruticosum flavum" (Forsskål, 1775) 23.Ex.10
Gossypium tomentosum Nutt. ex Seem. 46.Ex.13
Graderia Benth. Rec.60B.Ex.1
Gramineae 18.5, 19.Ex.2, 53.Ex.9
Graphis meridionalis Nakan. 45.Ex.2
Grislea L. Rec.50E.Ex.1
Guttiferae 18.5
Gymnadenia R. Br. H.6.Ex.1, H.6.Ex.3
×*Gymnanacamptis* Asch. & Graebn. H.6.Ex.1

Hedysarum 20.Ex.1
Helianthemum Mill. 49.Ex.4
– *aegyptiacum* (L.) Mill. 49.Ex.4
– *italicum* var. *micranthum* Gren. & Godr. 11.Ex.13
– *penicillatum* Thibaud ex Dunal 11.Ex.13
– – var. *micranthum* (Gren. & Godr.) Grosser 11.Ex.13
Helicosporium elinorae Linder 34.Ex.13
Helleborus niger L. 23.Ex.4
Hemerocallis L. 62.Ex.1
– *flava* (L.) L. 11.Ex.4
– *fulva* (L.) L. 11.Ex.4
– *lilioasphodelus* L. 11.Ex.4
– – var. *flava* L. 11.Ex.4
Hemisphace (Benth.) Opiz 32.Ex.5, 46.Ex.10
Hepaticae 13.1(c), 13.Ex.1, 53.Ex.12
Heracleum sibiricum L. 11.Ex.20
– – subsp. *lecokii* (Godr. & Gren.) Nyman 11.Ex.20
– – subsp. *sibiricum* 11.Ex.20
– *sphondylium* L. 11.Ex.20
– – subsp. *sibiricum* (L.) Simonk. 11.Ex.20
– – "subsp. *lecokii*" 11.Ex.20
Hesperomecon Greene 62.Ex.4
Hetaeria alta Ridl., not *"alba"* 60.Ex.2
Heuchera
– ×*tiarelloides* Lemoine & E. Lemoine H.11.Ex.1
×*Heucherella tiarelloides* (Lemoine & E. Lemoine) H. R. Wehrh. H.11.Ex.1
×*Holttumara* anon. H.8.Ex.2
Hordelymus (K. Jess.) K. Jess. 20A.Ex.1, 52.Ex.12, H.3.Ex.2, H.8.Ex.1
×*Hordelymus* Bachteev & Darevsk. non (K. Jess.) K. Jess. H.3.Ex.2, H.8.Ex.1
×*Hordeopyron* Simonet, not *"Hordeopyrum"* H.9.Ex.1
Hordeum L. Rec.20A.Ex.1, H.3.Ex.2H.8.Ex.1, H.9.Ex.1
– subg. *Hordelymus* K. Jess 52.Ex.12
Hyacinthus non-scriptus L., not *"non scriptus"* 23.Ex.11
Hydrocoleum glutinosum (C. Agardh) ex Gomont 46.Ex.19
Hydrophyllum Rec.60G.1(c)
Hymenocarpos Savi 62.Ex.2
Hyoseridinae Less. 19.Ex.4
Hyoseris L. 19.Ex.4
"Hypericum" (Tournefort, 1700) 10.Ex.3
Hypericum aegypticum L. 10.Ex.3
– *elodes* L. 10.Ex.3
Hypholoma (Fr. : Fr.) P. Kumm. 32.Ex.8
Hypnum crassinervium Wilson 46.Ex.8
Hypomyces chrysospermus Tul. 59.Ex.7

"Ibidium" (Salisbury, 1812) 34.Ex.8
Ifloga 20.Ex.1
Impatiens 20.Ex.1
– *noli-tangere* L., not *"noli tangere"* 23.Ex.1, 23.Ex.11
Indigofera longipedunculata Y. Y. Fang & C. Z. Zheng, not *"longipednnculata"* 60.Ex.4
Ionopsis Kunth H.6.Ex.6
Iria (Pers.) Hedw. 53.Ex.9
Iridaea cordata var. *splendens* (Setch. & N. L. Gardner) I. A. Abbott 7.Ex.3
– *splendens* (Setch. & N. L. Gardner) Papenf. 7.Ex.3
Iridophycus splendens Setch. & N. L. Gardner 7.Ex.3
Iris L. 53.Ex.9
Isoëtes 60.6

Juncus bufonius "var. *occidentalis*" (Hermann, 1975) 33.Ex.10
– *"sphaerocarpus"* auct. Am., non Nees 33.Ex.10
Juniperus L. 52.Ex.7

Kedarnatha P. K. Mukh. & Constance 42.Ex.2
– *sanctuarii* P. K. Mukh. & Constance 42.Ex.2
Kernera Rec.60B.1(b)
Kratzmannia Opiz 32.Ex.4
Kyllinga Rottb. 20.Ex.9

Labiatae 18.5, 53.Ex.1
"Labyrinthodyction" (Valkanov, 1969) 45.Ex.6
Lactuca L. 19.Ex.4
Lactuceae Cass. 19.Ex.4
Laelia Lindl. H.6.Ex.5, H.6.Ex.6
Lamiaceae 18.5
Lamium L. 18.5
"Lanceolatus" (Plumstead, 1952) 20.Ex.4
Lapageria Ruiz & Pav. H.9.Ex.1

Lapeirousia Pourr. Rec.60B.Ex.1
– *erythrantha* var. *welwitschii* (Baker) Geerinck & al. Rec.46C.Ex.2
Lasiobelonium corticale (Pers.) Raitv. 33.Ex.7
Lasiosphaeria elinorae Linder 34.Ex.13
Lecanidion Endl. 15.Ex.2
Lecanora campestris "f. *pseudistera*" (Grummann, 1963) 33.Ex.8
– *pseudistera* Nyl. 33.Ex.8
Leguminosae 18.5, 19.7, 53.Ex.9
Lemanea Sirodot, non Bory 46.Ex.12, 48.Ex.1
Lepidocarpaceae 3.Ex.2
Lepidocarpon D. H. Scott 3.Ex.2
Lepidodendrales 3.Ex.1
Leptogium cyanescens (Rabenh.) Körb., not "(Ach.) Körb.", nor "(Schaer.) Körb." 58.Ex.3
Leptonia (Fr. : Fr.) P. Kumm. 11.Ex.15
"Leptostachys" 20.4.Note.2
Lespedeza Michx. 60.Ex.6
Lesquerella lasiocarpa (Hook. ex A. Gray) S. Watson 11.Ex.22
– – subsp. *berlandieri* (A. Gray) Rollins & E. A. Shaw 11.Ex.22
– – subsp. *lasiocarpa* 11.Ex.22
– – var. *berlandieri* (A. Gray) Payson 11.Ex.22
– – var. *hispida* (S. Watson) Rollins & E. A. Shaw 11.Ex.22
Leucadendron 14.Ex.9
×*Leucadenia* Schltr. H.6.Ex.3
Leucodon nitidulus Hook. f. & Wilson 46.Ex.11
Leucorchis E. Mey. H.6.Ex.3
Lichen debilis Sm. 46.Ex.9
– *gracilis* L. 48.Ex.3
Lilium tianschanicum N. A. Ivanova ex Grubov 46.Ex.15
Linaria Mill. 11.Ex.7
– *spuria* (L.) Mill. 11.Ex.7
– *"linaria"* 23.Ex.3
Lindera Thunb., non Adans. Rec.50C.Ex.1
Linum multiflorum Lam. 58.Ex.1
– *radiola* L. 58.Ex.1
Liquidambar 20.Ex.1
Lithocarpus polystachyus (Wall. ex A. DC.) Rehder 46.Ex.14
"Lobata" (Chapman, 1952) 20.Ex.4
Lobelia spicata Lam. 26.Ex.1
– – "var. *originalis*" (McVaugh, 1936) 24.Ex.3
– – var. *spicata* 26.Ex.1
Loranthus (sect. *Ischnanthus*) *gabonensis* Rec.21A.Ex.1
– *macrosolen* Steud. ex Rich., not "Steud." 32.Ex.2
Lotus L. 62.Ex.1
Lupinus Tourn. ex L. 46.Ex.18
Luzuriaga Ruiz & Pav. 14.Ex.5
Lycium odonellii F. A. Barkley, not *"o'donellii"* 60.Ex.16
Lycoperdon atropurpureum Vittad., not *"atro-purpureum"* 60.Ex.13
Lycopersicon esculentum Mill. 14.Ex.1
– *lycopersicum* (L.) H. Karst. 14.Ex.1
Lycopodium L. 13.Ex.2
– *clavatum* L. 13.Ex.2
– *inundatum* L. 26.Ex.4
– – var. *inundatum* 26.Ex.4
– – var. *bigelovii* Tuck. 26.Ex.4
"Lycopsis" (Ray, 1724) 7.Ex.5
Lyngbya Gomont 53.Ex.10
– *glutinosa* C. Agardh 46.Ex.19
Lyngbyea Sommerf. 53.Ex.10
Lysimachia hemsleyana Oliv. Rec.23A.2, 53.Ex.9
– *hemsleyi* Franch. Rec.23A.2, 53.Ex.9
Lythrum intermedium Ledeb. 11.Ex.3
– *salicaria* L. 11.Ex.3
– – var. *glabrum* Ledeb. 11.Ex.3
– – var. *intermedium* (Ledeb.) Koehne 11.Ex.3

Macrothyrsus Spach 11.Ex.5
Magnolia foetida (L.) Sarg. 11.Ex.2
– *grandiflora* L. 11.Ex.2
– *virginiana* var. *foetida* L. 11.Ex.2
Mahonia Nutt. 14.Ex.3
Malpighia L. 22.Ex.3
– sect. *Apyrae* DC. 22.Ex.3
– subg. *Homoiostylis* Nied. 22.Ex.3
– sect. *Malpighia* 22.Ex.3
– subg. *Malpighia* 22.Ex.3
– *glabra* L. 22.Ex.3
Maltea B. Boivin H.6.Ex.4
Malvastrum bicuspidatum subsp. *tumidum* S. R. Hill 34.Ex.11
– – var. *tumidum* S. R. Hill 34.Ex.11

Malvineae 17.Ex.2
Manihot Mill. 20.Ex.1, 62.Ex.8
Martia Spreng. Rec.60B.Ex.1
Martiusia Schult. & Schult. f. Rec.60B.Ex.1
Maxillaria mombachoënsis A. H. Heller ex J. T. Atwood 46.Ex.24
Mazocarpon M. J. Benson 3.Ex.2
Medicago orbicularis (L.) Bartal. 49.Ex.1
– *polymorpha* L. 26.Ex.2
– – var. *orbicularis* L. 49.Ex.1
– – var. *hispida* L. 26.Ex.2
Melilotus Rec.60G.1(c), 62.Ex.1
Meliola 33.Ex.8
– *albiziae* Hansford & Deighton, not *"albizziae"* Rec.60H.Ex.1
Meliosma Rec.60G.1(c)
Mentha 51.Ex.1
– *aquatica* L. H.2.Ex.1, H.11.Ex.3
– *arvensis* L. H.2.Ex.1
– ×*piperita* f. *hirsuta* Sole H.12.Ex.1
– – nothosubsp. *piperita* H.11.Ex.3
– – nothosubsp. *pyramidalis* (Ten.) Harley H.11.Ex.3
– ×*smithiana* R. A. Graham H.3.Ex.1
– *spicata* L. H.2.Ex.1
– – subsp. *spicata* H.11.Ex.3
– – subsp. *tomentosa* (Briq.) Harley H.11.Ex.3
Mesembryanthemum L., not *"Mesembrianthemum"* 60.Ex.1
– sect. *Minima* Haw. 34.Ex.3
Mespilodaphne mauritiana Meisn. 55.Ex.1
Mespilus 33.Ex.5
– *arbutifolia* var. *nigra* Willd. 33.Ex.5
Metasequoia Hu & W. C. Cheng, non Miki 11.Ex.24
– *disticha* (Heer) Miki 11.Ex.24
– *glyptostroboides* Hu & W. C. Cheng 11.Ex.24
Micromeria benthamii Webb & Berthel. H.10.Ex.3
– ×*benthamineolens* Svent. H.10.Ex.3
– *pineolens* Svent. H.10.Ex.3
Micropteryx poeppigiana Walp. 34.Ex.6
Mimosa cineraria L. 53.Ex.17
– *cinerea* L. 1753, No. 10, non L. 1753 No. 25 53.Ex.17
"Minthe" 51.Ex.1
Minuartia 11.Ex.10
– *stricta* (Sw.) Hiern 11.Ex.10
Monarda ciliata L. 33.Ex.2
Monochaete Döll 53.Ex.9
Monochaetum (DC.) Naudin 53.Ex.9
Monotropeae D. Don Rec.19A.Ex.2
Monotropoideae (D. Don) A. Gray Rec.19A.Ex.2
Montia parvifolia (DC) Greene 25.Ex.1
– – subsp. *flagellaris* (Bong.) Ferris 25.Ex.1
– – subsp. *parvifolia* 25.Ex.1
Mouriri subg. *Pericrene* 6.Ex.2
Mucor chrysospermus (Bull.) Bull. 59.Ex.7
Musci 13.1.(b), 13.Ex.1, 13.Ex.2
Mussaenda frondosa L., not *"fr. [fructu] frondoso"* 23.Ex.13
Mycosphaerella aleuritidis S. H. Ou59.Ex.5
Myogalum boucheanum Kunth 34.Ex.5
Myosotis L. 47.Ex.2, Rec60G.3(b)
Myrcia laevis O. Berg, non G. Don 7.Ex.1
– *lucida* McVaugh 7.Ex.1
Myrtaceae 53.Ex.11

Napaea L. 53.Ex.9, 60.Ex.12
"Napea" 60.Ex.12
Narcissus pseudonarcissus L., not *"Pseudo Narcissus"* 23.Ex.12
Nasturtium Mill. 14.Ex.4
Nasturtium R. Br., *nom cons.,* non Mill. 14.Ex.4
– *"nasturtium-aquaticum"* 23.Ex.3
Nelumbo 18.Ex.2
Nelumbonaceae 18.Ex.2
Neoptilota Kylin 36.Ex.4
Neotysonia phyllostegia (F. Muell.) Paul G. Wilson 46.Ex.6
Nepeta "×*faassenii*" "Bergmans", non Lawrence" 40.Ex.1
Neves-armondia K. Schum. 20.Ex.7
Nolanea (Fr. : Fr.) P. Kumm. 11.Ex.15
Nostocaceae 13.1(e)
Nothotsuga Hu ex C. N. Page 46.Ex.23

Odontoglossum Kunth H.6.Ex.6
Oedogoniaceae 13.1(e)
Oenothera depressa Greene H.4.Ex.1
– ×*hoelscheri* Renner ex Rostański H.4.Ex.1
– *rubricaulis* Kleb. H.4.Ex.1
– ×*wienii* Renner ex Rostański H.4.Ex.1

Omphlaria nummularia Durieu & Mont. 42.Ex.1
Oncidium Sw. H.6.Ex.6
Oplopanax (Torr. & A. Gray) Miq. 62.Ex.3
Opuntia Mill. 32.Ex.7
– *ficus-indica* (L.) Mill. 32.Ex.7
– *vulgaris* Mill. 32.Ex.7
"*Opuntia vulgo herbariorum*" (Bauhin & Cherler, 1650-1651) 32.Ex.7
×*Orchicoeloglossum mixtum* Asch. & Graebn. H.11.Ex.2
Orchis L. 62.Ex.1
– *fuchsii* Druce H.11.Ex.2
Ormocarpum P. Beauv. 62.Ex.2
Ornithogalum
– *boucheanum* (Kunth) Asch. 34.Ex.5
– "*undulatum* hort. Bouch." (Kunth, 1843) 34.Ex.5
Orobanche artemisiae 51.Ex.1
– *artemisiepiphyta* 51.Ex.1
– *columbariae* 51.Ex.1
– *columbarihaerens* 51.Ex.1
– *rapum* 51.Ex.1
– *sarothamnophyta* 51.Ex.1
"sectio *Orontiaceae*" (Brown, 1810) 33.Ex.11
Osbeckia L. 53.Ex.2
Ostrya virginiana (Mill.) K. Koch Rec.60D.Ex.1
Ottoa Rec.60B.1(a)

Palmae 18.5, 53.Ex.12
Panax nossibiensis Drake 44.Ex.1
Papaver rhoeas 23.Ex.1
Papilionaceae 18.5, 19.7
Papilionoideae 19.7
Parietales 17.Ex.1
Parmelia cyanescens (Pers.) Ach., non Schaer. 58.Ex.3
Patellaria Fr. : Fr., non Hedw. 15.Ex.2
Paullinia paullinioides Radlk. 10.Ex.2
Pavia Mill. 11.Ex.5
Pecopteris (Brongn.) Sternb. 3.Ex.1
Peltophorum (Vogel) Benth. 53.Ex.9
Peltophorus Desv. 53.Ex.9
Penicillium brefeldianum B. O. Dodge 59.Ex.3
Peperomia san-felipensis J. D. Smith 60.Ex.14
Peponia Grev. 53.Ex.9
Peponium Engl. 53.Ex.9
Pereskia opuntiiflora DC., not "*opuntiaeflora*" 60.Ex.11
Peridermium balsameum Peck 23.Ex.4
Pernettya Gaudich. 11.Ex.26
Petalodinium Cachon & Cachon-Enj. 45.Ex.5
Petrophiloides Bowerb. 11.Ex.23
Petrosimonia brachiata (Pall.) Bunge 51.Ex.4
– *oppositifolia* (Pall.) Litv. 51.Ex.4
Peyrousea DC. Rec.60B.Ex.1
Peziza corticalis Pers., not "Mérat", nor "Fr." 33.Ex.7
"*Phaelypea*" (Browne, 1756) 42.Ex.4
×*Philageria* Mast. H.9.Ex.1
Philesia Comm. ex Juss. H.9.Ex.1
Philgamia Baill. 42.Ex.5
– *hibbertioides* Baill. 42.Ex.5
Phippsia (Trin.) R. Br. H.6.Ex.4
Phlox divaricata L. subsp. *divaricata* H.3.Ex.3
– – subsp. *laphamii* (A. W. Wood) Wherry H.3.Ex.3
– *drummondii* 'Sternenzauber' 28.Ex.1
– *pilosa* subsp. *ozarkana* Wherry H.3.Ex.3
Phlyctidia Müll. Arg. 43.Ex.2
– "*andensis*" (Müller, 1880) 43.Ex.2
– *boliviensis* (Nyl.) Müll. Arg. 43.Ex.2
– "*brasiliensis*" (Müller, 1880) 43.Ex.2
– "*hampeana*" (Müller, 1880) 43.Ex.2
– *ludoviciensis* Müll. Arg. 43.Ex.2
– "*sorediiformis*" (Müller, 1880) 43.Ex.2
Phlyctis andensis Nyl. 43.Ex.2
– *boliviensis* Nyl. 43.Ex.2
– *brasiliensis* Nyl. 43.Ex.2
– *sorediiformis* Kremp. 43.Ex.2
Pholiota (Fr. : Fr.) P. Kumm. 33.Ex.13
Phoradendron Nutt., not "*Phoradendrum*" 60.Ex.1
Phyllachora annonicola Chardon, not "*anonicola*" Rec.60H.Ex.1
Phyllanthus L. emend. Müll. Arg. Rec.47A.Ex.1
Phyllerpa prolifera var. *firma* Kütz., not "var. *Ph. firma*" 24.Ex.5
Physconia Poelt 10.Ex.4
Physospermum Cuss. 29.Ex.1
Phyteuma L. 62.Ex.1
Picea abies (L.) H. Karst. 52.Ex.4

– *excelsa* Link 52.Ex.4
Pinaceae 53.Ex.12
Pinus abies L. 52.Ex.4
– *excelsa* Lam. 52.Ex.4
– *mertensiana* Bong. 7.Ex.2
Piptolepis Benth. 42.Ex.3
– *phillyreoides* Benth. 42.Ex.3
Piratinera Aubl. 34.Ex.9
"Pirus mairei" Rec.50F.Ex.1
Pisocarpium Link 62.Ex.2
Planera aquatica J. F. Gmel., not "(Walter) J. F. Gmel." 43.Ex.3
Platycarya Siebold & Zucc. 11.Ex.23
Plectranthus L'Hér. 14.Ex.7
– *fruticosus* L'Hér. 14.Ex.7
– *punctatus* (L. f.) L'Hér. 14.Ex.7
Pleuripetalum T. Durand 53.Ex.5
Pleuropetalum Hook. f. 53.Ex.5
Plumbaginaceae 18.Ex.1
Plumbago, Plumbaginis 18.Ex.1
Pluteus Fr. 15.Ex.3
– *atricapillus* (Batsch) Fayod 15.Ex.3
– *cervinus* (Schaeff.) P. Kumm. 15.Ex.3
Poa L. 18.5, 19.Ex.2
Poaceae Barnhart 18.5, 19.Ex.2
Poëae R. Br. 19.Ex.2
"Polifolia" (Buxbaum, 1721) 60.Ex.11
Polycarpaea Lam. 62.Ex.2
Polycarpon L. 62.Ex.2
Polycnemum oppositifolium Pall. 51.Ex.4
Polygonales 17.Ex.2
Polygonum pensylvanicum L. Rec.60D.Ex.1
Polypodium australe Fée 52.Ex.13
– *filix-femina* L., not *"F. femina"* 23.Ex.14
– *filix-mas* L., not *"F. mas"* 23.Ex.14
– ×*font-queri* Rothm. 52.Ex.13
– *fragile* L., not *"F. fragile"* 23.Ex.14
– ×*shivasiae* Rothm. 52.Ex.13
– *vulgare* nothosubsp. *mantoniae* (Rothm.) Schidlay H.3.Ex.1
– – subsp. *prionodes* (Asch.) Rothm. 52.Ex.13, H.2.Ex.1
– – subsp. *vulgare* 52.Ex.13,. H.2.Ex.1
Polypogon Desf. H.2.Ex.1, H.6.Ex.1
– *monspeliensis* (L.) Desf. H.2.Ex.1
Pooideae Asch. 19.Ex.2
Populus ×*canadensis* var. *marylandica* (Poir.) Rehder H.12.Ex.1
– – var. *serotina* (R. Hartig) Rehder H.12.Ex.1
Porella pinnata L. 13.Ex.1
Potamogeton 18.Ex.1
Potamogetonaceae 18.Ex.1
Potentilla atrosanguinea Lodd. ex D. Don H.10.Ex.1
– *"atrosanguinea-pedata"* (Maund, 1833) H.10.Ex.1
– *pedata* Nestl. H.10.Ex.1
×*Potinara* Charlesworth & Co. H.6.Ex.5
Primula sect. *Dionysiopsis* Pax 11.Ex.6
Protea L. 1771, *nom. cons.,* non L. 1753 14.Ex.9, Rec.50E.Ex.1
– *cynaroides* (L.) L. 14.Ex.9
Protodiniferaceae Kof. & Swezy, not *"Protodiniferidae"* 45.Ex.7
×*Pseudadenia* P. F. Hunt H.6.Ex.3
Pseudelephantopus Rohr, *nom. cons.,* not *"Pseudo-elephantopus"* 60.Ex.15
"Pseudoditrichaceae" (Steere & Iwatsuki, 1974) 41.Ex.1
"Pseudoditrichum" (Steere & Iwatsuki, 1974) 41.Ex.1
–*"mirabile"* (Steere & Iwatsuki, 1974) 41.Ex.1
Pseudorchis Ség. H.6.Ex.3
Pseudo-salvinia Piton, not *"Pseudosalvinia"* 60.Ex.15
Psilotum truncatum R. Br. 52.Ex.8
– "*truncatum*" auct., or R. Br. pro parte 52.Ex.8
Pteridium aquilinum subsp. *caudatum* (L.) Bonap. Rec.26A.Ex.2
– – var. *caudatum* (L.) Sadeb. Rec.26A.Ex.2
Pteridophyta 13.1(a), 53.Ex.12
Pteridopsida 3.Ex.1
Pteris caudata L. Rec.26A.Ex.2
Ptilostemon Cass. 11.Ex.8
– sect. *Cassinia* Greuter H.7.Ex.1
– nothosect. *Platon* Greuter H.7.Ex.1
– sect. *Platyrhaphium* Greuter H.7.Ex.1
– nothosect. *Plinia* Greuter H.7.Ex.1
– sect. *Ptilostemon* H.7.Ex.1
– *chamaepeuce* (L.) Less. 11.Ex.8
– *muticus* Cass. 11.Ex.8
Puccinellia Parl. H.6.Ex.4
Puccinia Pers. : Pers. 59.Ex.2
×*Pucciphippsia* Tzvelev H.6.Ex.4
Pulsatilla montana subsp. *australis* (Heuff.) Zămelis 49.Ex.6

– – subsp. *dacica* Rummelsp. 49.Ex.6
– – var. *serbica* W. Zimm., not "(W. Zimm.) Rummelsp." 49.Ex.6
Pyroleae D. Don Rec.19A.Ex.2
Pyroloideae (D. Don) A. Gray Rec.19A.Ex.2
Pyrus 33.Ex.5
– *calleryana* Decne. Rec.50F.Ex.1
– *mairei* H. Lév., not "*Pirus*" Rec.50F.Ex.1

Quercus alba L. H.10.Ex.4
– ×*deamii* Trel. H.10.Ex.4
– *macrocarpa* Michx. H.10.Ex.4
– *muehlenbergii* Engelm. H.10.Ex.4
– *polystachya* A. DC. 46.Ex.14
Quisqualis L. 20.Ex.7

"*Radicula*" (Hill, 1756) 20.Ex.2
Radicula Moench 20.Ex.2
Radiola linoides Roth 58.Ex.1
– "*radiola*" (Karsten, 1882) 58.Ex.1
Raphidomonas F. Stein 16.Ex.1
Raphidophyceae Chadef. ex P. C. Silva 16.Ex.1
Rauia Nees & Mart. 53.Ex.12
Rauhia Traub 53.Ex.12
Ravenelia cubensis Arthur & J. R. Johnst. 59.Ex.4
Ravensara Sonn. 55.Ex.1
Renanthera Lour. H.8.Ex.2
Rhamnus L. 62.Ex.1
– sect. *Pseudofrangula* Grubov Rec.22A.Ex.1
– subg. *Pseudofrangula* (Grubov) Brizicky Rec.22A.Ex.1
– *alnifolia* L'Hér. Rec.22A.Ex.1
– *vitis-idaea* Burm. f., not "*vitis idaea*" 23.Ex.11
"*Rhaptopetalaceae*" (Pierre, 1897) 34.Ex.7
Rhaptopetalum Oliv. 34.Ex.7
Rheedia kappleri Eyma 9.Ex.1
Rheum "×*cultorum*" (Thorsrud & Reisaeter, 1948) 40.Ex.2
Rhizoctonia microsclerotia Matz 59.Ex.6
Rhododendreae Brongn. 19.Ex.3
Rhododendroideae Endl. 19.Ex.3
Rhododendron L. 19.Ex.3, 20.Ex.1, 22.Ex.4
– sect. *Anthodendron* 22.Ex.4
– subg. *Anthodendron* (Rchb.) Rehder 22.Ex.4
– subg. *Pentanthera* G. Don 22.Ex.4
– *luteum* Sweet 22.Ex.4
Rhodomenia Grev. 14.Ex.11
Rhodophyllaceae 18.Ex.1
Rhodophyllidaceae 18.Ex.1
Rhodophyllis, Rhodophyllidos 18.Ex.1
Rhodophyllus, Rhodophylli 18.Ex.1
Rhodophyta 53.Ex.12
Rhodora L. 19.Ex.3
Rhodoreae D. Don 19.Ex.3
Rhodymenia Grev., not "*Rhodomenia*" 14.Ex.11
Rhynchostylis Blume H.6.Ex.6
Richardia L. 51.Ex.5
Richardsonia Kunth 51.Ex.5
Ricinocarpos sect. *Anomodiscus* 21.Ex.1
×*Rodrettiopsis* Moir H.6.Ex.6
Rodriguezia Ruiz & Pav. H.6.Ex.6
Rorippa Scop. 14.Ex.4
Rosa L. 18.Ex.1, 19.Ex.1, 20.Ex.1, 46.Ex.1
– *canina* L. H.3.Ex.3
– *gallica* L. 46.Ex.1
– – var. *eriostyla* R. Keller 46.Ex.1
– – var. *gallica* 46.Ex.1
– *glutinosa* var. *leioclada* H. Christ 24.Ex.6
– *jundzillii* f. *leioclada* Borbás 24.Ex.6
– *pissardii* Carrière, not "*pissardi*", nor "*pissarti*" 60.Ex.17
– ×*toddiae* Wolley-Dod, not "×*toddii*" 60.Ex.18
– *webbiana* Rec.60C.1(d)
Rosaceae Juss. I18.Ex.1, 19.Ex.1, 46.Ex.1, 53.Ex.12
Roseae DC. 19.Ex.1
Rosoideae Endl. 19.Ex.1
Rubia L. 53.Ex.9
Rubus L. 53.Ex.9
– *amnicola* Blanch., not "*amnicolus*" 23.Ex.4
– *fanjingshanensis* L. T. Lu ex Boufford & al. 46.Ex.16
– *quebecensis* L. H. Bailey Rec.60D.Ex.1
Rutaceae 53.Ex.12
Sacheria Sirodot 48.Ex.1
Sadleria hillebrandii Rob. 33.Ex.9
– "*pallida*" (sensu Hillebrand, 1888), non Hook. & Arn. 33.Ex.9
Salicaceae 18.Ex.1

Salix 18.Ex.1
– sect. *Argenteae* W. D. J. Koch 49.Ex.7
– sect. *Glaucae* Pax 49.Ex.7
– subsect. *Myrtilloides* C. K. Schneid., not "(C. K. Schneid.) Dorn" 49.Ex.7
– *aurita* L. H.2.Ex.1
– *caprea* L. H.2.Ex.1
– ×*capreola* Andersson H.3.Ex.1
– *glaucops* Andersson (pro hybr.) 50.Ex.2
– *humilis* Marshall 11.Ex.21
– – var. *microphylla* Fernald 11.Ex.21
– – var. *tristis* Griggs 11.Ex.21
– *myrsinifolia* Salisb. 52.Ex.3
– *"myrsinites"* sensu Hoffm., non L. 52.Ex.3
– *tristis* Aiton 11.Ex.21
– – var. *microphylla* Andersson 11.Ex.21
– – var. *tristis* 11.Ex.21
Salvia sect. *Hemisphace* Benth. 32.Ex.5, 46.Ex.10
– *"africana coerulea"* (Linnaeus, 1753) 23.Ex.10
– *grandiflora* subsp. *willeana* Holmboe, subsp. *"S. willeana"* 24.Ex.4
– – *oxyodon* Webb & Heldr. 30.Ex.1
Sapium 11.Ex.17
– subsect. *Patentinervia* 21.Ex.1
Saxifraga aizoon [var. *aizoon* subvar. *brevifolia* f. *multicaulis*] subf. *surculosa* Engl. & Irmsch. 24.Ex.1
Scandix pecten-veneris L., not "*pecten* ♀" 23.Ex.2
Scenedesmus armatus var. *brevicaudatus* Pankow, non (L. S. Péterfi) E. H. Hegew. 53.Ex.15
– – f. *brevicaudatus* L. S. Péterfi 53.Ex.15
– *carinatus* var. *brevicaudatus* Hortob. 53.Ex.15
"Schaenoides" (Rottbøll, 1772) 20.Ex.9
Schiedea "gregoriana" (Degener, 1936) 36.Ex.2
– *kealiae* Caum & Hosaka 36.Ex.2
Schoenoxiphium 46.Ex.3
– *altum* Kukkonen 46.Ex.3
Schoenus 20.Ex.9
Scilla peruviana L. 51.Ex.3
Scirpoides Ség. 41.Ex.2, 43.Ex.4
"Scirpoides" (Rottbøll, 1772) 20.Ex.9, 43.Ex.4
– *"paradoxus"* (Rottbøll, 1772) 43.Ex.4
Scirpus 20.Ex.9
– sect. *Pseudoëriophorum* Jurtzev, not sect. *"Pseudo-eriophorum"* 60.Ex.13
– *cespitosus* L., not *"caespitosus"* 60.Ex.1
Sclerocroton 11.Ex.17
– *integerrimus* Hochst. 11.Ex.17
– *reticulatus* Hochst. 11.Ex.17
Scleroderma 18.Ex.1
Sclerodermataceae 18.Ex.1
Scyphophora ecmocyna Gray 48.Ex.3
Scytanthus Hook. 53.Ex.5
Scytopetalaceae Engl. 34.Ex.7
"Scytopetalum" (Pierre, 1897) 34.Ex.7
Sebastiano-schaueria Nees 20.Ex.7
Sebertia Engl., not "Pierre ex Baill." *34.Ex.1*
– "*acuminata* Pierre (ms.)" (Baillon, 1891) 34.Ex.1
Selaginella 31.Ex.3
Selenicereus (A. Berger) Britton & Rose H.6.Ex.1
×*Seleniphyllum* Rowley H.6.Ex.1
Senecio napaeifolius (DC.) Sch. Bip., not *"napeaefolius"* 53.Ex.9, 60.Ex.12
– *napifolius* MacOwan 53.Ex.9
Sepedonium chrysospermum (Bull.) Fr. 59.Ex.7
Serratula chamaepeuce L. 11.Ex.8
Sersalisia R. Br. 34.Ex.1
– *? acuminata* Baill. 34.Ex.1
Sesleria Rec.60B.1(b)
Sicyos L. 62.Ex.1
Sida retusa L. 9.Ex.2
Sigillariaceae 3.Ex.2
Silene L. 11.Ex.12
– *behen* L. 11.Ex.12
– *cucubalus* Wibel 11.Ex.12
– *vulgaris* Garcke 11.Ex.12
Siltaria Traverse 3.Ex.2
Simarouba Aubl. 53.Ex.10
Simaruba Boehm. 53.Ex.10
Skytanthus Meyen 53.Ex.5
Sloanea Rec.60B.1(a)
Smilax "caule inermi" (Aublet, 1775) 23.Ex.5
Smithia Scop. 14.Ex.10
Smithia Aiton, *nom. cons.*, non Scop. 14.Ex.10
Solanum ferox 52.Ex.9
– *indicum* L. 52.Ex.9

– *insanum* 52.Ex.9
– *lycopersicum* L. 14.Ex.1
– *melongena* var. *insanum* Prain, *"insana"* 24.Ex.2
– *saltense* C. V. Morton 53.Ex.11
– *saltiense* S. Moore 53.Ex.11
– *torvum* Sw. 52.Ex.9
– *tuberosum* var. *murukewillu* Ochoa, not *"muru'kewillu"* 60.Ex.16
Solidago L. 11.Ex.25
×*Solidaster* H. R. Wehrh. 11.Ex.25
Sophora tomentosa subsp. *occidentalis* Brummitt 46.Ex.7
×*Sophrolaeliocattleya* Hurst H.6.Ex.6
Sophronitis Lindl. H.6.Ex.5, H.6.Ex.6
Spartium biflorum Desf. 11.Ex.9
Spathiphyllum solomonense Nicolson, not *"solomonensis"* Rec.50F.Ex.3
Spergula stricta Sw. 11.Ex.10
Spermatites Miner 3.Ex.1
Spermatophyta 13.1(a)
Sphagnaceae 13.1(b), 13.1(c)
Spondias mombin 23.Ex.1
Stachys L. 62.Ex.1
– ×*ambigua* Sm. (pro sp.) 50.Ex.1
– *palustris* subsp. *pilosa* (Nutt.) Epling Rec.26A.Ex.1
– – var. *pilosa* (Nutt.) Fernald Rec.26A.Ex.1
Staphylea, not *"Staphylis"* 51.Ex.1
Stenocarpus R. Br. 62.Ex.2
Steyerbromelia discolor L. B. Smith & H. Rob. 46.Ex.7
Stigmaria Brongn. 3.Ex.1
Stillingia 11.Ex.17
– *integerrima* (Hochst.) Baill. 11.Ex.17
Strychnos L. 62.Ex.1
"Suaeda" (Forsskål, (1775) 43.Ex.1
– *"baccata"* (Forsskål, 1775) 43.Ex.1
– *"vera"* (Forsskål, 1775) 43.Ex.1
Swainsona formosa (D. Don) Joy Thomps. 11.Ex.19
Symphostemon Hiern 53.Ex.9
Symphyostemon Miers 53.Ex.9
Synthlipsis berlandieri A. Gray 11.Ex.22
– var. *berlandieri* 11.Ex.22
– var. *hispida* S. Watson 11.Ex.22

Talinum polyandrum Hook., non Ruiz & Pav. 58.Ex.2
Tamus, not *"Tamnus"* 51.Ex.1
Taonabo Aubl. 62.Ex.6
– *dentata* Aubl. 62.Ex.6
– *punctata* Aubl. 62.Ex.6
Tapeinanthus Boiss. ex Benth., non Herb. 41.Ex.2, 53.Ex.1
Taraxacum Zinn, not *"Taraxacvm"* 60.Ex.7
Taxus baccata var. *variegata* Weston 28.Ex.1
– – 'Variegata' (cv. Variegata) 28.Ex.1
Tephroseris (Rchb.) Rchb. 49.Ex.3
– sect. *Eriopappus* Holub 49.Ex.3
Tersonia cyathiflora (Fenzl) A. S. George ex J. W. Green 46.Ex.17
"Thamnos", *"Thamnus"* 51.Ex.1
Thea L. 13.Ex.3
Thuspeinanta T. Durand 41.Ex.2, 53.Ex.1
Tiarella cordifolia L. H.11.Ex.1
Tilia 10.Ex.4
Tillaea 51.Ex.1
Tillandsia bryoides Griseb. ex Baker 9.Ex.3
Tillia 51.Ex.1
Tithymalus "jaroslavii" (Poljakova, 1953) 34.Ex.10
Tmesipteris elongata P. A. Dang. 52.Ex.8
– *truncata* (R. Br.) Desv. 52.Ex.8
Torreya Arn., *nom. cons.*, non Raf. 53.Ex.3
Triaspis mossambica A. Juss., not *"mozambica"* 60.Ex.1
Trichipteris kalbreyeri (Baker) R. M. Tryon 33.Ex.6
Tricholomataceae Pouzar 18.Ex.7
"Tricholomées" Roze 18.Ex.7
Trifolium indicum L., not *"M. indica"* 23.Ex.14
Trilepisium Thouars 41.Ex.3
– *madagascariense* DC. 41.Ex.3
Triticum L. H.8.Ex.1
– *aestivum* L. H.3.Ex.3
– *dicoccoides* (Körn.) Körn. H.3.Ex.3
– *laxum* Fr. H.5.Ex.1
– *speltoides* (Tausch) Gren. ex K. Richt. H.3.Ex.3
– *tauschii* (Coss.) Schmalh. H.3.Ex.3
×*Tritordeum* Asch. & Graebn. H.8.Ex.1
Tropaeolum majus L. 23.Ex.4
Tsuga 7.Ex.2
– *heterophylla* (Raf.) Sarg. 7.Ex.2
– *mertensiana* (Bong.) Carrière 7.Ex.2
Tuber F. H. Wigg. : Fr. 20.Ex.3

– *gulosorum* F. H. Wigg. 20.Ex.3

Ubochea Baill. Rec.60B.Ex.1
Uffenbachia Fabr., not "*Vffenbachia*" 60.Ex.7
Ulmus racemosa Thomas, non Borkh. Rec.50C.Ex.1
Umbelliferae 18.5
Uredinales 13.1(d)
Uredo Pers. : Pers. 59.Ex.2
– *cubensis* (Arthur & J. R. Johnst.) Cummins 59.Ex.4
– *pustulata* Pers., not "*Vredo pvstvlata*" 60.Ex.8
Uromyces fabae 23.Ex.1
Urtica "dubia?" (Forsskål, 1775) 23.Ex.6
Urvillea Kunth 53.Ex.9, Rec.60B.Ex.1
Ustilaginales 13.1(d), 17.Ex.2, 59.1
Utricularia inflexa Forssk. 26.Ex.5
– – var. *stellaris* (L. f.) P. Taylor 26.Ex.5
– *stellaris* L. f. 26.Ex.5
– – var. *coromandeliana* A. DC. 26.Ex.5
– – var. *stellaris* 26.Ex.5
Uva-ursi Duhamel, not "*Uva ursi*" (Miller, 1754) 20.Ex.6

Vaccinieae D. Don Rec.19A.Ex.2
Vaccinioideae (D. Don) Endl. Rec.19A.Ex.2
Valantia L., not "*Vaillantia*" 60.Ex.9
Valeriana sect. *Valerianopsis* 21.Ex.1
Vanda W. Jones ex R. Br. H.6.Ex.6, H.8.Ex.2
– *lindleyana* Rec.60C.1(c)
×*Vascostylis* Takakura H.6.Ex.6
Verbascum sect. *Aulacosperma* Murb. 53.Ex.16
– *lychnitis* L. H.10.Ex.2
– "*nigro-lychnitis*" (Schiede, 1825) H.10.Ex.2
– *nigrum* L. 23.Ex.4, H.10.Ex.2
– ×*schiedeanum* W. D. J. Koch H.10.Ex.2
Verbena hassleriana Rec.60C.1(d)
Verbesina alba L. 11.Ex.18
– *prostrata* L. 11.Ex.18
Veronica anagallis-aquatica L., not "*anagallis*" 23.Ex.2, 60.Ex.14
Vexillifera Ducke 6.Ex.1
– *micranthera* 6.Ex.1
Viburnum ×*bodnantense* 'Dawn' 28.Ex.1
– *ternatum* Rehder 46.Ex.2
Vicia L. 18.5
Vinca major L. 23.Ex.4
Vincetoxicum 51.Ex.1
Viola hirta L. 24.Ex.6
– "*qualis*" (Krocker, 1790) 23.Ex.6
– *tricolor* var. *hirta* Ging. 24.Ex.6
Vulpia myuros (L.) C. C. Gmel. 26.Ex.3
– – "subsp. *pseudo-myuros* (Soy.-Will.) Maire & Weiller" 26.Ex.3

Wahlenbergia Roth 11.Ex.1
Waltheria americana L. 11.Ex.16
– *indica* L. 11.Ex.16
Weihea Spreng. 14.Ex.2
×*Wilsonara* Charlesworth & Co. H.6.Ex.6
Wintera Murray 18.Ex.4
Winteraceae Lindl. 18.Ex.4

Xanthoceras Bunge 62.Ex.5
"*Xanthoxylon*", "*Xanthoxylum*" Rec.50F.Ex.2
Xerocomus Quél. Rec.62A.Ex.1
Xylomataceae Fr., "ordo *Xylomaceae*" 18.Ex.3

Zanthoxylum caribaeum var. *floridanum* (Nutt.) A. Gray, not "*Xanthoxylum*" Rec.50F.Ex.2
– *cribrosum* Spreng., not "*Xanthoxylon*" Rec.50F.Ex.2
Zygophyllum billardierei DC., not "*billardierii*" 60.Ex.10

SUBJECT INDEX

The references in this index are not to pages but to the Articles, Recommendations, etc. of the *Code,* as follows: Div. = Division; Pre. = Preamble; Prin. = Principles; arabic numerals = Articles or, when followed by a letter, Recommendations; Ex. = Examples; N. = Notes; fn. = footnotes; H. = App. I (hybrids); App. = other Appendices.

For ease of reference, a few sub-indices have been included under the following headings: Abbreviations, Definitions, Epithets, Publications, Transcriptions (including related subjects), and Word elements.

For plant names in Latin appearing in the body of the *Code* plus App. I, refer to the special Index of scientific names on p. 354-368. A separate Index to App. IIIA (nomina generica conservanda et rejicienda) is provided on p. 336-353

Abbreviation, authors' name 46A.1-5
– herbarium name 37.N.1
– personal name 60B.N.1

Abbreviations:
anam. nov. (anamorphe nova) 59A.1
auct. (auctorum) 50D
comb. nov. (combinatio nova) 7.4
emend. (emendavit) 47A
& (et) 46C.1
& *al.* (et alii) 46C.2
excl. gen. (excluso genere, exclusis generibus) 47A
excl. sp. (exclusa specie, exclusis speciebus) 47A
excl. var. (exclusa varietate, exclusis varietibus) 47A
loc. cit. (loco citato) 33A
m. (mihi) 46D
mut. char. (mutatis characteribus) 47A
n- (notho-) H.3.1
nob. (nobis) 46D
nom. alt. (nomen alternativum) 19.Ex.2+4, 19.7
nom. cons. (nomen conservandum) 50E.1
nom. nud. (nomen nudum) 50B
op. cit. (opere citato) 33A
p. p. (pro parte) 47A
pro hybr. (pro hybrida) 50.Ex.2
pro sp. (pro specie) 50.Ex.1
pro syn. (pro synonymo) 50A

Abbreviations (cont.)
s. ampl. (sensu amplo) 47A
s. l. (sensu lato) 47A
s. str. (sensu stricto) 47A
St. (Saint), in epithet 60C.4(d)
stat. nov. (status novus) 7.4

Absence of a rule Pre.9
Abstracting journals 30A
Acceptance of name 34.1(a), 45.1
Adjective, as epithet 23.1+5+6(a+c), 24.2, 60.8, 60C.1(c-d), 60D
– – plural 21.2, 21B.1-2
– as name 20A.1(f)
– – plural 18.1, 19.1
Agreement, grammatical, see under Gender
– in number 60.N.3
Agricultural plants, designation 28.N.1
Algae, Committee for Div.III.2
– fossil vs. non-fossil 11.7
– homonymy with bacterial name 54.N.1
– illustration 39, 39A
– Latin description or diagnosis 36.1-2
– living culture from holotype 8B.1
– names of divisions, classes and subclasses 16.2, 16A.3(a)
– originally assigned to non-plant group 45.5
– starting points 13.1(e)
Alliance, instead of order 17.2

Alteration of circumscription, author citation 47, 47A
– of rank, author citation 49.1
– – priority 11.2
– of status, hybrid/non-hybrid 50
Alternative names 34.2
– different ranks 34.2
– families 10.6, 11.1, 18.5-6
– pleomorphic fungi 34.N.1
– subfamily 10.6, 11.1, 19.7
– valid publication 34.2
Ambiguity, avoidance Pre.1
Amendment, see Modification
American code of botanical nomenclature 10.Ex.6
Ampersand (&) 46C.1-2
Anagram 20.Ex.1, 60B.N.1
Analysis 42.4
– equivalence for non-vascular plants 44.2
– for valid publication 42.3, 44.1
Anamorph 3.N.1, 59.1
– alternative names 7.9, 34.N.1
– binary names admissible 59.5
– name, binary 59.5
– – new 59A.1
– – not alternative 34.N.1
– – priority 13.6, 59.1+4
– – type 7.9, 59.1-3+6
– – use permitted 59.5
– new taxon/new combination 59.6, 59A.2
Apostrophe, deletion 60.10
Appendix, I, see Hybrid
– II 14.1+5+N.1, 18.Ex.7, 60.Ex.5
– III 14.1+3+N.1-2, 50E.1, 53.Ex.10
– IV 56.1
– V 20.N.2, 32.8
Arbitrary formation, epithet 23.2
– generic name 20.1, 62.3
Article(s), in personal name 60C.4(c)
– of the *Code* Pre.3
Ascomycetes, pleomorphic 59.1
Ascription, definition 46.3
– effect on author citation 46.2+4
Asexual form, see Anamorph
Author, citation (see also Citation) 46
– – alteration of diagnostic characters 47
– – autonyms 22.1, 26.1, 46.1
– – basionym 49.1
– – change of rank 19A.Ex.1-2, 49.1
– – followed by "in" 46.N.1
– – homonyms 50C
– – hybrid names 50
– – incorrect form 33.3
– – internal evidence 46.6
– – omission 22.1, 26.1, 46.1
– – parenthetical 49.1
– – pre-starting point authors 46.5
– – unchanged 16A.4, 17.3, 18.4, 19.6, 24.4, 32.6, 47
– – with "&" or "& al." 46C.1
– – with "ex" 46.4-5+N.2
– names, abbreviation 46A.1-5+N.1
– – own to be used 46D
– – romanization 46B.1-2
Authors of proposals Div.III.4
Autograph, indelible 30.1, 32.2
Automatic typification, see under Typification
Autonym 6.8
– establishment 22.3, 26.3
– in new combination 11.N.2
– infraspecific taxa 26.1, 55.2
– no author name 22.1, 26.1, 46.1
– priority 11.6
– subdivisions of genus 22.1, 55.1
– valid publication 32.7
Available, epithet 11.5, 58.2
– name 14.10, 58.1
– printed matter 31.1
– under zoological Code 45.Ex.4.fn.
Avowed substitute, see *Nomen novum*

Back-cross H.4.1
Bacterial nomenclature, see *International code of nomenclature of bacteria*
Basidiomycetes, pleomorphic 59.1
Basionym 33.2
– author citation 49
– legitimate 52.3
– reference 32.4-5, 33.2-4, 33A, 41
– – full and direct 32.4, 33.2+4, 33A
– – indirect 32.4-5, 41
– – mere cross-reference to bibliography 33A
– type 7.4
– without indication of rank 35.2
Bibliographic citation, error 33.3
– use of "in" 46.N.1
Bigeneric hybrid H.6.2

Binary name 23.1
– as epithet 24.4
Binding decision, on confusability 53.4
Blue-green algae Pre.7
Bona fide botanists 7A
Bryophyta, Committee for Div.III.2
Bureau of Nomenclature Div.III.3-4

Canonization, prefix indicating 60C.4(d)
Capital initial letter 20.1, 21.2, 60.2, 60F
Catalogue 30.3
Caulis, not a generic name 20.Ex.5
Change of name, proper reasons Pre.9
Choice between names, see under Priority
Choice of type, see Lectotype and Designation
Circumscription Prin.IV, 6.5, 11.1
– alteration, citation 47, 47A
– anticipation 34.1(b)
– causing nomenclatural superfluity 52.1
– nothotaxon H.4
Citation (see also Author citation)
– altered circumscription 47A
– basionym 32.4-5, 33.2-4, 33A, 41
– bibliographic error 33.3
– collection data 37.3
– date of publication 45B.1
– holotype 37.2
– homonym 50C
– invalid name 50A
– lectotype 37.2
– *loc. cit.*, avoid use 33.A.1
– misapplied name 50D
– *nomen conservandum* 50E.1
– *nomen nudum* 50B
– *op. cit.*, avoid use 33A
– orthographical variant 61.N.1
– single element 37.3
– synonym 34.1, 50A
Class *(classis)*, name 16.1-2, 16A.3
– rank 3.1
Classical usage, see Tradition
Coal-ball 8A.3
Code, editing Div.III.2
– modification Div.III
Cohors, instead of order 17.2
Collection, public 7A, 8B.1
Collection data, citation 37.3
– illustrated material 8A.2
Collective epithet, definition H.3.3+N.2
Combination 6.7
– based on rejected name 14.4+7+10, 56.1
– binary 21.1, 23.1, 24.4
– ternary 24.1
– under conserved later homonym 55.3
– valid publication 33.1-2
Committees, nomenclature 14.12+14, 14A, 53.4, 56.2, Div.III.2
Component, fungal, in lichens 13.1(d)
Compound, correctable epithets 60.8
– generic names, gender 62.2
– names and epithets 60G.1
Condensed formula H.6.1-2+4, H.7.1
– as epithet H.7.1, H.8.1
– commemorating person H.6.3-4, H.6A, H.8.2
– equivalent to H.6.3-4, H.6A, H.8.2, H.9.N.1
– form H.6.2-4, H.7.1
– no type 37.1, H.9.N.1
– nothogeneric name 20.N.1, H.6, H.6A, H.8-9
– – more than two parental genera H.6.3-4
– – valid publication 32(c), 37.1, 40.N.1, H.9
– parental names H.8.1, H.9.1
Conflict with protologue 9.13(b), 9A.5, 10.2+5
Confused name, not to be used 57
Confusingly similar names 53.3-4, 61.5
– binding decision 53.4
– treated as homonyms 53.3-4, 61.5
Connecting vowel 60.Ex.12, 60G.1(a2)
Conservation (see also Deposited material, Preservation) 14, 14A, App.II-III
– aims 14.1-2
– citation 50E
– combination under conserved homonym 55.3
– date, effect on competing names 14.N.3
– – not affecting priority 14.5-7
– extent 14.4-5+10
– family name 14.1+4-5, App.IIA-B
– – based on illegitimate generic name 18.3
– gender 14.11
– General Committee approval 14.14
– generic name 14.1+4, App.IIIA
– illegitimate name 6.4, 14.1
– junior homonym 14.6+10+14, 45.4, 53.1

- limitation of priority 11.3
- lists permanently open 14.12
- name against itself 14.N.1+fn.
- no entry to be deleted 14.13
- orthography 14.11, 60.N.1
- overrides sanctioning 15.6
- proposal 14.12+14, 14A
- species name 14.1+4+N.2, App.IIIB
- type 14.3+8-9, 48.N.2

Correct, grammar Pre.1
- name Prin.IV, 6.5, 11.1+3-4
- – choice 11.3-7+N.1 14.5-7, 15.3-5
- – of nothotaxon H.4-5, H.8
- – of pleomorphic fungus 59.1
- – potentially 52.3, H.5.N.1

Correction of spelling, see Orthography

Correction slip 30A

Cultivar epithet 28.N.2

Cultivated plants Pre.8, 28
- from the wild 28.1

Culture 8.2
- collection 8B.1
- from type 8B.2

Custom (see also Tradition), author abbreviation 46A.5
- established Pre.10, 23.8
- prevailing Pre.1

Date, of autonym 32.7
- of name, definition 45.1
- – ease of verification 46.1
- – Linnaean generic 13.4, 41.N.1
- of publication 31.1-2, 31A, 45B.1, 45C.1
- – starting point works 13.1+5
- unchanged 16A.4, 17.3, 18.4, 19.6, 24.4, 32.6

Definitions:

Alternative names 34.2
Analysis 42.4
Anamorph 59.1
Ascription 46.3
Autonym 6.8
Available name 45.Ex.4.fn.
Basionym 33.2
Binary combination 23.1
Collective epithet H.3.3+N.2
Combination 6.7
Correct name 6.5, 11.4
Date of name 45.1
Descriptio generico-specifica 42.1

Definitions (cont.)

Diagnosis 32.3
Duplicate 9.3.fn.
Effective publication 6.1, 29
Epithet 6.7, 21.2, 23.1-2, 24.2
Epitype 9.7
Ex-type, ex-holotype, ex-isotype 8B.2
Final epithet 11.4.fn.
Form-genera 3.3
Forma specialis 4.N.3
Fossil plant Pre.7.fn.
Holomorph 59.1
Holotype 9.1
Homonyn 53.1
Hybrid formula H.2.1
Illegitimate name 6.4
Indelible autograph 30.2
Indirect reference 32.5
Isosyntype 9.9
Isotype 9.3
Later homonym 53.1
Lectotype 9.2
Legitimate name 6.3
Monotypic genus 42.2
Name 6.6
Name below rank of genus 11.4
Name of infraspecific taxon 24.1
Name of species 23.1
Neotype 9.6
Nomen novum 7.3
Nomen rejiciendum 56.1
Nomenclatural synonym 14.4, App.IIIA
Nomenclatural type 7.2
Non-fossil plant Pre.7.fn.
Non-valid publication 34.1
Nothogenus 3.2
Nothomorph H.12.2.fn.
Nothospecies 3.2
Nothotaxon H.3.1
Objective synonym 14.4.fn.
Opus utique oppressum 32.8, App.V
Original material 9.9.fn.
Original spelling 60.2
Orthographical variant 61.2
Page reference 33.N.1
Paratype 9.5
Plant Pre.1.fn.
Protologue 9.4,footN.1
Provisional name 34.1(b)
Pseudocompound 60G.1

Definitions (cont.)
Registration 32.2
Sanctioned name 7.8,13.1(d),15.1
Subdivision of family 4.N.1
Subdivision of genus 4.N.1
Subjective synonym 14.4.fn.
Superfluous name 52.1
Syntype 9.4
Tautonym 23.4
Taxon 1
Taxonomic synonym 14.4, App.IIIA
Teleomorph 59.1
Type 7.2
Valid publication 6.2, 32.1

Delivery, printed matter to carrier 31A
Deposited material, access policy 7A
– specification of herbarium 9.14, 37.5+N.1
Descriptio generico-specifica 42.1
Description, in addition to diagnosis 36A
– or diagnosis, ascription 46.2+4
– – combined generic and specific 41.3(c), 42.1
– – English 36.3
– – Latin 36.1-3, 36A
– – pre-Linnaean 32A
– – provision 32.1, 32B+E, 41.1-3, 42.1
– – published before 1753 32A
– – reference 32.1(c)+4-5, 32A, 33.2, 36, 38, 39, 41
– – – full and direct 32.4, 33.2+4, 33A
– – – indirect 32.4-5, 41
– – – not acceptable 42.1
Descriptive, name 16A.1, 17.1
– phrase 23.6-7
Designation of type (see also Lectotype and Unitary) 10.1+5-6, 10A
– effective, requirements 7.10-11
– mandatory 37.1-5, 37A
– supersedable 10.5
Desmidiaceae, starting point 13.1(e)
Diacritical signs 46B.2, 60.6, 60B.1(d), 60C.3
Diaeresis 60.6
Diagnosis (see also Description) 32.3
– accompanying description 36A
Diagnostic characters, alteration 47, 47A
Direction of cross H.2A.1
Disadvantageous change of name 56.1

Division *(divisio)* or phylum, name 16.1-2, 16A
– rank 3.1, 16.N.1
Doubt Pre.10, 34.1, 52.N.1
Duplicate, definition 9.3.fn.

Editorial Committee Div.III.2
Effective publication, date 31.1
– definition 6.1, 29
– indelible autograph 30.1
– valid publication 32.1
Element, citation 37.3, 52.2
– conflicting with description 9A.5
– heterogeneous 9.10, 9A.5
– inclusion 52.2
– – with doubt 52.N.1
English, description or diagnosis 36.3
Ennoblement, prefix indicating 60C.4
Ephemeral printed matter 30A
Epithet (see also Adjective, Substantive) 6.7, 21.2, 23.1-2, 24.2
– avoidance 23A.2-3, 58A, 60C.3, H.10A
– compound 60G.1
– considered as hybrid formula H.10.3
– cultivar 28.N.2
– definite association with genus or species name 33.1
– derived from, generic name 60F
– – geographical name 23A.1-2, 60.7, 60D
– – host plant name 60H
– – illegitimate name 58.3, 58A
– – personal name 23A.1-2, 60.7, 60C.1-4, 60F
– – vernacular name 60.7, 60F
– etymology 60I.1
– final 11.4.fn.
– fungal name, derived from host 60H
– – identical for anamorph and holomorph 59.N.1
– hyphenation 60.9
– inadmissible 21.3, 23.4+6, 24.3-4, 58.2
– initial letter 60F
– nothotaxon H.10.3, H.10A, H.11.2
– original spelling 60E
– pleomorphic fungi 59.N.1, 59A.2
– recommended spelling 60E
– sequence in hybrid formula H.2A.1
Epithets (see also Index to scientific names):
albomarginatus 60G.1(b)

Epithets (cont.)
alexandri 60C.2
anagallis-aquatica 23.2, 60.Ex.14
atropurpureus 60G.1(b)
augusti 60C.2
austro-occidentale 60.Ex.14
balansana, -um, -us 60C.1(c)
beatricis 60C.2
billardierei 60.Ex.10
brauniarum 60C.1(b)
brienianus 60C.4(b)
candollei 60C.4(d)
"cannaefolius" 60G.1(b)
caricaefolius 60G.1(b)
ceylanica 53.Ex.8
chinensis 53.Ex.8, 60E.Ex.1
clusianus, clusii 23A.1
dahuricus 23A.1
dubuyssonii 60C.4(c)
fedtschenkoi 60C.1(a)
"genuinus" 9A.3, 24.3
geppiorum 60.Ex.20
glazioui 60C.1(a)
hectoris 60C.2
heteropodus, heteropus 53.Ex.8
hilairei 60C.4(d)
hookerorum 60C.1(a)
iansonii 60.Ex.16
iheringii 60C.4(e)
jamacaru 60.Ex.6
jussieui 60C.4(d)
laceae 60C.1(a)
lacryma-jobi 60.Ex.14
lafarinae 60C.4(c)
lecardii 60C.1(b)
leclercii 60C.4(c)
linnaei 60C.2
logatoi 60C.4(c)
macfadyenii 60C.4(a)
macgillivrayi 60C.4(a)
mackenii 60C.4(a)
macnabii 60C.4(a)
macrocarpon, macrocarpum 53.Ex.8
macrostachys, macrostachyus 53.Ex.8
"mandacaru" 60.Ex.6
martii 60C.2+4(e)
martini 60C.2
matthewsiae, matthewsii 60.Ex.19
munronis 60C.2
murukewillu 60.Ex.16

Epithets (cont.)
napaeifolia 60.Ex.12
napaulensis, nepalensis, nipalensis 53.Ex.8
nidus-avis 60G.1(b)
nova 23.Ex.7
novae-angliae 60.Ex.14
obrienii 60C.4(b)
odonellii 60.Ex.16
okellyi 60C.4(b)
opuntiiflora 60.Ex.11
"originalis", "originarius" 24.3
pissardii 60.Ex.17
poikilantha, poikilanthes 53.Ex.8
polifolia 60.Ex.11
polyanthemos, polyanthemus 53.Ex.8
porsildiorum 23A.1
pteroides, pteroideus 53.Ex.8
quinquegona 51.Ex.2
remyi 60C.4(d)
richardsonis 60C.2
saharae 23A.1
san-felipensis 60.Ex.14
sanctae-helenae 60C.4(d)
sancti-johannis 60C.4(d)
scopolii 60C.1(a)
silvatica 60.Ex.1
sinensis 53.Ex.8, 60E.Ex.1
steenisii 60C.4(e)
strassenii 60C.4(e)
sylvatica 60.Ex.1
toddiae 60.Ex.18
trachycaulon, trachycaulum 53.Ex.8
trianae 60C.1(a)
trinervis, trinervius 53.Ex.8
tubaeflorus 60G.1(b)
"typicus" 9A.3, 24.3
uva-ursi 60.Ex.14
vanbruntiae 60C.4(e)
vanderhoekii 60C.4(e)
vechtii 60C.4(e)
"veridicus", "verus" 24.3
verlotiorum 60C.1(b)
vonhausenii 60C.4(e)
wilsoniae 60C.1(b)
zeylanica 53.Ex.8

Epitype, definition 9.7
Error, application of name on transfer 7.4
– bibliographic citation 33.3
– correctable orthographical 60.1+8-11

Established custom, see Custom
et (&) 46C.1
Etymology 60I.1
Euphony Pre.1
"ex" in author citation 46.4-5+N.2
Examples in the *Code* Pre.3
Exclusion of type, see under Type
Exsiccata 30.4+N.1
ex-type *(ex typo)*, ex-holotype *(ex holotypo)*, ex-isotype *(ex isotypo)* 8B.2

Family *(familia)*, name 18
– – alternative 10.6, 11.1, 18.5-6
– – based on illegitimate generic name 18.3
– – conservation 14.1+4+5, App.IIA-B
– – correction of termination 18.4
– – form 18.1
– – type 10.6
– – valid publication 41.1
– rank 3.1
– – change 19A.1-2
– – termed order 18.2
– subdivision of, see Subdivision of family 4.N.1
Fancy name, in cultivars 28.N.2
Female symbol H.2A.1
Figure, see Illustration
Final epithet 11.4+fn.+N.2, 26.1-2, 27
First, see Priority
Folium, not a generic name 20.Ex.5
Forestry plants, designation 28.N.1
Form *(forma)*, rank 4.1
Forma specialis, definition 4.N.3
Form-genus (see also Anamorph) 3.3-4, 7.9, 11.1
Fossil plant Pre.7.fn., 13.3
– Committee for Div.III.2
– description or diagnosis 36.1+3
– form-genus 3.3, 7.9, 11.1
– fragmentary specimen 3.3
– name, priority 11.7
– – type 7.9, 8A.3, 8.4-5, 9A.6
– – valid publication 36.1+3, 38
– – – illustration required 38
– starting point 13.1(f)
– vs. non-fossil 11.7, 13.3
Fungi (see also Anamorph, Holomorph, Teleomorph), Committee for Div.III.2
– epithet, derived from name of host 60H
– *formae speciales* 4.N.3
– homonymy with bacterial name 54.N.1
– host 32E, 60H
– indication of sanctioned status 50E.2
– lichen-forming 13(1)d, 59.1
– living culture from holotype 8B.1
– names of higher taxa 16A.1-2+3(b)
– parasitic 4.N.3, 32E,' 60H
– pleomorphic 59, 59A
– – form-taxon (anamorph) 3.4, 11.1, 59.1, 53.3-5
– – generic name pertaining to the wrong morph 51, 59.3
– – holomorph including correlated anamorphs 25, 34.N.1, 59.1+4
– – morph, nomenclatural type 59.1-3+6, 59A
– – new morph 59.6, 59.A
– – priority of names 13.6
– – typification of names 7.8-9+N.1, 59.2-3+6
– starting point 13.1(d)

Gasteromycetes, starting point 13.1(d)
Gender, agreement in 21.2, 23.5, 24.2, 60.N.3
– generic name 62, 62A
– – arbitrarily formed name 62.3
– – assigned by author 62.1+3
– – botanical tradition 62.1+N.1
– – compound 62.2
– – conservation 14.11
– – correction of epithets 60.N.3
– – feminine when commemorating person 20A.1(i)
– – irrespective of original author 62.2+4
– – not apparent 62.3
– – when genus is divided 62A
General Committee 14.12+14, 14A, 32.8-9, 32F, 53.4, 56.2, Div.III.2+4
Generic, see Genus
Genitive, see under Substantive
Genus (genera) (see also Nothogenus) 3, 20, 20A
– anamorphic 59.3
– holomorphic 59.3
– monotypic 37.2, 41.1(a)+2
– name 20, 20A
– – adjective used as noun 20A.1(f), 62.3
– – as autonym epithet 22.1-2+4
– – biverbal 20.2

– – capital initial letter 20.1
– – coinciding with technical term 20.2
– – commemorating person 20A.1(i), 60B
– – composed arbitrarily 20.1, 62.3
– – conservation
10.4+N.2, 14.1+3-4, 60.N.1-2, App.IIIA
– – form 20.1
– – – advisable 20A.1(a+e+i)
– – – not advisable 20A.1(b-d+f-h+j)
– – former, as epithet 60F
– – gender, see Gender
– – hyphenated 20.3, 60.N.2
– – illegitimate 55.1
– – in Linnaean works 13.4, 41.N.1
– – not regarded as such 20.4, 44.N.1
– – type 10.1-5+N.1-2, 14.8
– – – designation 10.1+5
– – – inclusion 10.2-3
– – – indication 37.1-4
– – valid publication 37.2+4, 41.2+N.1-2, 42, App.V
– – vernacular 62.3
– rank 3.1
– – raised section or subgenus 21B.3, 49
– subdivision, see Subdivision of genus
Geographical names, in epithets 23A.1+3(j), 60.7, 60D
– use of "St." 60C.4
Grammatical correctness Pre.1
Greek, gender of nouns 62
– personal names 60C.2
– transliteration 60A
– word elements 60G

Handwritten material 29, 30.2
Hepaticae, starting point 13.1(c)
Herbarium, abbreviation 37.N.1
– access policy 7A
– author's institution 9A.4
– sheet 8.1, 9.10, 9A.3
– to be specified 37.5
Hierarchy of ranks 2-5
– subordinate taxa 25, 34.1(d)
Holomorph 25, 59.1
– name, binary 59.2
– – correct 59.1
– – not alternative 34.N.1
– – priority 13.6, 59.1+4
– – type 59.1-3+6
– new taxon/new combination 59.6, 59A.2
Holotype *(holotypus)* (see also Type) 9.1, 10.N.1
– automatic 7.3-6
– collection data 8A.2, 37.3
– definite indication 7.5
– designation 7,5, 9.1+4
– destroyed 9.9+11
– duplicate 9.3
– equivalent in modern language 37.4
– exclusion 48.1
– identification ambiguous 9.7
– illustration 8.1+3, 8A.1-2, 9.1
– inclusion, in named taxon 10.2-3, 22.1-3, 22A, 26, 26A, 37.2, 52.2+N.1-2
– – in other taxon 48.1
– – single element 37.3
– indication 9.9, 37.1-4, 37A
– living 8.2, 8B.1-2
– location 9A.4, 37.5
– lost 9.9+11
– missing 9.2+6+9+11
– more than one taxon 9.2+9-10+N.1-2
– permanently preserved 8.2-3+Ex.1, 8A.2, 8B.2
– – in public herbarium 7A
– previously published species name 37.2
– rediscovered 9.13
– specimen data 8A.2, 37.3
– supporting epitype 9.7
Homonym (see also Confusingly similar names) 14.10, 22.6, 53.1
– by conservation 14.9
– by exclusion of type 48.1
– choice between simultaneous 53.6
– citation 46.3, 50C
– conserved later 14.9-10, 45.4, 53.1, 55.3
– disregard hybrid status H.3.4
– earlier 14.6+10, 45.4, 53.N.1
– equal priority 53.6
– illegitimate 21.N.1, 24.N.2, 53.1-2+N.1
– infraspecific taxa, same species 24.N.2, 53.5
– later 14.10, 22.6, 45.4, 53.1-2+N.1
– names likely to be confused 53.3
– of conserved name 14.10
– rejected, earlier 14.10
– – later 45.4
– – legitimate 14.10
– sanctioned 45.4, 53.1+3
– subdivisions of same genus 21.N.1, 53.5

– taxon not treated as plants 54.1
– unranked 35.2
Horticultural plants, designation 28.N.1
Host name 32E, 60H
– specific form 4.N.3
Hybrid Pre.8, 3.2, 4.4, 20.N.1, 28.2, 40, 50, H.1-12
– anticipation of existence H.9.N.2
– arising in cultivation 28.2
– change to non-hybrid status 50
– condensed formula, see Condensed formula
– "Cultivated Code" Pre.8, 28.N.1-2
– formula 23.6(d), H.2, H.2A, H.4.1, H.10.3
– – definition H.2.1
– – more informative H.10B.1
– genus, see Nothogenus and Condensed formula
– multiplication sign H.1-2,H.3.1+4,H.3A
– name, see under Nothotaxon
– parental taxa, see under Nothotaxon
– prefix notho (n-) H.1, H.3.1+4
– rank, see under Nothotaxon
– species, see Nothospecies
– statement of parentage 52.N.3
– – of secondary importance H.10.N.1
– status, indication of 50, H.1.1, H.2
– taxon, see Nothotaxon
– variety, see Nothomorph
Hyphen, in compound epithet 23.1, 23A.3(d), 60.9+N.2
– in generic name 20.3, 60.N.2
– in hybrid designation H.10.3

Illegitimate name 6.4, 52-54
– adoption of epithet 58.3, 58A
– becoming legitimate later 6.4
– by conservation 6.4, 14.1
– by sanctioning 6.4, 53.2
– family 18.3
– genus 18.3, 55.1
– homonyms 21.N.1, 24.N.2, 53.1-2, 54, 58.1(b)
– hybrid 52.N.3
– species 55.2
– subdivision of family 19.5
– type 7.5
Illustration, advisable 32D.1
– algae 39, 39A
– as type 8.1+3, 8A, 9.1-2+6-7+fn.+14, 10.4, 37.3+5
– equivalent to description 41.N.2, 42.3, 44.2
– fossil plant 8.4, 9A.6, 38
– of type 8.4, 39A
– original material 9.7.fn.
– scale 32D.3
– specimen used 8A.1-2, 32D.2, 39A
– with analysis 41.N.2, 42.3-4, 44.1-2
Improper, see Incorrect
"in" in citation 46.N.1
Inadmissible, see under Epithet
Inclusion of type 22.1-3, 22A, 26, 26A, 37.2, 52.1-2
Incorrect (see also Orthography, correction), Latin termination 16.A.4, 17.3, 18.4, 19.6, 32.6
– name 52.3, H.5.N.1
Indelible autograph 30.2
Index herbariorum 37.N.1
Index kewensis 33.4
Index of fungi 33.4+Ex.8
Indication, of rank, see Rank
– of type, see Type
Indirect reference 32.4-5
Infrageneric, see Epithet, Infraspecific, Species, Subdivision of genus
Infraspecific, autonym 26
– epithet (see also Epithet) 24, 24A-B, 26, 26A, 27
– – binary combination instead of 24.4
– – cultivar 28.N.2
– – form 24.2
– – grammatical agreement 24.2
– – inadmissible 24.3-4, 58.2
– – nothotaxon H.10.3, H.10A, H.11.2
– – to be avoided 24A.1, 24B.2, 26A.1-3
– – to be retained 24A, 24B.1, 26A.3
– – under illegitimate species name 55.2
– ranks 4.1-2
– – change 24B.2
– – not clearly indicated 35.1-2
– – single 35.3
– name 24, 26-27
– – form 24.1
– – homonymous within species 53.5
– – legitimate 55.2
– – valid publication 41.3, 43.1, 44.1
– taxon 24-27, H.10-12

– – including type of species name 26.1-2, 26A, 27
– – rank not indicated 35.3
Institution, see Collection and Herbarium
Institutional votes Div.III.4(b)
Intentional latinization 60.7
Intercalated ranks 4.3
Intergeneric hybrid, see Nothogenus
Interim designation 23.Ex.8
Internal evidence in publication 35.4, 46.6
International, Association for Plant Taxonomy 32.2, Div.III.2+4
– Botanical Congress 14.14+Ex.7, 32.1+9, 53.4, Div.III
– – Bureau of Nomenclature Div.III.3-4
– – decisions 14.4, 32.9, 53.4, 56.2, Div.III.2
– – plenary session Div.III.1
– – Tokyo 14.N.1.fn.
– *code of nomenclature for cultivated plants* Pre.8, 28.N.1, H.3.N.2, H.4.N.1
– *code of nomenclature of bacteria* Pre.7.fn., Prin.I., 14.4.fn., 54.N.1
– *code of zoological nomenclature* Prin.I., 14.4.fn., 45.Ex.4-7+Ex.4.fn.
– Commission for the Nomenclature of Cultivated Plants Pre.8
– Union of Biological Sciences Div.III.1.fn.
Invalid, see Name, not validly published
Isosyntype 9.9
Isotype 9.3+9

Kingdom *(regnum)*, rank 3.1

Later homonym, see Homonym
Latin, and latinization, accepted usage 60E
– – geographical names23A.1-2, 60.7, 60D
– – personal names 23A.1-2, 60.7, 60C.-21
– – vernacular names 60.7, 60F, 62.3
– description or diagnosis 36, 36A
– termination 32.6, 60B, 60C.1
– transliteration to, see Romanization
– word elements 60G
Lectotype (see also Designation of type) 9.2, 10.N.1
– designation 9.9, 9A
– – effective, requirements 7.10-11, 9.14
– – first to be followed 9.13
– destroyed 9.11
– identification ambiguous 9.7
– illustration 8.3, 8A.1, 9.2+14
– inclusion, in named taxon 26.2, 37.2, 52.2(b)+N.1-2
– location to be specified 9.14
– names of fossil species 9A.6
– precedence over neotype 9.9
– preserving current usage 9A.5
– previously designated 9.11, 26.2, 52.2(b)
– previously published species name 37.2
– supersedable 9.13
– supporting epitype 9.7
Legitimate name 6.3, 51, 52.3+N.1-3
– by conservation 14.1
– epithet under illegitimate or rejected name 55.1-3
– maintenance 51
– nothotaxon H.4.1
– priority 11.3-5, 45.4
Letters, foreign to classical Latin 60.4, 60B.1, 60C.1
– initial 20.1, 21.2, 60.3, 60F
– used interchangeably 60.5
Lichen-forming fungi 13.1(d)
Ligatures 60.4+6
Limitation of priority, see under Priority
Linnaean symbols 23.3, 60.Ex.14
Lyophilization 8.Ex.1

Mail vote Div.III.4
Male symbol H.2A.1
Manuscript 29.1, 30.2
– names 23A.3(i), 34A
– notes 9A.3
Mechanical methods of typification 9A.2, 10.5(b)
Microfilm 29
mihi, as author citation 46D
Misapplied name 7.4, 33.N.2, 48.N.1, 50D, 57, 59.3
Misplaced term 33.5
Modification of *Code* Pre.6, Div.III
– Appendix II-III 14.12-14, 14A
– Appendix IV 56.2
– Appendix V 32.8-9, 32F
Monotypic genus 42.1-2
Morph, see Fungi, pleomorphic
Morphology, technical term 20.2
Multiplication sign H.1-2, H.3.1+4, H.3A
Musci, starting point 13.1(b)

Mythical persons 60F

Name (see also Adjective, Author, Nomenclature, Personal name Substantive) 6.6, 12
– alternative 34.2+N.1
– – of family 10.6, 11.1, 18.5-6
– – of subfamily 10.6, 11.1, 19.7
– avowed substitute 7.3, 33.2
– based on generic name 7.1, 10.7, 16.1-2, 17.1+3, 18.1+3, 19.5
– class or subclass 16.1-2, 16A.3
– compound 60G
– confused 57
– confusingly similar 53.3-4, 61.5
– conserved, see Conservation
– contrary to rules Pre.4, Pre.9
– correct 6.5, 11, 14.5-7, 15.3-5, 59.1, H.4
– current usage 57
– derived, from Greek 60A
– – from person's name 60B
– etymology 60I
– euphony Pre.1
– fancy 28.N.2
– – termed order 18.2
– first syllable 60.3
– illegitimate, see Illegitimate
– initial letter 20.1, 21.2, 60.3
– misapplied 7.4, 33.N.2, 48.N.1, 50D, 57, 59.3
– new, in modern language works 45A
– not to be adopted 32C.1
– not validly published 32.8, 33.1-2+4-5, 34.1-2, 35.1, 43, App.V
– of division or phylum 16.1-2+N.1, 16A.1
– of family 10.6, 18, 19A.1
– of form-taxon 3.4, 7.9, 11.1, 59.3+5
– of genus 10.1-5, 20, 20A, 21B.3
– of hybrid, see Hybrid
– of infraspecific taxon 8-9, 24, 24B, 25-26, 26A, 27
– of order or suborder 17.1-3, 17A, 18.2, 19.2
– of species 8-9, 23
– of subdivision, of family 10.6, 19, 19A
– – of genus 10.1-2+5, 21, 21A-B, 22, 22A
– – (subphylum) 16.1-2, 16A.2
– of subfamily, termed suborder 19.2
– of suprafamilial taxon 10.7, 11.9, 16.1+N.2, 16B
– of taxon of lower rank than variety 26A.2
– orthography, errors 23.7, 45.3, 50F, 60.1+3+8-11, 61.1+N.1
– – variants 61
– rank, see Rank
– reasons for change Pre.9
– regularity Pre.1
– rejected, see Rejection
– spelling 13.4, 14.11, 45.3, 50F, 60, 60A-J, 61
– stability Pre.1, 14.2, 56.1
– superfluous 52.1+3
– type, see Type
– useless creation Pre.1
– with question mark 34.1, 52.N.1
– words, not generic names 20.4
Natural order *(ordo naturalis)* 18.2
nec, in homonym citation 50C
Neotype 9.6
– designation 9.9, 9B
– – effective, requirements 7.10-11, 9.14
– – first to be followed 9.13
– identification ambigous 9.7
– illustration 8.3, 8A.1, 9.6+14
– inclusion, in named taxon 26.2, 52.2(b)+N.1-2
– location to be specified 9.14
– precedence of lectotype 9.9
– preserving usage 9.11
– previously designated 9.7, 26.2, 52.2(b)
– supersedable 9.12-13
– supporting epitype 9.7
New combination 33.1-3
– authorship 46.4
– basionym citation 33.2-4, 33A
– type 7.4
– valid publication 33.1-4
Newspaper, non-scientific 30.3
Nixus, instead of order 17.2
nobis, as author citation 46D
Nomen conservandum (see also Conservation) 14, 14A, 50E.1, App.II-III
Nomen novum 7.3
– authorship 46.4
– by error 33.Ex.9-10, 59.6
– replacing illegitimate name 58.1+3
– type 7.3
– valid publication 33.2+N.2
Nomen nudum 50B
Nomen rejiciendum, see Rejected name

Nomen specificum legitimum 23.6
Nomen triviale 23.7
Nomen utique rejiciendum 56, App.IV
Nomenclatural synonym 14.4, App.IIIA-B
Nomenclatural type, see Type
Nomenclature, botanical Pre.1-2+4
– contrary to rules Pre.9
– – disadvantageous change 14.1, 56.1
– – independence Prin.I
– – principles Prin.I-VI
– – stability 14.2
– Committees 14.12, 14A, 32.8, 32F.1, 53.4, 56.2, Div.III.2
– Section Div.III.1-3+4(b)
– – officers Div.III.3
– – voting Div.III.4(b)
non, in homonym citation 50C
Non-algal taxon, originally assigned to non-plant group 45.5
Non-fossil, see Fossil
Non-vascular plant, figure 42.3, 44.2
– type of name 8.1
Nostocaceae, starting point 13.1(e)
"notho-", etymology H.1.fn.
– prefix H.1, H.3.1+4
Nothogenus 3.2
– name, see Condensed formula
– rank 3.2, H.1, H.4-5, H.12
Nothomorph H.12.2+fn.
Nothospecies 3.2
– epithet H.3.3
– from different genera H.11.1
– name H.11.1
– rank 3.2, 40, 50
Nothotaxon (see also Hybrid) H.3.1
– circumscription H.4.1
– name (see also Condensed formula) Pre.8, 28.2, 40, H.1-12
– – author citation 50
– – correct H.4-5, H.8, H.11-12
– – legitimate 52.N.3
– – validly published 32.1(c), 37.1, 40, H.9, H.10.1(a)+3, H.12.2
– parental taxa 52.N.3, H.2, H.2A, H.3.2, H.4-5, H.5A, H.6.2+4, H.8-9, H.10.3+N.1, H.10A, H.11, H.12.1+N.1
– rank (see also Nothospecies, Nothogenus, Nothomorph) 3.2, 4.4, 40, 50, H.1, H.3.1, H.4-5, H.12
– – appropriate H.5.1-2
– – inappropriate H.5.N.1
– – infraspecific 4.4, H.3.1, H.4.N.1, H.10, H.10A-B, H.11.2, H.12
– – subdivision of genus H.7, H.9
– variation H.4.N.1, H.12
Noun, se Substantive

Objective synonym 14.4.fn.
Oedogoniaceae, starting point 13.1(e)
Opus utique oppressum 32.8, App.V
Order *(ordo),* name 16.1-2, 17.1, 17A
– rank 3.1
– – intended as family 18.2
– – termed otherwise 17.2
– relative, of ranks 5, 33.5
Organisms treated as plants Pre.7
Original material 9.7.fn.
– identification ambiguous 9.7
– not extant 9.9
– rediscovered 9.13
Original spelling 60.2, 61.1
– indication of 50F
– correction 45.3, 60.1
– retention 60.1, 60B.1(d), 60C.3, 60E, 61.1
– standardization 60.5-6+8-11, 60H
– variant 61.3
Orthography 60-61
– conservation 14.11
– correction 23.7, 45.3, 60.1+3+5-6+8-11, 60c.1, 60G.1(b), 60H, 61.1
– epithets 60.8-11, 60A-I
– error 23.7, 60.1+8-11, 60G.1(b), 61.1
– Linnaean, generic names 13.4
– – phrase-like epithets 23.7
– standardization 60.5-6+8-11, 60H
– variants 61.2
– – citation 50F, 61.N.1
– – confusingly similar 61.5
– – correction 61.4
– – in original publication 61.3
– – not validly published 61.1

Page reference 33.N.1
Pagination 45C
Parasite 4.N.3
– host name 32E, 60H
Paratype 9.5
Parentage, see Nothotaxon, parental taxa
Parentheses 21A, 49-50

Particles, in personal names 46A.1, 60C.4
Patronymic prefix 60C.4(a-b)
Periodical, date 45C
- popular 30A
- separates 31.2, 45C
Person, mythical 60F
Personal name (see also Author), anagram 60.N.1
- diacritical signs 60B.1
- Greek or Latin 60C.2
- in epithets 23A.1-2, 60C
- in generic name 20A.1(i), 60B
- in nothogeneric name H.6.3-4, H.6A
- intentional latinization 60.7
- romanization 46B
- well-established latinized form 60C.2
Phrase name 23.6-7
Phylum, see Division
Plants (see also Cultivated and Fossil) Pre.7
- not originally so treated 45.5
Pleomorphic fungi, see Fungi, pleomorphic
Pleonasm 23A.3(e)
Polyploid H.3.Ex.3
Popular periodicals 30A
Position Prin.IV, 6.5, 11.1
- anticipation 34.1(b)
Prefix (see also Word elements) 16A.2
- in personal name 60B.N.1, 60C.4
- notho- (n-) H.1, H.3.1+4
- sub- 4.1
Preparation 8.1, 9.10
Preservation (see also Deposited material) 7A, 8.3
- impossible 8.3, 8A.2
President, International Association for Plant Taxonomy Div.III.2(1)
- Nomenclature Section Div.III.3(1)
Pre-starting-point, author 46.5
- publication 32A
Previous editions of the *Code* Pre.11
Principles Pre.2, Prin.I-VI
Printed matter (see also Publication), accompanying exsiccata 30.4+N.1
- delivery to carrier 31A
- ephemeral 30A
Priority, equal 11.5, 53.6
- of autonyms 11.6
- of choice 11.5+N.1, 53.6, 61.3, 62.3
- of designation of type 7.10, 9.13, 10.5
- of homonyms 45.4, 53.6
- of names 11, 16B
- - anamorphs 59.4
- - fossil vs. non-fossil 11.7
- - higher taxa 11.9, 16.N.1, 16B
- - holomorphs 59.4
- - hybrids 11.8
- - legitimate 45.3
- - limitation 11.2-4+6-7, 13-15, 59.4
- - - to rank 11.2
- - unaffected by, date of conservation 14.N.3
- - - date of sanctioning 15.N.1
- - unranked taxa 35.2
- principle Prin.III
Prokaryots Pre.7.fn.
Pronunciation, difficult in Latin 20A.1
Proposal, to amend *Code* Div.III.4
- to conserve name 14.12+14, 14A
- to reject name 14.14, 14A, 56.1-2
- to suppress work 32.8-9
Protologue 9.4.fn.
- conflict with 9.13(b), 9A.3, 10.5(a)
- generic 10.2+4
- guide in lectotypification 9A.2
- page reference 33.N.1
Provisional name 23.Ex.8, 34.1
Pseudocompound 60G.1
Pteridophyta, Committee for Div.III.2
- starting point 13.1(a)
Public meeting 29
Publication (see also Simultaneous) 35.4, 46.6
- abstracting journals 30A
- date 31, 31A, 45
- effective, see Effective publication
- ephemeral printed matter 30A
- in parts 35.4, 45B
- indelible autograph 30.1
- independently of exsiccata 30.N.1
- non-scientific newspaper 30.3
- popular periodical 30A
- printed herbarium labels 30.4
- seed-exchange list 30.3
- separates 31.2, 45C
- trade catalogue 30.3
- valid, see Valid publication

Publications:
Agosti, *Re Bot. Tract.* (1770) App.V
Aiton, *Hort. Kew.* (1789) 46.Ex.21

Publications (cont.)
Bornet & Flahault, *Rév. Nostoc. Hét.* (1886-1888) 13.1(e)
Britton & Brown, *Ill. Fl. N. U.S.* (1896-1898) 46.Ex.20
– & – *Ill. Fl. N. U.S.*, ed.2 (1913) 10.Ex.6, 46.Ex.20
Brummitt & Powell, *Auth. Pl. Names* (1992) 46A.N.3
Buc'hoz, *Dict. Univ. Pl. France* (1770-1771) App.V
– *Tournefortius Lothar.* (1764) App.V
– *Traité Hist. Pl. Lorraine* (1762-1770) App.V
Don, see Sweet
Donati, *Ausz. Natur-Gesch. Adriat. Meer.* (1753) App.V
Ehrhart, J. B., *Oecon. Pflanzenhist.* (1753-1762) App.V
Ehrhart, J. F., "Index phytophylacii ehrhartiani" (1789) App.V
– *Phytophylacium Ehrhart.* (1780-1785) App.V
Feuillée, *Beschr. Arzen. Pfl.* (1756-1757; ed. 2, 1766) App.V
Forsskål, *Fl. Aegypt.-Arab.* (1775) 23.Ex.6+8+10, 43.Ex.1
Fries, E. M., *Elench. Fung.* (1828)13.1(d)
– *Syst. Mycol.* (1821-1832) 13.1(d), 33.6
Fries, T. M., *Lich. Arct.* (1861) 31.Ex.2
Gandoger, *Fl. Eur.* (1883-1891) App.V
Garsault, *Descr. Vertus Pl.* (1764-1767) App.V
– *Expl. Abr. Pl.* (1765) App.V
– *Fig. Pl. Méd.* (1764) App.V
Gilibert, *Exerc. Bot.* (1782) App.V
– *Exerc. Phyt.* (1792) App.V
– *Fl. Lit. Inch.* (1782) App.V
– *Syst. Pl. Eur.* (1785-1787) App.V
Gleditsch, *Obs. Pneumonanthe* (1753) App.V
Gomont, *Monogr. Oscill.* (1892) 13.1(e), 46.Ex.19
Guettard, *Ve Mém. Glandes Pl.* (1715) App.V
Haller, *Enum. Pl. Gotting.*(1753) App.V
Hedwig, *Sp. Musc. Frond.* (1801) 13.1(b)
Heister, *Descr. Nov. Gen.* (1753) App.V
Hill, *Brit. Herb.* (1756-1757) App.V

Publications (cont.)
– [entries in] *New Dict. Arts Sci.* (1753-1754) App.V
– [entries in] *Suppl. Chambers's Cyclopaedia* (1753) App.V
– *Useful Fam. Herb.* (1754, etc.) App.V
Hirn, *Monogr. Oedogon.* (1900) 13.1(e)
Krocker, *Fl. Siles.*, (1787) 23.Ex.6
Kummer, *Führer Pilzk.* (1871) 32.Ex.8
Léveillé, *Fl. Kouy-Tchéou* (1914-1915) 30.Ex.2
Linnaeus, *Amoen. Acad.*, 4 (1759) 45.Ex.1
– *Fl. Monsp.* (1756) 45.Ex.1
– *Gen. Pl.*, ed.5 (1754) 13.4
– *Herb. Amboin.* (1754) 34.Ex.2
– *Sp. Pl.* (1753) 13.1(a+c-e,)+4-5, 33.Ex.1, 41.N.
– *Sp. Pl.*, ed.2 (1762-1763) 13.4, 41.N.1
Lundell & Nannfeldt, *Fung. Exs. Suec.* (1934-1947) 30.Ex.3
Michaux, *Fl. Bor.-Amer.* (1805-1806) 30.Ex.1
Miller, *Gard. Dict.*, ed.8 (1768) 32.Ex.8, 33.Ex.1
Necker, *Elem. Bot.* (1790, 1791, 1808) App.V
Persoon, *Syn. Meth. Fung.* (1801) 13.1(d)
Ralfs, *Brit. Desmid.* (1848) 13.1(e)
Rumphius & Burman, *Herb. Amboin. Auctuar.* (1755) App.V
Schlotheim, *Petrefactenkunde* (1820) 13.1(f)
Secretan, *Mycogr. Suisse* (1833) App.V
Séguier, *Pl. Veron.* (1745-1754) 41.Ex.2
Sternberg, *Vers. Fl. Vorwelt,* 1 (1820) 13.1(f)
Steudel, *Nomencl. Bot.* (1821-1824) 33.Ex.1
Stickman, see Linnaeus (1754)
Swartz, *Prodr.* (1788) 52.Ex.9
Sweet, *Hort. Brit.*, ed.3 (1839) 32.Ex.3
Trew & al., *Herb. Blackwell.* (1747-1773) App.V
Walter, *Fl. Carol.* (1788) 20.Ex.8
Webb & Heldreich, *Cat. Pl. Hisp.* (1850) 30.Ex.1
Willdenow, *Sp. Pl.* (1797-1803) 31.Ex.1

Publishing author, see Author
Purpose of giving a name Pre.1

Question mark 34.1, 52.N.1
Quotation marks 50F

Radix, not a generic name 20.Ex.5
Rank Pre.1, Prin.IV, 2-5, 6.5, 11.1
– alteration, author citation 49.1
– anticipation 34.1(b)
– appropriate, for hybrid H.5
– basic 2
– – prefix sub- 4.2
– further intercalated 4.3
– hierarchy 2
– inappropriate, for hybrid H.5.N.1
– indication 21.1, 21A, 24.1, 35.4
– not indicated 35.1-3,
– nothotaxon
3.2, 4.4, 50, H.1.1, H.3.1, H.5, H.12
– particular Prin.IV
– principal 3.1-2
– priority outside 11.2
– relative order 5, 33.5-6
– secondary 4.1
– simultaneous different 34.2
– single infraspecic 35.3
– term denoting 3.1, 4.1-2, 21.1, 24.1
– – misplaced 33.5-6
Rapporteur-général Div.III.2(1+8)+3(3)
Recommendations, in the *Code* Pre.3+5
Recorder, Nomenclature
Section Div.III.3(2)
Reference, page 33.2+N.1
– to previous description or
diagnosis 32.1(c)+4-5, 32A, 36, 41
– – direct or indirect 32.1(c)+4-5, 41
– – full and direct 32.4, 33.2+4, 45.1
– to previous illustration 38-39
Registration 32.2
– requirement for valid publication 32.1
– re-submission 45.2
Regnum, see Kingdom
Reihe, instead of order 17.2
Rejected name, combination based on 14.4
Rejection, of name 56, App.IV
– – against conserved name 14.4+6-7+10
– – as illegitimate 45.4, 52-54, 58
– – authorized pending decision 14.14
– – disallowed 51
– – not affecting legitimacy 14.10
– – overriding sanctioning 15.6
– – proposal 14.14, 14A, 56.1-2

explicar....

– of proposal 57
Relative order of ranks 5, 33.5-6
Replaced synonym 33.2+N.2, 58.1+3
Restoration, of rejected name 14.6-7
Retention of name authorized 14.14
Retroactivity of rules Prin.VI
Romanization of author names 46B.2
– of Greek 60A
Rules, absence or doubt Pre.10
– in the *Code* Pre.3+4
– retroactivity Prin.VI

Sanctioning 13.1(d), 15
– competing names 15.3-4
– date not affecting priority 15.N.1
– homonym 15.1-2+N.1, 53.1-2, 55.3
– – earlier not illegitimate 15.2
– illegitimate name 6.4, 52.1
– indication by author citation 50E.2
– overridden by conservation or
rejection 15.6
– typification of names 7.5+8
– works 13.1(d)
Scale of figure 32D.3
Scientific plant names in Latin Pre.1, Prin.V
Secretary, International Association for
Plant Taxonomy Div.III.2
Section *(sectio)* (see also Subdivision of
genus) 4.1, 21.1
– change in rank 21B.3
– epithet, from personal name 60B
– – preferably a substantive 21B.1
– – same as for subgenus 22A
Seed-exchange list 30.3
Selection of type, see Designation and
Lectotype
Separates 31.2, 45C
Series (see also Subdivision of
genus) 4.1, 21.1
– epithet, a plural adjective 21B.1
Sex, of persons 60.N.3, 60C.1(a-b)
Sexual symbols H.2A.1
Signs, see Symbols
Simultaneous publication (see also
Priority of choice) 11.5-6,
13.1(e)+5, 34.2+N.1, 53.6, 59.N.1, 61.3
Small herbaceous plants, type 8.1
Special form *(forma specialis)* 14.N.3
Species (see also Notho-
species) 3.1-2, 23, 23A

– anamorphic 59.3+N.1,59A
– epithet (see also Epithet) 6.7, 23.1-2, 23A
– – adjectival 23.5, 60C.1(c-d), 60D
– – compound 60.8-9, 60G
– – etymology 60I
– – form 23.1-3+5, 23A, 60.8-11, 60C-H
– – grammatical agreement 23.5, 32.6, 60.11, 60C.1
– – inadmissible 23.4+6, 58.2
– – initial letter 60F
– – Linnaean 23.3+7-8
– – spelling, see Orthography
– – termination 23.5, 23A.2(a), 32.6, 60.11, 60C.1-2, 60D
– – to adopt 23A.1&3(a)
– – to avoid 23A.2+3(b-j), 60C.2
– – two or more words 23.1
– – under illegitimate generic name 55.1
– – with apostrophe 60.10, 60C.4(a-c)
– – with hyphen 23.1, 60.9, 60G.N.1
– – with symbol 23.3
– holomorphic 59.1-2+N.1,59A
– name 23
– – conserved 10.N.2, 14.1+4+N.2
– – equivalent to type 10.1
– – illegitimate 52, 53.1, 55.2
– – legitimate 55.1
– – not regarded as such 23.6
– – type 8-9, 8A-9B, 10.N.2, 37.3-5, 37A
– – – typifying name of supraspecific taxon 10.1-4+6, 22.5-6
– – valid publication 23.3-7, 32.1(d), 33.1-4, 37.3-5, 41.3, 42, App.V
– raised infraspecific taxon 24B.2
– rank 2, 3.1-2
– tautonym 23.4
– unitary designation 20.4
Specimen (see also Collection and Type) 8, 8A
– cited in protologue 9.4
– fossil 8.4-5, 8A.3
– illustrated 8A.1, 32D.2, 39A
– impossible to preserve 8A.2
– reference to detail 37.3
Spelling, see Original spelling, Orthography
Spermatophyta, Committee for Div.III.2
– starting point 13.1(a)
Sphagnaceae, starting point 13.1(b-c)
Spina, not a generic name 20.Ex.5
Spiritus asper 60A.2
"St." in epithets 60C.4(d)
Stability of names Pre.1, 14.1-2. 56.1
Standard species 7.Ex.7
Starting points, nomenclatural 13.1
– – taxonomic position of type 13.2
– – valid publication of names 13.1, 32.1(a)
Status, hybrid vs. non-hybrid 50
– of name 6.1-6, 12
Stem augmentation (see also Word elements) 60C.1(b)
Stratigraphic relations 13.3
sub-, in rank designation 4.1
Subclass *(subclassis),* name 16.1, 16A.3
– rank 4.2
Subdivision of family 4.N.1
– including type of family name 19.4
– name 19, 19A
– – illegitimate 19.5
– – termination 19.1+3+6, 19.A.1, 32.6
– – type 10.6
– – valid publication 19.5, 32.1, 41.1
Subdivision of genus 4.N.1
– autonyms 22.1-4+N.1
– – type 7.6, 10A
– change in rank 21B.3
– epithet 21, 21A-B
– – capital initial letter 21.2
– – condensed formula H.7.1, H.8.1
– – etymology 60I
– – form 21.2-3, 21B.2, 60B
– – from constituent species 22.5
– – grammatical agreement 21.2
– – inadmissible 21.3
– – parenthetical, in species name 21A
– – same in different ranks 22A
– – to adopt 21B.1+3, 22A, 60B
– – to avoid 21B.2
– – under illegitimate generic name 55.1
– homonyms 21.N.1, 53.5
– hybrid 40.N.1, H.7, H.8.1, H.9
– name 21.1
– – illegitimate 21.N.1, 53.5
– – legitimate 55.1
– – type 10.1-3+5+N.1, 22.5, 22A
– – valid publication 32.1, 33.1, 41.2, H.9
– nomenclaturally typical 10A, 22.1-2
Subdivision *(subdivisio)* or subphylum, name 16.1, 16A.2

– rank 4.2
Subfamily *(subfamilia)* (see also Subdivision of family), name 19.1
– rank 4.2
– – termed suborder 19.2
Subforma (see also Infraspecific taxon) 4.2
Subfossil type 11.7
Subgenus (see also Subdivision of genus), change in rank 21B.3
– epithet, preferably substantive 21B.1
– – from personal name 60B
– – same in different ranks 22A
– rank 4.2, 21.1
Subjective synonym 14.4.fn.
Suborder *(subordo)*, name 16.1, 17.1
– – improper Latin terminatin 17.3, 32.6
– rank 4.2
– – intended as subfamily 19.2
Subordinate taxa 25, 34.1(d)
Subphylum, see Subdivision
Subregnum 4.2
Subsection *(subsectio)* (see also Subdivision of genus) 4.2
– epithet a plural adjective 21B.1
Subseries (see also Subdivison of genus) 4.2
– epithet a plural adjective 21B.1
Subspecies (see also Infraspecific taxon) 4.2
– epithet, same in varietal rank 26A.1-2
– – maintenance in infravarietal ranks 26A.3
Substantive (see also Gender), as epithet 21.2, 21B.1-2, 23.1+6,(a), 60.N.3
– – genitive 23A.1-2, 60C.1(a-b)
– as name 20.1
– – plural adjective 18.1, 19.1
Subtribe *(subtribus)* (see also Subdivision of family), name 19.3
– rank 4.2
Subvariety *(subvarietas)* (see also Infraspecific taxon) 4.2
Suffix, see Word elements
Sum of subordinate taxa 25
Superfluous name 52.1
– basionym legitimate 52.3
– illegitimate 52.1
– not illegitimate 52.3+N.1-2
Supplementary ranks 4.3
Suppressed works 32.8, App.V
– approval by General Committee 32.9
– proposal under study 32F.1
Symbols (see also Multiplication sign) 23.3
– female 23.Ex.2, H.2A.1
– Linnaean 23.3, 23.Ex.2
– male H.2A.1
Synonym, citation as 34.1, 50A
– name- or epithet-bringing 33.2
– nomenclatural 14.4, App.IIIA-B
– objective 14.4.fn.
– regardless of multiplication sign H.3.4
– regardless of prefix "notho-" H.3.4
– replaced 33.2
– subjective 14.4.fn.
– taxonomic 14.4, 53.N.1, App.IIIA-B
Syntype(s) 9.4, 10.N.1
– designated as lectotype 9.9
– duplicate 9.8
– inclusion of all 52.2

Tautonym 23.4., 58.2
Taxon (taxa) Pre.1, Prin.I-V, 1.1
– change in rank 19A, 21B.3, 49, 24B.2
– not treates as plants Prin.I, 32.N.1, 45.5, 54
– one correct name Prin.IV, 11.1
– parental, see under Hybrid
– subordinate 25, 34.1(d)
– transferred 49-50
– treated as plants Pre.7, Prin.I, 45.5, 54
– unranked 35.1-2
Taxonomic, group, see Taxon
– position, see Position
– rank, see Rank
– synonym 14.4, 53.N.1, App. IIIA-B
Technical term, morphology 20.2
Teleomorph 59.1
– diagnosis or description 59.2
Termination (see also Word elements) 16.2, 16A, 17.1, 18.1, 19.1+3, 19A.1, 20A.1(a), 60.11, 60B-D, H.6.3-4, H.6A, H.8.2
– contrary to rules 60.11
– correction when improper 16A.4, 17.3, 18.4, 19.6, 60.7+11
– epithet from geographical name 60D
– – from personal name 60.11, 60B-C
– incorrect but name validly published 32.6
– Latin if possible 20A.1
Ternary name 24.1

Tetraploid H.3.Ex.3
Tokyo Congress 14.N.1.fn.
Trade catalogue 30.3
Tradition, botanical 62.1+2(b-c)+N.1
– classical 60A.1-2, 60C.2, 60E, 60G.1

Transcriptions (etc.):
ae, for ä, æ, è, é, or ê 60.6
ao, for å 60.6
diacritical signs 60.6, 60C.3
e, for è, é, or ê 60.6
h, for spiritus asper 60A.1
i & j used interchangeably 60.1, 60.5
k, permissible in Latin plant names 60.4
letters 60.5
letters foreign to classical Latin 60.4, 60C.3
Linnaean symbols 23.3-4
ligature 60.4
n, for ñ 60.6
oe, for ö, ø, or œ 60.6
spiritus asper 60A.2
ss, for ß 60.4
u & v used interchangeably 60.1, 60.5
ue, for ü 60.6
w, permissible in Latin plant names 60.4
x, used for × H.3A
y, permissible in Latin plant names 60.4

Transfer, hybrid/non-hybrid 50, H.10.2
– to other genus or species 49
Transliteration to Latin, see Romanization
Tribe *(tribus)* (see also Subdivision of family), in Fries's *Systema* 33.6
– name 19.3
– rank 4.1
Trigeneric hybrid, name H.6.4
Type *(typus)* (see also Holotype, Lectotype, Neotype, etc.) Prin.II, 7-8+10
– acceptance, by typifying author 7.11
– anamorphic 59.3, 59A.1
– automatic 7.3-6, 10.7, 16.1, 17.1, 22.5
– collection data 8A.1-2, 37.3
– conservation (see also Type, preservation) 10.4+N.2, 14.9, App.IIIA-B
– correction of term 9.8
– definite indication 7.5
– definition 7.2
– deposit 7A, 8B.1, 9A.4, 37.5
– designation, see Designation
– duplicate 9.3+fn.+9
– equivalent in modern language 7.11, 7.Ex.7, 37.4
– exclusion 47, 48, 52.2
– from context of description 7.7, 9.N.1
– identification ambiguous 9.7
– illustration 8.1+3, 8A.1-2, 9.1-2+6-7
– inclusion, in named taxon 10.2-3, 22.1-3, 22A, 26, 26A, 37.2, 52.1-2
– – in other taxon 48.1
– – single element 37.2-3
– indication 9.9, 22.5, 24.3. 37.1-4, 37A
– interpretative 9.7
– living 8.2+Ex.1, 8B.1-2
– location 7A, 8B.1, 9A.4, 37.5
– mechanical designation 9A.2, 10.5
– missing 9.2+6+9+11
– mis-use of term 9.8
– more than one individual 8.1
– more than one taxon 9.9, 9A.5
– not always typical of taxon 7.2
– not conspecific with original material 10.2
– of alternative name 10.6, 18.5, 19.8
– of autonym 7.6
– of avowed substitute 7.3
– of basionym 7.4
– of condensed formula 37.1, H.9.N.1
– of name, conserved 10.4+N.2, 14.3+8-9, 48.N.2, App.IIIA-B
– – erroneously applied 7.4
– – family 7.1, 10.6
– – – alternative 10.6, 18.5
– – – included in subdivision of family 19.4
– – form-genus 7.9
– – fossil taxon 7.9, 8.4-5, 13.3
– – fungal anamorph 7.9+N.1, 59.3, 59A
– – fungal teleomorph 59.1-2, 59A
– – genus 10.1-5, 10A, 14.3
– – illegitimate 7.5
– – rejected 14.3
– – nothotaxon 37.1, H.9.N.1, H.10.N.1
– – sanctioned 7.5+8
– – species or infraspecific taxon 8-9, 8A-9B
– – subdivision, of family 7.1, 10.6
– – – of genus 10.1-5+N.3, 10A, 22.5
– – subfamily, alternative 10.6, 19.8
– – suprafamilial taxon 7.1, 10.7, 16.1, 17.1

– – validated by reference 7.7
– – with later starting point 7.7
– of new combination 7.4
– of orthographic variants 61.2+5
– of sanctioned name 7.8
– original 48.1+N.2
– preservation, impossible 8.3, 8A.2
– – permanent 8.1-3+Ex.1, 8B.2
– – place 7A, 8B.1, 9A.4, 37.5
– previously designated 9.11, 22.2, 26.2, 52.2(b)
– rediscovered 9.13
– required 37.1
– serious conflict with protologue 9.13(b), 10.5(a)
– single specimen 8.1
– specimen data 8A.1, 37.3
– standard species 7.Ex.7
– stratigraphic relations 13.3
– taxonomic position 13.2, 59.3, H.10.N.1
– teleomorphic 59.1-2, 59A.1
Type-scripts 29
Typification, see Designation, Lectotype, Neotype
– principle Prin.II, 10.7
Typography 60.2
– error 60.1, 61.1
Typus, see type

Unitary designation of species 20.4
Unpublished, material 29
– names 23A.3(i), 34A
Unranked taxa 35.1-3
Uredinales, starting point 13.1(d)
Usage, see Custom and Tradition
– to be followed pending decision 14A, 32F
Ustilaginales, starting point 13.1(d)
– pleomorphic fungi 59.1

Valid publication 6.2, 12.1, 32-45, H.9
– advisable 45A
– correction of orthography 45.3
– date 45.1+3+5, 45B-C
– – for names of taxa not originally treated as plants 45.5
– – unaffected by conservation 14.N.3
– – unaffected by sanctioning 15.N.1
– – unaffected by spelling change 45.3
– despite taxonomic doubt 34.1
– not by mere mention of subordinate taxa 34.1(d)
– not by reference to general indices 33.4
– not of name, cited as synonym 34.1(c)
– – proposed in anticipation 34.1, H.9.N.2
– – provisonal 34.1(b)
– not when misplaced term denotes rank 33.5
– of autonym 22.3, 26.3, 32.7
– of avowed substitute 32.2-3+N.2
– of basionym 32.2, 33A
– of combination 33.1, 43.1
– of name, algae 36.2, 39, 45.4
– – alternative 34.2+N.1
– – family 18.3, 41.1
– – fossil plant 36.3, 38
– – genus 37.2, 41.2, 42.1+3
– – – Linnaean 13.4, 41.N.1
– – hybrid 40, H.9, H.10.1
– – infraspecific taxon 37.3-5, 41.3, 44.1
– – monotypic new genus 42
– – non-vascular plant 42.3
– – nothogenus H.9
– – nothospecies or lower ranked hybrid H.10.1
– – species 37.3, 41.3, 42.1+3, 44.1
– – subdivision of family 19.5, 41.1
– – subdivision of genus 37.2, 41.2
– – taxon not originally treated as plant 32.N.1, 45.5
– of new combination 32.1-3, 35.1
– of orthographical variant 61.1
– requirements 32.1, 45.1
– – acceptance of name 34.1(a), 45.1
– – association of epithet with name 33.1
– – citation of basionym or replaced synonym 33.2
– – – despite bibliographic error 33.3
– – – despite incorrect author citation 33.3
– – citation of type data 37.2-5, 37A
– – compliance with provisions on form of name 32.1(b), 32.6
– – date reference 33.2, 45B-C
– – description or diagnosis 32.1(c)+3, 41, 42.1
– – – English 36.3
– – – Latin 36, 36A
– – effective publication 32.1(a)
– – illustration 38-39, 39A, 41.N.2, 42.3, 44.2

– – – with analysis 42.3-4, 44
– – indication of rank 35.1
– – indication of type 37, 37A
– – page and plate reference 33.2+N.1
– – reference to description or diagnosis 7.7, 32.1(c), 32A, 36, 41
– – – direct or indirect 32.1(c)+4-5, 41
– – – full and direct 32.4, 33.2+4, 45.1
– – reference to illustration 38-39
– – registration 32.1, 45.2
– – specification of type herbarium 37.5
– – statement of parentage H.9
– starting points 13.1
Variant, see under Orthography
– cultivated 4.N.1, 28.N.1-2
Variety *(varietas)* (see also Infraspecific taxon and Nothomorph) 4.1, H.12.2+fn.
– epithet, same in subspecies rank 26A.1-2
– – maintenance in infravarietal ranks 26A.3
– single infraspecific rank 35.3
Vernacular name 60.7, 60.F, 62.3,
Vice-rapporteur Div.III.3(4)
Vote, final Div.III.4(b)
– preliminary Div.III.4(a)
Vowel, connecting 60G.1(a2)
– final 60B.1(a), 60C.1(a+c)
– transcription 60.5-6

Wild plants in cultivation 28.1
Wood, fossil 8A.3
Word, Greek or Latin 60G
– last in compound, gender 62.2
– not epithet or name 20.4(a), 43.N.1
– standing independently 60.9
Word element, omitted in suprafamilial name 16.2

Word elements:
-a 60B.1(a-b), 60C.1(a+c)
-aceae 17.1, 18.1, 19.1, 19A.1
-achne 62.2(b)
-ae 18.1, 60.11, 60C.1(a), 60G.1(a1)
-ae- 60.Ex.12
-ales 17.1
-an- 60C.1(c-d)
-ana 60C.1(c)
-anthes 62.4
-anthos, -anthus 62.2(c)
-anum 60C.1(c)
-anus 60.11, 60C.1(c), 60D

Word elements (cont.)
-ara H.6.3-4, H.6A.1, H.8.2
-as 18.1, 60G.1(a1)
-aster (-asteris), -astrum (-astri) 61.Ex.1
-carpa, -aea, -ium, -on, -os, -um, -us 62.Ex.2
-ceras 62.2(c)
-cheilos, -chilos, -chilus 62.2(c)
-chlamys 62.2(b)
-clado- 16.2
-cocco- 16.2
-codon 60.2(a)
-daphne 62.2(b)
-dendron 62.2(c)
-e 60C.1(a)
-ea 60B.1(a)
-eae 19.3+7, 19A.1
-ensis 60D
-eos 18.1, 60G.1(a)
-er 60B.1(b), 60C.1(a-b)
-es 18.1, 60G.1(a)
eu- 21.3
-gaster 62.2(b)
-i 18.1, 60.11, 60C.1(a), 60G.1(a1)
-i- 60.Ex.12, 60C.1(b+d), 60G.1(a2)
-ia 60B.1(b)
-iae 60.11, 60C.1(b)
-ianus, -iana, -ianum 60.11, 60C.1(d)
-iarum 60C.1(b)
-icus 60D
-idae 16A.3(c)
-ii 60.11, 60C.1(b)
-inae 17.1, 19.3, 19A.1
-ineae 17.1
-inus 60D
-iorum 60C.1(b)
-is 18.1, 60G.1(a1)
-ites 62.4
-mecon 62.2(b)
-monado- 16.2
-myces 62.2(a)
-mycetes 16A.3(b)
-mycetidae 16A.3(b)
-mycota 16A.1
-mycotina 16A.2
-n- 60C.1(c)
-nema 62.2(c)
-nemato- 16.2
-nus, -na, -num 60C.1(c), 60D
-o- 60G.1(a2)

Word elements (cont.)
-odes 62.4
-odon 62.2(a)
-oideae 19.1+7, 19A.1
-oides 62.4
-opsida 16A.3
-orum 60C.1(a)
-os 18.1, 60G.1(a1)
-osma 62.2(b)
-ou 18.1
-ous 18.1, 60G.1(a)
-panax 62.2(a)
-phyceae 16.2, 16A.3(a)
-phycidae 16A.3(e)
-phyta 16.2, 16A.1

Word elements (cont.)
-phytina 16A.2
-pogon 62.2(a)
-rum 60C.1(a)
-sancta, sanctus 60C.4(d)
-stemon 62.2(a)
-stigma 62.2(c)
-stoma 62.2(c)
-us 18.1, 60B.1(c), 60G.1(a)

Work, see Publication

Zoological nomenclature, see *International code of zoological nomenclature*